Control Engineering

Control Engineering

Edited by **Ashley Potter**

NY RESEARCH PRESS
New York

Published by NY Research Press,
23 West, 55th Street, Suite 816,
New York, NY 10019, USA
www.nyresearchpress.com

Control Engineering
Edited by Ashley Potter

International Standard Book Number: 978-1-63238-522-2 (Hardback)

Printed in the United States of America.

Contents

Preface

This book has been an outcome of determined endeavour from a group of educationists in the field. The primary objective was to involve a broad spectrum of professionals from diverse cultural background involved in the field for developing new researches. The book not only targets students but also scholars pursuing higher research for further enhancement of the theoretical and practical applications of the subject.

This book traces the progress of the field of control engineering and highlights some of its key concepts and applications. It elucidates new theories and techniques in a multidisciplinary approach. Control engineering is a branch of engineering that implements the laws of control theory to design and manufacture systems that are used to control the machines and to monitor their performance in order to enhance their efficiency. A control engineer uses the elements of mathematics and engineering to make the systems work proficiently and smoothly. This text includes detailed explanations of various approaches and techniques of this branch. The topics introduced herein are of utmost significance and are bound to provide incredible insights to readers. This book will prove to be an essential guide for students, scientists, engineers, researchers and all those who are interested in control systems engineering.

It was an honour to edit such a profound book and also a challenging task to compile and examine all the relevant data for accuracy and originality. I wish to acknowledge the efforts of the contributors for submitting such brilliant and diverse chapters in the field and for endlessly working for the completion of the book. Last, but not the least; I thank my family for being a constant source of support in all my research endeavours.

Editor

Anticollocated Backstepping Observer Design for a Class of Coupled Reaction-Diffusion PDEs

Antonello Baccoli and Alessandro Pisano

Department of Electrical and Electronic Engineering, University of Cagliari, 09123 Cagliari, Italy

Correspondence should be addressed to Antonello Baccoli; antonello.baccoli@diee.unica.it

Academic Editor: Ai-Guo Wu

The state observation problem is tackled for a system of n coupled reaction-diffusion PDEs, possessing the same diffusivity parameter and equipped with boundary sensing devices. Particularly, a backstepping-based observer is designed and the exponential stability of the error system is proven with an arbitrarily fast convergence rate. The transformation kernel matrix is derived in the explicit form by using the method of successive approximations, thereby yielding the observer gains in the explicit form, too. Simulation results support the effectiveness of the suggested design.

1. Introduction

Model-based control and advanced process monitoring of Distributed-Parameter Systems (DPSs), governed by Partial Differential Equations (PDEs), typically require full state information. However, the available measurements of DPS are typically located on the boundary of the spatial domain that motivates the need of the state observer [1, 2].

For linear infinite-dimensional systems the Luenberger observer theory was established by replacing matrices with linear operators [2–4], and the observer design was confined to determining a gain operator that stabilizes the associated error dynamics. In contrast to finite dimensional systems, finding such a gain operator is not trivial even numerically because operators were not generally represented with a finite number of parameters.

Design methods, which are not relying on any discretization or finite-dimensional approximation (thereby preserving the infinite-dimensional representation of the system during the entire design process) and which are yielding the observer gains in the explicit form, have only recently been investigated. In this context, the backstepping method appears to be a particularly effective systematic design approach which can be applied for a broad class of systems governed by PDEs [5, 6]. Basically, the backstepping approach relies on the application of an invertible Volterra integral transformation mapping, a predefined exponentially stable target system, into the observer error dynamics.

For systems governed by parabolic PDEs defined on a one-dimensional (1D) spatial domain, a systematic observer design approach using boundary sensing is introduced in [6]. Recently, the backstepping-based observer design was presented in [7] for reaction-diffusion processes with spatially varying reaction coefficient and a certain weighted average of the state over the spatial domain as measured output. In [8, 9], backstepping-based observer design was addressed for reaction-diffusion processes evolving in multidimensional spatial domains. In [10], the backstepping-based design for parabolic processes was applied by adopting a nonconventional target system for the error dynamics, embedding certain discontinuous output injection terms.

More recently, high-dimensional systems of *coupled* PDEs were considered in the backstepping-based boundary control and observer design settings. The most intensive efforts of the current literature seem however to be oriented towards coupled hyperbolic processes of the transport type [11–15]. In [14], a 2×2 linear hyperbolic system was stabilized by a single observer-based boundary control input, with an additional feature that an unmatched disturbance, generated by an *a priori* known exosystem, was rejected. Both the controller and the observer were designed by following the backstepping approach. In [12] a state estimator in a semi-infinite three-dimensional (3D) domain is presented for a coupled model of magnetohydrodynamic flow, and Fourier transform methods were applied to put the system in a form, where the 1D backstepping method is applicable. In [15], a backstepping-based observer was designed for a system of two diffusion-convection-reaction processes coupled through the corresponding boundary conditions. In [13], a 2×2 system of coupled linear heterodirectional hyperbolic systems was stabilized by a backstepping-based observer-controller under some boundedness restriction on the spatially dependent coupling coefficients. In [11], observer-controller design was studied for a system of $n+1$ coupled first-order heterodirectional hyperbolic linear PDEs (n of which featured rightward convecting transport and one leftward) with a single boundary input. Some specific results concerning the backstepping-based output feedback boundary stabilization of parabolic coupled PDEs have been presented in the literature. In [16] the controller/observer design for the linearized 2×2 model of thermal-fluid convection has been treated.

In this work, the observer design is developed for a class of n coupled diffusion-reaction PDEs in the 1D spatial domain $x \in [0,1]$. The task of the present paper is to generalize some results presented in [6], where explicit backstepping observers were developed for a scalar unstable reaction-diffusion equation. Here a generalization is made for a set of n reaction-diffusion processes, which are coupled through the corresponding reaction terms. The motivation to this investigation comes from chemical processes [17] where coupled temperature-concentration parabolic PDEs were involved to describe system dynamics. This generalization is shown to be far from being trivial because the underlying backstepping-based treatment gives rise to more complex development of finding out an explicit form of the observer gains in the form of matrix Bessel series, and, furthermore, it turns out to be unfeasible in the general case where each process possesses its own diffusivity parameter. In this work we therefore address the simplified case where all processes possess the same diffusivity value, and we postpone the more general case for further investigations (see Remark 2). The present paper can be considered as the observer design counterpart of our recent work [18], where the stabilizing boundary controller design problem was addressed for a similar class of systems differing only in the boundary conditions from that considered in the present work. Subsequently, in [19], the stabilizing boundary control design problem in the general case of different diffusivity parameters was addressed and solved.

Particularly, in the present context, two output injections are needed in the observer dynamics (one distributed along the spatial domain and another one located at the uncontrolled boundary).

The structure of the paper is as follows. After introducing some useful notation in Section 1.1, Section 2 states the problem to be investigated and introduces the proposed observer structure with the underlying backstepping transformation and (matrix) kernel PDE. In Section 3, the explicit solution of the kernel PDE is derived. In Section 4, the proposed observer design is summarized and the main result of this paper is presented. Section 5 discusses supporting simulation results, and Section 6 collects some concluding remarks and future perspectives of this research.

1.1. Notation. The notation used throughout is fairly standard. $L_2(0,1)$ stands for the Hilbert space of square integrable scalar functions $z(\zeta)$ on $(0,1)$ with the corresponding norm

$$\|z(\cdot)\|_2 = \sqrt{\int_0^1 z^2(\zeta)\,d\zeta}. \tag{1}$$

Throughout the paper the notation

$$[L_2(0,1)]^n = \underbrace{L_2(0,1) \times L_2(0,1) \times \cdots \times L_2(0,1)}_{n \text{ times}} \tag{2}$$

is also utilized and

$$\|Z(\cdot)\|_{2,n} = \sqrt{\sum_{i=1}^{n} \|z_i(\cdot)\|_2^2} \tag{3}$$

stands for the corresponding norm of a generic vector function $Z(\zeta) = [z_1(\zeta), z_2(\zeta), \ldots, z_n(\zeta)] \in [L_2(0,1)]^n$.

With reference to a generic real-valued symmetric matrix W of dimension n, $\sigma_1(W)$ denotes the smallest eigenvalue of W. Finally, $I_{n\times n}$ stands for the identity matrix of dimension n.

2. Problem Formulation and Backstepping Transformation

The following n-dimensional system of coupled reaction-diffusion processes, equipped with Neumann-type boundary conditions and governed by the boundary-value problem

$$Q_t(x,t) = \theta Q_{xx}(x,t) + \Lambda Q(x,t), \tag{4}$$

$$Q_x(0,t) = 0, \tag{5}$$

$$Q(1,t) = U(t), \tag{6}$$

is under study. Hereinafter,

$$Q(x,t) = [q_1(x,t), q_2(x,t), \ldots, q_n(x,t)]^T \in [L_2(0,1)]^n \tag{7}$$

is the vector collecting the state of all systems,

$$U(t) = [u_1(t), u_2(t), \ldots, u_n(t)]^T \in \mathfrak{R}^n \tag{8}$$

is the boundary input vector, $\Lambda = \{\lambda_{ij}\} \in \mathfrak{R}^{n\times n}$ is a real-valued square matrix, and $\theta \in \mathfrak{R}^+$ is a positive scalar. The open-loop system (4)–(6) (with $U(t) = 0$) may possess arbitrarily many unstable eigenvalues when the symmetric part $(\Lambda + \Lambda^T)/2$ of matrix Λ possesses sufficiently large positive eigenvalues. For system (4)–(6) of n coupled reaction-diffusion processes, the following observer

$$\begin{aligned} \widehat{Q}_t(x,t) &= \theta\widehat{Q}_{xx}(x,t) + \Lambda\widehat{Q}(x,t) \\ &\quad + G(x)\left[Q(0,t) - \widehat{Q}(0,t)\right], \\ \widehat{Q}_x(0,t) &= M\left[Q(0,t) - \widehat{Q}(0,t)\right], \\ \widehat{Q}(1,t) &= U(t) \end{aligned} \tag{9}$$

is proposed with $G(x)$ being a nth order square matrix of observer gain functions and $M \in \mathfrak{R}^{n,n}$ being a square matrix of constant observer gains. The error variable

$$\widetilde{Q}(x,t) = Q(x,t) - \widehat{Q}(x,t) \tag{10}$$

is then governed by the error system

$$\widetilde{Q}_t(x,t) = \theta\widetilde{Q}_{xx}(x,t) + \Lambda\widetilde{Q}(x,t) - G(x)\widetilde{Q}(0,t), \tag{11}$$

$$\widetilde{Q}_x(0,t) = -M\widetilde{Q}(0,t), \tag{12}$$

$$\widetilde{Q}(1,t) = 0. \tag{13}$$

To design the observer gains $G(x)$ and M, the backstepping approach is involved to find out an invertible transformation

$$\widetilde{Q}(x,t) = \widetilde{Z}(x,t) - \int_0^x P(x,y)\widetilde{Z}(y,t)\,dy, \tag{14}$$

where $P(x,y)$ is a $n\times n$ matrix kernel function whose elements are denoted as $p_{ij}(x,y)$, $i,j = 1,2,\ldots,n$, which maps the error system (11)–(13) into the exponentially stable (the exponential stability properties of the target error system (15)–(17) will be investigated in Theorem 4) target error dynamics

$$\widetilde{Z}_t(x,t) = \theta\widetilde{Z}_{xx}(x,t) - \overline{C}\widetilde{Z}(x,t), \tag{15}$$

$$\widetilde{Z}_x(0,t) = 0, \tag{16}$$

$$\widetilde{Z}(1,t) = 0. \tag{17}$$

The following lemma is in order.

Lemma 1. *The error system (11)–(13) is transferred by (14) into the target error dynamics (15)–(17) provided that the design terms M and $G(x)$ are selected as*

$$G(x) = \theta P_y(x,0), \tag{18}$$

$$M = P(0,0), \tag{19}$$

where $P(x,y)$ is a solution to the kernel PDE

$$\begin{aligned} &P_{xx}(x,y) - P_{yy}(x,y) \\ &\quad = -\frac{1}{\theta}\left[P(x,y)\overline{C} + \Lambda P(x,y)\right], \end{aligned} \tag{20}$$

$$P(x,x) = \frac{\Lambda + \overline{C}}{2\theta}(x-1), \tag{21}$$

$$P(1,y) = 0. \tag{22}$$

Proof. Employing the Leibnitz differentiation rule, the spatial differentiation of (14) results in

$$\begin{aligned} \widetilde{Q}_x(x,t) &= \widetilde{Z}_x(x,t) - P(x,x)\widetilde{Z}(x,t) \\ &\quad - \int_0^x P_x(x,y)\widetilde{Z}(y,t)\,dy, \end{aligned} \tag{23}$$

$$\begin{aligned} \widetilde{Q}_{xx}(x,t) &= \widetilde{Z}_{xx}(x,t) - \left[\frac{d}{dx}P(x,x)\right]\widetilde{Z}(x,t) \\ &\quad - P(x,x)\widetilde{Z}_x(x,t) - P_x(x,x)\widetilde{Z}(x,t) \\ &\quad - \int_0^x P_{xx}(x,y)\widetilde{Z}(y,t)\,dy. \end{aligned} \tag{24}$$

In turn, the temporal differentiation of (14) and recursive integration by parts yield

$$\begin{aligned} \widetilde{Q}_t(x,t) &= \widetilde{Z}_t(x,t) - \int_0^x P(x,y)\widetilde{Z}_t(y,t)\,dy \\ &= \widetilde{Z}_t(x,t) - P(x,x)\theta\widetilde{Z}_x(x,t) \\ &\quad + \theta P(x,0)\widetilde{Z}_x(0,t) + \theta P_y(x,x)\widetilde{Z}(x,t) \\ &\quad - \theta P_y(x,0)\widetilde{Z}(0,t) \\ &\quad - \theta\int_0^x P_{yy}(x,y)\widetilde{Z}(y,t)\,dy \\ &\quad + \int_0^x P(x,y)\overline{C}\widetilde{Z}(y,t)\,dy. \end{aligned} \tag{25}$$

By evaluating (14) at $x = 0$ and $x = 1$ and considering (17), one derives that

$$\widetilde{Q}(0,t) = \widetilde{Z}(0,t), \tag{26}$$

$$\widetilde{Q}(1,t) = -\int_0^1 P(1,y)\widetilde{Z}(y,t)\,dy. \tag{27}$$

Substituting (14), (16) and (24)–(26) into (11) and performing lengthy but straightforward computations yield

$$\begin{aligned} &\widetilde{Z}_t(x,t) - \theta\widetilde{Z}_{xx}(x,t) + \overline{C}\widetilde{Z}(x,t) \\ &\quad = -\left\{\theta\left[\frac{d}{dx}P(x,x)\right] + \theta P_y(x,x) + \theta P_x(x,x)\right. \\ &\quad \left. - \Lambda - \overline{C}\right\}\widetilde{Z}(x,t) + \left[P(x,x)\theta - \theta P(x,x)\right]\widetilde{Z}_x(x,t) \\ &\quad + \left[\theta P_y(x,0) - G(x)\right]\widetilde{Z}(0,t) + \int_0^x \left[\theta P_{yy}(x,y)\right. \\ &\quad \left. - \theta P_{xx}(x,y) - P(x,y)\overline{C} - \Lambda P(x,y)\right] \\ &\quad \cdot \widetilde{Z}(y,t)\,dy. \end{aligned} \tag{28}$$

By evaluating (23) at $x = 0$ and considering (16), it follows that

$$\widetilde{Q}_x(0,t) = -P(0,0)\widetilde{Z}(0,t). \tag{29}$$

Substituting (29) and (26)-(27) into (12) and (13), one derives the conditions

$$[M - P(0,0)]\widetilde{Z}(0,t) = 0,$$
$$\int_0^1 P(1,y)\widetilde{Z}(y,t)\,dy = 0. \tag{30}$$

Clearly, to obtain the target error PDE (15) the right hand side of (28) should be identically zero. To meet this requirement, it suffices to employ relations (30) and exploit the identity $(d/dx)P(x,x) = P_x(x,x) + P_y(x,x)$, thereby obtaining both the kernel boundary-value problem

$$\theta\left(P_{xx}(x,y) - P_{yy}(x,y)\right) = -P(x,y)\overline{C} - \Lambda P(x,y), \tag{31}$$

$$2\theta\frac{d}{dx}P(x,x) = \Lambda + \overline{C}, \tag{32}$$

$$P(1,y) = 0 \tag{33}$$

and the observer gain design conditions in the form of (18)-(19). Integrating (32) with respect to x and considering (19) result in

$$P(x,x) = \frac{1}{2\theta}\left(\Lambda + \overline{C}\right)x + P(0,0) = \frac{1}{2\theta}\left(\Lambda + \overline{C}\right)x + M. \tag{34}$$

Evaluating (34) at $x = 1$ yields

$$P(1,1) = \frac{1}{2\theta}\left(\Lambda + \overline{C}\right) + M. \tag{35}$$

By evaluating (33) at $y = 1$ it is concluded that $P(1,1) = 0$, thus getting from (35) that

$$M = -\frac{1}{2\theta}\left(\Lambda + \overline{C}\right). \tag{36}$$

Considering (34) and (36), one finally rewrites (31)–(33) in the form of (20)–(22). Lemma 1 is proven. □

Remark 2. The present paper is confined to the case in which all the coupled PDEs (4) possess the same diffusivity parameter θ. The reason behind this is that in the more general case where each process has its own diffusivity θ_i $(i = 1, 2, \ldots, n)$, the corresponding "generalized" version

$$\Theta\left(P_{xx}(x,y) - P_{yy}(x,y)\right) = -P(x,y)\overline{C} - \Lambda P(x,y),$$
$$\Theta\frac{d}{dx}P(x,x) + \Theta P_x(x,x) + \Theta P_y(x,x) = \Lambda + \overline{C}, \tag{37}$$
$$P(x,x)\Theta = \Theta P(x,x),$$
$$P(1,y) = 0$$

of (20)–(22), where $\Theta = \operatorname{diag}(\theta_i)$, sets an overdetermined boundary-value problem that has no solution, unless specific constraints are imposed on the matrix $\overline{C}$ and on the form of the kernel matrix $P(x,y)$. This topic calls for further investigation and will be published elsewhere.

3. Solving the Kernel PDE (20)–(22)

For later use, the following result is reproduced.

Theorem 3. *Problem (20)–(22) possesses a solution*

$$P(x,y) = -\sum_{n=0}^{\infty}\frac{2(1-x)\left((1-y)^2 - (1-x)^2\right)^n}{n!\,(n+1)!} \cdot \left(\frac{1}{4\theta}\right)^{n+1}\left[\sum_{i=0}^{n}\binom{n}{i}\Lambda^i\left(\Lambda + \overline{C}\right)\overline{C}^{n-i}\right] \tag{38}$$

which is of class C^∞ in the domain $0 \le y \le x \le 1$.

Proof. By the invertible change of variables

$$\overline{x} = 1 - y,$$
$$\overline{y} = 1 - x, \tag{39}$$

one transforms (20)–(22) into

$$\overline{P}_{\overline{xx}}(\overline{x},\overline{y}) - \overline{P}_{\overline{yy}}(\overline{x},\overline{y}) = \frac{1}{\theta}\left[\overline{P}(\overline{x},\overline{y})\overline{C} + \Lambda\overline{P}(\overline{x},\overline{y})\right],$$
$$\overline{P}(\overline{x},\overline{x}) = -\frac{\Lambda + \overline{C}}{2\theta}\overline{x}, \tag{40}$$
$$\overline{P}(\overline{x},0) = 0.$$

Following [20], the existence of a solution to problem (40) can be shown by transforming it into an integral equation using the change of the variables

$$\xi = x + y,$$
$$\eta = x - y. \tag{41}$$

Setting

$$H(\xi,\eta) = \overline{P}(x,y) = \overline{P}\left(\frac{\xi+\eta}{2}, \frac{\xi-\eta}{2}\right) \tag{42}$$

the relations

$$\overline{P}_x = H_\xi + H_\eta,$$
$$\overline{P}_{xx} = H_{\xi\xi} + 2H_{\xi\eta} + H_{\eta\eta},$$
$$\overline{P}_y = H_\xi - H_\eta, \tag{43}$$
$$\overline{P}_{yy} = H_{\xi\xi} - 2H_{\xi\eta} + H_{\eta\eta}$$

are obtained, and the matrix kernel boundary-value problem (40), written in the new coordinates, takes the form

$$H_{\xi\eta}(\xi,\eta) = \frac{1}{4\theta}H(\xi,\eta)\overline{C} + \frac{1}{4\theta}\Lambda H(\xi,\eta), \quad (44)$$

$$H(\xi,0) = -\frac{1}{4\theta}\left(\Lambda+\overline{C}\right)\xi, \quad (45)$$

$$H(\xi,\xi) = 0. \quad (46)$$

Integrating (44) with respect to η from 0 to η and considering the relation $H_{\xi}(\xi,0) = -(1/4\theta)(\Lambda+\overline{C})$, which follows from (45), one obtains

$$H_{\xi}(\xi,\eta) = -\frac{1}{4\theta}\left(\Lambda+\overline{C}\right) + \frac{1}{4\theta}\int_0^{\eta}\left[H(\xi,s)\overline{C} + \Lambda H(\xi,s)\right]ds. \quad (47)$$

Integrating (47) with respect to ξ from η to ξ yields

$$\int_{\eta}^{\xi} H_{\tau}(\tau,\eta)\,d\tau = \int_{\eta}^{\xi} -\frac{1}{4\theta}\left(\Lambda+\overline{C}\right)d\tau + \frac{1}{4\theta}\int_{\eta}^{\xi}\left\{\int_0^{\eta}\left[H(\tau,s)\overline{C} + \Lambda H(\tau,s)\right]ds\right\}d\tau \quad (48)$$

which can further be manipulated to

$$H(\xi,\eta) - H(\eta,\eta) = -\frac{1}{4\theta}\left(\Lambda+\overline{C}\right)(\xi-\eta) + \frac{1}{4\theta}\int_{\eta}^{\xi}\left\{\int_0^{\eta}\left[H(\tau,s)\overline{C} + \Lambda H(\tau,s)\right]ds\right\}d\tau. \quad (49)$$

An explicit form of $H(\eta,\eta)$ is subsequently derived. For this purpose, (46) is used to obtain

$$H(\eta,\eta) = 0. \quad (50)$$

By substituting (50) into (49) one derives an integral equation for $H(\xi,\eta)$:

$$H(\xi,\eta) = -\frac{1}{4\theta}\left(\Lambda+\overline{C}\right)(\xi-\eta) + \frac{1}{4\theta}\int_{\eta}^{\xi}\left\{\int_0^{\eta}\left[H(\tau,s)\overline{C} + \Lambda H(\tau,s)\right]ds\right\}d\tau. \quad (51)$$

The method of successive approximations is then applied to show that (51) has a smooth solution. Let us start with an initial approximation

$$H^0(\xi,\eta) = 0 \quad (52)$$

and set up the recursive formula for (51) as follows:

$$H^{n+1}(\xi,\eta) = -\frac{1}{4\theta}\left(\Lambda+\overline{C}\right)(\xi-\eta) + \frac{1}{4\theta}\int_{\eta}^{\xi}\left\{\int_0^{\eta}\left[H^n(\tau,s)\overline{C} + \Lambda H^n(\tau,s)\right]ds\right\}d\tau. \quad (53)$$

Provided that this recursion converges, solution $H(\xi,\eta)$ can be represented as

$$H(\xi,\eta) = \lim_{n\to\infty} H^n(\xi,\eta). \quad (54)$$

Let

$$\Delta H^n(\xi,\eta) = H^{n+1}(\xi,\eta) - H^n(\xi,\eta) \quad (55)$$

stand for the difference between two consecutive terms. Then, the recursion

$$\Delta H^0(\xi,\eta) = H^1(\xi,\eta) = -\frac{1}{4\theta}\left(\Lambda+\overline{C}\right)(\xi-\eta), \quad (56)$$

$$\Delta H^{n+1}(\xi,\eta) = \frac{1}{4\theta}\int_{\eta}^{\xi}\left\{\int_0^{\eta}\left[\Delta H^n(\tau,s)\overline{C} + \Lambda\Delta H^n(\tau,s)\right]ds\right\}d\tau \quad (57)$$

is correspondingly concluded from (52)-(53) and (54) is alternatively represented as

$$H(\xi,\eta) = \sum_{n=0}^{\infty}\Delta H^n(\xi,\eta). \quad (58)$$

Since variables ξ and η lie in the bounded domain $0 \le \eta \le \xi \le 2$, one can apply (56) to show that

$$\left\|\Delta H^0(\xi,\eta)\right\| \le \frac{1}{\theta}\left(\|\Lambda\| + \left\|\overline{C}\right\|\right) = N. \quad (59)$$

In order to apply the mathematical induction method, suppose that

$$\left\|\Delta H^n(\xi,\eta)\right\| \le N^{n+1}\frac{(\xi+\eta)^n}{n!}. \quad (60)$$

Then, by employing (57), (59), and (60) one arrives at

$$\begin{aligned}\left\|\Delta H^{n+1}(\xi,\eta)\right\| &\le \frac{1}{4\theta}\left(\|\Lambda\| + \left\|\overline{C}\right\|\right) \cdot \frac{N^{n+1}}{n!}\left|2\int_0^{\eta}\int_0^{\tau}(\tau+s)^n\,ds\,d\tau + \int_{\eta}^{\xi}\int_0^{\eta}(\tau+s)^n\,ds\,d\tau\right| \\ &= \frac{N^{n+2}}{4n!}\left|2\int_0^{\eta}\int_0^{\tau}(\tau+s)^n\,ds\,d\tau + \int_{\eta}^{\xi}\int_0^{\eta}(\tau+s)^n\,ds\,d\tau\right|.\end{aligned} \quad (61)$$

It is readily shown (cf. [21], equation (2.14)) that the next estimate

$$\left|2\int_0^\eta\int_0^\tau(\tau+s)^n\,ds\,d\tau+\int_\eta^\xi\int_0^\eta(\tau+s)^n\,ds\,d\tau\right| \leq 4\frac{(\xi+\eta)^{n+1}}{(n+1)} \quad (62)$$

holds. Therefore, combining (61) and (62), one gets

$$\left\|\Delta H^{n+1}(\xi,\eta)\right\| \leq N^{n+2}\frac{(\xi+\eta)^{n+1}}{(n+1)!}. \quad (63)$$

Thus, by mathematical induction, (63) holds for all $n \geq 0$. It then follows from the Weierstrass M-test that the series (58) converges absolutely and uniformly in $0 \leq \eta \leq \xi \leq 2$. By (56)-(57), it follows that

$$\Delta H^1(\xi,\eta) = -\frac{\xi^2\eta+\xi\eta^2}{2}\left(\frac{1}{4\theta}\right)^2\left[(\Lambda+\overline{C})\Lambda+\overline{C}(\Lambda+\overline{C})\right]. \quad (64)$$

Iterating on the computations, one observes the pattern which leads to the following formula:

$$\Delta H^n(\xi,\eta) = -\frac{(\xi\eta)^n(\xi-\eta)}{n!\,(n+1)!}\left(\frac{1}{4\theta}\right)^{n+1} \cdot\left[\sum_{i=0}^{n}\binom{n}{i}\Lambda^i(\Lambda+\overline{C})\overline{C}^{n-i}\right]. \quad (65)$$

The solution to the integral equation (51) is therefore given by the next series expansion

$$H(\xi,\eta) = -\sum_{n=0}^{\infty}\frac{(\xi\eta)^n(\xi-\eta)}{n!\,(n+1)!}\left(\frac{1}{4\theta}\right)^{n+1} \cdot\left[\sum_{i=0}^{n}\binom{n}{i}\Lambda^i(\Lambda+\overline{C})\overline{C}^{n-i}\right] \quad (66)$$

which is absolutely and uniformly converging.

Converting (66) into the original x, y variables, one obtains the series expansion (38) for the Kernel matrix $P(x, y)$ which solves the kernel boundary-value problem (40). Straightforward inspection reveals that (38) is infinitely times continuously differentiable. Returning back to the original (x, y) variables, one obtains (38). Theorem 3 is thus proven. □

3.1. Inverse Transformation. Transformation (14) is a matrix Volterra integral equation of the second type. Since $P(x, y)$ is continuous by Theorem 3, there exists a continuous inverse kernel $L(x, y)$ (see, e.g., [11, 22] for the scalar case which is straightforwardly extended to the present vector case) such that

$$\widetilde{Q}(x,t) = \widetilde{Z}(x,t) + \int_0^x L(x,y)\widetilde{Z}(y,t)\,dy \quad (67)$$

implicitly defined on $T = \{(x, y) \in R^2 : 0 \leq y \leq x \leq 1\}$ by

$$L(x,y) = P(x,y) + \int_y^x L(x,s)P(s,y)\,ds. \quad (68)$$

Relation (68) can in fact be easily derived by substituting (14) into (67) and performing straightforward manipulations of the resulting integral equation. The method of successive approximations can be then applied to show that (68) gives rise to a unique $R(x, y)$, which has as much regularity as $P(x, y)$ has. Detailed computations, which follow similar steps as those carried out in the proof of Theorem 3, are skipped for brevity.

4. Main Result

Taking advantage of the explicit solution (38) to the kernel boundary-value problem (20)–(22), the explicit representation

$$M = -\frac{\Lambda+\overline{C}}{2\theta},$$
$$G(x) = \theta\sum_{n=0}^{\infty}\frac{4n(1-x)(2x-x^2)^{n-1}}{n!\,(n+1)!}\left(\frac{1}{4\theta}\right)^{n+1} \cdot\left[\sum_{i=0}^{n}\binom{n}{i}\Lambda^i(\Lambda+\overline{C})\overline{C}^{n-i}\right] \quad (69)$$

of the observer gains is straightforwardly derived by specifying (18)-(19) accordingly.

The stability features of the target error dynamics (15)–(17) are going to be studied. The following result is in force.

Theorem 4. *If the design matrix $\overline{C}$ is selected such that its symmetric part $\overline{C}_s = (\overline{C}+\overline{C}^T)/2$ is positive definite, then system (15)–(17) is exponentially stable in the space $[L_2(0,1)]^n$ with the convergence rate specified by*

$$\left\|\widetilde{Z}(\cdot,t)\right\|_{2,n} \leq \left\|\widetilde{Z}(\cdot,0)\right\|_{2,n} e^{-\sigma_1(\overline{C}_s)t}. \quad (70)$$

Proof. Consider the Lyapunov function $V(t) = (1/2)\int_0^1 \widetilde{Z}^T(\xi, t)\widetilde{Z}(\xi,t)\,d\xi = (1/2)\|\widetilde{Z}(\cdot,t)\|_{2,n}^2$. The corresponding time derivative along the solutions of (15)–(17) is given by

$$\dot{V}(t) = \int_0^1 \widetilde{Z}^T(\xi,t)\Theta\widetilde{Z}_{xx}(\xi,t)\,d\xi - \int_0^1 \widetilde{Z}^T(\xi,t)\overline{C}\widetilde{Z}(\xi,t)\,d\xi. \quad (71)$$

Integrating by parts taking into account the BCs (16) and (17) and exploiting the diagonal form of matrix Θ yield

$$\int_0^1 \widetilde{Z}^T(\xi,t)\Theta\widetilde{Z}_{xx}(\xi,t)\,d\xi = \widetilde{Z}^T(\chi,t)\Theta\widetilde{Z}_x(\chi,t)\Big|_{\chi=0}^{\chi=1} - \int_0^1 \widetilde{Z}_x^T(\xi,t)\Theta\widetilde{Z}_x(\xi,t)\,d\xi \leq -\theta_m\left\|\widetilde{Z}_x(\cdot,t)\right\|_{2,n}^2, \quad (72)$$

where $\theta_m = \min_{1\le i\le n}\theta_i > 0$. Since $\sigma_1(\overline{C}_s)$ is assumed to be positive then exploiting the trivial inequality $\widetilde{Z}^T(\xi,t)\overline{C}\widetilde{Z}(\xi,t) \ge \sigma_1(\overline{C}_s)^T\widetilde{Z}(\xi,t)\widetilde{Z}(\xi,t)$ and employing (72), one manipulates (71) to derive

$$\begin{aligned}\dot{V}(t) &\le -\theta_m \left\|\widetilde{Z}_\xi(\cdot,t)\right\|_{2,n}^2 - 2\sigma_1\left(\overline{C}_s\right)V(t)\\ &\le -2\sigma_1\left(\overline{C}_s\right)V(t),\end{aligned} \tag{73}$$

thereby concluding the exponential stability of the target error dynamics in the space $[L_2(0,1)]^n$ with a convergence rate obeying the estimate (70). Theorem 4 is proven. □

The next theorem specifies the proposed observer design and summarizes the main result of this paper.

Theorem 5. *The observer (9), with gains M and $G(x)$ set as in (69) and with matrix $\overline{C}$ being selected such that its symmetric part $\overline{C}_s = (\overline{C}+\overline{C}^T)/2$ is positive definite, reconstructs the state of system (4)–(6) with an arbitrarily fast convergence rate in accordance with*

$$\left\|\widetilde{Q}(\cdot,t)\right\|_{2,n} \le A\left\|\widetilde{Q}(\cdot,0)\right\|_{2,n} e^{-\sigma_1(C_s)t}, \tag{74}$$

where A is a positive constant independent of $\widetilde{Q}(\xi,0)$.

Proof. In Lemma 1 and Theorem 3, it was shown that the error system (11)–(13) is transferred, by means of (14), into the target error dynamics (15)–(17) provided that the gains M and $G(x)$ are selected as in (18)-(19) where solution $P(x,y)$ to kernel PDE (20)–(22) is given by (38). Specifying (18)-(19) in light of the actual form of solution (38), it straightforwardly results in (69), where $P(0,0)$ is derived by specifying (21) at $x=0$ and $P_y(x,0)$ is readily obtained by differentiating (38) with respect to y at $y=0$.

The asymptotic stability features of (15)–(17), subject to the design requirement that the arbitrary design parameter $\overline{C}_s = (\overline{C}+\overline{C}^T)/2$ is positive definite, were demonstrated in Theorem 4. In particular, according to (70), the corresponding convergence rate can be made arbitrarily fast by a proper selection of the $\overline{C}$ matrix.

From now on, we follow [20] to derive analogous convergence properties for the original system (4)–(6) as well. Observing that $\xi+\eta=x$, one derives from (58)–(60) that $\|P(x,y)\| \le Ne^{2Nx}$, and the same bound can be derived for the norm of the inverse transformation kernel matrix $L(x,y)$ as well; that is, $\|L(x,y)\| \le Ne^{2Nx}$. A straightforward generalization of [20, Th 4] yields that those two boundedness relations, coupled together, establish the equivalence of norms of $\widetilde{Z}(x,t)$ and $\widetilde{Q}(x,t)$ in $[L_2(0,1)]^n$ which means that there exists a positive constant A independent of $\widetilde{Q}(\xi,0)$ such that the estimate (74) is in force as a direct consequence of (70). Theorem 5 is proven. □

5. Simulation Results

5.1. Academic Example. To validate the proposed observer, system (4)–(6) of coupled reaction-diffusion processes is specified for simulation purposes with $n=3$ and with parameters

$$\theta = 2,\qquad \Lambda = \begin{bmatrix}1 & 2 & 3\\ 4 & 5 & 3\\ 2 & 5 & 1\end{bmatrix}. \tag{75}$$

The initial conditions are set to $q_1(x,0) = q_2(x,0) = q_3(x,0) = 2\sin(\pi x) + 2\sin(3\pi x)$. For solving the underlying PDEs, a standard finite-difference approximation method is used by discretizing the spatial solution domain $x\in[0,1]$ into a finite number of N uniformly spaced solution nodes $x_i = ih$, $h = 1/(N+1)$, $i = 1,2,\ldots,N$. The value $N=40$ is then used. The resulting 40th order discretized system is subsequently solved by fixed-step Runge-Kutta ODE4 method with step $T_s = 10^{-4}$.

The unstable behaviour of the plant subject to the open-loop input vector $U(t) = [5\sin t, 10\sin 2t, 15\sin 3t]^T$ is displayed in Figure 1, which for certainty shows the diverging spatiotemporal evolution of the states $q_1(x,t)$ and $q_3(x,t)$.

Observer (9), (69) has been implemented by selecting the design matrix $\overline{C} = 10I_{3\times3}$ and by specifying the initial conditions at $\hat{q}_1(x,0) = \hat{q}_2(x,0) = \hat{q}_3(x,0) = 0$. Figure 2 displays the spatiotemporal evolution of the observed states $\hat{q}_1(x,t)$ and $\hat{q}_3(x,t)$, which clearly mimic the corresponding actual states. Figure 3 shows the temporal evolution of the norm $\|\widetilde{Q}(\cdot,t)\|_{2,3}$, which tends to zero exponentially, thus confirming the correct functioning of the proposed observer and supporting the theoretical analysis.

5.2. Application Example. To provide a more valuable validation of the proposed scheme, we consider the coupled temperature-concentration dynamics of a Chemical Tubular Reactor (CTR) at low fluid superficial velocities, when convection terms become negligible, dealt with in [17]. After a suitable transformation, the next dimensionless model was derived

$$\begin{aligned}\frac{\partial x_1}{\partial t} &= D_1\frac{\partial^2 x_1}{\partial \xi^2} + k_0\delta\left(1-x_2\right)e^{-\gamma/(1+x_1)},\\ \frac{\partial x_2}{\partial t} &= D_2\frac{\partial^2 x_2}{\partial \xi^2} + k_0\left(1-x_2\right)e^{-\gamma/(1+x_1)},\\ x_{1\xi}(0,t) &= x_{2\xi}(0,t) = 0,\\ x_1(1,t) &= u_1(t),\\ x_2(1,t) &= u_2(t),\end{aligned} \tag{76}$$

where the states x_1 and x_2 denote the normalized temperature and concentration, respectively, and the underlying physical parameters take the values

$$\begin{aligned}D_1 &= D_2 = 0.167,\\ \delta &= 0.5,\\ k_0 &= 2.426\cdot 10^7,\\ \gamma &= 20.\end{aligned} \tag{77}$$

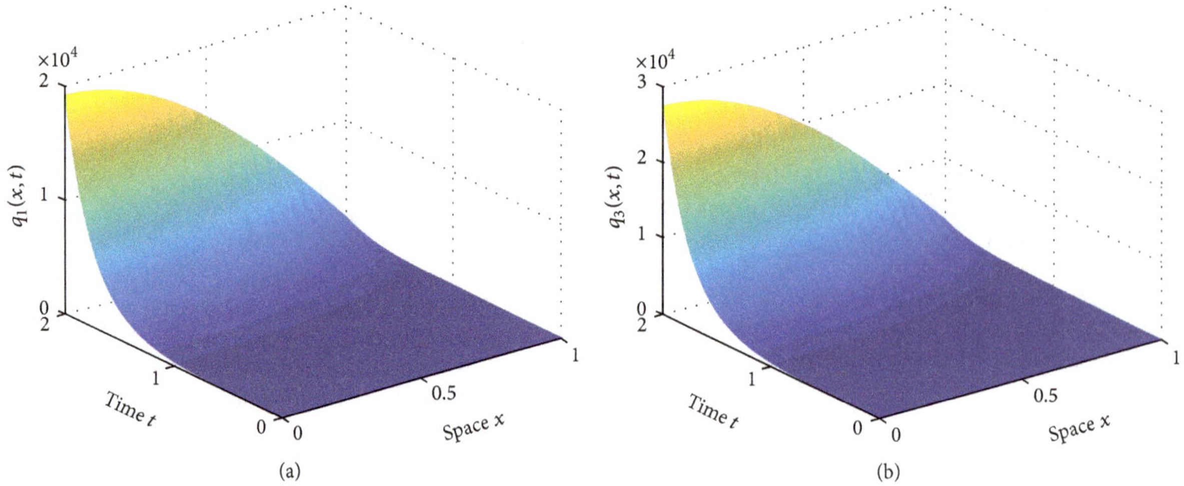

FIGURE 1: Spatiotemporal evolution of $q_1(x,t)$ (a) and $q_3(x,t)$ (b).

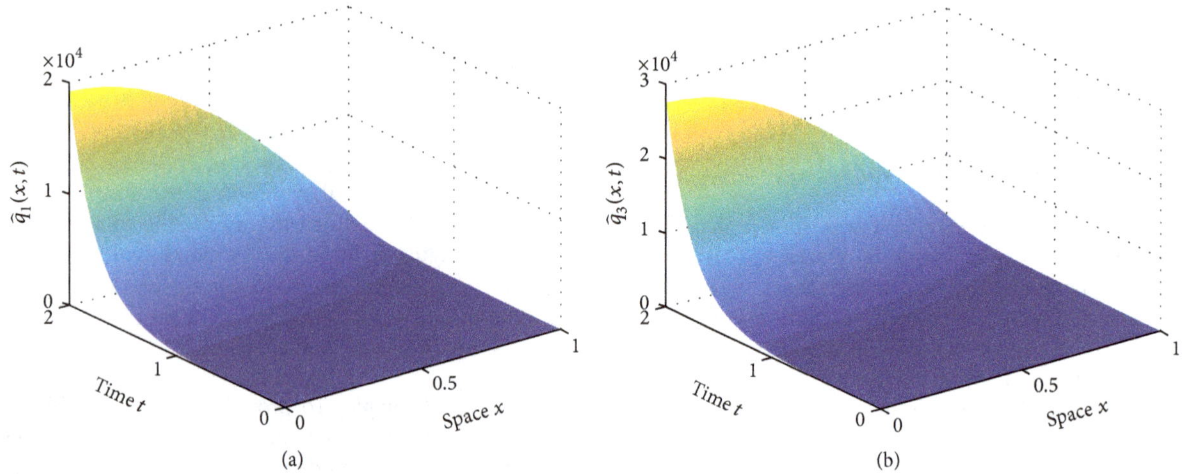

FIGURE 2: Spatiotemporal evolution of $\hat{q}_1(x,t)$ (a) and $\hat{q}_3(x,t)$ (b).

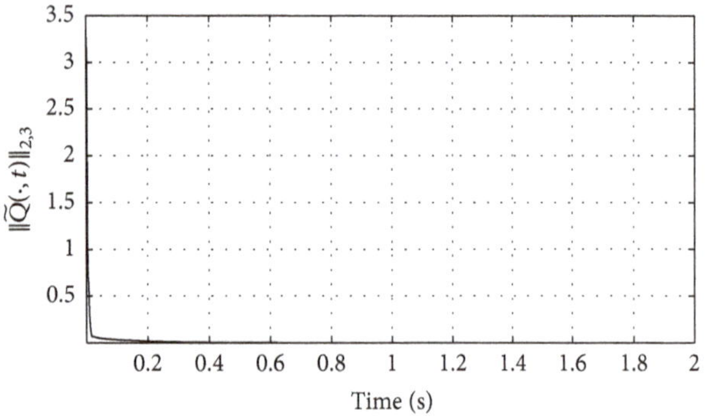

FIGURE 3: Temporal evolution of the norm $\|\widetilde{Q}(\cdot,t)\|_{2,3}$.

Its linearization around the constant profiles

$$x_1^*(\xi,t) = 0.1,$$
$$x_2^*(\xi,t) = 0.98 \quad (78)$$

gives rise to model (4)–(6) with the following diffusivity and reaction parameters:

$$\theta = 0.167,$$
$$\Lambda = \begin{bmatrix} 1.018 & 0.154 \\ 2.037 & 0.308 \end{bmatrix}. \quad (79)$$

The open-loop control input $U(t) = [5\sin t, 10\sin 2t]^T$ was selected. The plant ICs are set to $x_1(x,0) = x_2(x,0) = 2\sin(\pi\xi) + 2\sin(3\pi\xi)$. The unstable open-loop behaviour of the plant state $x_2(\xi,t)$ is displayed in Figure 4(a). Observer (9), (69) has been implemented by selecting the design matrix $\overline{C} = 20I_{2\times 2}$ and by specifying ICs $\hat{x}_1(\xi,0) = \hat{x}_2(\xi,0) = 0$. Figure 4(b) shows that the observer is able to correctly reconstruct the unstable profile of the plant state $x_2(\xi,t)$. Figure 5 shows the temporal evolution of the norm $\|\widetilde{Q}(\cdot,t)\|_{2,2}$, which confirms the correct functioning of the observer for the estimation of the state variable $x_2(\xi,t)$, too.

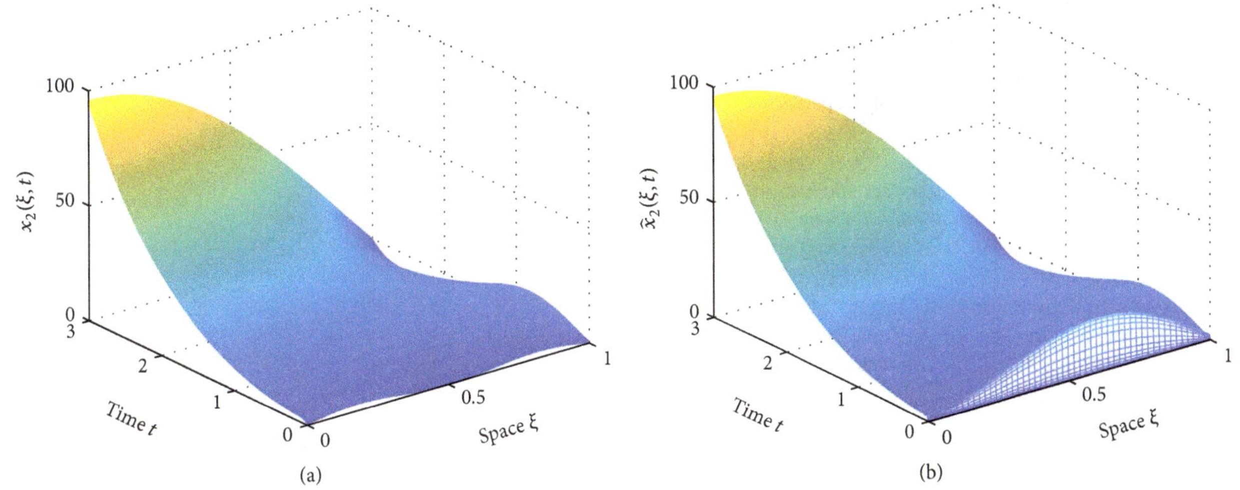

FIGURE 4: Spatiotemporal evolution of $x_2(\xi, t)$ (a) and $\hat{x}_2(\xi, t)$ (b).

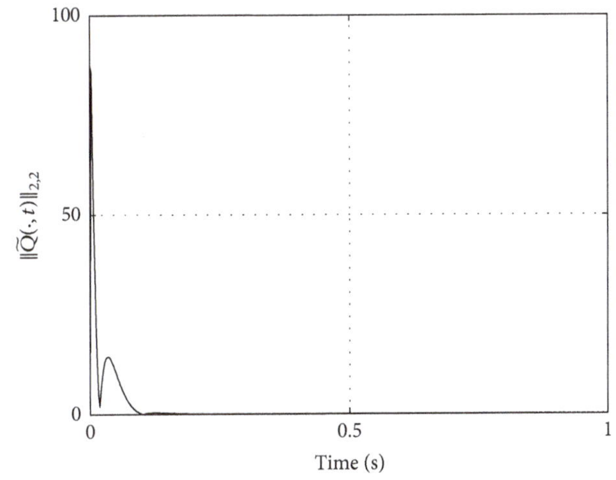

FIGURE 5: Temporal evolution of the norm $\|\widetilde{Q}(\cdot, t)\|_{2,2}$.

6. Conclusions

The backstepping-based anticollocated observer design of a system of n coupled parabolic linear PDEs has been tackled, and an explicit representation of the underlying observer gains has been derived which allows one to enforce an arbitrarily fast exponential decay of the observation error dynamics in the space $[L_2(0, 1)]^n$. The extension to the case of different diffusivities and spatially dependent parameters and the observer-based output feedback design of a stabilizing controller $U(t)$ are among the most interesting future lines of related investigations that will be pursued in our future work.

Conflict of Interests

The authors declare that there is no conflict of interests regarding the publication of this paper.

Acknowledgments

The research leading to these results has received funding from the Research Project "Modeling, Control and Experimentation of Innovative Thermal Storage Systems," funded by Sardinia Regional Government under Grant Agreement no. CRP-60193, and from the Research Project "RODEO—RObust Decentralised Estimation fOr Large-Scale Systems," funded by the Italian Ministry for Foreign Affairs under Grant Agreement PGR00152.

References

[1] A. Vande Wouwer and M. Zeitz, "State estimation in distributed parameter systems," in *Encyclopedia of Life Support Systems (EOLSS)*, H. Unbehauen, Ed., Control Systems, Robotics and Automation, EOLSS Publishers, Oxford, UK, 2001.

[2] Z. Hidayat, R. Babuska, B. De Schutter, and A. Nunnez, "Observers for linear distributed-parameter systems: a survey," in *Proceedings of the IEEE International Symposium on Robotic and Sensors Environments (ROSE '11)*, pp. 166–171, IEEE, Montreal , Canada, September 2011.

[3] R. F. Curtain and H. Zwart, *An Introduction to Infinite-Dimensional Linear Systems Theory*, vol. 21 of *Texts in Applied Mathematics*, Springer, New York, NY, USA, 1995.

[4] T. Lasiecka and R. Triggiani, *Control Theory for Partial Differential Equations: Continuous and Approximation Theories, I Abstract Parabolic Systems*, Cambridge University Press, Cambridge, UK, 2000.

[5] M. Krstic and A. Smyshlyaev, *Boundary Control of PDEs: A Course on Backstepping Designs*, SIAM Advances in Design and Control, SIAM, Philadelphia, Pa, USA, 2008.

[6] A. Smyshlyaev and M. Krstic, "Backstepping observers for a class of parabolic PDEs," *Systems & Control Letters*, vol. 54, no. 7, pp. 613–625, 2005.

[7] D. Tsubakino and S. Hara, "Backstepping observer design for parabolic PDEs with measurement of weighted spatial averages," *Automatica*, vol. 53, pp. 179–187, 2015.

[8] L. Jadachowski, T. Meurer, and A. Kugi, "State estimation for parabolic PDEs with varying parameters on 3-dimensional

spatial domains," in *Proceedings of the 18th IFAC World Congress*, pp. 13338–13343, Milano, Italy, September 2011.

[9] L. Jadachowski, T. Meurer, and A. Kugi, "Backstepping observers for linear PDEs on higher-dimensional spatial domains," *Automatica*, vol. 51, pp. 85–97, 2015.

[10] R. Miranda, I. Chairez, and J. Moreno, "Observer design for a class of parabolic PDE via sliding modes and backstepping," in *Proceedings of the 11th International Workshop on Variable Structure Systems (VSS '10)*, pp. 215–220, Mexico City, Mexico, June 2010.

[11] F. Di Meglio, R. Vazquez, and M. Krstic, "Stabilization of a system of coupled first-order hyperbolic linear pdes with a single boundary input," *IEEE Transactions on Automatic Control*, vol. 58, no. 12, pp. 3097–3111, 2013.

[12] R. Vazquez, E. Schuster, and M. Krstic, "Magnetohydrodynamic state estimation with boundary sensors," *Automatica*, vol. 44, no. 10, pp. 2517–2527, 2008.

[13] R. Vazquez, M. Krstic, and J.-M. Coron, "Backstepping boundary stabilization and state estimation of a 2×2 linear hyperbolic system," in *Proceedings of the 50th IEEE Conference Decision and Control and European Control Conference (CDC-ECC '11)*, pp. 4937–4942, Orlando, Fla, USA, December 2011.

[14] O. M. Aamo, "Disturbance rejection in 2×2 linear hyperbolic systems," *IEEE Transactions on Automatic Control*, vol. 58, no. 5, pp. 1095–1106, 2013.

[15] S. Moura, J. Bendtsen, and V. Ruiz, "Observer design for boundary coupled PDEs: application to thermostatically controlled loads in smart grids," in *Proceedings of the 52nd IEEE Conference on Decision and Control (CDC '13)*, pp. 6286–6291, Firenze, Italy, December 2013.

[16] R. Vazquez and M. Krstic, "Boundary observer for output-feedback stabilization of thermal-fluid convection loop," *IEEE Transactions on Control Systems Technology*, vol. 18, no. 4, pp. 789–797, 2010.

[17] Y. Orlov and D. Dochain, "Discontinuous feedback stabilization of minimum-phase semilinear infinite-dimensional systems with application to chemical tubular reactor," *IEEE Transactions on Automatic Control*, vol. 47, no. 8, pp. 1293–1304, 2002.

[18] A. Baccoli, Y. Orlov, and A. Pisano, "On the boundary control of coupled reaction-diffusion equations having the same diffusivity parameters," in *Proceedings of the IEEE 53rd Annual Conference on Decision and Control (CDC '14)*, pp. 5222–5228, Los Angeles, Calif, USA, December 2014.

[19] A. Baccoli, A. Pisano, and Y. Orlov, "Boundary control of coupled reaction-diffusion processes with constant parameters," *Automatica*, vol. 54, pp. 80–90, 2015.

[20] A. Smyshlyaev and M. Krstic, "Closed-form boundary state feedbacks for a class of 1-D partial integro-differential equations," *IEEE Transactions on Automatic Control*, vol. 49, no. 12, pp. 2185–2202, 2004.

[21] B.-Z. Guo and J.-J. Liu, "Sliding mode control and active disturbance rejection control to the stabilization of one-dimensional Schrödinger equation subject to boundary control matched disturbance," *International Journal of Robust and Nonlinear Control*, vol. 24, no. 16, pp. 2194–2212, 2014.

[22] R. Vazquez, *Boundary control laws and observer design for convective, turbulent and magnetohydrodynamic flows [Ph.D. thesis]*, University of California, San Diego, Calif, USA, 2006.

2

Model Predictive Control for Load Frequency Control with Wind Turbines

Yi Zhang,[1,2] Xiangjie Liu,[2] and Yujia Yan[2]

[1]*Department of Electrical Engineering, North China University of Science and Technology, Tangshan 063000, China*
[2]*The State Key Laboratory of Alternate Electrical Power System with Renewable Energy Sources, North China Electric Power University, Beijing 102206, China*

Correspondence should be addressed to Yi Zhang; zhangyizhouzhao@163.com

Academic Editor: Onur Toker

Reliable load frequency (LFC) control is crucial to the operation and design of modern electric power systems. Considering the LFC problem of a four-area interconnected power system with wind turbines, this paper presents a distributed model predictive control (DMPC) based on coordination scheme. The proposed algorithm solves a series of local optimization problems to minimize a performance objective for each control area. The scheme incorporates the two critical nonlinear constraints, for example, the generation rate constraint (GRC) and the valve limit, into convex optimization problems. Furthermore, the algorithm reduces the impact on the randomness and intermittence of wind turbine effectively. A performance comparison between the proposed controller with and that without the participation of the wind turbines is carried out. Good performance is obtained in the presence of power system nonlinearities due to the governors and turbines constraints and load change disturbances.

1. Introduction

Wind energy is considered as a promising and encoring renewable energy alternative for power generation owing to environmental and economical benefits. The world market of wind installation set a new record in the year of 2014 and reached a total size of 51 GW [1]. Nowadays, due to the interconnection of more distributed generators, especially wind turbines that are committed to grid operation, electric power system has become more complicated than ever.

Power system LFC incorporating WTGs can be a quite challengeable issue. The output power of WTGs varies with wind speed fluctuation [2]. This wind power fluctuation imposed additional power imbalance to the power system and may cause frequency deviation from the nominal value [3]. Significant frequency deviations may lead to the disconnection of some loads and generations and even can lead to whole power system oscillations. Previous studies [4–6] provide extensive overviews of the primary and secondary frequency control strategies of power systems with wind power plants.

LFC, secondary frequency control, has been performed by integrating the area control error (ACE), which acts on the load reference settings of the governors. LFC tasks are maintaining tie-line power flow and system frequency close to nominal value for the multiarea interconnected power system [7]. As a fundamental characteristic of electric power operations, frequency of the system deviates from its nominal value due to generation-demand imbalance. Conventional generators, in which the turbine rotational speed is nearly constant, provide inertia and governor response against frequency deviations; however, the speed of a wind turbine is not synchronous with the grid and is usually controlled to track the maximum power point. It implies that the wind turbines will have less time to react to the power imbalance, probably resulting in lager frequency deviations.

Thus, it is thus necessary to establish the optimal profile of the WTGs power surge in coordination with the characteristics of conventional plants to achieve a more economical and reliable operation of power system. With the large amount of realistic constraints, for example, generation rate constraints

(GRCs) in the conventional units, the pitch angle, and generator torque constraints in WTGs, the LFC becomes a large-scale, distributed, multiconstraints optimization problem.

Recently, a few attempts studied the idea of wind turbines in the issue of LFC [8–10]. Two types of wind farm models are derived and demonstrated to portray the capability of set-point tracking under automatic generation control (AGC) [8]. This inference leads to the development of a simplified wind farm model that is specially designed for the set-point control in the power system study. However, the durability of inertia effect depends on the allowable rotor speed range. An adaptive fuzzy logic structure was used to propose a new LFC scheme in the interconnected large-scale power system in the presence of wind turbines [9]. The performance against sudden load change and wind power fluctuations in different wind power penetration rates is confirmed by simulation. A flatness-based method to control frequency and power flow for multiarea power system with wind turbine is presented in [10]. And, practical constraints such as generator ramping rates of wind turbine generator can be considered in designing the controllers. As abovementioned reference, the control schemes are designed for each area to maintain the frequency at nominal value and to keep power flows near scheduled values. However, local controller in each area does not work cooperatively towards satisfying systemwide control objectives. In addition, the control scheme [8–10] mentioned above could yield unsatisfactory performance since the effects of nonlinearities such as generation rate constraint and generation ramping rate were not considered.

Model predictive control (MPC), also called receding horizon control, was originally developed to be an effective method for processing industrial control. It transforms the control problem into a finite horizon optimal control problem that can also satisfy multivariable constraints on the input, output, and state variables. In the power industry, MPC has been successfully used in controlling power plant steam-boiler generation processes [11–13]. In power system control, MPC was first developed to be an economic-oriented LFC [14], which generates the control action based on the open-loop optimization method over a finite horizon. MPC has subsequently been developed to realize the constrained optimal algorithm for LFC problem. In [15], the constraint handling ability of MPC is employed to effectively account for the generation rate constraints (GRCs) but without the analysis of closed-loop stability and robustness. Recently, MPC has been successfully used in LFC design of multiarea power system with wind turbines [16]. However, each area controller is designed independently and the communication between the local controllers is not considered. On the other hand, with the size and capacity of wind farms increasing in recent years, traditional centralized MPCs encounter many difficulties due to limitations in exchanging information with large-scale, geographically extensive control areas. In order to deal with these issues, advanced distributed control strategies have to be investigated and implemented.

Developing decentralized/distributed LFC structures can be an effective way of solving this problem. The decentralized model predictive control scheme for the LFC of multiarea interconnected power system is presented in [17]. However, the local controller does not consider generation rate constraint that is only imposed on the turbine in the simulation. In the distributed MPC (DMPC), the benefits from using a decentralized structure are partially preserved, and the plant-wide performance and stability are improved through coordination [18, 19]. In [20], feasible cooperation-based MPC method is used in distributed LFC instead of centralized MPC. It is noted that the range of load change used in the cases is very large and inappropriate for the LFC issue.

This paper studies the effect of merging the wind turbines on the system frequency of multiarea power system. The first control area includes an aggregated wind turbine model (which consists of 60 wind turbine units) beside the thermal power plant. According to the distributed LFC structure, the dynamics model of the four-area interconnected power system is established. In our scheme, the overall power system is decomposed into four areas and each area has its own local MPC controller. These areas-based MPCs exchange their measurement and predictions by communication and incorporate the information from other controllers into their local objective so as to coordinate with each other. The controllers calculate the optimal control signal while respecting constraints over the wind turbines output frequency deviation and the load change. Not only do the effects of the physical constraints conclude generating rate constraints (GRCs) and the limit of governor position in conventional power plant, but also the wind speed constraints in wind turbines are considered. Comparisons of response to step load change, computational burden, and robustness have been made between DMPC, centralized MPC, and decentralized MPC. The results confirm the superiority of the proposed DMPC technique.

The remainder of the paper is organized as follows. Modeling of wind turbines participation in LFC is presented in Section 2, and the proposed DMPC algorithm is presented in Section 3. Section 4 presents the application of the algorithm in a four-area interconnected power system. The conclusions are presented in Section 5.

2. Distributed Model of Hybrid Power System

Figure 1 illustrates the interconnected power system consisting of four control areas connected by tie-lines, which consists of thermal power plant, variable speed wind turbines (VSWTs), and hydro power plant. In area 1, wind turbine is taken into consideration as it can provide a new solution to the contradiction between economic development and environment pollution. Area 4 is the thermal power plant, while area 2 and area 3 are hydro power plants.

Detailed compositions of each area are shown in Figures 2–4. In addition, area 1 includes an aggregated wind turbine model which consists of 40 VSWT units while the capacity of thermal plant is 600 MW. The variables and parameters are listed in Table 1. In each control area, a change in local demand (load) alters the nominal frequency. The DMPC in each control area i manipulates the load reference set-point $\Delta P_{\mathrm{ref},i}$ to drive the frequency deviations Δf_i and tie-line power flow deviations $\Delta P_{\mathrm{tie},ij}$ to zero.

TABLE 1: Power system variables and parameter.

Parameter/variable	Description	Unit
ω_r	Angular velocity of rotor	rad/s
ω_g	Angular velocity of high speed shaft and generator	rad/s
T_g	Generator reaction torque	Nm
T_r	Aerodynamic torque	Nm
K_s	Total stiffness on low speed shaft	Nm/rad
J_r	Inertia of the rotor (low speed shaft and gearbox)	Kgm^2
J_g	Inertia of the rotor (high speed shaft and gearbox)	Kgm^2
n_{gear}	Exchange ratio	Null
η_{gear}	Efficiency of the gear box	%
τ_β	Actuator time constant	s
K_β	Actuator gain	Hz/p.u.MW
v_m	Wind speed	m/s
θ_{ref}	Pitch demand	rad
θ	Pitch angle	rad
P_e^{ref}	Power demand	p.u.MW
P_e	The output of wind turbine	p.u.MW
$\Delta f_i(t)$	Frequency deviation	Hz
$\Delta P_{gi}(t)$	Generator output power deviation	p.u.MW
$\Delta X_{gi}(t)$	Governor valve position deviation	p.u.
$\Delta X_{ghi}(t)$	Governor valve servomotor position deviation	p.u.
$\Delta P_{tie,i}(t)$	Tie-line active power deviation	p.u.MW
$\Delta P_{di}(t)$	Load disturbance	p.u.MW
K_{Pi}	Power system gain	Hz/p.u.MW
K_{ri}	Reheat turbine gain	Hz/p.u.MW
T_{Pi}	Power system time constant	s
T_{ri}	Reheat turbine time constant	s
T_{Wi}	Water starting time	s
T_{ii}, T_{Ri}	Hydro governor time constants	s
T_{Gi}	Thermal governor time constant	s
T_{Ti}	Turbine time constant	s
K_{Sij}	Interconnection gain between control areas	p.u.MW
K_{Bi}	Frequency bias factor	p.u.MW/Hz
R_i	Speed drop due to governor action	Hz/p.u.MW
ACE_i	Area control error	p.u.MW

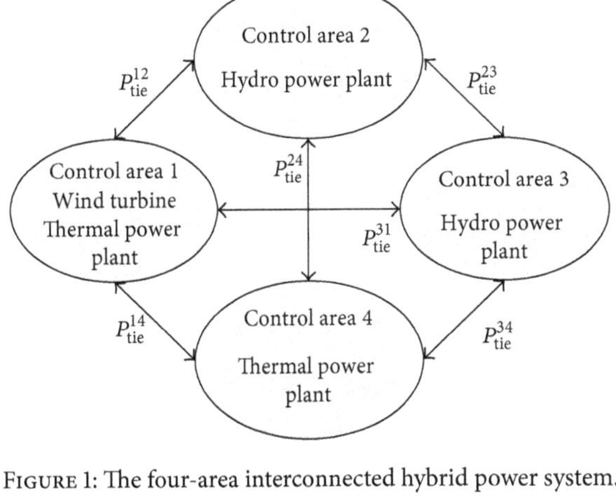

FIGURE 1: The four-area interconnected hybrid power system.

2.1. Wind Turbine Model. A wind turbine is an installation for converting kinetic energy extracted from wind to electrical energy. Figure 5 illustrates the basic model structure of a wind turbine and the interactions between the different dynamic components in the model. The whole wind turbine can be divided into four subsystems: aerodynamics subsystem, mechanical subsystem, electrical, and actuator subsystem [21].

The linearization model for the variable speed wind turbine in Figure 6 can be represented by

$$\Delta\dot{\varphi}_\varepsilon = \Delta\omega_r - \frac{1}{n_{\text{gear}}}\Delta\omega_g, \quad (1a)$$

$$\Delta\dot{\omega}_r = -\frac{K_s}{J_r}\Delta\varphi_\varepsilon + \frac{1}{J_r}\Delta T_r, \quad (1b)$$

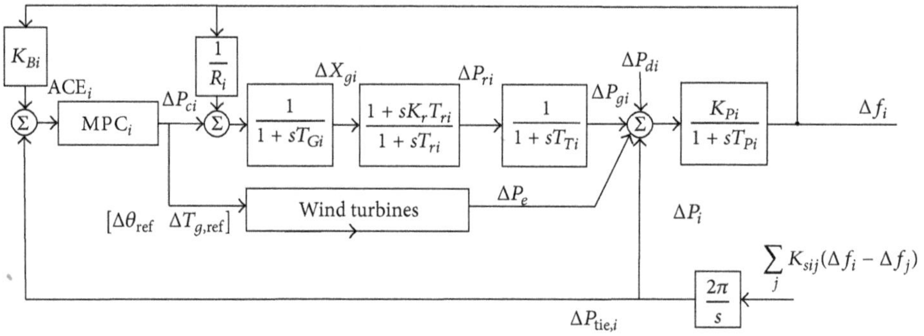

FIGURE 2: Block diagram of a thermal power plant and wind turbines ($i = 1$).

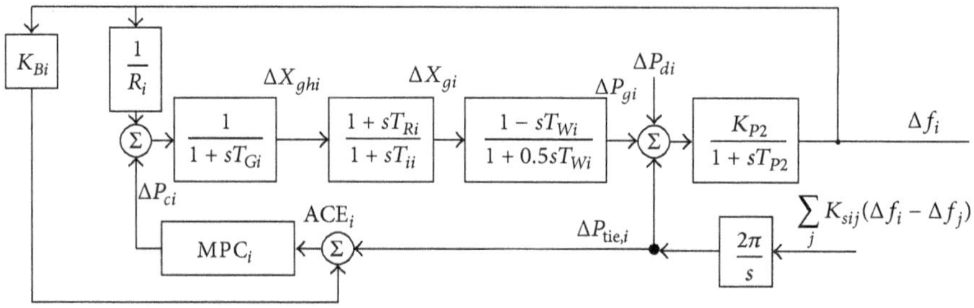

FIGURE 3: Block diagram of a hydro power plant ($i = 2, 3$).

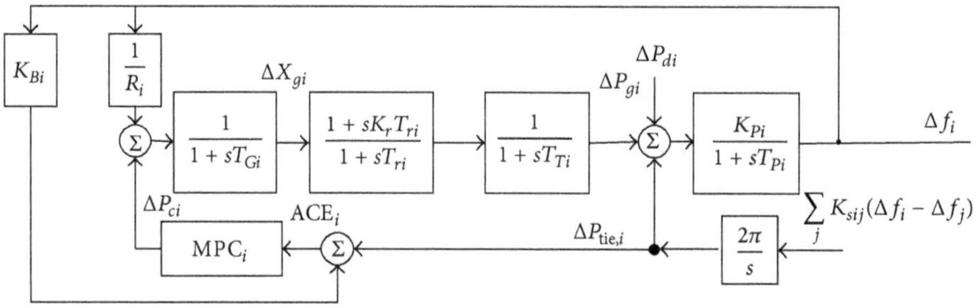

FIGURE 4: Block diagram of a thermal power plant ($i = 4$).

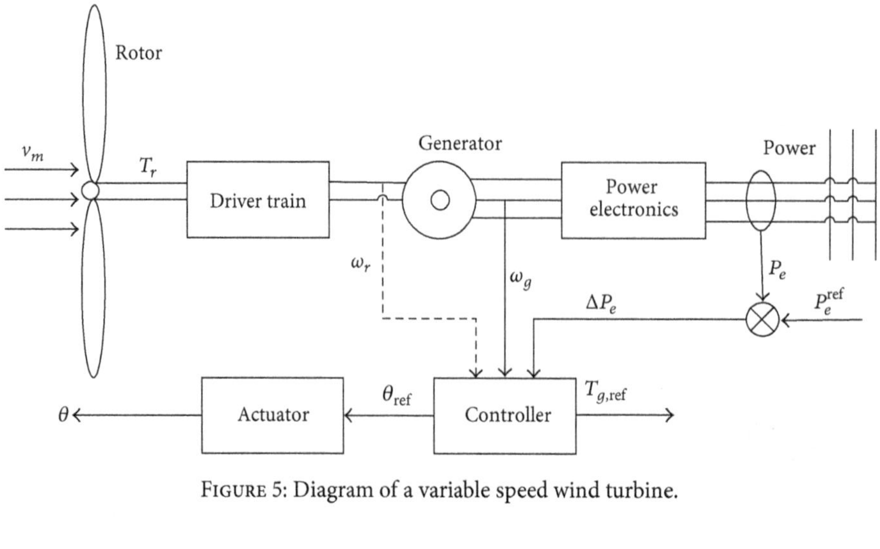

FIGURE 5: Diagram of a variable speed wind turbine.

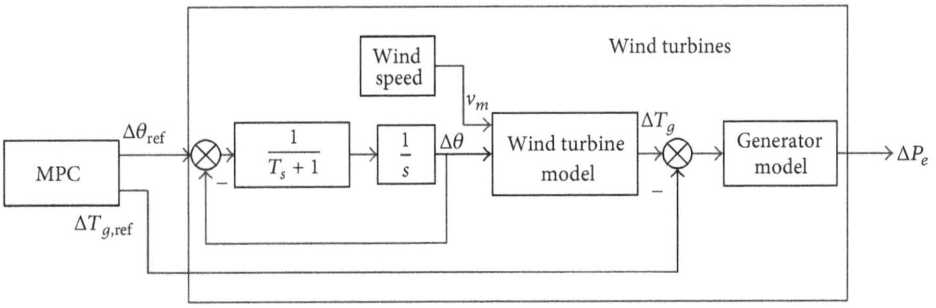

FIGURE 6: Diagram of wind power plant in area 1.

$$\Delta\dot{\omega}_g = \frac{\eta_{\text{gear}} K_s}{n_{\text{gear}}}\Delta\varphi_\varepsilon - \frac{1}{J_r}\Delta T_g, \quad (1c)$$

$$\Delta\dot{\theta} = -\frac{1}{\tau_\beta}\Delta\theta + \frac{K_\beta}{\tau_\beta}\Delta\theta_{\text{ref}}. \quad (1d)$$

The generator reaction torque T_g and the reference pitch angle θ_{ref} are used as indicator of the input of VSWT, as $u_e = [\Delta\beta_{\text{ref}}\ \ \Delta T_g]^T \in R^2$. Moreover, η is the efficiency of the generator and ω_g and T_g are used as indicator of the output power as $P_e = \eta\omega_g T_g \in R^1$, where ω_g is the angular velocity of generator shaft. A generalized representation of the state-space model of the variable speed turbine can be described as

$$\dot{x}_e(t) = A(v_m)x_e(t) + B_1(v_m)\omega(t) + B_2u_e(t), \quad (2a)$$

$$z_e(t) = Cx_e(t) + D_1\omega(t) + D_2u_e(t) \quad (2b)$$

with

$$A(v_m) = \begin{bmatrix} 0 & 1 & -\frac{1}{n_{\text{gear}}} & 0 \\ -\frac{K_s}{J_r} & \frac{1}{J_r}\frac{\partial T_r}{\partial\omega_r}\Big|_{\text{op}} & 0 & \frac{1}{J_r}\frac{\partial T_r}{\partial\theta}\Big|_{\text{op}} \\ \frac{\eta_{\text{gear}}K_s}{n_{\text{gear}}J_g} & 0 & 0 & 0 \\ 0 & 0 & 0 & -\frac{1}{\tau_\theta} \end{bmatrix},$$

$$B_1(v_m) = \begin{bmatrix} 0 \\ \frac{1}{J_r}\frac{\partial T_r}{\partial v}\Big|_{\text{op}} \\ 0 \\ 0 \end{bmatrix},$$

$$B_2 = \begin{bmatrix} 0 & 0 \\ 0 & 0 \\ 0 & -\frac{1}{J_g} \\ \frac{K_\beta}{\tau_\beta} & 0 \end{bmatrix},$$

$$\begin{aligned} C &= \begin{bmatrix} 0 & 0 & 1 & 0 \\ 0 & 0 & 0 & 0 \end{bmatrix}, \\ D_1 &= \begin{bmatrix} 0 \\ 0 \end{bmatrix}, \\ D_2 &= \begin{bmatrix} 0 & 0 \\ 0 & 1 \end{bmatrix}, \\ x_e &= [\Delta\varphi_\varepsilon\ \ \Delta\omega_r\ \ \Delta\omega_g\ \ \Delta\theta]^T, \\ u_e &= [\Delta\theta_{\text{ref}}\ \ \Delta T_{g,\text{ref}}]^T, \\ z_e &= [\Delta\omega_g\ \ \Delta T_g]^T, \\ y_e &= P_e = \eta\omega_g T_g. \end{aligned} \quad (3)$$

2.2. Four-Area Power System with Wind Turbine. Denoting that the control area i ($i = 1,2,3,4$) is to be interconnected with the control area j, $j \neq i$, through a tie-line, a linear continuous time-varying model of control area i can be written as

$$\begin{aligned} \dot{x}_i &= A_{ii}x_i + B_{ii}u_i + F_{ii}d_i + \sum_{i\neq j}\left(A_{ij}x_j + B_{ij}u_j + F_{ij}d_j\right), \\ y_i &= C_{ii}x_i, \end{aligned} \quad (4)$$

where $x_i \in R^n$, $u_i \in R^m$, $d_i \in R^k$, and $y_i \in R^l$ are the state vector, the control signal vector, the disturbance vector, and the vector of output of control area i, respectively. $x_j \in R^p$, $u_j \in R^q$, and $d_j \in R^s$ are the state vector, the control signal vector, and the disturbance vector of neighbor control area, respectively. Matrices A_{ii}, B_{ii} C_{ii}, and F_{ii} represent appropriate system matrices of control area i, and A_{ij}, B_{ij}, and F_{ij} represent the matrices of interaction variables between area i and area j. Tie-line power for area i is represented by

$$\begin{aligned} \Delta P_{\text{tie},i} &= \sum_{\substack{j=1 \\ j\neq i}}^{4}\Delta P_{\text{tie}}^{ij} = \sum_{\substack{j=1 \\ j\neq i}}^{4}K_{sij}\left(\Delta f_i - \Delta f_j\right), \\ \Delta P_{\text{tie}}^{ij} &= -\Delta P_{\text{tie}}^{ji}. \end{aligned} \quad (5)$$

The state, disturbance, and output vectors for area i are defined by

$$x_i = [\Delta f_i \;\; \Delta P_{\mathrm{tie},i} \;\; \Delta P_{gi} \;\; \Delta X_{gi} \;\; \Delta \varphi_\varepsilon \;\; \Delta \omega_r \;\; \Delta \omega_g \;\; \Delta \theta]^T \quad (i = 1),$$

$$x_i = \left[\Delta f_i \;\; \Delta P_{\mathrm{tie},i} \;\; \Delta P_{gi} \;\; \Delta X_{gi} \;\; \Delta X_{ghi}\right]^T \quad (i = 2, 3),$$

$$x_i = [\Delta f_i \;\; \Delta P_{\mathrm{tie},i} \;\; \Delta P_{gi} \;\; \Delta X_{gi} \;\; \Delta P_{ri}(t)]^T \quad (i = 4),$$

$$d_i = \Delta P_{di} \quad (i = 1, 2, 3, 4),$$

$$u_1 = [\Delta P_{c1} \;\; \Delta \theta_{\mathrm{ref}} \;\; \Delta T_g]^T,$$

$$y_i = \mathrm{ACE}_i = [K_{Bi} \Delta f_i + \Delta P_{\mathrm{tie},i}] \quad (i = 1, 2, 3, 4). \tag{6}$$

The state, control, and disturbance matrices for area 1 are as follows:

$$A_{11} = \begin{bmatrix} -\frac{1}{T_{P1}} & -\frac{K_{P1}}{T_{P1}} & \frac{K_{P1}}{T_{P1}} & 0 & 0 & 0 & 0 & 0 \\ \sum_j K_{sij} & 0 & 0 & 0 & 0 & 0 & 0 & 0 \\ 0 & 0 & -\frac{1}{T_{T1}} & 0 & \frac{1}{T_{T1}} & 0 & 0 & 0 \\ \frac{1}{T_{G1}} & 0 & 0 & -\frac{1}{T_{G1}} & 0 & 0 & 0 & 0 \\ 0 & 0 & 0 & 0 & 0 & 1 & -\frac{1}{n_{\mathrm{gear}}} & 0 \\ 0 & 0 & 0 & 0 & -\frac{K_s}{J_r} & \frac{1}{J_r}\frac{\partial T_r}{\partial \omega_r}\Big|_{\mathrm{op}} & 0 & \frac{1}{J_r}\frac{\partial T_r}{\partial \theta}\Big|_{\mathrm{op}} \\ 0 & 0 & 0 & 0 & \frac{\eta_{\mathrm{gear}} K_s}{n_{\mathrm{gear}} J_g} & 0 & 0 & 0 \\ 0 & 0 & 0 & 0 & 0 & 0 & 0 & -\frac{1}{\tau_\theta} \end{bmatrix}, \tag{7}$$

$$B_{11} = \begin{bmatrix} 0 & 0 & 0 & \frac{1}{T_{G1}} & 0 & 0 & 0 & 0 \\ 0 & 0 & 0 & 0 & 0 & 0 & 0 & \frac{K_\beta}{\tau_\theta} \\ 0 & 0 & 0 & 0 & 0 & 0 & -\frac{1}{J_g} & 0 \end{bmatrix}^T,$$

$$F_{11} = \begin{bmatrix} -\frac{K_{P1}}{T_{P1}} & 0 & 0 & 0 & 0 & 0 & 0 & 0 \\ 0 & 0 & 0 & 0 & 0 & \frac{1}{J_r}\frac{\partial T_r}{\partial v_m}\Big|_{\mathrm{op}} & 0 & 0 \end{bmatrix}^T,$$

$$C_{11} = [K_{b1} \;\; 1 \;\; 0 \;\; 0 \;\; 0 \;\; 0 \;\; 0 \;\; 0].$$

However for hydro plants in areas 2 and 3 they are as follows:

$$A_{22} = A_{33} = \begin{bmatrix} -\frac{1}{T_{Pi}} & -\frac{K_{Pi}}{T_{Pi}} & \frac{K_{Pi}}{T_{Pi}} & 0 & 0 \\ \sum_j K_{Sij} & 0 & 0 & 0 & 0 \\ 2\alpha & 0 & -\frac{2}{T_{Wi}} & 2\kappa & 2\beta \\ -\alpha & 0 & 0 & -\frac{1}{T_{2i}} & -\beta \\ -\frac{1}{T_{1i}R_i} & 0 & 0 & 0 & -\frac{1}{T_{1i}} \end{bmatrix},$$

$$B_{22} = B_{33} = \left[0 \;\; 0 \;\; -2R_i\alpha \;\; R_i\alpha \;\; \frac{1}{T_{1i}}\right]^T,$$

$$C_{22} = C_{33} = [K_{Bi} \;\; 1 \;\; 0 \;\; 0 \;\; 0],$$

$$F_{22} = F_{33} = \left[-\frac{K_{pi}}{T_{pi}} \;\; 0 \;\; 0 \;\; 0 \;\; 0\right]^T, \tag{8}$$

where $\alpha = T_{Ri}/T_{1i}T_{2i}R_i$, $\beta = (T_{Ri} - T_{1i})/T_{1i}T_{2i}$, and $\kappa = (T_{2i} + T_{Wi})/T_{2i}T_{Wi}$.

However for thermal power plants in area 4 they are as follows:

$$A_{44} = \begin{bmatrix} -\frac{1}{T_{Pi}} & -\frac{K_{Pi}}{T_{Pi}} & \frac{K_{Pi}}{T_{Pi}} & 0 & 0 \\ \sum_j K_{Sij} & 0 & 0 & 0 & 0 \\ 0 & 0 & -\frac{1}{T_{Ti}} & 0 & \frac{1}{T_{Ti}} \\ -\frac{1}{T_{Gi}R_i} & 0 & 0 & -\frac{1}{T_{Gi}} & 0 \\ -\frac{K_{ri}}{T_{Gi}R_i} & 0 & 0 & \frac{1}{T_{ri}} - \frac{K_{ri}}{T_{Gi}} & -\frac{1}{T_{ri}} \end{bmatrix}, \tag{9}$$

$$B_{44} = \begin{bmatrix} 0 & 0 & 0 & \frac{1}{T_{Gi}} & 0 \end{bmatrix}^T,$$

$$C_{44} = \begin{bmatrix} K_{Bi} & 1 & 0 & 0 & 0 \end{bmatrix},$$

$$F_{44} = \begin{bmatrix} -\frac{K_{pi}}{T_{pi}} & 0 & 0 & 0 & 0 \end{bmatrix}^T.$$

The interaction matrices between the four control areas are as follows:

$$A_{ij} = \begin{bmatrix} 0 & 0 & 0 & 0 & 0 & 0 & 0 & 0 \\ -K_{Sij} & 0 & 0 & 0 & 0 & 0 & 0 & 0 \\ 0 & 0 & 0 & 0 & 0 & 0 & 0 & 0 \\ 0 & 0 & 0 & 0 & 0 & 0 & 0 & 0 \\ 0 & 0 & 0 & 0 & 0 & 0 & 0 & 0 \end{bmatrix} \quad (i = 1,\ j = 2, 3, 4),$$

$$A_{ij} = \begin{bmatrix} 0 & 0 & 0 & 0 & 0 & 0 & 0 & 0 \\ -K_{Sij} & 0 & 0 & 0 & 0 & 0 & 0 & 0 \\ 0 & 0 & 0 & 0 & 0 & 0 & 0 & 0 \\ 0 & 0 & 0 & 0 & 0 & 0 & 0 & 0 \end{bmatrix} \quad (i = 1,\ j = 2, 3,\ i \neq j),$$

$$A_{ij} = \begin{bmatrix} 0 & 0 & 0 & 0 & 0 \\ -K_{Sij} & 0 & 0 & 0 & 0 \\ 0 & 0 & 0 & 0 & 0 \\ 0 & 0 & 0 & 0 & 0 \end{bmatrix} \quad (i = j = 2, 3, 4,\ i \neq j),$$

$$B_{ij} = 0_{8\times4},$$

$$F_{ij} = 0_{8\times2}$$

$$(i = 1,\ j = 2, 3, 4,\ i \neq j),$$

$$B_{ij} = 0_{5\times1},$$

$$F_{ij} = 0_{5\times1}$$

$$(i = j = 2, 3, 4,\ i \neq j). \tag{10}$$

The GRCs for the thermal plants are $|\Delta\dot{P}_{gi}| \leq 0.0017$ p.u.MW/s and the hydro units are $|\Delta\dot{P}_{gi}| \leq 0.045$ p.u.MW/s. In addition, the load disturbance is constrained to $|\Delta\dot{P}_{d_i}| \leq 0.3$.

3. Distributed Model Predictive Controller

3.1. Distributed Model Predictive Controller. The block diagram of the DMPC scheme for a four-area interconnected power system is illustrated in Figure 7. Though there exists large amount of variables in the interconnected power system, the 30 state variables expressed in (1a), (1b), (1c), and (1d) concerning the frequency, the generator output power, the governor valve (servomotor) position, the tie-line active power, the wind power, and the 4 load disturbance ΔP_{di} are crucial to LFC problem. They can be measured or estimated directly by the local controller. The DMPC in each area exchange control information through the power line communication, which is a sole networking technology with high reliability that can provide high speed communication to power grids applications [22].

Distributed MPC. The partitioned discrete-time model for control area i of the continuous-time four-area interconnected power system ((1a), (1b), (1c), and (1d)) can be expressed as follows:

$$\begin{aligned} x_i(k+1) &= \overline{A}_{ii}x_i(k) + \overline{B}_{ii}u_i(k) + \overline{F}_{ii}d_i(k) \\ &\quad + \sum_{i\neq j}\left(\overline{A}_{ij}x_j(k) + \overline{B}_{ij}u_j(k) + \overline{F}_{ij}d_j(k)\right), \end{aligned} \tag{11}$$

$$y_i(k) = \overline{C}_{ii}x_i(k),$$

where $\overline{A}_{ii}$, $\overline{B}_{ii}$, $\overline{C}_{ii}$, $\overline{F}_{ii}$, $\overline{A}_{ij}$, $\overline{B}_{ij}$, and $\overline{F}_{ij}$ represent the discrete new matrices obtained from original matrices in (4) based on the Zero-Order Hold (ZOH) method.

Assume that the state variables $x_i(k)$ and the disturbance D_i can be measured or estimated directly by the controller in area i at sampling time k. Optimizations and exchange of variables are termed iterate. The iteration number is denoted by p.

For DMPC, the optimal state-input trajectory (x_i, u_i) for each area i, $i = 1, 2, 3, 4$ at iterate p is obtained as the solution to the optimization problem:

$$\min_{u_i(k+n|k)} \quad J_i(k) \tag{12}$$

$$J_i(k) = \sum_{n=0}^{N}\left[x_i^T(k+n \mid k)\, Q_i x_i(k+n \mid k) + u_i^T(k+n \mid k)\, R_i u_i(k+n \mid k)\right] \tag{13}$$

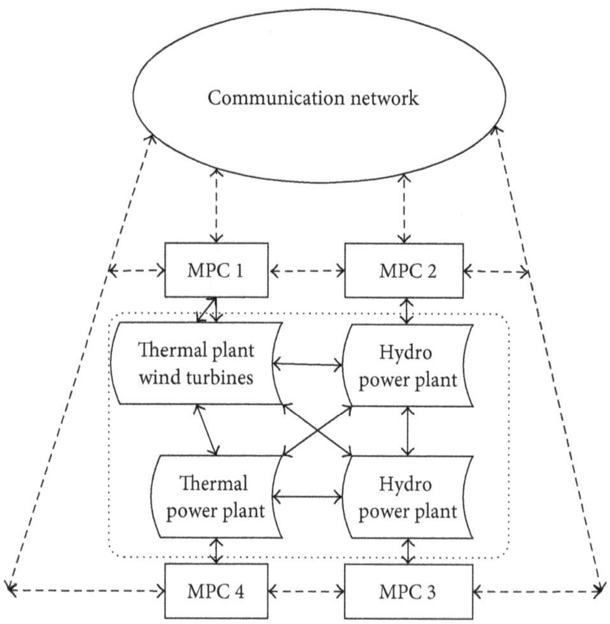

FIGURE 7: Block diagram of DMPC for power system with wind turbines.

$$\text{Subject to} \quad \|x_{i3}(k+n \mid k)\|_2 \le 0.0017, \quad i = 1,4 \tag{14a}$$

$$\|x_{i3}(k+n \mid k)\|_2 \le 0.0045, \quad i = 2,3 \tag{14b}$$

$$\|x_{i4}(k+n \mid k)\|_2 \le \sigma_i, \quad i = 1,2,3,4. \tag{14c}$$

For notational convenience, we drop the k dependence of $x_i(k)$, $u_i(k)$, $i = 1,2,3,4$. It is shown in [20] that each $\overline{x}_i$ can be expressed as

$$\overline{x}_i = \overline{E}_{ii}\overline{u}_i + \overline{f}_{ii}x_i(k) + \overline{\beta}_{ii}d_i(k) + \sum_{i \ne j}\left(\overline{E}_{ij}\overline{u}_j + \overline{g}_{ij}\overline{x}_j + \overline{f}_{ij}x_j(k) + \overline{\beta}_{ij}d_j(k)\right) \tag{15}$$

with

$$\overline{x}_i = \left[x_i(k+1 \mid k)^T \; x_i(k+2 \mid k)^T \; \cdots \; x_i\left(k+N_p \mid k\right)^T\right]^T,$$
$$\overline{u}_i = \left[u_i(k \mid k)^T \; u_i(k+1 \mid k)^T \; \cdots \; u_i(k+N_c-1 \mid k)^T\right]^T. \tag{16}$$

Let N_c denote the control horizon and let N_p denote the predictive horizon. $\overline{x}_i$ is no more a vector but a matrix after iteration obtained from original equation (4). The matrices in (15) have detailed expressions as follows:

$$\overline{E}_{ii} = \begin{bmatrix} \overline{B}_{ii} & 0 & \cdots & 0 \\ \overline{A}_{ii}\overline{B}_{ii} & \overline{B}_{ii} & \cdots & 0 \\ \vdots & \vdots & & \vdots \\ \overline{A}_{ii}^{N-1}\overline{B}_{ii} & \overline{A}_{ii}^{N-2} & \cdots & 0 \end{bmatrix},$$

$$\overline{E}_{ij} = \begin{bmatrix} \overline{B}_{ij} & 0 & \cdots & 0 \\ \overline{A}_{ii}\overline{B}_{ij} & \overline{B}_{ij} & \cdots & 0 \\ \vdots & \vdots & & \vdots \\ \overline{A}_{ii}^{N-1}\overline{B}_{ij} & \overline{A}_{ii}^{N-2} & \cdots & 0 \end{bmatrix},$$

$$\overline{f}_{ii} = \begin{bmatrix} \overline{A}_{ii} \\ \overline{A}_{ii}\overline{A}_{ii} \\ \vdots \\ \overline{A}_{ii}^{N-1}\overline{A}_{ii} \end{bmatrix},$$

$$\overline{f}_{ij} = \begin{bmatrix} \overline{A}_{ij} \\ \overline{A}_{ii}\overline{A}_{ij} \\ \vdots \\ \overline{A}_{ii}^{N-1}\overline{A}_{ij} \end{bmatrix},$$

$$\overline{\beta}_{ii} = \begin{bmatrix} \overline{F}_{ii} \\ \overline{A}_{ii}\overline{F}_{ii} \\ \vdots \\ \overline{A}_{ii}^{N-1}\overline{F}_{ii} \end{bmatrix},$$

$$\overline{\beta}_{ij} = \begin{bmatrix} \overline{F}_{ij} \\ \overline{A}_{ii}\overline{F}_{ij} \\ \vdots \\ \overline{A}_{ii}^{N-1}\overline{F}_{ij} \end{bmatrix},$$

$$\overline{g}_{ij} = \begin{bmatrix} 0 & 0 & 0 & \cdots & 0 \\ \overline{A}_{ij} & 0 & 0 & \cdots & 0 \\ \overline{A}_{ii}\overline{A}_{ij} & \overline{A}_{ij} & 0 & \cdots & 0 \\ \vdots & \vdots & \vdots & \cdots & \vdots \\ \overline{A}_{ii}^{N-2}\overline{A}_{ij} & \overline{A}_{ii}^{N-3}\overline{A}_{ij} & \cdots & \overline{A}_{ij} & 0 \end{bmatrix}, \tag{17}$$

where $\overline{E}_{ii}$, $\overline{f}_{ii}$, $\overline{\beta}_{ii}$, $\overline{E}_{ij}$, $\overline{f}_{ij}$, $\overline{\beta}_{ij}$, and $\overline{g}_{ij}$ are the new matrices obtained from $\overline{A}_{ii}$, $\overline{B}_{ii}$, $\overline{C}_{ii}$, $\overline{F}_{ii}$, $\overline{A}_{ij}$, $\overline{B}_{ij}$, and $\overline{F}_{ij}$ after iteration.

Combining the models in (15) gives the following system of equations:

$$\Lambda\tilde{x} = \varepsilon\tilde{u} + \mu\tilde{x}(k) + \phi d(k) \tag{18}$$

with

$$\Lambda = \begin{bmatrix} I & -\overline{g}_{12} & -\overline{g}_{13} & -\overline{g}_{14} \\ -\overline{g}_{21} & I & -\overline{g}_{23} & -\overline{g}_{24} \\ -\overline{g}_{31} & -\overline{g}_{32} & I & -\overline{g}_{34} \\ -\overline{g}_{41} & -\overline{g}_{42} & -\overline{g}_{43} & I \end{bmatrix},$$

$$\varepsilon = \begin{bmatrix} \overline{E}_{11} & \overline{E}_{12} & \overline{E}_{13} & \overline{E}_{14} \\ \overline{E}_{21} & \overline{E}_{22} & \overline{E}_{23} & \overline{E}_{24} \\ \overline{E}_{31} & \overline{E}_{32} & \overline{E}_{33} & \overline{E}_{34} \\ \overline{E}_{41} & \overline{E}_{42} & \overline{E}_{43} & \overline{E}_{44} \end{bmatrix},$$

$$\mu = \begin{bmatrix} \overline{f}_{11} & \overline{f}_{12} & \overline{f}_{13} & \overline{f}_{14} \\ \overline{f}_{21} & \overline{f}_{22} & \overline{f}_{23} & \overline{f}_{24} \\ \overline{f}_{31} & \overline{f}_{32} & \overline{f}_{33} & \overline{f}_{34} \\ \overline{f}_{41} & \overline{f}_{42} & \overline{f}_{43} & \overline{f}_{44} \end{bmatrix},$$

$$\phi = \begin{bmatrix} \overline{\beta}_{11} & \overline{\beta}_{12} & \overline{\beta}_{13} & \overline{\beta}_{14} \\ \overline{\beta}_{21} & \overline{\beta}_{22} & \overline{\beta}_{23} & \overline{\beta}_{24} \\ \overline{\beta}_{31} & \overline{\beta}_{32} & \overline{\beta}_{33} & \overline{\beta}_{34} \\ \overline{\beta}_{41} & \overline{\beta}_{42} & \overline{\beta}_{43} & \overline{\beta}_{44} \end{bmatrix},$$

$$\tilde{x} = \left[\overline{x}_1 \;\; \overline{x}_2 \;\; \overline{x}_3 \;\; \overline{x}_4\right]^T,$$

$$\tilde{u} = \left[\overline{u}_1 \;\; \overline{u}_2 \;\; \overline{u}_3 \;\; \overline{u}_4\right]^T. \tag{19}$$

Since matrix Λ is invertible, we can write it as

$$\begin{aligned}\overline{x}_i &= \overline{E}_{ii}\overline{u}_i + \overline{f}_{ii}x_i(k) + \overline{\beta}_{ii}d_i(k) \\ &\quad + \sum_{i\neq j}\left(\overline{E}_{ij}\overline{u}_j + \overline{f}_{ij}x_j(k) + \overline{\beta}_{ij}d_j(k)\right)\end{aligned} \tag{20}$$

in which

$$\begin{aligned}\overline{E}_{ij} &= \Lambda^{-1}\varepsilon, \\ \overline{f}_{ij} &= \Lambda^{-1}\mu, \\ \overline{\beta}_{ij} &= \Lambda^{-1}\phi.\end{aligned} \tag{21}$$

To do so, we eliminate the unknown matrix $\overline{x}_j$ because we have knowledge of $x_j(k)$ since it is just a vector at time k.

In the distributed MPC algorithm, for subsystem i, the control signal $\overline{U}_i$ is designed at each time interval $k \geq 0$. By solving the following optimization problem denoted by J_i, it is usually defined as

$$J_i = \min_{\overline{u}_i}\frac{1}{2}\overline{u}_i^T\Phi_i\overline{u}_i^T + \left(\gamma_i + \Gamma_i + \sum_{i\neq j}H_{ij}\overline{u}_j\right)^T\overline{u}_i \tag{22}$$

in which

$$\mathbb{Q}_i = \operatorname{diag}\overbrace{(\omega_i Q_i, \ldots, \omega_i Q_i)}^{N_p},$$

$$\mathbb{R}_i = \operatorname{diag}\overbrace{(\omega_i R_i, \ldots, \omega_i R_i)}^{N_c},$$

$$\Phi_i = \mathbb{R}_i + \overline{E}_{ii}^T\mathbb{Q}_i\overline{E}_{ii} + \sum_{\substack{j=1\\ j\neq i}}^{4}\overline{E}_{ji}^T\mathbb{Q}_j\overline{E}_{ji},$$

$$\gamma_i = \overline{E}_{ii}^T\mathbb{Q}_i\overline{g}_{ii} + \sum_{\substack{j=1\\ j\neq i}}^{4}\overline{E}_{ji}^T\mathbb{Q}_j\overline{g}_{ji},$$

$$
\begin{aligned}
g_{ii} &= f_{ii}x_i(k) + \sum_{j=1}^{4} f_{ij}x_j(k),\\
\Gamma_i &= E_{ii}^T \mathbb{Q}_i \rho_i + \sum_{j=1}^{4} E_{ji}^T \mathbb{Q}_j \rho_j,\\
\rho_i &= \beta_{ii}d_i(k) + \sum_{j=1}^{4} \beta_{ij}d_j(k),\\
H_{ij} &= \overline{E}_{ii}^T \mathbb{Q}_i \overline{E}_{ij} + \sum_{\substack{j=1\\ j\neq i}}^{4} \overline{E}_{ji}^T \mathbb{Q}_j \overline{E}_{ji}.
\end{aligned}
\tag{23}
$$

At time interval k, (22) is implemented based on the future states and manipulated variables. The first input in the optimal sequence is injected into the processes, and the procedure is repeated at subsequent time intervals.

$Q_i \geq 0, R_i \geq 0$ are symmetric weighting matrices and $\omega_i > 0, \sum_{i=1}^{4} \omega_i = 1$.

Define $\eta_i = \gamma_i + \Gamma_i + \sum_{j\neq i} H_{ij}\overline{u}_j$.

Then (22) is rewritten as

$$J_i = \min_{\overline{u}_i} \frac{1}{2}\overline{u}_i^T \Phi_i u_i^T + \eta_i^T \overline{u}_i. \tag{24}$$

3.2. Constraint Handling. The two crucial nonlinearities, for example, the GRCs and the valve position limits of the governor, have been considered as the state constraints in the designed DMPC, as shown in Figures 8 and 9.

In power system, the GRC can be expressed as $\Delta\dot{P}_g(k)_{\min} \leq \Delta\dot{P}_g(k) \leq \Delta\dot{P}_g(k)_{\max}$, and then the constraints on ΔP_g can be expressed as follows:

$$
\begin{aligned}
&T\left(\Delta\dot{P}_g(k)\right)_{\min} + \Delta P_g(k-1) \leq \Delta P_g(k)\\
&\leq T\left(\Delta\dot{P}_g(k)\right)_{\max} + \Delta P_g(k-1),
\end{aligned}
\tag{25}
$$

$$
\begin{aligned}
&\Delta\overline{P}_g\\
&= \left[\Delta P_g(k+1\mid k)\ \ \Delta P_g(k+2\mid k)\ \cdots\ \Delta P_g\left(k+N_p\mid k\right)\right]^T.
\end{aligned}
\tag{26}
$$

Since $\Delta P_{gi} = X_{i3}$, the state form can be expressed as

$$\Delta\overline{P}_g = S_i\overline{x}_i, \tag{27}$$

where $S_i = \mathrm{diag}(\overbrace{\omega_i S_{ii}, \ldots, \omega_i S_{ii}}^{N_p})$.

When $i = 1, 4$, $S_{ii} = [0\ \ 0\ \ 1\ \ 0\ \ 0]$, and when $i = 2, 3$, $S_{ii} = [0\ \ 0\ \ 1\ \ 0\ \ 0]$, with (25) and (27), the constraints on $\Delta P_g(k)$ are expressed as $\overline{N}_i \leq S_i\overline{x}_i \leq \overline{M}_i$.

Define

$$
\begin{aligned}
\overline{N}_i &= \left[\overbrace{N_i\ \ N_i\ \cdots\ N_i}^{N_p}\right]^T,\\
\overline{M}_i &= \left[\overbrace{M_i\ \ M_i\ \cdots\ M_i}^{N_p}\right]^T,
\end{aligned}
\tag{28}
$$

where N_i and M_i are obtained from (15).

Consider the constraints on $\Delta P_g(k)$:

$$
\begin{bmatrix} S_iE_{ii}\\ -S_iE_{ii}\end{bmatrix}\overline{u}_i \leq
\begin{bmatrix}
\overline{M}_i - S_i\left\{f_{ii}x_i(k) + \beta_{ii}d_i(k) + \sum_{j\neq i}\left[E_{ij}\overline{u}_j + f_{ij}x_j(k) + \beta_{ij}d_j(k)\right]\right\}\\
-\overline{N}_i + S_i\left\{f_{ii}x_i(k) + \beta_{ii}d_i(k) + \sum_{i\neq j}\left[E_{ij}\overline{u}_j + f_{ij}x_j(k) + \beta_{ij}d_j(k)\right]\right\}
\end{bmatrix}.
\tag{29}
$$

Define

$$
\begin{aligned}
\Psi_i &= \begin{bmatrix} S_iE_{ii}\\ -S_iE_{ii}\end{bmatrix},\\
\Pi_i &= \begin{bmatrix}
\overline{M}_i - S_i\left\{f_{ii}x_i(k) + \beta_{ii}d_i(k) + \sum_{j\neq i}\left[E_{ij}\overline{u}_j + f_{ij}x_j(k) + \beta_{ij}d_j(k)\right]\right\}\\
-\overline{N}_i + S_i\left\{f_{ii}x_i(k) + \beta_{ii}d_i(k) + \sum_{i\neq j}\left[E_{ij}\overline{u}_j + f_{ij}x_j(k) + \beta_{ij}d_j(k)\right]\right\}
\end{bmatrix}.
\end{aligned}
\tag{30}
$$

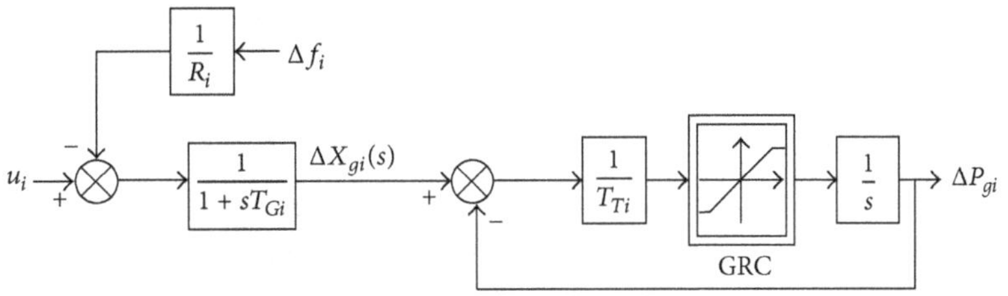

FIGURE 8: Thermal power plant with GRC.

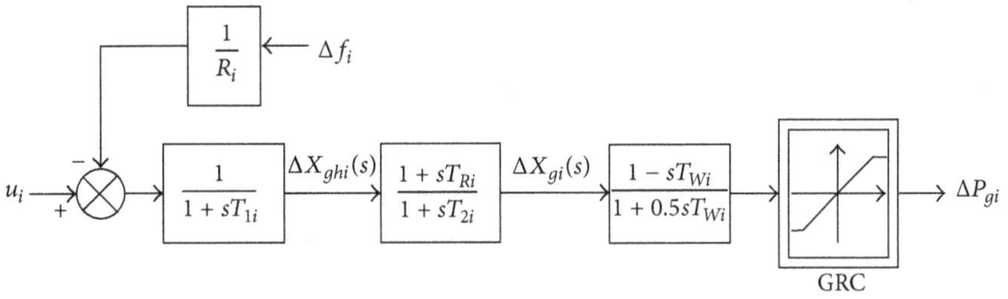

FIGURE 9: Hydro power plant with GRC.

Then, distributed MPC algorithm (24) for multiple-interconnected system can be transformed into the following optimization problem with GRC constraints:

$$\begin{aligned} J_i &= \min_{\overline{u}_i} \quad \frac{1}{2}\overline{u}_i^T \Phi_i \overline{u}_i^T + \eta_i^T \overline{u}_i \\ &\text{Subject to} \quad \Psi_i \overline{u}_i \leq \Pi_i. \end{aligned} \tag{31}$$

3.3. The DMPC Algorithm

Step 1 (initialization). The constant matrices R_i, R_j and Q_i, Q_j, at control interval $k = 0$, are given. Choose the specified error tolerance ε_i. Set iteration $p = 0$.

Step 2 (communication). The controller in each subsystem i exchanges its previous predictions $\overline{x}_i(k)$, $\overline{x}_j(k)$, set $\overline{u}_i^0(k)$, and $\overline{u}_j^0(k)$ at initial instant.

Step 3 (optimization and iteration).

While $p < p_{\max}$,

$\overline{u}_i^{*(p)}$ is solved by the optimal problem (31)

If $\|u_i^{(p)} - u_i^{(p-1)}\| \leq \varepsilon_i \ \forall i \in \{1, 2, 3, 4\}$

Break

End if

Exchange the solutions $\overline{u}_i^p$ and $\overline{u}_j^p$ and set $p = p + 1$

If $\varepsilon_i = 0 \ \forall i \in \{1, 2, 3, 4\}$

Break

End if

End while

Step 4 (assignment and prediction). Send out $u_i(k) = \overline{u}_i(k)$. Otherwise, $u_i(k) = \overline{u}_i(k-1)$. Predict the future states.

Step 5 (implementation). Set $k = k + 1$, and repeat Step 1.

4. Simulation Results

In this section, the four-area power system stability is analyzed, and the performances of the proposed DMPC have been tested in case of wind turbines participation at nominal parameters. The simulation of the proposed DMPC scheme is also verified by two cases. The performance and the implementation of the proposed DMPC are compared with other two types of typical LFC scheme.

As comparison, we design the centralized MPC and decentralized MPC controller for four-area interconnected power system, respectively. The four-area interconnected power system can be described as

$$\begin{aligned} x(k+1) &= Ax(k) + Bu(k) + Fd(k), \\ y(k+1) &= Cx(k), \end{aligned} \tag{32}$$

where

$$A = \begin{bmatrix} A_{11} & A_{12} & A_{13} & A_{14} \\ A_{21} & A_{22} & A_{23} & A_{24} \\ A_{31} & A_{32} & A_{33} & A_{34} \\ A_{41} & A_{42} & A_{43} & A_{44} \end{bmatrix},$$

$$B = \begin{bmatrix} B_{11} & B_{12} & B_{13} & B_{14} \\ B_{21} & B_{22} & B_{23} & B_{24} \\ B_{31} & B_{32} & B_{33} & B_{34} \\ B_{41} & B_{42} & B_{43} & B_{44} \end{bmatrix},$$

$$C = \begin{bmatrix} C_{11} & 0 & 0 & 0 \\ 0 & C_{22} & 0 & 0 \\ 0 & 0 & C_{33} & 0 \\ 0 & 0 & 0 & C_{44} \end{bmatrix},$$

$$F = \begin{bmatrix} F_{11} & 0 & 0 & 0 \\ 0 & F_{22} & 0 & 0 \\ 0 & 0 & F_{33} & 0 \\ 0 & 0 & 0 & F_{44} \end{bmatrix},$$

$$x = \left[x_1^T \ x_2^T \ x_3^T \ x_4^T\right]^T,$$

$$u = \left[u_1^T \ u_2^T \ u_3^T \ u_4^T\right]^T,$$

$$y = \left[y_1^T \ y_2^T \ y_3^T \ y_4^T\right]^T,$$

$$d = \left[d_1^T \ d_2^T \ d_3^T \ d_4^T\right]^T \quad (33)$$

with constraints (12), (13), (14a), (14b), and (14c) for each control area. In centralized MPC framework, the MPC for overall system (32) solves the following optimization problem:

$$\min_{u(k+n|k)} J(k) \quad (34)$$

$$J(k) = \sum_{n=0}^{N} \left[x^T(k+n \mid k) Q x(k+n \mid k) + u^T(k+n \mid k) R u(k+n \mid k)\right] \quad (35)$$

subject to (14a), (14b), and (14c).

The weighting matrices Q and R in objective function (35) are chosen as $R = \text{diag}(1,1,1,1)$ and

$$Q = \text{diag}\begin{pmatrix} 1000, 0, 0, 1000, 1000, 0, 0, 1000, 1000, \\ 0, 0, 1000, 1000, 0, 0, 1000 \end{pmatrix}. \quad (36)$$

In the decentralized modeling framework, it is assumed that the interaction between the control areas is negligible. Subsequently, the decentralized model for each control area is

$$x_i(k+1) = A_{ii}x_i(k) + B_{ii}u_i(k) + F_{ii}d_i(k), \quad y_i(k+1) = C_{ii}x_i(k) \quad (37)$$

with the system matrices and constraints (12), (13), (14a), (14b), and (14c) for each control area denoted as in Section 2. In decentralized MPC framework, each control area based MPC solves the following optimization problem:

$$\min_{u_i(k+n|k)} J_i(k) \quad (38)$$

$$J_i(k) = \sum_{n=0}^{N} \left[x_i^T(k+n \mid k) Q_i x_i(k+n \mid k) + u_i^T(k+n \mid k) R_i u_i(k+n \mid k)\right] \quad (39)$$

subject to (14a), (14b), and (14c).

The weighting matrices Q_i and R_i in objective function (39) are chosen as $R_1 = R_2 = R_3 = R_4 = 1$ and

$$Q_1 = Q_2 = Q_3 = Q_4 = \text{diag}(1000, 0, 0, 1000). \quad (40)$$

Choose the prediction horizon of the centralized MPC, decentralized MPC, and RDMPC to be $N = 15$, choose the control horizon to be $N_c = 10$, and choose the sample time $T_s = 0.1$ and $\lambda = 0.1$. Consider GRC for the thermal power plants in area 1 and area 4 to be $|\Delta \dot{P}_g^i| \le r = 0.1$ p.u.MW/min = 0.0017 p.u.MW/s and GRC for the hydro power plants in area 2 and area 3 to be $|\Delta \dot{P}_g^i| \le r = 2.7$ p.u.MW/min = 0.045 p.u.MW/s. In addition, area 1 includes an aggregated wind turbine model which consists of 30 wind turbine units of 2 MW rated VSWTs while the capacity of thermal plant is 600 MW. The wind turbine parameters and operating points [23] are indicated as follows:

Operating point: 80 MW; *wind speed*: 12 m/s.

$T_g = 3781.9$ Nm; $\omega_g = 105$ rad/s; $\omega_r = 2.6869$ rad/s.

$K_s = 7.87e6$ Nm/rad; $n_{\text{gear}} = 1 : 28.7$; $\eta_{\text{gear}} = 97.5\%$.

$J_r = 2.8675$ kgm^2; $J_g = 54.5432$ kgm^2.

$R_3 = 3.3$ Hz/p.u.MW; $R_4 = 3$ Hz/p.u.MW.

The parameters for the thermal and hydro plants used in the simulation are listed as follows:

$K_{P1} = 120$ Hz/p.u.MW; $K_{P2} = 115$ Hz/p.u.MW.

$K_{P3} = 80$ Hz/p.u.MW; $K_{P4} = 75$ Hz/p.u.MW.

$T_{P1} = 20$ s; $T_{P2} = 20$ s; $T_{P3} = 13$ s; $T_{P4} = 15$ s.

$R_1 = 2.4$ Hz/p.u.MW; $R_2 = 2.5$ Hz/p.u.MW.

$R_3 = 3.3$ Hz/p.u.MW; $R_4 = 3$ Hz/p.u.MW.

$K_{B1} = 0.425$ p.u.MW/Hz; $K_{B2} = 0.409$ p.u.MW/Hz.

$K_{B3} = 0.316$ p.u.MW/Hz; $K_{B4} = 0.347$ p.u.MW/Hz.

$T_{G1} = 0.08$ s; $T_{G2} = 0.1$ s; $T_{G3} = 0.08$ s; $T_{G4} = 0.2$ s.

$T_{T1} = T_{T4} = 0.3$ s; $T_{r1} = T_{r4} = 10$ s; $T_{R2} = 0.6$ s.

$T_{R3} = 0.513$ s; $T_{22} = 5$ s; $T_{23} = 10$ s; $T_{W2} = 1$ s; $T_{W3} = 2$ s.

$K_{S12} = -K_{S21} = 0.545$ p.u.MW.

$K_{S23} = -K_{S32} = 0.444$ p.u.MW.

$K_{S13} = -K_{S31} = 0.545$ p.u.MW.

$K_{S14} = -K_{S41} = 0.5$ p.u.MW.

$K_{S24} = -K_{S42} = 0.545$ p.u.MW.

$K_{S34} = -K_{S43} = 0.545$ p.u.MW.

Case 1 (response to step load change without wind turbines participation). Wind turbine is present but it does not provide any power support in the event of grid frequency deviation. An event is simulated in which a system shown in

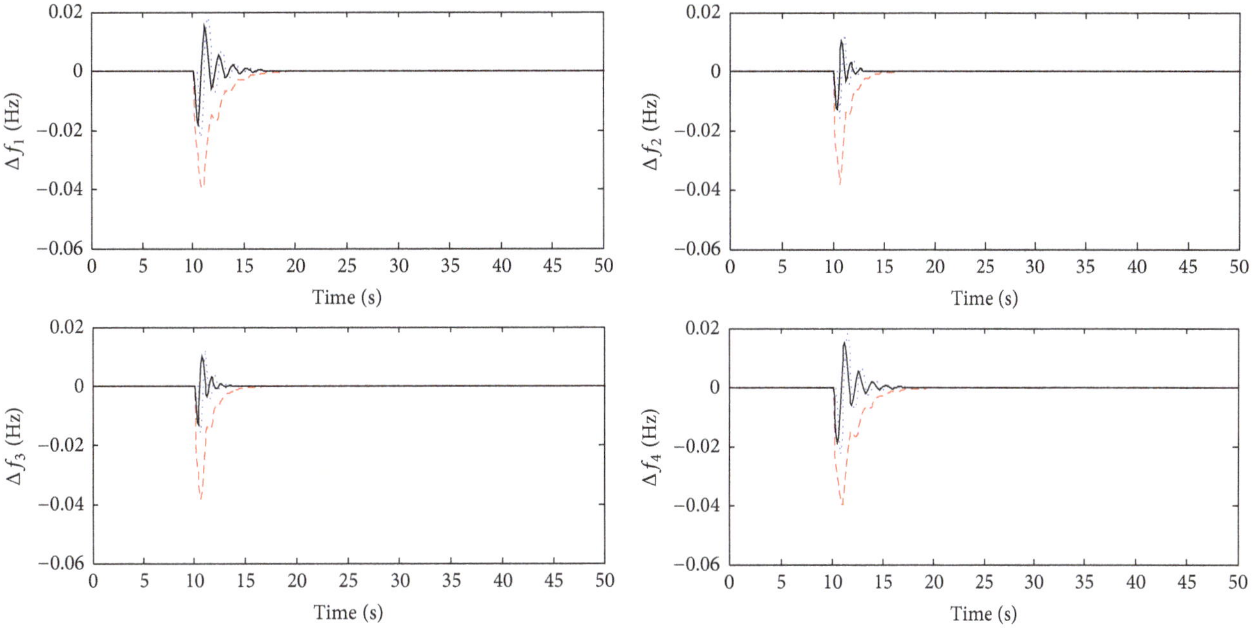

FIGURE 10: Response of frequency deviation to step load disturbance in Case 1: distributed MPC (solid line), centralized MPC (dotted line), and decentralized MPC (dashed line).

TABLE 2: Cost of the different strategies.

Strategy	Cost [20]
Centralized MPC	0.10
Decentralized MPC	0.083
Distributed MPC	0.078

Figure 1 is subjected to step load disturbances as give in (41) at $t = 10$ s. Consider

$$\Delta P_{d1} = \Delta P_{d2} = \Delta P_{d3} = \Delta P_{d4} = 0.1. \quad (41)$$

Figure 10 shows the simulation results of distributed MPC, centralized MPC, and decentralized MPC without wind turbine participation and only conventional integrator systems. The relative performance of distributed MPC, centralized MPC, and decentralized MPC rejecting the load disturbance in each area in Figure 10 is denoted by solid, dotted, and dashed lines, respectively. It has been noticed that the closed-loop trajectory of distributed MPC obtained by algorithm is little fast and almost indistinguishable from the closed-loop trajectory of centralized MPC. It successfully improves the dynamic response of area frequencies compared with decentralized MPC.

The control costs defined by [20] for different strategies are listed in Table 2. It is obviously seen that the DMPC controller needs nearly as much CPU time as decentralized MPC controller and significantly less CPU time than centralized MPC controllers. The proposed DMPC algorithm has significant computational advantages when compared to centralized MPC while achieving the best performance.

Case 2 (response to step load change with wind turbines participation). Wind turbine is present and it will provide active power support in the event of grid frequency deviation. An event is simulated in which a system shown in Figure 1 is subjected to step load disturbances as give in (41) at $t = 10$ s. Mean wind speed is assumed to be 17 m/s in area 1.

In Figures 11 and 12, the behavior for the frequency is presented for Case 2 where the wind turbines are participating in load frequency control. The results from top to the bottom in Figure 11 are the frequency deviations for area 1 to area 4 and in Figure 12 are six tie-lines power change. In simulation, it is obvious that both the DMPC and the centralized MPC converge rapidly and drive the local frequency changes and tie-line power deviation to zero. The wind turbines that have participated in the interconnected power system do not affect the performance of the power system under distributed MPC and centralized MPC while satisfying all the physical constraints, for example, the GRC, the limit of the governors, and load step change constraints. However, with decentralized MPC, the rapid convergence cannot be guaranteed in the presence of wind turbines in area 1. This confirms the performance advantage of the proposed distributed model predictive control algorithm.

Figure 13 shows the dynamic response of active power deviation ΔP_e and rotor speed ω_g of wind turbine while participating in the load frequency control. When the control is activated, the frequency deviation becomes zero which consequently eliminated the additional active power deviation ΔP_e and wind turbine is driven to operate again at the optimal rotor speed ω_g. It may be noted here that an increase in power step on top of the converter further reduces the rotor speed, thereby transferring more kinetic power to reduce the frequency dip. As shown in this figure, the distributed MPC

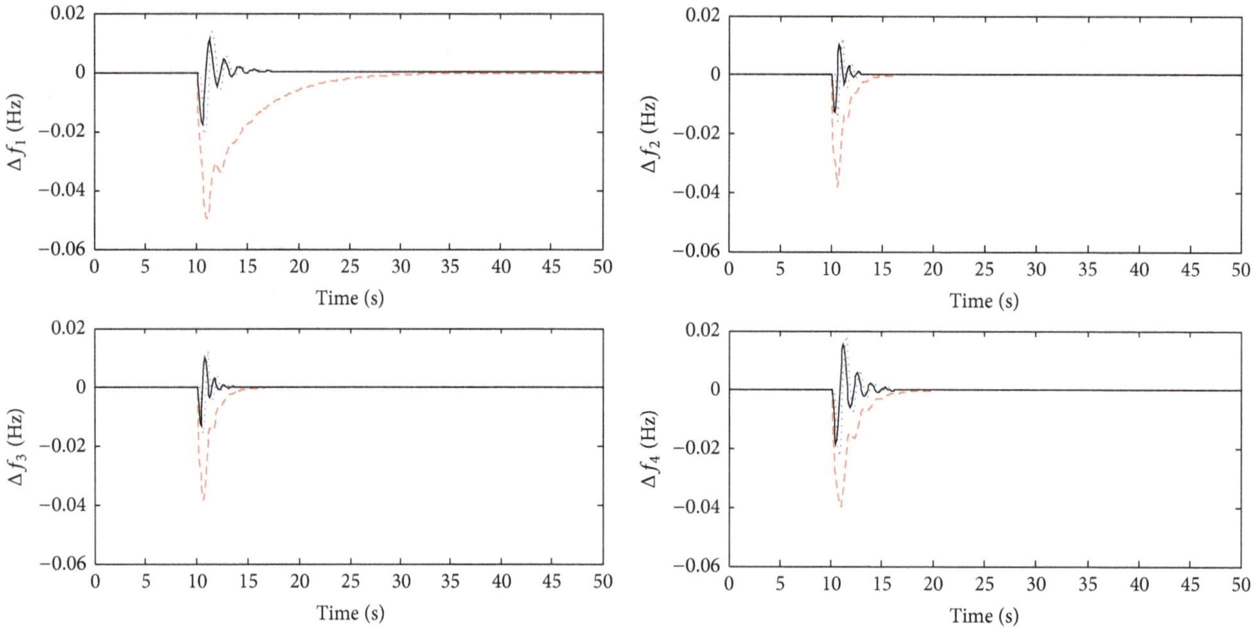

Figure 11: Response of frequency deviation to step load disturbance in Case 2: distributed MPC (solid line), centralized MPC (dotted line), and decentralized MPC (dashed line).

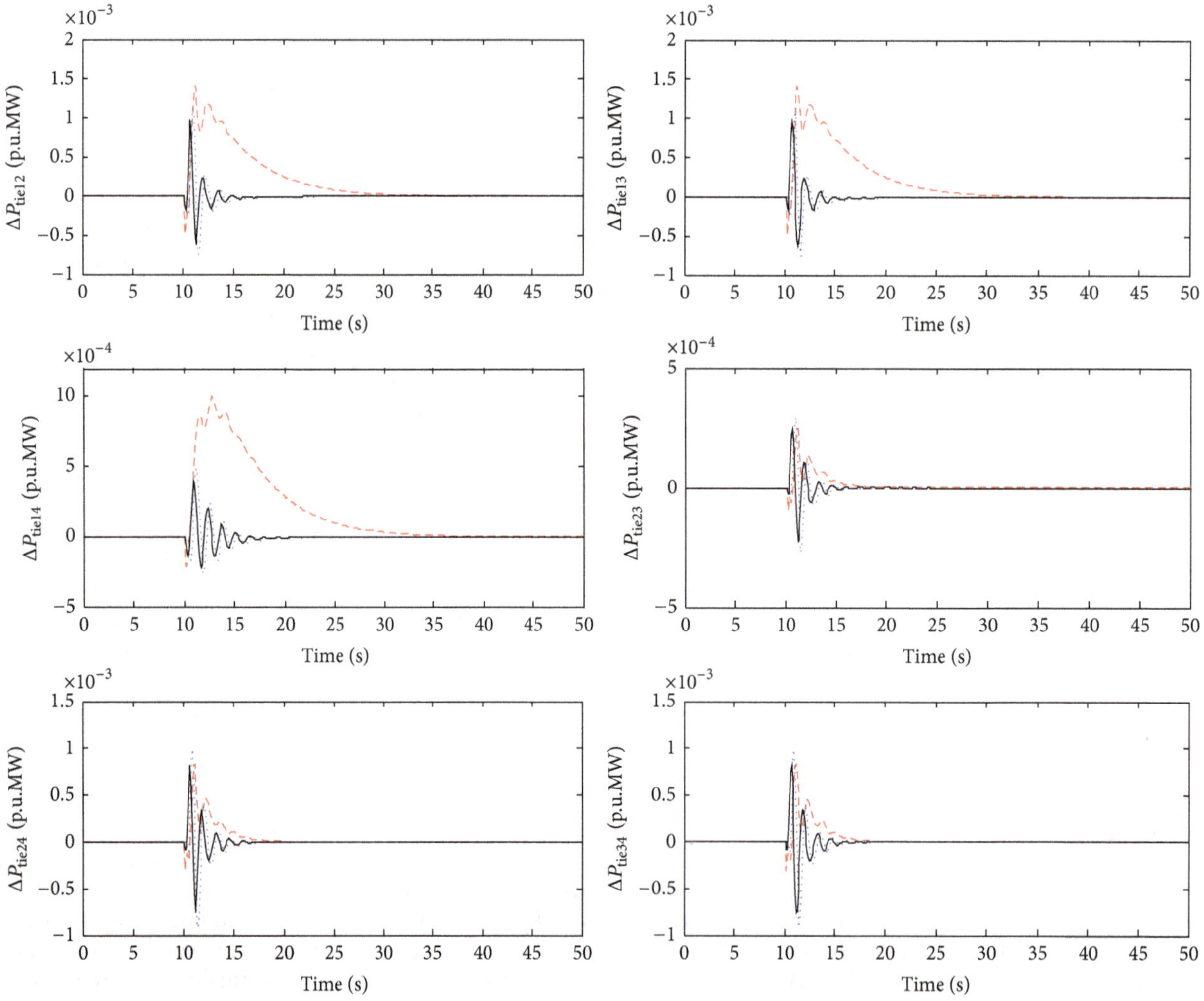

Figure 12: Response of tie-line active power deviation in Case 2: distributed MPC (solid line), centralized MPC (dotted line), and decentralized MPC (dashed line).

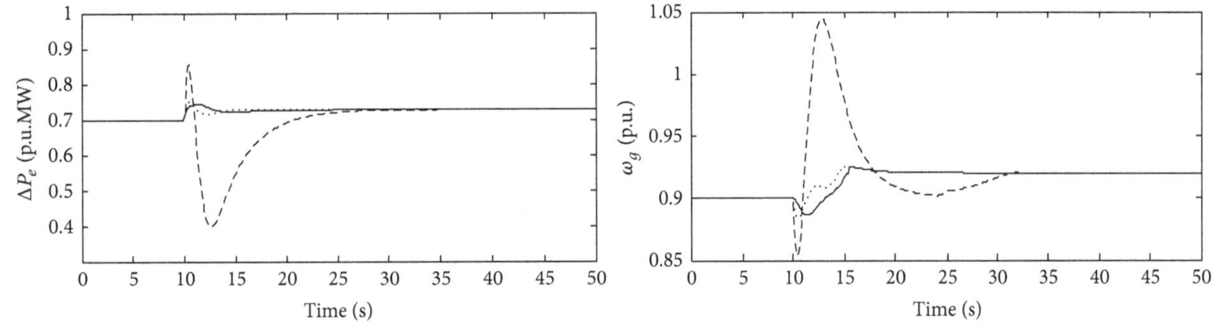

FIGURE 13: Wind turbine response of electrical power and rotor speed in Case 2: distributed MPC (solid line), centralized MPC (dotted line), and decentralized MPC (dashed line).

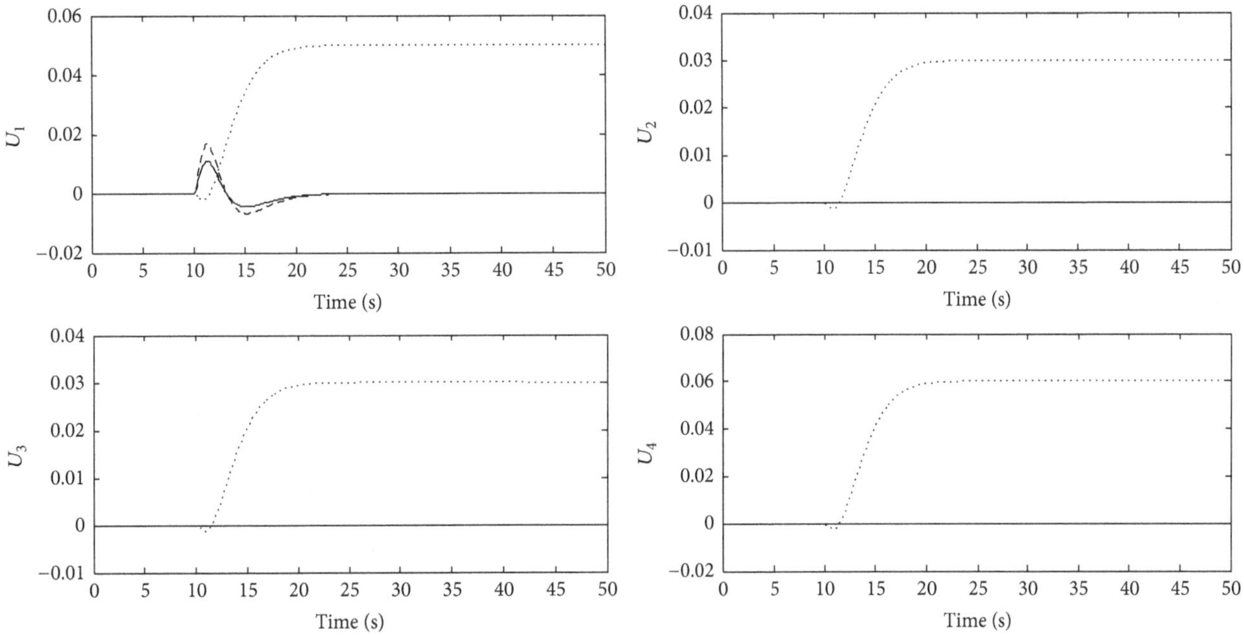

FIGURE 14: Control signal of distributed MPC in Case 2: $\Delta\theta_{\text{ref}}$ in area 1 (solid line), ΔP_{ci} in four areas (dotted line), and ΔT_g in area 1 (dashed line).

in the presence of wind turbine has desirable performance in comparison to centralized MPC and decentralized MPC.

The distributed MPC control actions as shown in Figure 14, $\Delta\theta_{\text{ref}}$, ΔP_{ci}, and ΔT_g in four areas are depicted as solid, dotted, and dashed line, respectively. $\Delta\theta_{\text{ref}}$ and ΔT_g are the control signals of wind turbine in area 1, and ΔP_{ci} is the control signal of traditional power plants in the four areas. Figure 15 shows the generating outputs of traditional plants.

5. Conclusions

In this paper, a DMPC scheme is presented for the LFC of a four-area interconnected power system with wind turbines. The state and input constraints including the valve position limit on the governor and the GRCs were incorporated into the system design. In our scheme, each control area has a local MPC controller, in which the four controllers coordinated with each other by exchanging their information. Comparisons of response to step load change and computational burden have been made between DMPC, centralized MPC, and decentralized MPC. The simulation results verified the reliability of the DMPC for achieving a performance that has advantages over the centralized MPC and distributed MPC in the presence of load changes. Moreover, the proposed DMPC scheme can guarantee a good performance under the wind turbines participation in LFC. Future work will be the extension of the proposed DMPC to different renewable energy contained LFC, since the greater utilization of intermittent renewable resources will induce greater power flow fluctuations.

Conflict of Interests

The authors declare that there is no conflict of interests regarding the publication of this paper.

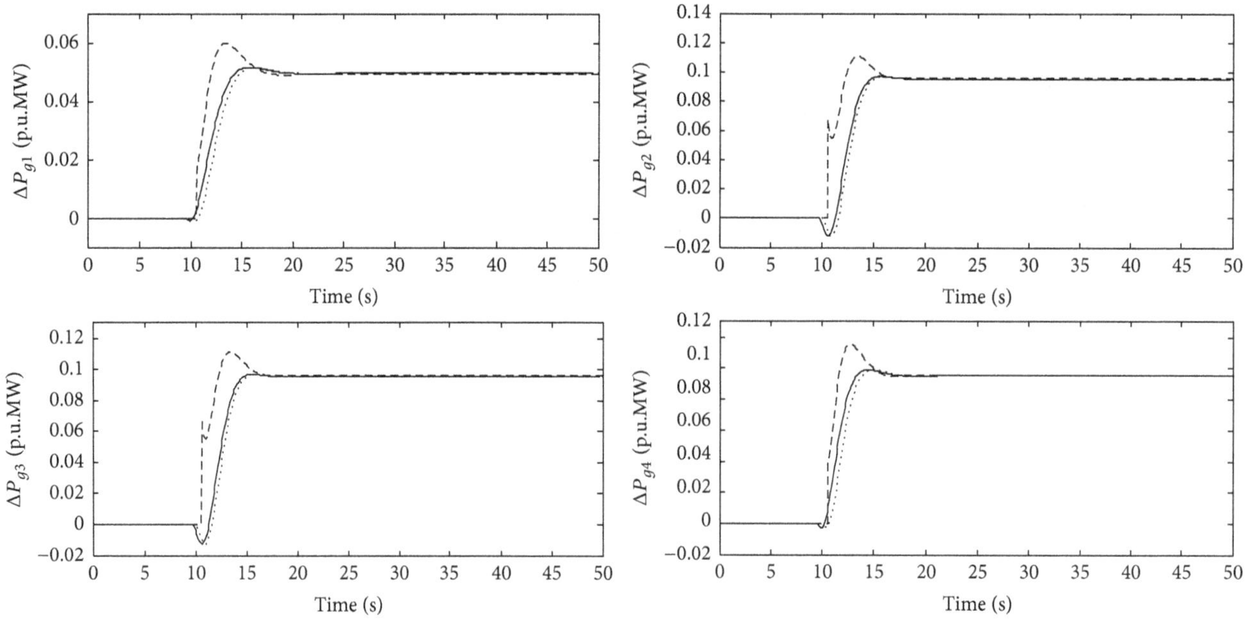

FIGURE 15: Response of generated power deviation in Case 2: distributed MPC (solid line), centralized MPC (dotted line), and decentralized MPC (dashed line).

Acknowledgments

This project was supported by National Natural Science Foundation of China under Grants 60974051 and 61273144, Natural Science Foundation of Beijing under Grant 4122071, Scientific Technology Research and Development Plan Project of Tangshan under Grant 13130298b, and Scientific Technology Research and Development Plan Project of Hebei under Grant z2014070.

References

[1] Global Wind Energy Council, *Global Wind Report on Annual Market*, Global Wind Energy Council, 2014.

[2] H. Bevrani, F. Daneshfar, and R. P. Daneshmand, "Intelligent power system frequency regulations concerning the integration of wind power units," in *Wind Power Systems: Applications of Computational Intelligence*, L. F. Wang, C. Singh, and A. Kusiak, Eds., Green Energy and Technology, pp. 407–437, Springer, Berlin, Germany, 2010.

[3] X. Yingcheng and T. Nengling, "Review of contribution to frequency control through variable speed wind turbine," *Renewable Energy*, vol. 36, no. 6, pp. 1671–1677, 2011.

[4] Y.-Z. Sun, Z.-S. Zhang, G.-J. Li, and J. Lin, "Review on frequency control of power systems with wind power penetration," in *Proceedings of the International Conference on Power System Technology*, pp. 1–8, IEEE, Hangzhou, China, October 2010.

[5] S. K. Pandey, S. R. Mohanty, and N. Kishor, "A literature survey on load-frequency control for conventional and distribution generation power systems," *Renewable and Sustainable Energy Reviews*, vol. 25, pp. 318–334, 2013.

[6] F. Díaz-González, M. Hau, A. Sumper, and O. Gomis-Bellmunt, "Participation of wind power plants in system frequency control: review of grid code requirements and control methods," *Renewable and Sustainable Energy Reviews*, vol. 34, pp. 551–564, 2014.

[7] H. Shayeghi, H. A. Shayanfar, and A. Jalili, "Load frequency control strategies: a state-of-the-art survey for the researcher," *Energy Conversion and Management*, vol. 50, no. 2, pp. 344–353, 2009.

[8] L.-R. Chang-Chien, C.-C. Sun, and Y.-J. Yeh, "Modeling of wind farm participation in AGC," *IEEE Transactions on Power Systems*, vol. 29, no. 3, pp. 1204–1211, 2014.

[9] H. Bevrani and P. R. Daneshmand, "Fuzzy logic-based load-frequency control concerning high penetration of wind turbines," *IEEE Systems Journal*, vol. 6, no. 1, pp. 173–180, 2012.

[10] M. H. Variani and K. Tomsovic, "Distributed automatic generation control using flatness-based approach for high penetration of wind generation," *IEEE Transactions on Power Systems*, vol. 28, no. 3, pp. 3002–3009, 2013.

[11] X. J. Liu, P. Guan, and C. W. Chan, "Nonlinear multivariable power plant coordinate control by constrained predictive scheme," *IEEE Transactions on Control Systems Technology*, vol. 18, no. 5, pp. 1116–1125, 2010.

[12] X.-J. Liu and C. W. Chan, "Neuro-fuzzy generalized predictive control of boiler steam temperature," *IEEE Transactions on Energy Conversion*, vol. 21, no. 4, pp. 900–908, 2006.

[13] X. J. Liu and X. B. Kong, "Nonlinear fuzzy model predictive iterative learning control for drum-type boiler–turbine system," *Journal of Process Control*, vol. 23, no. 8, pp. 1023–1040, 2013.

[14] D. Rerkpreedapong, N. Atic, and A. Feliachi, "Economy oriented model predictive load frequency control," in *Proceedings of the Large Engineering Systems Conference on Power Engineering*, pp. 12–16, IEEE, Montreal, Canada, May 2003.

[15] X. Liu, X. Kong, and X. Deng, "Power system model predictive load frequency control," in *Proceedings of the American Control Conference (ACC '12)*, pp. 6602–6607, June 2012.

[16] T. H. Mohamed, J. Morel, H. Bevrani, and T. Hiyama, "Model predictive based load frequency control design concerning

wind turbines," *International Journal of Electrical Power & Energy Systems*, vol. 43, no. 1, pp. 859–867, 2012.

[17] T. H. Mohamed, H. Bevrani, A. A. Hassan, and T. Hiyama, "Decentralized model predictive based load frequency control in an interconnected power system," *Energy Conversion and Management*, vol. 52, no. 2, pp. 1208–1214, 2011.

[18] Y. Zheng, S. Li, and H. Qiu, "Networked coordination-based distributed model predictive control for large-scale system," *IEEE Transactions on Control Systems Technology*, vol. 21, no. 3, pp. 991–998, 2013.

[19] E. Camponogara and H. F. Scherer, "Distributed optimization for model predictive control of linear dynamic networks with control-input and output constraints," *IEEE Transactions on Automation Science and Engineering*, vol. 8, no. 1, pp. 233–242, 2011.

[20] A. N. Venkat, I. A. Hiskens, J. B. Rawlings, and S. J. Wright, "Distributed MPC strategies with application to power system automatic generation control," *IEEE Transactions on Control Systems Technology*, vol. 16, no. 6, pp. 1192–1206, 2008.

[21] M. Mirzaei, N. K. Poulsen, and H. H. Niemann, "Robust model predictive control of a wind turbine," in *Proceedings of the American Control Conference (ACC '12)*, pp. 114–119, Toronto, Canada, June 2012.

[22] M. Yigit, V. C. Gungor, G. Tuna, M. Rangoussi, and E. Fadel, "Power line communication technologies for smart grid applications: a review of advances and challenges," *Computer Networks*, vol. 70, pp. 366–383, 2014.

[23] M. Ma, H. Chen, X. Liu, and F. Allgöwer, "Moving horizon H^{∞} control of variable speed wind turbines with actuator saturation," *IET Renewable Power Generation*, vol. 8, no. 5, article 498, 2014.

3

Wind Characteristics of Three Meteorological Stations in China

Yang Yang,[1] Yao Gang,[1] Wang Rong,[1] and Wang Hengyu[2]

[1] *College of Civil Engineering, Chongqing University, Chongqing 400044, China*
[2] *Huadian New Energy Development Co., Gansu 730000, China*

Correspondence should be addressed to Yao Gang; yaocqu@vip.sina.com

Academic Editor: Petko Petkov

With rapid economic development of China, demand for energy is growing rapidly. Many experts have begun to pay attention on exploiting wind energy. Wind characteristics of three meteorological stations in China were analyzed to find out if or not it is possible to build a wind farm in this paper. First of all, studies about the wind characteristics and potential wind energy were summarized. Then ways of collecting and manipulating wind data were introduced. Wind-generation potential was assessed by the method of Weibull distribution. Wind shear exponent, extreme wind speed in 50 years, and turbulence intensity were calculated. The wind characteristics were summarized and assessment of wind-generation potential was given. At last, the wind was simulated with autoregressive method by Matlab software.

1. Introduction

With rapid economic development of China, demand for energy is growing rapidly [1]. Electricity produced by consuming coal had taken up of 76% in the whole electricity. This led to the spring of 2013 when the fog hazy weather lasted for a long time in China.

China is one of the countries which have a large number of wind resources and make use of the wind resource in the early days. According to statistics [2], density of wind energy in China is about 100 W/m^2 on average, and the total wind energy amount is 3226 GW. The available development and utilization wind energy is about 253 GW on land. The wind energy utilization has been growing at a fast pace around all over the world. But the utilization is held back by the shortage of wind data which are sufficient and credible. The wind speeds' distribution is not only significant to the wind farm designers but also to other power generators. Nowadays, vast quantities of information is offered to researchers by research of wind characteristic, investors, and wind farm planners that pay close attention to the renewable energy's evolution.

Gansu province which is the fifth on the potential wind energy in China has 237 million KW potential wind energy in theory [3]. Jiuquan region in Gansu is rich of wind energy, and its annual average wind density is above 150 W/m^2. Nowadays, energy policy of China is committed to developing a kind of new energy which is abundant and free pollution with low cost. So considering the exploitation of wind energy resources in Jiuquan is required.

In the last five years, lots of researches had been done in China on potential wind energy and wind characteristics. Song et al. [4] investigated major measures of observation data and computational precision based on field measurement and analyses of a number of wind power facilities, in particular over complex terrain. Du et al. [5, 6] proposed an improved method which had the advantage of less uncertainty and less error in computing the wind speed of the representative year in wind farm and presented changes of the wind shear index basing on the measured data of three 70 meter-high wind towers of two consecutive years in Inner Mongolia. Peng et al. [7] studied the wind shear exponent for wind resource assessment and found that the wind shear exponent in different vertical layers calculated at different speed bins had the highest precision. Chen et al. [8] studied the wind characteristics in northwest and southeast of China by analyzing the observational data obtained from towers which were representative in the two areas. Fang et al. [9] analyzed the wind power in northeast of China. The method and model for wind energy resource investigation in cold area had been established in his study. Wen et al. [10] analyzed

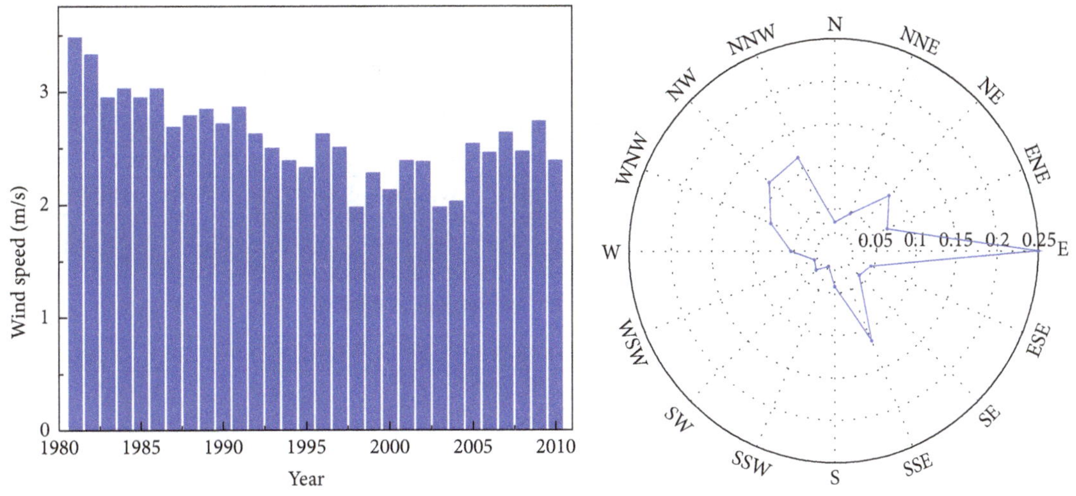

(a) Wind speed (b) Wind direction

FIGURE 1: Annual average wind speed and wind direction during 1981–2010 year in Guazhou meteorological station.

the reserves and distribution characteristics of wind speed and wind energy at 70 m height with statistical method, basing on observational data at 70 m height from 18 wind towers in Fujian coastal areas. Tong [11] put forward a new method of wind speed forecasting-wavelet decomposition autoregressive combined with wind speed change parting forecast method by bringing wind speed change characteristics types into wind speed prediction. Shang and Xie [12] developed logistic models according to the development characteristics of each province to predict development trend of wind power in 2013~2020, and predicting the main problems and measures should be taken by different provinces in the future. Bassyouni et al. [13] collected eleven years' wind data in Jeddah to determine the wind characteristics including the daily, monthly, and annual wind speed, wind probability density distribution, and shape (k) and scale (c) parameters at 10 m height. Alam et al. [14] studied the variability of fluctuating nature of the wind both in time and spatial domain. By utilizing daily mean values of wind speed from different meteorological stations, dynamic nature of the wind at nine stations was analyzed in conjunction with wavelet transform and fast Fourier transform power spectrum techniques. Wu et al. [15] developed a prediction system to predict wind power with a method of combining statistical model and physical model. The inlet condition of the wind farm is forecasted by the autoregressive model in this system. The proposed prediction system was tested by the data from Wattle Point Wind Farm in Australia, and this system was effective for power output prediction of wind farm. Santamaría-Bonfil et al. [16] proposed a hybrid methodology based on Support Vector Regression for wind speed forecasting with autoregressive model, and results showed that forecasts made with this method were more accurate for medium short term wind power forecasting and wind speed forecasting than other models.

China has sufficient potential wind energy especially in the northwest region. But development of wind resource is under the expected standard until now. This paper aims to make sure of the wind's characteristics and simulate the wind with autoregressive method.

2. Wind Speed Data Analysis

2.1. Description of Wind Data Collection System. Guazhou meteorological station which belonged to national principal station was built in 1951 with a 1171 m altitude above sea level. The wind speed data from 1981 to 2010 was collected at Guazhou meteorological station. Measurement data was kept with EL and EL15-2/2A anemometer.

Wind farm was about 67 km away the northeast of Guazhou meteorological station, which had a similar geographic and geomorphic condition with Guazhou meteorological station. Therefore, quantitative data of Guazhou meteorological station could be used in analyzing the characteristics of wind.

The annual average wind speed and wind direction during 1981–2010 year in Guazhou was given in Figure 1. It can be calculated that 2.55 m/s was the average speed in 30 years. Average wind speed kept stability and east wind direction was the prevailing wind direction. The month average wind speed during 1981–2010 year in Guazhou was given in Figure 2. The highest average speed was 3.4 m/s in March and the smallest was 2.0 m/s in October. It evidenced that wind in Guazhou kept stability in each month. Therefore it is possible to build a wind farm in this region.

More than fifty meteorological stations with 70 m height had been built in Jiuquan region to develop the wind recourse. There were three meteorological stations in the survey region, and hence all of them were selected to analyze the wind characteristics of the wind farm. The numbers of meteorological stations were 8158#, 7411#, and 8601#. Table 1 gave the basic condition of the three meteorological stations.

Table 1: Site specific information of meteorological stations considered in this paper.

Number	H (m)	Period	Coordinate	Elevation (m)	Wind tower configuration	Instrument device
8158#	70	2008.09.01–2011.04.28	N: 40°48′39″ E: 96°31′20″	1566	Wind speed: 10, 30, 50, 70 direction: 10, 70	NRG
7411#	70	2008.09.01–2011.04.27	N: 40°44′28″ E: 96°31′18″	1494	Wind speed: 10, 30, 50, 70 direction: 10, 70	NRG
8601#	70	2008.08.25–2011.04.28	N: 40°45′33″ E: 96°24′46″	1500	Wind speed: 10, 30, 50, 70 direction: 10, 70	NRG

Table 2: Valid data's rate of three meteorological stations.

Time	2008.9.01–2011.4.28	2008.9.01–2008.12.31	2009.1.01–2009.12.31	2010.1.01–2010.12.31	2011.1.01–2011.4.28
8158#	93.64%	95.16%	95.83%	89.76%	97.31%
7411#	96.47%	97.20%	96.71%	95.28%	98.68%
8601#	89.32%	95.33%	80.09%	95.59%	96.74%

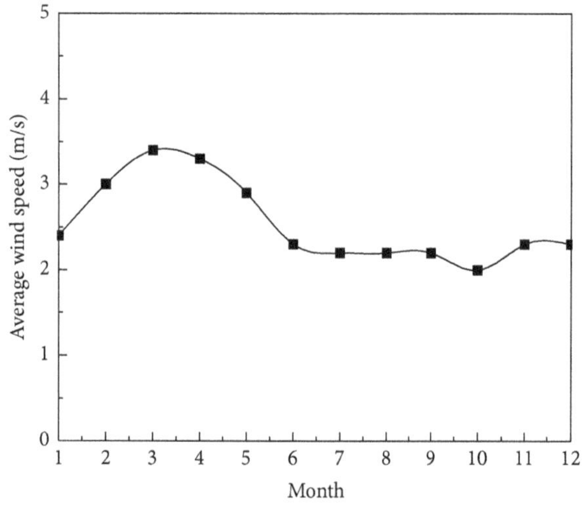

Figure 2: Month average wind speed during 1981–2010 year in Guazhou meteorological station.

2.2. Manipulation of Wind Data. In order to get the assessment which was most close to the actual wind power generation potential in the wind farm, the collected wind data were separately checked by all of wind data and the wind data which were measured in 2009 year. NRG anemoscopes were used in the meteorological stations, and sampling frequency of wind speed was 1 Hz. Data of wind speed and wind direction was output each 10 min. Sampling frequency of extreme wind speed was 2 Hz. Collection of wind direction kept the same pace with the wind speed, and all of wind direction records were moment sampling values. The anemoscopes could store the wind data collected in three months completely, and the staff got the wind data every two months. According to [17], the wind data were checked with data integrity, data scope, data dependency, and trend of wind speed. The unreasonable and missing data were deleted. At last, all the unreasonable data were collected and the rationality of data was judged again. The valid data were picked out and returned to the original data set. Table 2 showed that the entire rate of valid data was beyond 80% during observation period.

In order to check the correlation between wind speed and wind direction, wind data collected from different meteorological stations were selected. Correlation between wind speed and wind direction was analyzed between 8601# and 8158#, 8601# and 7411#, and 8158# and 7411# meteorological stations. Table 3 showed the related equation and correlation coefficient between 8601# and 8158#, 8601# and 7411#, and 8158# and 7411# at 70 m height in 2009 year. The wind data between 8601# and 7411# was taken as an example, the total correlation coefficient was 0.859, and the related equation was $Y = 0.869X + 0.623$. The 8601# and 7411# mainly wind correlation coefficient was above 0.75. The wind speed had a good correlation coefficient in three meteorological stations. It can be seen that wind power generation potential in the wind farm was almost the same.

From the checked results of wind data, the 8601#, 8158#, and 7411# meteorological stations can meet the requirement of design with more than 95% effective data. The monthly average wind speed data in 2009 collected from three meteorological stations at different heights were compared in Figure 3. It showed that wind speed data in 2009 collected from three meteorological stations at the same height had a good agreement, and hence wind speed data in 2009 had a certain representation. From Figure 3, it also can be seen that the wind speed was concentrated mainly on 5 m/s–9 m/s. Invalid wind speed under 3 m/s was rare, and there was no destructive wind all over the year. With the small change of the wind speed, this wind farm can generate electricity through the whole year.

3. Wind Characteristic Analysis

3.1. Weibull Parameter of Wind Speed. In the studied areas, frequency distribution of wind speed is a principal element to evaluate potential wind power. Potential wind power in this wind farm must be determined by the Weibull distribution. The range of wind data is wide. As a consequence, keeping a few important parameters which could certify the behavior of wind speed data with a wide range is requisite. Using a distribution function is the easiest method for the procedure, and the most widely used function is Weibull function.

Table 3: Correlation coefficient of three meteorological stations.

(a) Correlation coefficient of 8601# and 8158# meteorological stations at 70 m height

The total correlation equation: $Y = 0.884X + 0.556$ Correlation coefficient: 0.859		
Quadrant	Correlation equation	Correlation coefficient
N	$Y = 0.605X + 1.824$	0.627
NNE	$Y = 0.776X + 0.414$	0.879
NE	$Y = 0.906X + 0.132$	0.8
ENE	$Y = 0.956X + 0.099$	0.843
E	$Y = 0.862X + 1.130$	0.824
ESE	$Y = 0.797X + 1.485$	0.746
SE	$Y = 0.787X + 1.096$	0.8
SSE	$Y = 0.605X + 1.370$	0.664
S	$Y = 0.532X + 1.369$	0.603
SSW	$Y = 0.481X + 1.371$	0.553
SW	$Y = 0.714X + 0.907$	0.676
WSW	$Y = 0.869X + 0.512$	0.829
W	$Y = 0.893X + 0.365$	0.871
WNW	$Y = 0.860X + 0.328$	0.855
NW	$Y = 0.834X + 0.497$	0.846
NNW	$Y = 0.840X + 1.090$	0.772

(b) Correlation coefficient of 8601# and 7411# meteorological stations at 70 m height

The total correlation equation: $Y = 0.869X + 0.623$ Correlation coefficient: 0.859		
Quadrant	Correlation equation	Correlation coefficient
N	$Y = 0.607X + 2.020$	0.635
NNE	$Y = 0.796X + 0.487$	0.882
NE	$Y = 0.820X + 0.731$	0.75
ENE	$Y = 0.917X + 0.173$	0.819
E	$Y = 0.876X + 0.623$	0.798
ESE	$Y = 0.785X + 0.967$	0.766
SE	$Y = 0.840X + 0.623$	0.849
SSE	$Y = 0.717X + 0.946$	0.769
S	$Y = 0.665X + 1.041$	0.721
SSW	$Y = 0.576X + 1.219$	0.626
SW	$Y = 0.800X + 0.719$	0.798
WSW	$Y = 0.902X + 0.441$	0.885
W	$Y = 0.918X + 0.429$	0.898
WNW	$Y = 0.910X + 0.561$	0.876
NW	$Y = 0.860X + 0.902$	0.846
NNW	$Y = 0.807X + 1.597$	0.772

(c) Correlation coefficient of 8158# and 7411# meteorological stations at 70 m height

(c) Continued.

The total correlation equation: $Y = 0.918X + 0.682$ Correlation coefficient: 0.932		
Quadrant	Correlation equation	Correlation coefficient
N	$Y = 0.786X + 0.671$	0.802
NNE	$Y = 0.959X + 0.114$	0.964
NE	$Y = 1.000X + 0.207$	0.933
ENE	$Y = 0.974X + 0.575$	0.949
E	$Y = 0.907X + 1.112$	0.943
ESE	$Y = 0.958X + 1.081$	0.926
SE	$Y = 0.903X + 0.800$	0.875
SSE	$Y = 0.743X + 0.983$	0.735
S	$Y = 0.669X + 0.878$	0.695
SSW	$Y = 0.573X + 1.129$	0.642
SW	$Y = 0.746X + 0.687$	0.737
WSW	$Y = 0.886X + 0.514$	0.858
W	$Y = 0.905X + 0.482$	0.909
WNW	$Y = 0.841X + 0.658$	0.873
NW	$Y = 0.878X + 0.377$	0.898
NNW	$Y = 0.921X + 0.261$	0.875

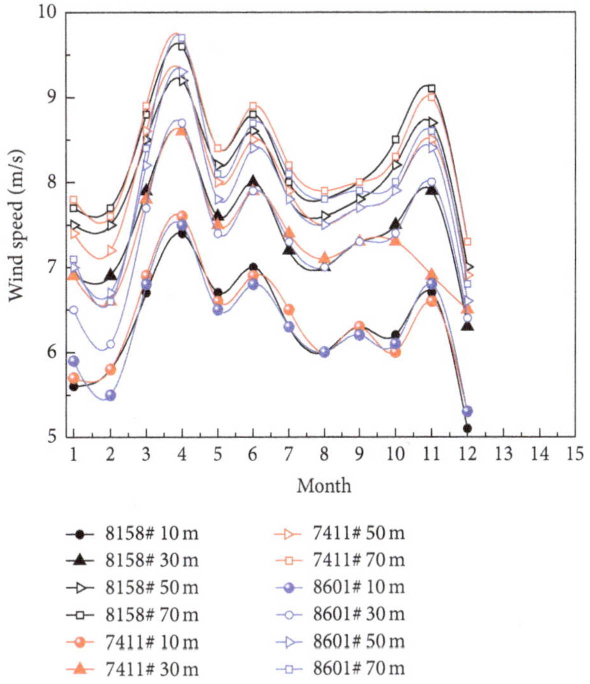

Figure 3: Wind speed of three meteorological stations at different heights in 2009 year.

The two-parameter distribution Weibull function could be showed in what follows:

$$f(V) = \frac{k}{c}\left(\frac{V}{c}\right)^{k-1} \times e^{-(V/c)^k}, \quad (1)$$

where k is a dimensionless parameter of Weibull shape, c is a Weibull scale parameter, and V is the wind speed, m/s. For the sake of assessing k and c, different kinds of ways have been come up with. Wind power can be assessed with the following:

$$W = \frac{1}{2}\rho v^3, \quad (2)$$

TABLE 4: The Weibull parameters at different heights.

Number	Parameter	10 m	30 m	50 m	65 m	70 m	75 m	80 m	90 m
8158#	c	6.4	7.9	8.6	8.8	8.8	8.9	9.0	9.1
	k	1.85	2.2	2.24	2.16	2.16	2.16	2.16	2.16
7411#	c	6.6	7.8	8.5	8.8	8.9	9.0	9.1	9.2
	k	1.92	2.14	2.21	2.15	2.15	2.16	2.16	2.16
8601#	c	6.5	7.8	8.3	8.4	8.5	8.5	8.6	8.8
	k	1.86	2.16	2.15	2.05	2.05	2.05	2.05	2.06

where W is the wind power density, ρ is average air density, and v is the average speed at 10 m height. Air density is calculated by

$$\rho = \frac{1.276}{1 + 0.00366t} \times \frac{p - 0.378e}{1000}, \tag{3}$$

where t is temperature on average, p is the air pressure on average, and e is the average vapor pressure. The air density in the wind farm is 1.045 kg/m^3.

The Weibull parameters were determined by using the software WASP10.0 in this paper. Table 4 showed the results of two Weibull parameters k and c at different height and different meteorological stations. Weibull parameters k and c ranged from 1.85 to 2.16, 6.4 to 9.1 separately. Both of k and c increased with the height under 50 m and kept nearly steady above the 50 m height. The wind speed frequency distribution Weibull fit curves at different height were described in order to check the calculate results. Table 5 showed density of wind power and wind speed at different height. Wind speeds in three meteorological stations were almost the same and had an error of no more than 4.6% at the same height. Average power density of wind ranged from 374 W/m^2 to 503 W/m^2 when the height of meteorological station changed from 50 m to 90 m height and had an error of no more than 9.9% at the same height. According to the standard [18], the grade of this wind farm was beyond three, and wind energy resources were abundant and steady.

3.2. Calculation of Wind Shear Exponent. Wind shear exponent is a key factor on wind turbine safety in wind resource assessment. Wind shear exponent in different vertical layers calculated in different speed bins has the highest precision in all methods, while the wind shear exponent calculated by the exponent law has the characteristic of high credibility and high stability. Thus exponent law was used in this paper. Wind shear exponent can be calculated by

$$a = \frac{\lg (v_2/v_1)}{\lg (z_2/z_1)}, \tag{4}$$

where v_2 is wind speed at z_2 height and m/s, v_1 is speed at z_1 height, m/s.

Table 6 showed the results of wind shear at different heights. When the height is above the 70 m height, the wind shear exponent is constant and independent of height. Therefore the wind shear exponent was assigned a value of 0.13 when the height was above 70 m height.

TABLE 5: Power density of wind and wind speed of three meteorological stations.

Number	Average wind speed (m/s)			Average power density of wind (W/m^2)		
	50 m	75 m	90 m	50 m	75 m	90 m
8158#	7.60	7.9	8.08	396	453	487
7411#	7.55	7.98	8.16	389	465	503
8601#	7.38	7.61	7.79	374	422	453

3.3. Extreme Wind Speed in 50 Years. Wind speed and turbulence intensity determine the grade of wind turbine generator system. The wind farm can be divided into three grades by extreme wind speed in 50 years, associating with three grades of wind turbine generator system. Extreme wind speed in 50 years determines the security and rationality of the wind farm. Table 7 showed the grade of wind farm.

Type I extreme value distribution method shown in (5) was used to assess the extreme wind speed in 50 years:

$$V_{50\text{-max}} = u - \frac{1}{\omega} \ln \left[\ln \left(\frac{250}{249} \right) \right], \tag{5}$$

where u is distribution location parameter, $u = 17.04$. ω is distribution scale parameter, and $\omega = 0.4892$.

Air density in this wind farm is 1.045 kg/m^3. Considering the change of air density, the extreme wind speed in 50 years should be modified with the condition of standard air density. Figure 4 showed the extreme wind speed in 50 years should be modified with the condition of standard air density.

3.4. Turbulence Intensity at Different Heights. Turbulence intensity function at the speed of 15 m/s could be calculated from

$$I_T = \frac{\delta}{V}, \tag{6}$$

where V is the average speed between 14.5 m/s and 15.5 m/s; σ is corresponding wind speed standard deviation. Table 8 showed the turbulence intensity at different heights. Turbulence intensity was decreased with the increasing of height. Range of turbulence intensity variation is 0.108 to 0.059. The C standard wind farm turbulence intensity is 0.12, so the turbulence intensity in this wind farm can meet the requirement [18].

TABLE 6: Results of wind shear at different heights.

Number	8158#				7411#				8601#			
H (m)	10	30	50	70	10	30	50	70	10	30	50	70
α	0.146	0.147	0.132	0.131	0.138	0.139	0.139	0.130	0.142	0.142	0.135	0.129

TABLE 7: Security grade of the wind farm.

IEC grade	I	II	III
Extreme wind speed in 50 years (m/s)	<70	<59.5	<52.5

4. Simulation of Wind Speed with Autoregressive Method

Simulation of wind speed is particularly important for wind farms because of cost-related issues, dispatch planning, and energy markets operations. Typically, wind farm energy production is estimated using a fixed weighted measure of the wind farm's nominal power and forecasts from historical atmospheric data. Further, it has been stated that wind speed is one of most important variables related to wind power generation. Autoregressive model is commonly used for time series forecasting wind speed since they are able to capture persistence in a time series. Therefore, wind speed was simulated with autoregressive method in this paper. Principle of autoregressive model can be expressed from

$$V(t) = \overline{v}(z) + v(t),$$

$$\begin{bmatrix} v^1(j\cdot t) \\ v^2(j\cdot t) \\ \vdots \\ v^M(j\cdot t) \end{bmatrix} = -\sum_{k=1}^{p} \psi_k \cdot \begin{bmatrix} v^1[(j-k)\cdot t] \\ v^2[(j-k)\cdot t] \\ \vdots \\ v^M[(j-k)\cdot t] \end{bmatrix} + \begin{bmatrix} N^1(j\cdot t) \\ N^2(j\cdot t) \\ \vdots \\ N^M(j\cdot t) \end{bmatrix}, \quad (7)$$

$$(j\cdot t = 0, 1, 2, 3, \dots, T,\ k \le j),$$

$$\overline{v}(z) = \overline{v}(10)\left(\frac{z}{10}\right)^{\alpha},$$

where $V(t)$ is simulation wind, $\overline{v}(z)$ is average wind, $v(t)$ is turbulent wind, and ψ_k is regression coefficient. $N(t)$ is a Gaussian function, p is the order of autoregressive mode, α can be found in Table 6, z is the height of wind, and $\overline{v}(10)$ is the average wind at the height of 10 m.

Simulation of wind speed with autoregressive method was realized by Matlab software. Time step was defined as 0.1, calculation step was defined as 1024, and operation order was 4. Wind in April 2009 at 70 m height was simulated according to davenport spectrum with autoregressive method. Figures 5 and 6 gave the turbulent and average wind in 400 seconds. From Figure 5, it can be seen that turbulent wind ranged from −1 m/s to 2 m/s at 70 m height, and it fluctuated around 0 m/s. From Figure 6, it can be seen that average wind ranged from 6.5 m/s to 9.5 m/s at 70 m height, and it fluctuated around 7.4 m/s which kept a good agreement with the measured wind speed data.

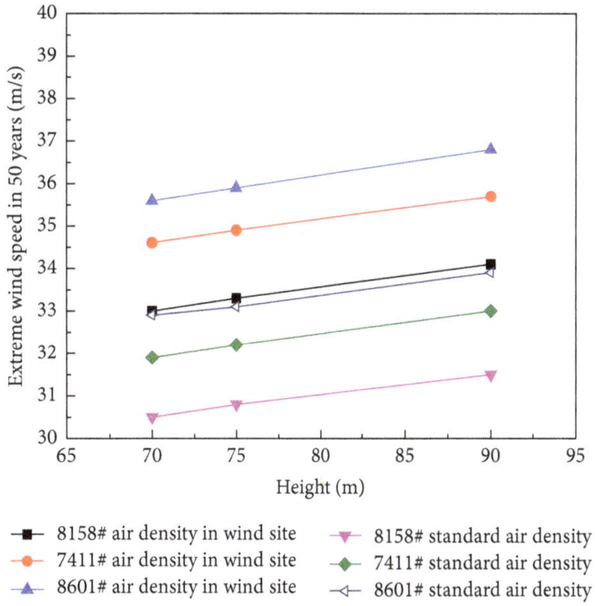

FIGURE 4: Comparison of wind speed between standard and actual air density.

TABLE 8: Turbulence intensity at different heights.

No.	70 m	50 m	30 m	10 m
8158#	0.059	0.073	0.093	0.109
7411#	0.059	0.074	0.093	0.109
8601#	0.062	0.076	0.089	0.108

Simulated power spectrum was compared with the standard one in Figure 7, the simulated power spectrum agreed well with the standard power spectrum. With the data collected, turbulent wind and average wind at different heights can be simulated accurately with autoregressive method.

5. Conclusions

(1) East wind is the prevailing wind direction in this wind farm. The wind speed is strong in winter and spring. By analyzing the rate of valid data and comprising wind speed data between meteorological stations, wind data collected in 2009 were used to assess wind-generation potential. The wind frequency mainly concentrated on 5 m/s–9 m/s. Invalid wind speed under

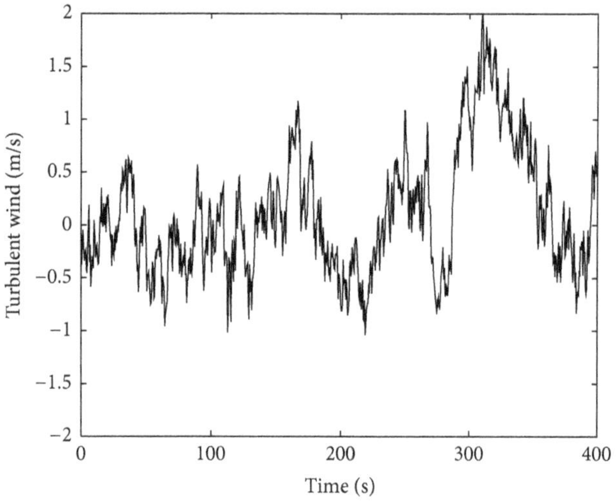

FIGURE 5: Turbulent wind in 400 seconds.

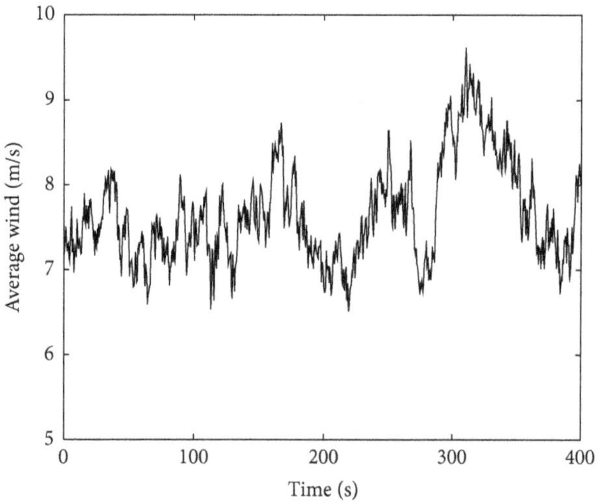

FIGURE 6: Average wind in 400 seconds.

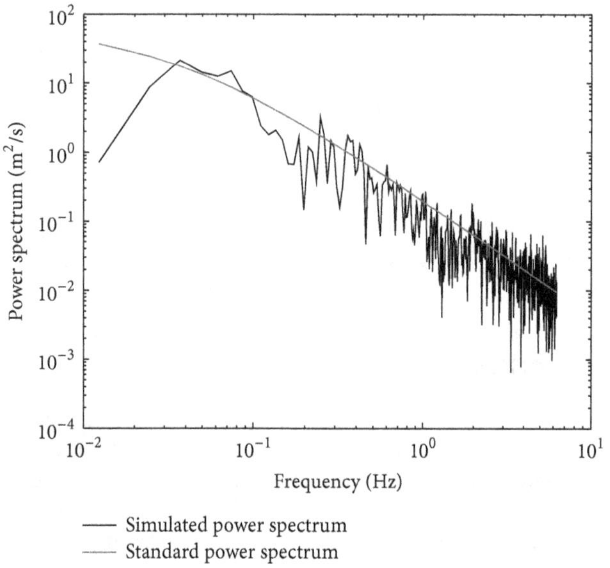

FIGURE 7: Simulated and standard power spectrum.

3 m/s was rare, and there was no destructive wind. With the small change of the wind speed, this wind farm can generate electricity through the whole year.

(2) Average wind power density at different heights was calculated. Average wind speed at 50 m height was 7.51 m/s and the density of wind power on average was 387 W/m^2. The grade of wind farm was beyond three.

(3) Under the condition of standard air density the maximum wind speed in 50 years is 36.4 m/s and 37.3 m/s at the height of 75 m and 90 m. Both of them are less than 37.5 m/s which is the reference wind speed in third grade of wind farm. And hence it is possible to build a wind farm of third grade in this region.

(4) Range of turbulence intensity variation is 0.108 to 0.059. The C standard wind farm turbulence intensity is 0.12, so the turbulence intensity in this wind farm can meet the requirement.

(5) With data collected and davenport spectrum, both wind speed at different heights and wind speed in this wind farm can be simulated accurately with autoregressive method.

Conflict of Interests

The authors declare that there is no conflict of interests regarding the publication of this paper.

Acknowledgments

This work was financially supported by science and technology project of Chongqing Municipal Construction Commission (20130844) and National Science Foundation of China (51578098).

References

[1] X. X. Wu, "Review of the energy work in 2013 and prediction of the energy work in 2014," *Macroeconomic Management*, vol. 19, no. 3, pp. 4–10, 2014.

[2] F. Wu and C. Fang, "Wind power resource appraisal and development stage regional division of China," *Journal of Natural Products*, vol. 24, no. 8, pp. 1413–1423, 2014.

[3] L. Shang and B. Xie, "Research on development trend of wind power in selected provinces of China," *Renewable Energy*, vol. 32, no. 2, pp. 191–196, 2014.

[4] L. L. Song, H. H. Huang, and S. Zhi, "Precision control in measurement and calculation of wind energy resource over wind power facilities," *Meteorological Monthly*, vol. 35, no. 3, pp. 73–81, 2009.

[5] Y. J. Du and C. Q. Feng, "Analysis on different computing method of wind speed of the representative year in wind farm," *Renewable Energy Resources*, vol. 28, no. 1, pp. 105–109, 2010.

[6] Y. J. Du and C. Q. Feng, "Application of wind shear index in the assessment of wind resources of wind farm," *Power System and Clean Energy*, vol. 26, no. 5, pp. 62–67, 2010.

[7] H. W. Peng, C. Q. Feng, and Z. G. Bao, "Study on the wind shear exponent for wind resource assessment," *Renewable Energy Resources*, vol. 28, no. 1, pp. 21–25, 2010.

[8] X. Chen, L. L. Song, and H. H. Huang, "Study on characteristics of wind energy resources in two tipical areas in China," *Acta Energiae Solaris Sinica*, vol. 32, no. 3, pp. 331–338, 2011.

[9] F. Fang, L. Yan, and Q. Y. Li, "Wind energy investigation and feasibility analysis of wind power in cold areas of China," *Journal of Northeast Agricultural University*, vol. 42, no. 8, pp. 66–76, 2011.

[10] M. Z. Wen, B. Wu, and X. F. Lin, "Distribution characteristics and assessment of wind energy resources at 70 m height over Fujian coastal areas," *Resources Science*, vol. 33, no. 7, pp. 1346–1352, 2011.

[11] J. L. Tong, *Research on change of wind speed and predicion of wind power in hexi corridor [Ph.D. thesis]*, Department of Applied Meteorology, Lanzhou University, Lanzhou, China, 2012.

[12] L. F. Shang and B. C. Xie, "Research on development trend of wind power in selected provinces of China," *Renewable Energy*, vol. 32, no. 2, pp. 191–196, 2014.

[13] M. Bassyouni, S. Gutub, U. Javaid et al., "Assessment and analysis of wind power resource using weibull parameters," *Energy, Exploration & Exploitation*, vol. 33, no. 1, pp. 105–122, 2015.

[14] M. M. Alam, S. Rehman, L. M. Al-Hadhrami, and J. P. Meyer, "Extraction of the inherent nature of wind speed using wavelets and FFT," *Energy for Sustainable Development*, vol. 22, no. 1, pp. 34–47, 2014.

[15] B. H. Wu, M. X. Song, K. Chen et al., "Wind power prediction system for wind farm based on auto regressive statistical model and physical model," *Journal of Renewable and Sustainable Energy*, vol. 6, no. 1, Article ID 0131011, 13 pages, 2014.

[16] G. Santamaría-Bonfil, A. Reyes-Ballesteros, and C. Gershenson, "Wind speed forecasting for wind farms: a method based on support vector regression," *Renewable Energy*, vol. 85, pp. 790–809, 2016.

[17] GB/T18710-2002, Methodology of wind energy resource assessment for wind farm.

[18] IEC6400-1-2005, International electro technical commission wind turbines.

4

Synthesis of Decentralized Variable Gain Robust Controllers with Guaranteed $\mathscr{L}_2$ Gain Performance for a Class of Uncertain Large-Scale Interconnected Systems

Shunya Nagai[1] and Hidetoshi Oya[2]

[1]*The Graduate School of Advanced Technology and Science, Tokushima University, 2-1 Minamijosanjima, Tokushima 770-8506, Japan*
[2]*The Institute of Technology and Science, Tokushima University, 2-1 Minamijosanjima, Tokushima 770-8506, Japan*

Correspondence should be addressed to Shunya Nagai; s-nagai@ee.tokushima-u.ac.jp

Academic Editor: Shengwei Mei

We consider a design problem of a decentralized variable gain robust controller with guaranteed $\mathscr{L}_2$ gain performance for a class of uncertain large-scale interconnected systems. For the uncertain large-scale interconnected system, the uncertainties and the interactions satisfy the matching condition. In this paper, we show that sufficient conditions for the existence of the proposed decentralized variable gain robust controller with guaranteed $\mathscr{L}_2$ gain performance are given in terms of linear matrix inequalities (LMIs). Finally, simple illustrative examples are shown.

1. Introduction

Due to the rapid development of industry in recent years, the controlled systems become more complex and such complex systems should be considered as large-scale interconnected systems (e.g., traffic systems and electric systems). As is well known, it is difficult to apply centralized control strategy to such systems because of physical constraints, calculation amount, and so on. Therefore, decentralized control problems for large-scale interconnected systems have been widely studied (see [1, 2] and references therein for details). The major problem of large-scale interconnected systems is how to deal with the interactions among subsystems.

On the other hand, robust control problem is one of the most important topics, and there is a large number of results for robust controller design and robust stability analysis (e.g., see [3, 4] and references therein). In particular, for uncertain linear systems, several quadratic stabilizing controllers have been suggested [5, 6] and the so-called robust $\mathscr{H}^\infty$ control problem has also been considered [7]. In most of the existing results, the proposed robust control systems have fixed gain controllers which are derived by considering worst case variations of uncertainties. In contrast with these, several design methods of variable gain controllers for uncertain continuous-time systems have been shown (e.g., [8–11]). These robust controllers are composed of a fixed gain controller and a variable gain one which are tuned by updating laws. In particular, in Oya and Hagino [11], the variable gain robust output feedback controller which achieves not only robust stability but also a specified $\mathscr{L}_2$ gain performance for a class of Lipschitz uncertain nonlinear systems has been proposed.

For large-scale interconnected systems with uncertainties, there are many existing results for decentralized robust control (e.g., [12–17]). In the work of Mao and Lin [12], for large-scale interconnected systems with unmodelled interaction, the aggregative derivation is tracked by using a model following technique with on-line improvement, and a sufficient condition for which the overall system when controlled by the completely decentralized control is asymptotically stable has been established. Furthermore, Gong [14] has proposed a decentralized robust controller which guarantees robust stability with prescribed degree of exponential convergence. Mukaidani et al. [15, 16] have also proposed decentralized guaranteed cost controllers for uncertain large-scale interconnected systems. In addition, we have suggested a decentralized variable gain robust controller which achieves not only robust stability but also satisfactory transient behavior for a class of uncertain large-scale interconnected systems [18, 19].

In this paper, on the basis of the existing results [11, 18, 19], we propose a decentralized variable gain robust controller with guaranteed $\mathscr{L}_2$ gain performance for a class of uncertain large-scale interconnected systems; that is, this study is an extension of our previous studies in this field. For the uncertain large-scale interconnected systems, uncertainties and interactions satisfy the matching condition. The proposed decentralized robust controller consists of a fixed gain matrix and a variable one determined by a parameter adjustment law. In this paper, LMI-based sufficient conditions for the existence of the proposed decentralized variable gain robust controller are derived. To put it in the concrete, the decentralized variable gain robust controller, that is, parameter adjustment law, is designed so that the effects of uncertainties and interactions are reduced.

This paper is organized as follows. In Section 2, we show the notations and useful lemmas used in this paper. In Section 3, the class of uncertain large-scale interconnected systems under consideration is presented. Section 4 contains the main results; that is, LMI-based sufficient conditions for the existence of the proposed decentralized variable gain robust controller with guaranteed $\mathscr{L}_2$ gain performance are derived. Finally, simple illustrative examples are included to show the effectiveness of the proposed decentralized robust controller.

2. Preliminaries

In this section, notations and useful and well-known lemmas (see [11, 20, 21] for details) which are used in this paper are shown.

In this paper, the following notations are adopted. For a matrix $\mathscr{X}$, the inverse of matrix $\mathscr{X}$ and the transpose of one are denoted by $\mathscr{X}^{-1}$ and $\mathscr{X}^T$, respectively. Additionally, $H_e\{\mathscr{X}\}$ and I_n mean $\mathscr{X} + \mathscr{X}^T$ and n-dimensional identity matrix, respectively, and the notation $\text{diag}(\mathscr{X}_1, \dots, \mathscr{X}_{\mathscr{M}})$ represents a block diagonal matrix composed of matrices $\mathscr{X}_i$ for $i = 1, \dots, \mathscr{M}$. For real symmetric matrices $\mathscr{X}$ and $\mathscr{Y}$, $\mathscr{X} > \mathscr{Y}$ (resp., $\mathscr{X} \geq \mathscr{Y}$), means that $\mathscr{X} - \mathscr{Y}$ is positive (resp., nonnegative) definite matrix. For a vector $\alpha \in \mathbb{R}^n$, $\|\alpha\|$ denotes standard Euclidian norm and, for a matrix $\mathscr{X}$, $\|\mathscr{X}\|$ represents its induced norm. The symbols "$\triangleq$" and "$\star$" mean equality by definition and symmetric blocks in matrix inequalities, respectively. Besides, $\mathscr{L}_2[0,\infty)$ is $\mathscr{L}_2$-space (i.e., the collection of all square integrable functions) defined on $[0,\infty)$, and, for a signal $f(t) \in \mathscr{L}_2[0,\infty)$, $\|f(t)\|_{\mathscr{L}_2}$ denotes its $\mathscr{L}_2$-norm.

Lemma 1. *For arbitrary vectors λ and ξ and the matrices $\mathscr{G}$ and $\mathscr{H}$ which have appropriate dimensions, the following inequality holds:*

$$H_e\left\{\lambda^T \mathscr{G} \Delta(t) \mathscr{H} \xi\right\} \leq 2\left\|\mathscr{G}^T \lambda\right\| \left\|\mathscr{H}\xi\right\|, \quad (1)$$

where $\Delta(t)$ with appropriate dimension is a time-varying unknown matrix satisfying $\|\Delta(t)\| \leq 1.0$.

Lemma 2 (Schur complement). *For a given constant real symmetric matrix Θ, the following items are equivalent:*

(1) $\Theta = \begin{pmatrix} \Theta_{11} & \Theta_{12} \\ \Theta_{12}^T & \Theta_{22} \end{pmatrix} > 0$;

(2) $\Theta_{11} > 0$ *and* $\Theta_{22} - \Theta_{12}^T \Theta_{11}^{-1} \Theta_{12} > 0$;

(3) $\Theta_{22} > 0$ *and* $\Theta_{11} - \Theta_{12} \Theta_{22}^{-1} \Theta_{12}^T > 0$.

3. Problem Formulation

Let us consider the uncertain large-scale interconnected system composed of $\mathscr{N}$ subsystems represented by

$$\begin{aligned} \frac{d}{dt} x_i(t) &= A_{ii}(t) x_i(t) + \sum_{\substack{j=1 \\ j \neq i}}^{\mathscr{N}} A_{ij}(t) x_j(t) + B_i u_i(t) \\ &\quad + \Gamma_{x_i} \omega_i(t), \\ z_i(t) &= C_{ii} x_i(t) + \Gamma_{z_i} \omega_i(t), \end{aligned} \quad (2)$$

where $x_i(t) \in \mathbb{R}^{n_i}$, $u_i(t) \in \mathbb{R}^{m_i}$, $z_i(t) \in \mathbb{R}^{p_i}$, and $\omega_i \in \mathbb{R}^{q_i}$ $(i = 1, \dots, \mathscr{N})$ are the vectors of the state, the control input, the controlled output, and the disturbance input for the ith subsystem, respectively. Besides, the disturbance input is assumed to be square integrable; that is, $\omega_i(t) \in \mathscr{L}_2[0,\infty)$. The matrices $A_{ii}(t)$ and $A_{ij}(t)$ in (2) are given by

$$\begin{aligned} A_{ii}(t) &= A_{ii} + B_i \Delta_{ii}(t) \mathscr{E}_{ii}, \\ A_{ij}(t) &= B_i \mathscr{D}_{ij} + B_i \Delta_{ij}(t) \mathscr{E}_{ij}; \end{aligned} \quad (3)$$

that is, the uncertainties and the interaction terms satisfy the matching condition [18, 19]. In (2) and (3), the matrices $A_{ii} \in \mathbb{R}^{n_i \times n_i}$, $A_{ij} \in \mathbb{R}^{n_i \times n_j}$, $B_i \in \mathbb{R}^{n_i \times m_i}$, $C_{ii} \in \mathbb{R}^{p_i \times n_i}$, $\Gamma_{x_i} \in \mathbb{R}^{n_i \times q_i}$, and $\Gamma_{z_i} \in \mathbb{R}^{p_i \times q_i}$ are known system parameters and the matrices $\mathscr{D}_{ij}$, $\mathscr{E}_{ii}$, and $\mathscr{E}_{ij}$ with appropriate dimensions represent the structure of interactions or uncertainties. Besides, the matrices $\Delta_{ii}(t) \in \mathbb{R}^{m_i \times r_i}$ and $\Delta_{ij}(t) \in \mathbb{R}^{m_i \times s_{ij}}$ denote unknown time-varying parameters satisfying the relations $\|\Delta_{ii}(t)\| \leq 1.0$ and $\|\Delta_{ij}(t)\| \leq 1.0$, respectively.

Since the ith subsystem is given by (2), we find that the overall system can be written as

$$\begin{aligned} \frac{d}{dt} x(t) &= \mathscr{A}(t) x(t) + \mathscr{B} u(t) + \Gamma_x \omega(t), \\ z(t) &= \mathscr{C} x(t) + \Gamma_z \omega(t), \end{aligned} \quad (4)$$

where $x(t) = (x_1^T(t), \dots, x_{\mathscr{N}}^T(t))^T$, $u(t) = (u_1^T(t), \dots, u_{\mathscr{N}}^T(t))^T$, $z(t) = (z_1^T(t), \dots, z_{\mathscr{N}}^T(t))^T$, and $\omega(t) = (\omega_1^T(t), \dots, \omega_{\mathscr{N}}^T(t))^T$ are the state, the control input, the controlled output, and the disturbance input of the overall system. Besides, the matrices

$\mathscr{A}(t) \in \mathbb{R}^{n\times n}$, $\mathscr{B} \in \mathbb{R}^{n\times m}$, $\mathscr{C} \in \mathbb{R}^{p\times n}$, $\Gamma_x \in \mathbb{R}^{n\times q}$, and $\Gamma_z \in \mathbb{R}^{p\times q}$ are given by

$$\mathscr{A}(t) = \begin{pmatrix} A_{11}(t) & A_{12}(t) & \cdots & A_{1\mathscr{N}}(t) \\ A_{21}(t) & A_{22}(t) & \cdots & A_{2\mathscr{N}}(t) \\ \vdots & \vdots & \ddots & \vdots \\ A_{\mathscr{N}1}(t) & A_{\mathscr{N}2}(t) & \cdots & A_{\mathscr{N}\mathscr{N}}(t) \end{pmatrix},$$

$$\mathscr{B} = \operatorname{diag}(B_1, B_2, \ldots, B_{\mathscr{N}}), \tag{5}$$

$$\mathscr{C} = \operatorname{diag}(C_{11}, C_{22}, \ldots, C_{\mathscr{N}\mathscr{N}}),$$

$$\Gamma_x = \operatorname{diag}(\Gamma_{x_1}, \Gamma_{x_2}, \ldots, \Gamma_{x_{\mathscr{N}}}),$$

$$\Gamma_z = \operatorname{diag}(\Gamma_{z_1}, \Gamma_{z_2}, \ldots, \Gamma_{z_{\mathscr{N}}}),$$

where n, m, p, and q are given by $n = \sum_{i=1}^{\mathscr{N}} n_i$, $m = \sum_{i=1}^{\mathscr{N}} m_i$, $p = \sum_{i=1}^{\mathscr{N}} p_i$, and $q = \sum_{i=1}^{\mathscr{N}} q_i$, respectively.

Now for the ith subsystem of (2), we define the following control input [18, 19]:

$$u_i(t) \triangleq F_i x_i(t) + \mathscr{L}_i(x_i, t)\, x_i(t). \tag{6}$$

In (6), $F_i \in \mathbb{R}^{m_i\times n_i}$ and $\mathscr{L}_i(x_i, t) \in \mathbb{R}^{m_i\times n_i}$ are the fixed compensation gain matrix and the variable one for the ith subsystem of (2). From (2), (3), and (6), the following closed-loop subsystem can be derived:

$$\begin{aligned} \frac{d}{dt}x_i(t) &= (A_{ii} + B_iF_i)\, x_i(t) + B_i\Delta_{ii}\mathscr{E}_{ii}x_i(t) \\ &\quad + B_i\sum_{\substack{j=1 \\ j\neq i}}^{\mathscr{N}} (\mathscr{D}_{ij} + \Delta_{ij}\mathscr{E}_{ij})\, x_j(t) \\ &\quad + B_i\mathscr{L}_i(x_i, t)\, x_i(t) + \Gamma_{x_i}\omega_i(t). \end{aligned} \tag{7}$$

Now we will give a definition of the decentralized variable gain robust control with guaranteed $\mathscr{L}_2$ gain performance.

Definition 3. For the uncertain large-scale interconnected system of (2), the control input of (6) is said to be a decentralized variable gain robust control with guaranteed $\mathscr{L}_2$ gain performance $\gamma^* > 0$ if the resultant closed-loop system of (7) is internally stable, and $\mathscr{H}_\infty$-norm of the transfer function from the disturbance input $\omega(t)$ to the controlled output $z(t)$ is less than or equal to a positive constant γ^*.

By using symmetric positive definite matrices $\mathscr{P}_i \in \mathbb{R}^{n_i\times n_i}$, we consider the following quadratic function:

$$\mathscr{V}(x, t) \triangleq \sum_{i=1}^{\mathscr{N}} \mathscr{V}_i(x_i, t), \tag{8}$$

where $\mathscr{V}_i(x_i, t)$ is

$$\mathscr{V}_i(x_i, t) \triangleq x_i^T(t)\,\mathscr{P}_i x_i(t). \tag{9}$$

Besides, we introduce the following Hamiltonian:

$$\begin{aligned} \mathscr{H}(x, t) &\triangleq \frac{d}{dt}\mathscr{V}(t) \\ &\quad + \sum_{i=1}^{\mathscr{N}} \left\{ z_i^T(t)\, z_i(t) - (\gamma_i^*)^2\, \omega_i^T(t)\,\omega_i(t) \right\}. \end{aligned} \tag{10}$$

Then we have the following lemma for the decentralized variable gain robust control with guaranteed $\mathscr{L}_2$ gain performance $\gamma^* > 0$ for the uncertain large-scale interconnected system of (2) and the control input of (6).

Lemma 4. *Consider the uncertain large-scale interconnected system of (2) and the control input of (6).*

For the quadratic function $\mathscr{V}(x, t)$ and the signals $z(t)$ and $\omega(t)$, if there exist symmetric positive definite matrices $\mathscr{P}_i$ ($i = 1, \ldots, \mathscr{N}$) and positive scalars γ^* which satisfy the inequality

$$\mathscr{H}(x, t) < 0, \tag{11}$$

then the control input of (6) is a decentralized variable gain robust control with guaranteed $\mathscr{L}_2$ gain performance γ^*, where γ^* is given by

$$\gamma^* = \max_i \gamma_i^* \quad (i = 1, \ldots, \mathscr{N}). \tag{12}$$

Proof. By integrating both sides of the inequality of (11) from 0 to ∞ with $x_i(0) = 0$, we obtain the following inequality from $\mathscr{V}(x, 0) = 0$. Consider

$$\begin{aligned} &\mathscr{V}(x, \infty) \\ &\quad + \sum_{i=1}^{\mathscr{N}} \left\{ \int_0^\infty z_i^T(t)\, z_i(t)\, dt - (\gamma_i^*)^2 \int_0^\infty \omega_i^T(t)\,\omega_i(t)\, dt \right\} \\ &< 0. \end{aligned} \tag{13}$$

We see from the inequality of (13) that the uncertain closed-loop subsystem of (7) is robustly stable (internally stable). Namely, robust stability of the uncertain closed-loop subsystem with $\omega(t) = 0$ is guaranteed and the $\mathscr{H}_\infty$-norm of the transfer function from the disturbance input $\omega(t)$ to the controlled output $z(t)$ is less than a positive constant γ^* because the inequality of (13) means the following relation:

$$\|z(t)\|_{\mathscr{L}_2} < \gamma^* \|\omega(t)\|_{\mathscr{L}_2}. \tag{14}$$

Thus the proof of Lemma 4 is accomplished. □

From the above discussion, our design objective in this paper is to determine the decentralized variable gain robust control input of (6) such that the overall system achieves not only internal stability but also guaranteed $\mathscr{L}_2$ gain performance $\gamma^* > 0$. That is to derive the symmetric positive definite matrices $\mathscr{P}_i \in \mathbb{R}^{n_i\times n_i}$, positive constants γ^*, the fixed gain matrices $F_i \in \mathbb{R}^{m_i\times n_i}$, and the variable one $\mathscr{L}_i(x_i, t) \in$

$\mathbb{R}^{m_i \times n_i}$ satisfying the inequality of (11) for all admissible uncertainties $\Delta_{ii}(t) \in \mathbb{R}^{m_i \times r_i}$ and $\Delta_{ij}(t) \in \mathbb{R}^{m_i \times s_{ij}}$ and the disturbance input $\omega_i(t) \in \mathscr{L}_2[0,\infty)$.

4. Decentralized Variable Gain Controllers

In this section, we show a design method of the decentralized variable gain robust controller with guaranteed $\mathscr{L}_2$ gain performance.

The following theorem shows sufficient conditions for the existence of the proposed decentralized control system.

Theorem 5. *Let one consider the large-scale interconnected system of (2) and the control input of (6).*

By using symmetric positive definite matrices $\mathscr{Y}_i \in \mathbb{R}^{n_i \times n_i}$, the matrices $\mathscr{W}_i \in \mathbb{R}^{m_i \times n_i}$, and positive scalars ϵ_i and γ_i which satisfy the LMIs,

$$\begin{pmatrix} H_e\{A_{ii}\mathscr{Y}_i + B_i\mathscr{W}_i\} & \Gamma_{x_i} + \mathscr{Y}_i C_{ii}^T \Gamma_{z_i} & \Lambda(\mathscr{Y}_i) \\ \star & \Gamma_{z_i}^T\Gamma_{z_i} - \gamma_i I_{q_i} & 0 \\ \star & \star & -\Omega(\epsilon_i) \end{pmatrix} < 0, \tag{15}$$

the fixed gain matrix $F_i \in \mathbb{R}^{m_i \times n_i}$ and the variable one $\mathscr{L}_i(x_i,t) \in \mathbb{R}^{m_i \times n_i}$ are determined as

$$F_i = \mathscr{W}_i\mathscr{Y}_i^{-1},$$
$$\mathscr{L}_i(x_i,t) \triangleq \begin{cases} -\dfrac{\zeta_i(x_i,t) + \eta_i(x_i,t)}{\left\|B_i^T\mathscr{P}_i x_i(t)\right\|^2} B_i^T\mathscr{P}_i \\ \mathscr{L}_i(x_i,t_\epsilon). \end{cases} \tag{16}$$

In (15) and (16), matrices $\Lambda(\mathscr{Y}_i)$ and $\Omega(\epsilon_i)$ and scalar functions $\zeta(x_i,t)$ and $\eta(x_i,t)$ are given by

$$\begin{aligned} \Lambda(\mathscr{Y}_i) &\triangleq \left(\mathscr{Y}_i C_{ii}^T\ \mathscr{Y}_i\mathscr{D}_{1i}^T\ \mathscr{Y}_i\mathscr{E}_{1i}^T\ \mathscr{Y}_i\mathscr{D}_{2i}^T\ \mathscr{Y}_i\mathscr{E}_{2i}^T\ \cdots\ \mathscr{Y}_i\mathscr{D}_{i-1i}^T\ \mathscr{Y}_i\mathscr{E}_{i-1i}^T\ \mathscr{Y}_i\mathscr{D}_{i+1i}^T\ \mathscr{Y}_i\mathscr{E}_{i+1i}^T\ \cdots\ \mathscr{Y}_i\mathscr{D}_{\mathcal{N}i}^T\ \mathscr{Y}_i\mathscr{E}_{\mathcal{N}i}^T\right), \\ \Omega(\epsilon_i) &\triangleq \operatorname{diag}\left(I_{p_i}, \epsilon_1 I_{m_1}, \epsilon_1 I_{s_1 i}, \epsilon_2 I_{m_2}, \epsilon_2 I_{s_2 i}, \ldots, \epsilon_{i-1} I_{m_{i-1}}, \epsilon_{i-1} I_{s_{i-1} i}, \epsilon_{i+1} I_{m_{i+1}}, \epsilon_{i+1} I_{s_{i+1} i}, \ldots, \epsilon_{\mathcal{N}} I_{m_{\mathcal{N}}}, \epsilon_{\mathcal{N}} I_{s_{\mathcal{N}} i}\right), \\ \zeta_i(x_i,t) &= \left\|B_i^T\mathscr{P}_i x_i(t)\right\| \left\|\mathscr{E}_{ii} x_i(t)\right\|, \\ \eta_i(x_i,t) &= \epsilon_i(\mathcal{N}-1)\left\|B_i^T\mathscr{P}_i x_i(t)\right\|^2. \end{aligned} \tag{17}$$

Note that t_ϵ in (16) is given by $t_\epsilon = \lim_{\epsilon>0,\epsilon\to 0}(t-\epsilon)$ [8].

Then the control input of (6) is the decentralized variable gain robust control with guaranteed $\mathscr{L}_2$ gain performance γ^*.

Proof. In order to prove Theorem 5, we consider the quadratic function $\mathscr{V}(x,t)$ of (8), the Hamiltonian $\mathscr{H}(x,t)$ of (10), and the inequality of (11).

For the quadratic function $\mathscr{V}_i(x_i,t)$ of (9), its time derivative along the trajectory of the resultant closed-loop subsystem of (7) can be computed as

$$\begin{aligned} \frac{d}{dt}\mathscr{V}_i(x_i,t) &= x_i^T(t)\left[H_e\left\{(A_{ii}+B_iF_i)^T\mathscr{P}_i\right\}\right]x_i \\ &+ H_e\left\{x_i^T(t)\mathscr{P}_iB_i\Delta_{ii}(t)\mathscr{E}_{ii}x_i(t)\right\} \\ &+ H_e\left\{x_i^T(t)\mathscr{P}_iB_i\sum_{\substack{j=1\\j\neq i}}^{\mathcal{N}}\left(\mathscr{D}_{ij}+\Delta_{ij}\mathscr{E}_{ij}\right)x_j(t)\right\} \\ &+ H_e\left\{x_i^T(t)\mathscr{P}_iB_i\mathscr{L}_i(x_i,t)x_i(t)\right\} \\ &+ H_e\left\{x_i^T(t)\mathscr{P}_i\Gamma_{x_i}\omega_i(t)\right\}. \end{aligned} \tag{18}$$

Besides, by using Lemma 1 and the well-known inequality

$$2\alpha^T\beta \le \delta\alpha^T\alpha + \frac{1}{\delta}\beta^T\beta \tag{19}$$

for any vectors α and β with appropriate dimensions and a positive scalar δ, we have the following relation for the function $\mathscr{V}_i(x_i,t)$:

$$\begin{aligned} \frac{d}{dt}\mathscr{V}_i(x_i,t) &\le x_i^T(t)\left[H_e\left\{(A_{ii}+B_iF_i)^T\mathscr{P}_i\right\}\right]x_i(t) \\ &+ 2\left\|B_i^T\mathscr{P}_ix_i(t)\right\|\left\|\mathscr{E}_{ii}x_i(t)\right\| \\ &+ 2\epsilon_i(\mathcal{N}-1)x_i^T(t)\mathscr{P}_iB_iB_i^T\mathscr{P}_ix_i(t) \\ &+ \frac{1}{\epsilon_i}\sum_{\substack{j=1\\j\neq i}}^{\mathcal{N}}x_j^T(t)\left(\mathscr{D}_{ij}^T\mathscr{D}_{ij}+\mathscr{E}_{ij}^T\mathscr{E}_{ij}\right)x_j(t) \\ &+ H_e\left\{x_i^T(t)\mathscr{P}_iB_i\mathscr{L}_i(x_i,t)x_i(t)\right\} \\ &+ H_e\left\{x_i^T(t)\mathscr{P}_i\Gamma_{x_i}\omega_i(t)\right\}. \end{aligned} \tag{20}$$

Firstly, we consider the case of $B_i^T\mathscr{P}_ix_i(t) \neq 0$. In this case, substituting the variable gain matrix of (16) into (20) and some algebraic manipulations derive the following inequality:

$$\frac{d}{dt}\mathscr{V}_i(x_i,t) \le x_i^T(t)\left[H_e\left\{(A_{ii}+B_iF_i)^T\mathscr{P}_i\right\}\right]x_i(t) + \frac{1}{\epsilon_i}\sum_{\substack{j=1\\ j\neq i}}^{\mathscr{N}} x_j^T(t)\left(\mathscr{D}_{ij}^T\mathscr{D}_{ij}+\mathscr{E}_{ij}^T\mathscr{E}_{ij}\right)x_j(t) + H_e\left\{x_i^T(t)\mathscr{P}_i\Gamma_{x_i}\omega_i(t)\right\}. \quad (21)$$

Additionally, one can see from (2) that the following relation holds:

$$\begin{aligned} z_i^T(t)z_i(t) - (\gamma_i^*)^2\omega_i^T(t)\omega_i(t) &= x_i^T(t)C_{ii}^TC_{ii}x_i(t) + H_e\left\{x_i^T(t)C_{ii}\Gamma_{z_i}\omega_i(t)\right\} \\ &\quad + \omega_i^T(t)\left(\Gamma_{z_i}^T\Gamma_{z_i} - \gamma_i I_{q_i}\right)\omega_i(t), \end{aligned} \quad (22)$$

where $(\gamma_i^*)^2 \triangleq \gamma_i$. Therefore, from (8), (10), (21), and (22), we can obtain the following relation for the Hamiltonian $\mathscr{H}(x,t)$:

$$\begin{aligned} \mathscr{H}(x,t) &\le \sum_{i=1}^{\mathscr{N}} x_i^T(t)\left[H_e\left\{(A_{ii}+B_iF_i)^T\mathscr{P}_i\right\}\right]x_i(t) \\ &\quad + \sum_{i=1}^{\mathscr{N}}\frac{1}{\epsilon_i}\sum_{\substack{j=1\\ j\neq i}}^{\mathscr{N}} x_j^T(t)\left(\mathscr{D}_{ij}^T\mathscr{D}_{ij}+\mathscr{E}_{ij}^T\mathscr{E}_{ij}\right)x_j(t) \\ &\quad + \sum_{i=1}^{\mathscr{N}} H_e\left\{x_i^T(t)\left(\mathscr{P}_i\Gamma_{x_i}+C_{ii}^T\Gamma_{z_i}\right)\omega_i(t)\right\} \\ &\quad + \sum_{i=1}^{\mathscr{N}} x_i^T(t)C_{ii}^TC_{ii}x_i(t) \\ &\quad + \sum_{i=1}^{\mathscr{N}} \omega_i^T(t)\left(\Gamma_{z_i}^T\Gamma_{z_i}-\gamma_i I_{q_i}\right)\omega_i(t). \end{aligned} \quad (23)$$

The inequality of (23) can also be rewritten as

$$\begin{aligned} \mathscr{H}(x,t) &\le \sum_{i=1}^{\mathscr{N}} x_i^T(t)\left[H_e\left\{(A_{ii}+B_iF_i)^T\mathscr{P}_i\right\}\right]x_i(t) \\ &\quad + \sum_{i=1}^{\mathscr{N}} x_i^T(t)\left\{\sum_{\substack{j=1\\ j\neq i}}^{\mathscr{N}}\frac{1}{\epsilon_j}\left(\mathscr{D}_{ji}^T\mathscr{D}_{ji}+\mathscr{E}_{ji}^T\mathscr{E}_{ji}\right)\right\}x_i(t) \\ &\quad + \sum_{i=1}^{\mathscr{N}} H_e\left\{x_i^T(t)\mathscr{P}_i\Gamma_{x_i}\omega_i(t)\right\} + \sum_{i=1}^{\mathscr{N}} x_i^T(t)C_{ii}^TC_{ii}x_i(t) \\ &\quad + \sum_{i=1}^{\mathscr{N}} H_e\left\{x_i^T(t)C_{ii}^T\Gamma_{z_i}\omega_i(t)\right\} + \sum_{i=1}^{\mathscr{N}} \omega_i^T(t)\left(\Gamma_{z_i}^T\Gamma_{z_i} - \gamma_i I_{q_i}\right)\omega_i(t) \\ &= \sum_{i=1}^{\mathscr{N}} x_i^T(t)\left[H_e\left\{(A_{ii}+B_iF_i)^T\mathscr{P}_i\right\} + C_{ii}^TC_{ii} + \sum_{\substack{j=1\\ j\neq i}}^{\mathscr{N}}\frac{1}{\epsilon_j}\left(\mathscr{D}_{ji}^T\mathscr{D}_{ji}+\mathscr{E}_{ji}^T\mathscr{E}_{ji}\right)\right]x_i(t) \\ &\quad + \sum_{i=1}^{\mathscr{N}} H_e\left\{x_i^T(t)\left(\mathscr{P}_i\Gamma_{x_i}+C_{ii}^T\Gamma_{z_i}\right)\omega_i(t)\right\} + \sum_{i=1}^{\mathscr{N}}\omega_i^T(t) \cdot \left(\Gamma_{z_i}^T\Gamma_{z_i}-\gamma_i I_{q_i}\right)\omega_i(t). \end{aligned} \quad (24)$$

Furthermore, some algebraic manipulations for (24) give the following inequality:

$$\mathscr{H}(x,t) \le \sum_{i=1}^{\mathscr{N}} \begin{pmatrix} x_i(t) \\ \omega_i(t) \end{pmatrix}^T \Psi_i(\mathscr{P}_i,\epsilon_i,\gamma_i)\begin{pmatrix} x_i(t) \\ \omega_i(t) \end{pmatrix}, \quad (25)$$

where $\Psi_i(\mathscr{P}_i,\epsilon_i,\gamma_i)$ is given by

$$\begin{aligned} \Psi_i(\mathscr{P}_i,\epsilon_i,\gamma_i) &\triangleq \left(\begin{array}{c|c} \Upsilon_i(\mathscr{P}_i,\epsilon_i) & \mathscr{P}_i\Gamma_{x_i}+C_{ii}^T\Gamma_{z_i} \\ \hline \star & \Gamma_{z_i}^T\Gamma_{z_i}-\gamma_i I_{q_i} \end{array}\right), \\ \Upsilon_i(\mathscr{P}_i,\epsilon_i) &\triangleq H_e\left\{(A_{ii}+B_iF_i)^T\mathscr{P}_i\right\} + C_{ii}^TC_{ii} + \sum_{i=1}^{\mathscr{N}}\frac{1}{\epsilon_i}\left(\mathscr{D}_{ji}^T\mathscr{D}_{ji}+\mathscr{E}_{ji}^T\mathscr{E}_{ji}\right). \end{aligned} \quad (26)$$

Hence, if the matrix inequality

$$\Psi_i(\mathscr{P}_i,\epsilon_i,\gamma_i) < 0 \quad (27)$$

holds, then the inequality of (11) for the Hamiltonian is satisfied.

Next we consider the case of $B_i^T\mathscr{P}_ix_i(t) = 0$. In this case, one can see from (18) and (22) and the definition of the control input of (6) and the variable gain matrix (16) that if the matrix inequality of (27) holds, then the inequality of (11) is also satisfied.

Finally, we consider the matrix inequality of (27). By introducing the matrices $\mathscr{Y}_i \triangleq \mathscr{P}_i^{-1}$ and $\mathscr{W}_i \triangleq F_i\mathscr{Y}_i$ and pre- and postmultiplying both sides of the matrix inequality of (27) by diag$(\mathscr{Y}_i, I_{q_i})$, we have the following inequality:

$$\Phi_i(\mathscr{Y}_i,\mathscr{W}_i,\epsilon_i,\gamma_i) = \left(\begin{array}{c|c} \Xi_i(\mathscr{Y}_i,\mathscr{W}_i,\epsilon_i) & \Gamma_{x_i}+\mathscr{Y}_iC_{ii}^T\Gamma_{z_i} \\ \hline \star & \Gamma_{z_i}^T\Gamma_{z_i}-\gamma_i I_{q_i} \end{array}\right) < 0, \quad (28)$$

where $\Xi_i(\mathscr{Y}_i,\mathscr{W}_i,\epsilon_i) \in \mathbb{R}^{n_i\times n_i}$ is given by

$$\Xi_i(\mathscr{Y}_i,\mathscr{W}_i,\epsilon_i) \triangleq H_e\left\{A_{ii}\mathscr{Y}_i + B_i\mathscr{W}_i\right\} + \mathscr{Y}_iC_{ii}^TC_{ii}\mathscr{Y}_i + \sum_{i=1}^{\mathscr{N}}\frac{1}{\epsilon_i}\mathscr{Y}_i\left(\mathscr{D}_{ji}^T\mathscr{D}_{ji}+\mathscr{E}_{ji}^T\mathscr{E}_{ji}\right)\mathscr{Y}_i. \quad (29)$$

Thus by applying Lemma 2 (Schur complement) to (28) we find that the matrix inequalities of (28) are equivalent to the LMIs of (15). In the LMIs of (15), scalar variables $\epsilon_i > 0$ and $\gamma_i > 0$ can arbitrarily be selected. Therefore, we find that the LMIs of (15) are always feasible; that is, there always exists the solution of the LMIs of (15). Namely, by solving the LMIs of (15), the fixed gain matrix is determined as $F_i = \mathscr{W}_i \mathscr{Y}_i^{-1}$ and the variable one is given by (16) and the proposed control input of (6) becomes a decentralized variable gain robust control with guaranteed $\mathscr{L}_2$ gain performance γ^* of (12). Therefore, the proof of Theorem 5 is accomplished. □

Remark 6. Although the uncertainties included in the large-scale interconnected system of (2) are structured ones, the proposed design method can also be applied to the systems with the parameter structured uncertainties [19]. Besides, in [19], the nominal system is introduced so as to generate the desired trajectory of the state and the control input. The proposed design method in this paper can be easily extended to such control problem.

Remark 7. The proposed decentralized variable gain robust controller can be obtained by solving LMIs of (15). Since LMIs of (15) define convex solution sets of $(\mathscr{Y}_i, \mathscr{W}_i, \epsilon_i, \gamma_i)$, thus various efficient convex optimization algorithms can be applied to test whether these LMIs are solvable and to generate particular solutions [22, 23]. In addition, these solutions parametrize the set of decentralized variable gain robust controllers with $\mathscr{L}_2$ gain performance. Namely, one can see that the result in Theorem 5 can easily be extended to the decentralized variable gain robust controller with suboptimal $\mathscr{L}_2$ gain performance (see Appendices for details).

Remark 8. The decentralized robust controller synthesis proposed in this paper is adaptable when some assumptions are satisfied. Namely, if the matching condition for uncertainties and interactions is satisfied, then the proposed decentralized variable gain robust controller is applicable; that is, the LMIs of (15) are always feasible (see [19]). On the other hand, for decentralized robust controllers with fixed gain matrices based on the existing results [14–16], the size of LMIs to be solved is $2n_i + \sum_{j=1,j\neq i}^{\mathscr{N}} n_j + 2p_i + q_i + r_i + \sum_{j=1,j\neq i}^{\mathscr{N}} s_{ij}$. However, the size of LMIs in the proposed design is equal to $n_i + 2q_i + \sum_{j=1,j\neq i}^{\mathscr{N}} (n_j + s_{ij})$. Moreover, the number of variables for LMIs of (15) is smaller than that of the decentralized robust controllers with fixed gain matrices. Therefore, one can see that the proposed decentralized robust controller design method is very useful (see also Remark 5 in [19]).

5. Numerical Examples

In order to demonstrate the efficiency of the proposed robust controller, we have run a simple example. The control problems considered here are not necessary practical. However, the simulation results stated below illustrate the distinct feature of the proposed decentralized robust controller.

In this example, we consider the uncertain large-scale interconnected systems consisting of three two-dimensional subsystems; that is, $\mathscr{N} = 3$. The system parameters are given as follows:

$$A_{11} = \begin{pmatrix} -1.0 & 1.0 \\ 0.0 & 1.0 \end{pmatrix},$$

$$A_{22} = \begin{pmatrix} 0.0 & 1.0 \\ -1.0 & -1.0 \end{pmatrix},$$

$$A_{33} = \begin{pmatrix} 1.0 & 0.0 \\ -1.0 & -3.0 \end{pmatrix},$$

$$B_1 = \begin{pmatrix} 0.0 \\ 1.0 \end{pmatrix},$$

$$B_2 = \begin{pmatrix} 1.0 \\ 1.0 \end{pmatrix},$$

$$B_3 = \begin{pmatrix} 1.0 \\ 0.0 \end{pmatrix},$$

$$\mathscr{E}_{11}^T = \begin{pmatrix} 1.0 \\ 0.0 \end{pmatrix},$$

$$\mathscr{E}_{22}^T = \begin{pmatrix} 2.0 \\ 1.0 \end{pmatrix},$$

$$\mathscr{E}_{33}^T = \begin{pmatrix} 2.0 \\ 2.0 \end{pmatrix},$$

$$\mathscr{D}_{12}^T = \begin{pmatrix} 1.0 \\ 2.0 \end{pmatrix},$$

$$\mathscr{D}_{13}^T = \begin{pmatrix} 2.0 \\ 1.0 \end{pmatrix},$$

$$\mathscr{D}_{21}^T = \begin{pmatrix} 1.0 \\ 0.0 \end{pmatrix},$$

$$\mathscr{D}_{23}^T = \begin{pmatrix} 1.0 \\ 1.0 \end{pmatrix},$$

$$\mathscr{D}_{31}^T = \begin{pmatrix} 2.0 \\ 1.0 \end{pmatrix},$$

$$\mathscr{D}_{32}^T = \begin{pmatrix} 0.0 \\ 2.0 \end{pmatrix},$$

$$\mathscr{E}_{12}^T = \begin{pmatrix} 2.0 \\ 1.0 \end{pmatrix},$$

$$\mathscr{E}_{13}^T = \begin{pmatrix} 2.0 \\ 2.0 \end{pmatrix},$$

$$\mathscr{E}_{21}^T = \begin{pmatrix} 1.0 \\ 0.0 \end{pmatrix},$$

$$\mathscr{E}_{23}^T = \begin{pmatrix} 0.0 \\ 3.0 \end{pmatrix},$$
$$\mathscr{E}_{31}^T = \begin{pmatrix} 1.0 \\ 2.0 \end{pmatrix},$$
$$\mathscr{E}_{32}^T = \begin{pmatrix} 3.0 \\ 1.0 \end{pmatrix},$$
$$\Gamma_{x_1} = \begin{pmatrix} 1.0 \\ 1.0 \end{pmatrix},$$
$$\Gamma_{x_2} = \begin{pmatrix} 1.0 \\ 0.0 \end{pmatrix},$$
$$\Gamma_{x_3} = \begin{pmatrix} 1.0 \\ 2.0 \end{pmatrix},$$
$$C_{11}^T = \begin{pmatrix} 1.0 \\ 0.0 \end{pmatrix},$$
$$C_{22}^T = \begin{pmatrix} 1.0 \\ 1.0 \end{pmatrix},$$
$$C_{33}^T = \begin{pmatrix} 1.0 \\ 1.0 \end{pmatrix},$$
$$\Gamma_{z_1} = 1.0,$$
$$\Gamma_{z_2} = 1.0,$$
$$\Gamma_{z_3} = 1.0. \quad (30)$$

Firstly, by using Theorem 5, we design the proposed decentralized variable gain robust controller. By solving LMIs of (15), we have positive definite matrices $\mathscr{Y}_i \in \mathbb{R}^{2\times 2}$, matrices $\mathscr{W}_i \in \mathbb{R}^{1\times 2}$, and positive scalars ϵ_i and γ_i given by

$$\mathscr{Y}_1 = \begin{pmatrix} 6.7486 \times 10^{-1} & -6.7790 \times 10^{-1} \\ \star & 1.8774 \end{pmatrix},$$
$$\mathscr{Y}_2 = \begin{pmatrix} 3.9813 \times 10^{-1} & 8.2896 \times 10^{-2} \\ \star & 7.9385 \times 10^{-1} \end{pmatrix},$$
$$\mathscr{Y}_3 = \begin{pmatrix} 1.5567 & -7.9700 \times 10^{-1} \\ \star & 1.2887 \end{pmatrix},$$
$$\mathscr{W}_1^T = \begin{pmatrix} -5.8074 \times 10^{-1} \\ -7.9480 \end{pmatrix},$$
$$\mathscr{W}_2^T = \begin{pmatrix} -2.9910 \\ -2.8262 \end{pmatrix},$$
$$\mathscr{W}_3^T = \begin{pmatrix} -7.3316 \\ -7.0692 \times 10^{-1} \end{pmatrix},$$

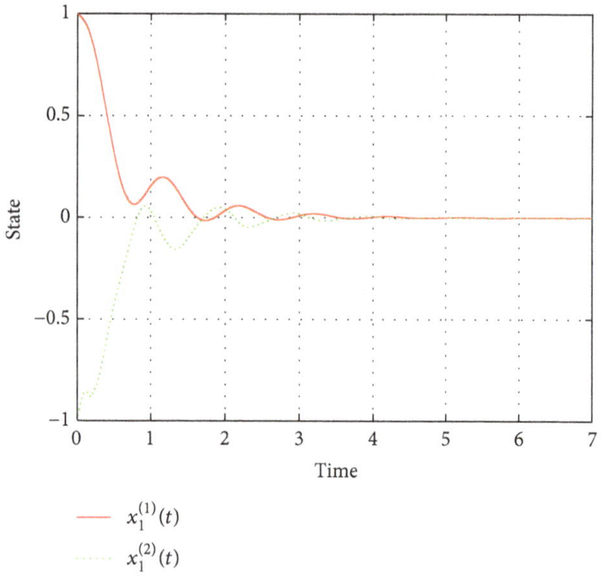

FIGURE 1: Time histories of $x_1(t)$.

$$\epsilon_1 = 3.8370,$$
$$\epsilon_2 = 7.2402,$$
$$\epsilon_3 = 2.4049,$$
$$\gamma_1 = 3.8259,$$
$$\gamma_2 = 3.4318,$$
$$\gamma_3 = 4.5212. \quad (31)$$

Thus the fixed gain matrices $F_i \in \mathbb{R}^{1\times 2}$ can be computed as

$$F_1 = \begin{pmatrix} -8.0232 & -7.1306 \end{pmatrix},$$
$$F_2 = \begin{pmatrix} -6.9218 & -2.8373 \end{pmatrix}, \quad (32)$$
$$F_3 = \begin{pmatrix} -7.3027 & -5.0650 \end{pmatrix}.$$

Furthermore, the positive scalars $\gamma_i^* = \sqrt{\gamma_i}$ can be obtained as

$$\gamma_1^* = 1.9560,$$
$$\gamma_2^* = 1.8525, \quad (33)$$
$$\gamma_3^* = 2.1263.$$

Therefore, the guaranteed $\mathscr{L}_2$ gain performance γ^* of (12) for the proposed controller is given by

$$\gamma^* = 2.1263. \quad (34)$$

The simulation result of this numerical example is shown in Figures 1–4. In this example, the initial value of the

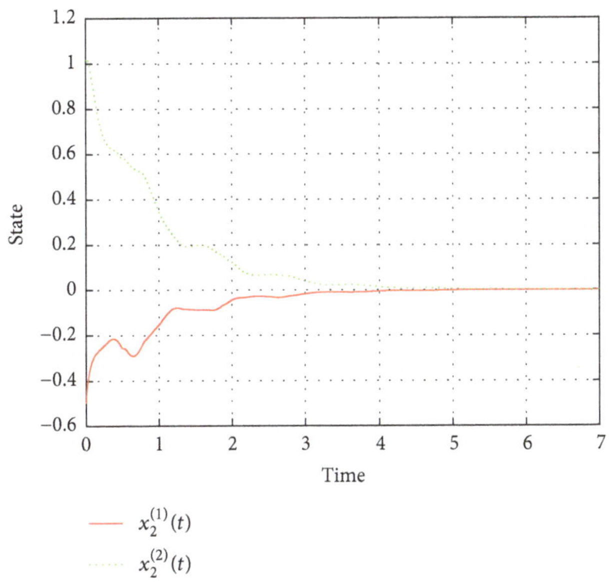

FIGURE 2: Time histories of $x_2(t)$.

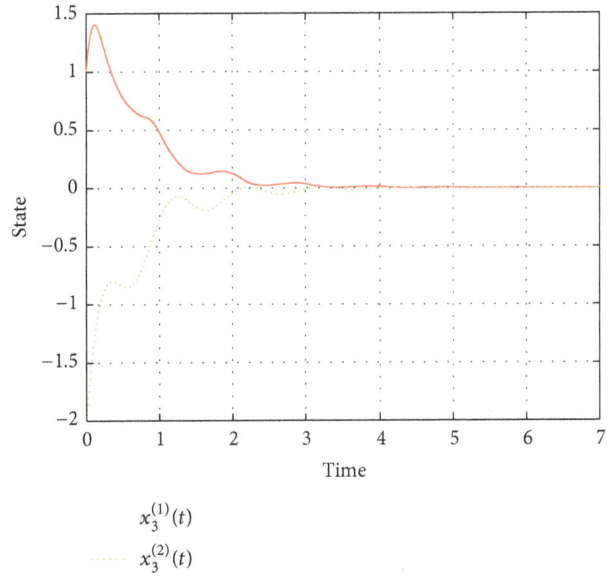

FIGURE 3: Time histories of $x_3(t)$.

uncertain large-scale system with system parameters of (30) is selected as follows:

$$x(0) = \left(1.0 \;\; -1.0 \,\middle|\, -0.5 \;\; 1.0 \,\middle|\, 1.0 \;\; -2.0 \right)^T. \tag{35}$$

In these figures, $x_i^{(l)}(t)$ denotes the lth element of the state $x_i(t)$ for the ith subsystem, respectively. Furthermore, unknown parameters and disturbance input are given as

$$\begin{aligned} \Delta_{ii}(t) &= \cos(5\pi t), \\ \Delta_{ij}(t) &= -\sin(2\pi t), \\ \omega_i(t) &= 2.0\exp(-t)\cos(5\pi t). \end{aligned} \tag{36}$$

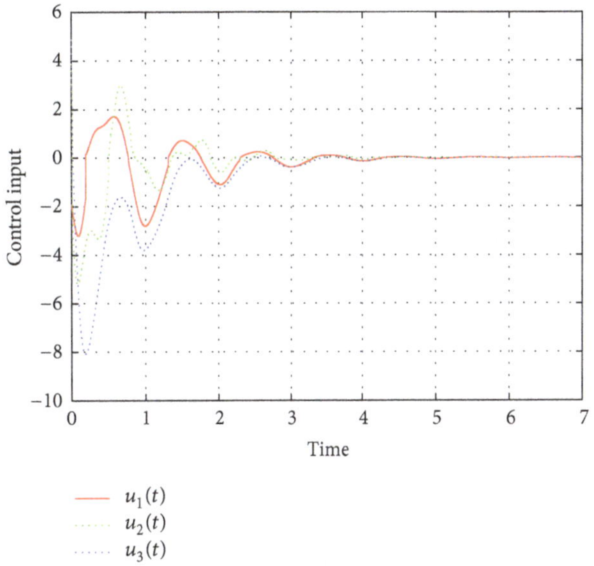

FIGURE 4: Time histories of $u_i(t)$.

From these figures, the proposed decentralized variable gain controller stabilizes the uncertain large-scale systems with system parameters of (30) in spite of uncertainties, interactions, and disturbance inputs. Therefore, the effectiveness of the proposed decentralized robust control system has been shown.

6. Conclusions

In this paper, on the basis of the result of [11, 18, 19], we have proposed a decentralized variable gain robust controller with guaranteed $\mathscr{L}_2$ gain performance for a large-scale interconnected system with uncertainties. For the uncertain large-scale interconnected systems, we have presented an LMI-based design method of the proposed decentralized variable gain robust controller. In addition, the effectiveness of the proposed decentralized robust controller has been shown by simple numerical examples. One can easily see that the result in this paper is an extension of the existing results [18, 19].

In the future research, we will extend the proposed controller synthesis to such a broad class of systems as large-scale systems with general uncertainties, uncertain large-scale systems with time delays, and so on.

Appendices

In this appendix, a decentralized variable gain robust controller with suboptimal $\mathscr{L}_2$ gain performance and the conventional fixed gain decentralized robust controller with $\mathscr{L}_2$ gain performance are presented.

A. Suboptimal Guaranteed $\mathscr{L}_2$ Gain Performance

In this section, we show a design method of a decentralized variable gain robust controller with suboptimal $\mathscr{L}_2$ gain performance.

Since the LMIs of (15) define a convex solution set, we consider minimizing the parameter γ_i because our interest is in establishing $\mathscr{L}_2$ gain performance. Furthermore, in the LMIs of (15), γ_i has no correlation with γ_j $(j \neq i)$. Thus our design problem can be replaced with the following constrained convex optimization problem (see [22, 23]):

$$\underset{\mathscr{Y}_i>0,\mathscr{W}_i,\epsilon_i>0,\gamma_i>0}{\text{Minimize}} \quad [\gamma_i] \quad \text{subject to} \quad (15). \tag{A.1}$$

If the optimal solution $\mathscr{Y}_i > 0$, $\mathscr{W}_i, \epsilon_i > 0$, and $\gamma_i > 0$ of the constrained optimization problem of (A.1) is obtained, then the control input of (6) with the fixed gain matrix F_i and variable one $\mathscr{L}_i(x_i, t)$ of (16) is the decentralized variable gain robust control with suboptimal $\mathscr{L}_2$ gain performance γ^* of (12).

Consequently, the following theorem for the decentralized variable gain robust controller with suboptimal $\mathscr{L}_2$ gain performance is obtained.

Theorem A.1. *Let one consider the uncertain large-scale interconnected system composed of $\mathscr{N}$ subsystems of (2) and the control input of (6).*

The control input of (6) is the decentralized variable gain robust control with suboptimal $\mathscr{L}_2$ gain performance γ^* of (12) provided that the constrained convex optimization problem of (A.1) is feasible.

One can easily see that although the solution of LMIs of (15) is not unique, the decentralized robust controller obtained by solving the convex optimization problem of (A.1) is unique.

Note that the constrained optimization problem of (A.1) can be solved by software such as MATLAB's LMI Control Toolbox and Scilab's LMITOOL.

B. An LMI-Based Design Method for the Conventional Fixed Gain Decentralized Robust Controller with $\mathscr{L}_2$ Gain Performance

In this section, the conventional fixed gain decentralized robust controller with $\mathscr{L}_2$ gain performance is presented.

Consider the uncertain large-scale interconnected system composed of $\mathscr{N}$ subsystems of (2). For the ith subsystem of (2), we define the following control input:

$$u_i(t) = K_i x_i(t), \tag{B.1}$$

where $K_i \in \mathbb{R}^{m\times n}$ is the fixed gain matrix for the ith subsystem of (2). From (2), (3), and (B.1), the closed-loop subsystem of (B.2) can be obtained as

$$\begin{aligned}\frac{d}{dt}x_i(t) &= (A_{ii} + B_iK_i)\, x_i(t) + B_i\Delta_{ii}\mathscr{E}_{ii}x_i(t) \\ &\quad + B_i\sum_{\substack{j=1\\ j\neq i}}^{\mathscr{N}} \left(\mathscr{D}_{ij} + \Delta_{ij}\mathscr{E}_{ij}\right) x_j(t) + \Gamma_{x_i}\omega_i(t).\end{aligned} \tag{B.2}$$

Namely, the control objective in this section is to design the fixed gain matrices $K_i \in \mathbb{R}^{m\times n}$ of (B.1) such that the overall closed-loop system composed of subsystems of (B.2) achieves not only internal stability but also guaranteed $\mathscr{L}_2$ gain performance $\gamma^* > 0$.

In order to design the fixed gain matrices $K_i \in \mathbb{R}^{m\times n}$, the quadratic function $\mathscr{V}(x,t)$ of (8) and the Hamiltonian $\mathscr{H}(x,t)$ of (10) are defined. Additionally, we consider the inequality of (11). The time derivative of the quadratic function $\mathscr{V}_i(x_i,t)$ of (9) along the trajectory of the closed-loop subsystem of (B.2) can be written as

$$\begin{aligned}\frac{d}{dt}\mathscr{V}_i(x_i,t) &= x_i^T(t)\left[H_e\left\{(A_{ii} + B_iK_i)^T\mathscr{P}_i\right\}\right]x_i(t) \\ &\quad + H_e\left\{x_i^T(t)\mathscr{P}_iB_i\Delta_{ii}(t)\mathscr{E}_{ii}x_i(t)\right\} \\ &\quad + H_e\left\{x_i^T(t)\mathscr{P}_iB_i\sum_{\substack{j=1\\ j\neq i}}^{\mathscr{N}}\mathscr{D}_{ij}x_j(t)\right\} \\ &\quad + H_e\left\{x_i^T(t)\mathscr{P}_iB_i\sum_{\substack{j=1\\ j\neq i}}^{\mathscr{N}}\Delta_{ij}(t)\mathscr{E}_{ij}x_j(t)\right\} \\ &\quad + H_e\left\{x_i^T(t)\mathscr{P}_i\Gamma_{x_i}\omega_i(t)\right\}.\end{aligned} \tag{B.3}$$

Besides, by using the well-known inequality of (19), we have the following relation for the function $\mathscr{V}_i(x_i,t)$:

$$\begin{aligned}\frac{d}{dt}\mathscr{V}_i(x_i,t) &\leq x_i^T(t)\left[H_e\left\{(A_{ii} + B_iK_i)^T\mathscr{P}_i\right\}\right]x_i(t) \\ &\quad + \epsilon_i x_i^T(t)\mathscr{P}_iB_iB_i^T\mathscr{P}_ix_i(t) \\ &\quad + \frac{1}{\epsilon_i}x_i^T(t)\mathscr{E}_{ii}^T\mathscr{E}_{ii}x_i(t) \\ &\quad + 2x_i^T(t)\mathscr{P}_iB_i\sum_{\substack{j=1\\ j\neq i}}^{\mathscr{N}}\mathscr{D}_{ij}x_j(t) \\ &\quad + \sum_{\substack{j=1\\ j\neq i}}^{\mathscr{N}}\epsilon_{ij}x_i^T(t)\mathscr{P}_iB_iB_i^T\mathscr{P}_ix_i(t) \\ &\quad + \sum_{\substack{j=1\\ j\neq i}}^{\mathscr{N}}\frac{1}{\epsilon_{ij}}x_j^T(t)\mathscr{E}_{ij}^T\mathscr{E}_{ij}x_j(t) \\ &\quad + H_e\left\{x_i^T(t)\mathscr{P}_i\Gamma_{x_i}\omega_i(t)\right\},\end{aligned} \tag{B.4}$$

where ϵ_i and ϵ_{ij} are positive constants. Moreover, one can see that the following relation holds:

$$\sum_{i=1}^{\mathcal{N}}\left\{\sum_{\substack{j=1\\ j\neq i}}^{\mathcal{N}}\left(x_j^T(t)x_j(t)-x_i^T(t)x_i(t)\right)\right\}=0. \quad \text{(B.5)}$$

Therefore from (B.4) and (22) we find the following relation for the Hamiltonian $\mathscr{H}(x,t)$:

$$\begin{aligned}\mathscr{H}(x,t)\le\sum_{i=1}^{\mathcal{N}}\Bigg\{ & x_i^T(t)\left[H_e\left\{(A_{ii}+B_iK_i)^T\mathscr{P}_i\right\}\right]x_i(t)\\ & +\epsilon_i x_i^T(t)\mathscr{P}_iB_iB_i^T\mathscr{P}_ix_i(t)+\frac{1}{\epsilon_i}x_i^T(t)\mathscr{E}_{ii}^T\mathscr{E}_{ii}x_i(t)\\ & +2x_i^T(t)\mathscr{P}_iB_i\sum_{\substack{j=1\\ j\neq i}}^{\mathcal{N}}\mathscr{D}_{ij}x_j(t)\\ & +\sum_{\substack{j=1\\ j\neq i}}^{\mathcal{N}}\epsilon_{ij}x_i^T(t)\mathscr{P}_iB_iB_i^T\mathscr{P}_ix_i(t)\\ & +\sum_{\substack{j=1\\ j\neq i}}^{\mathcal{N}}\frac{1}{\epsilon_{ij}}x_j^T(t)\mathscr{E}_{ij}^T\mathscr{E}_{ij}x_j(t)\\ & +H_e\left\{x_i^T(t)\mathscr{P}_i\Gamma_{x_i}\omega_i(t)\right\}+x_i^T(t)C_{ii}^TC_{ii}x_i(t)\\ & +2x_i^T(t)C_{ii}\Gamma_{z_i}\omega_i(t)+\omega_i^T(t)\left(\Gamma_{z_i}^T\Gamma_{z_i}-\gamma_iI_{q_i}\right)\\ & -\sum_{\substack{j=1\\ j\neq i}}^{\mathcal{N}}\left(x_j^T(t)x_j(t)-x_i^T(t)x_i(t)\right)\Bigg\}.\end{aligned} \quad \text{(B.6)}$$

Furthermore, some algebraic manipulations for (B.6) give the following inequality:

$$\mathscr{H}(x,t)\le\begin{pmatrix}x_i(t)\\ x_1(t)\\ \vdots\\ x_{\mathcal{N}}(t)\\ \omega_i(t)\end{pmatrix}^T\Phi\left(\mathscr{P}_i,\epsilon_i,\epsilon_{ij},\gamma_i\right)\begin{pmatrix}x_i(t)\\ x_1(t)\\ \vdots\\ x_{\mathcal{N}}\\ \omega_i(t)\end{pmatrix}, \quad \text{(B.7)}$$

where $\Phi(\mathscr{P}_i,\epsilon_i,\epsilon_{ij},\gamma_i)$ is given by

$$\Phi\left(\mathscr{P}_i,\epsilon_i,\epsilon_{ij},\gamma_i\right)\triangleq\left(\begin{array}{c|c|c}\Theta_i\left(\mathscr{P}_i,\epsilon_i,\epsilon_{ij}\right) & \Pi_i\left(\mathscr{P}_i\right) & \mathscr{P}_i\Gamma_{x_i}\\ \hline \star & \mathscr{Q}_i & 0\\ \hline \star & \star & \Gamma_{z_i}^T\Gamma_{z_i}-\gamma_iI_{q_i}\end{array}\right),$$

$$\begin{aligned}\Theta_i\left(\mathscr{P}_i,\epsilon_i,\epsilon_{ij}\right)&\triangleq H_e\left\{(A_{ii}+B_iK_i)^T\mathscr{P}_i\right\}\\ &+\epsilon_i\mathscr{P}_iB_iB_i^T\mathscr{P}_i+\frac{1}{\epsilon_i}\mathscr{E}_{ii}^T\mathscr{E}_{ii}+\sum_{\substack{j=1\\ j\neq i}}^{\mathcal{N}}\epsilon_{ij}\mathscr{P}_iB_iB_i^T\mathscr{P}_i\\ &+C_{ii}^TC_{ii}+(\mathcal{N}-1)I_n,\\ \Pi_i\left(\mathscr{P}_i\right)&\triangleq\left(\mathscr{P}_iB_i\mathscr{D}_{i1}\ \ \mathscr{P}_iB_i\mathscr{D}_{i2}\ \cdots\ \mathscr{P}_iB_i\mathscr{D}_{i\mathcal{N}}\right),\\ \mathscr{Q}_i&\triangleq\operatorname{diag}\left(-I_{n_1}+\frac{1}{\epsilon_{i1}}\mathscr{E}_{i1}^T\mathscr{E}_{i1},-I_{n_2}+\frac{1}{\epsilon_{i2}}\mathscr{E}_{i2}^T\mathscr{E}_{i2},\ldots,\right.\\ &\left.-I_{n_{\mathcal{N}}}+\frac{1}{\epsilon_{i\mathcal{N}}}\mathscr{E}_{i\mathcal{N}}^T\mathscr{E}_{i\mathcal{N}}\right).\end{aligned} \quad \text{(B.8)}$$

Hence, if the following inequality holds, then

$$\Phi\left(\mathscr{P}_i,\epsilon_i,\epsilon_{ij},\gamma_i\right)<0. \quad \text{(B.9)}$$

Additionally, by using $\mathcal{S}_i\triangleq\mathscr{P}_i^{-1}$ and $\mathscr{W}_i\triangleq K_i\mathcal{S}_i$ and pre- and postmultiplying both sides of the matrix inequality of (B.9) by $\operatorname{diag}(\mathcal{S}_i,I_{n_i^\star},I_{q_i})$, one can obtain

$$\left(\begin{array}{c|c|c}\Theta_i^\star\left(\mathcal{S}_i,\mathscr{W}_i,\epsilon_i,\epsilon_{ij}\right) & \Pi_i^\star & \Gamma_{x_i}\\ \hline \star & \mathscr{Q}_i & 0\\ \hline \star & \star & \Gamma_{z_i}^T\Gamma_{z_i}-\gamma_iI_{q_i}\end{array}\right)<0, \quad \text{(B.10)}$$

$$\begin{aligned}\Theta_i^\star\left(\mathcal{S}_i,\mathscr{W}_i,\epsilon_i,\epsilon_{ij}\right)&\triangleq H_e\left\{A_{ii}^T\mathcal{S}_i+B_i\mathscr{W}_i\right\}+\epsilon_iB_iB_i^T+\frac{1}{\epsilon_i}\mathcal{S}_i\mathscr{E}_{ii}^T\mathscr{E}_{ii}\mathcal{S}_i\\ &+\sum_{\substack{j=1\\ j\neq i}}^{\mathcal{N}}\epsilon_{ij}B_iB_i^T+\mathcal{S}_iC_{ii}^TC_{ii}\mathcal{S}_i+(\mathcal{N}-1)\mathcal{S}_i\mathcal{S}_i,\end{aligned} \quad \text{(B.11)}$$

$$\Pi_i^\star\triangleq\left(B_i\mathscr{D}_{i1}\ \ B_i\mathscr{D}_{i2}\ \cdots\ B_i\mathscr{D}_{i\mathcal{N}}\right). \quad \text{(B.12)}$$

Moreover, by applying Lemma 2 (Schur complement) to (B.10), we obtain

$$\left(\begin{array}{c|c|c|c} \Theta_i^{\star\star}\left(\mathcal{S}_i, \mathcal{W}_i, \epsilon_i, \epsilon_{ij}\right) & \Pi_i^\star & \Gamma_{x_i} & \Lambda_i\left(\mathcal{S}_i\right) \\ \hline \star & -I_{n_i^\star} & 0 & 0 \\ \hline \star & \star & \Gamma_{z_i}^T\Gamma_{z_i} - \gamma_i I_{q_i} & 0 \\ \hline \star & \star & \star & \Omega_i^\star\left(\epsilon_i, \epsilon_{ij}\right) \end{array}\right) < 0, \tag{B.13}$$

$$\begin{aligned} \Theta_i^{\star\star}\left(\mathcal{S}_i, \mathcal{W}_i, \epsilon_i, \epsilon_{ij}\right) &\triangleq H_e\left\{A_{ii}\mathcal{S}_i + B_i\mathcal{W}_i\right\} + \epsilon_i B_i B_i^T \\ &+ \sum_{\substack{j=1 \\ j\neq i}}^{\mathcal{N}} \epsilon_{ij} B_i B_i^T, \end{aligned} \tag{B.14}$$

$$\Lambda_i\left(\mathcal{S}_i\right) \triangleq \left(\mathcal{S}_i C_{ii}^T \;\; \mathcal{S}_i \;\; \mathcal{S}_i\mathcal{E}_{ii}^T \;\; \mathcal{E}_{i1}^T \;\; \mathcal{E}_{i2}^T \;\; \cdots \;\; \mathcal{E}_{i\mathcal{N}}^T\right),$$

$$\Omega_i^\star\left(\epsilon_i, \epsilon_{ij}\right) \triangleq -\operatorname{diag}\left(I_{p_i}, \frac{1}{\mathcal{N}-1} I_{n_i}, \epsilon_i I_{r_i}, \epsilon_{i1} I_{s_{i1}}, \ldots, \epsilon_{i\mathcal{N}} I_{i\mathcal{N}}\right).$$

Namely, by solving the LMIs of (B.13), the fixed gain matrix is determined as $K_i = \mathcal{W}_i \mathcal{S}_i^{-1}$.

Conflict of Interests

The authors declare that there is no conflict of interests regarding the publication of this paper.

Acknowledgments

The authors would like to thank Professor Shengwei Mei, Tsinghua University, and the anonymous reviewers for their valuable and helpful comments that greatly contributed to this paper.

References

[1] M. Ikeda and D. D. Sijjak, "Decentralized stabilization of linear time-varying systems," *IEEE Transactions on Automatic Control*, vol. 25, no. 1, pp. 106–107, 1980.

[2] D. D. Sijjak, *Decentralized Control of Complex Systems*, vol. 184 of *Mathematics in Science and Engineering*, Academic Press, New York, NY, USA, 1991.

[3] K. Zhou, J. C. Doyle, and K. Glover, *Robust and Optimal Control*, Prentice Hall, 1996.

[4] K. Zhou, *Essentials of Robust Control*, Prentice Hall, 1998.

[5] I. R. Petersen, "A Riccati equation approach to the design of stabilizing controllers and observers for a class of uncertain linear systems," *IEEE Transactions on Automatic Control*, vol. 30, no. 9, pp. 904–907, 1985.

[6] W. E. Schmitendorf, "Designing stabilizing controllers for uncertain systems using the Riccati equation approach," *IEEE Transactions on Automatic Control*, vol. 33, no. 4, pp. 376–379, 1988.

[7] P. Shi and S.-P. Shue, "Robust H_∞ control for linear discrete-time systems with norm-bounded nonlinear uncertainties," *IEEE Transactions on Automatic Control*, vol. 44, no. 1, pp. 108–111, 1999.

[8] M. Maki and K. Hagino, "Robust control with adaptation mechanism for improving transient behaviour," *International Journal of Control*, vol. 72, no. 13, pp. 1218–1226, 1999.

[9] H. Oya and K. Hagino, "Robust control with adaptive compensation input for linear uncertain systems," *IEICE Transactions on Fundamentals of Electronics, Communications and Computer Sciences*, vol. E86-A, no. 6, pp. 1517–1524, 2003.

[10] H. Oya and Y. Uehara, "Synthesis of variable gain controllers based on LQ optimal control for a class of uncertain linear systems," in *Proceedings of the UKACC International Conference on Control (CONTROL '12)*, pp. 87–91, Cardiff, UK, September 2012.

[11] H. Oya and K. Hagino, "L_2 gain performance for a class of lipschitz uncertain nonlinear systems via variable gain robust output feedback controllers," *Journal of Control Science and Engineering*, vol. 2013, Article ID 432034, 7 pages, 2013.

[12] C. J. Mao and W.-S. Lin, "Decentralized control of interconnected systems with unmodelled nonlinearity and interaction," *Automatica*, vol. 26, no. 2, pp. 263–268, 1990.

[13] C. J. Mao and J.-H. Yang, "Decentralized output tracking for linear uncertain interconnected systems," *Automatica*, vol. 31, no. 1, pp. 151–154, 1995.

[14] Z. Gong, "Decentralized robust control of uncertain interconnected systems with prescribed degree of exponential convergence," *IEEE Transactions on Automatic Control*, vol. 40, no. 4, pp. 704–707, 1995.

[15] H. Mukaidani, Y. Takato, Y. Tanaka, and K. Mizukami, "The guaranteed cost control for uncertain large-scale interconnected systems," in *Proceedings of the 15th IFAC World Congress*, Barcelona, Spain, 2002.

[16] H. Mukaidani, M. Kimoto, and T. Yamamoto, "Decentralized guaranteed cost control for discrete-time uncertain large-scale systems using fuzzy control," in *Proceedings of the IEEE World Congress on Computational Intelligence*, pp. 3099–3105, Vancouver, Canada, 2006.

[17] H. Oya and K. Hagino, "Trajectory-based design of robust non-fragile controllers for a class of uncertain linear continuous-time systems," *International Journal of Control*, vol. 80, no. 12, pp. 1849–1862, 2007.

[18] S. Nagai and H. Oya, "Decentralized variable gain robust controllers for a class of uncertain large-scale interconnected systems," in *Proceedings of the 33rd IASTED International Conference on Modelling, Identification and Control*, pp. 199–204, Innsbruck, Austria, 2014.

[19] S. Nagai and H. Oya, "Synthesis of decentralized variable gain robust controllers for large-scale interconnected systems with structured uncertainties," *Journal of Control Science and Engineering*, vol. 2014, Article ID 848465, 10 pages, 2014.

[20] F. R. Gantmacher, *The Theory of Matrices*, vol. 1, Chelsea Publishing Company, New York, NY, USA, 1960.

[21] S. Boyd, L. El Ghaoui, E. Feron, and V. Balakrishnan, *Linear Matrix Inequalities in System and Control Theory*, SIAM Studies in Applied Mathematics, 1994.

[22] H. Oya, K. Hagino, and M. Matsuoka, "Observer-based guaranteed cost control scheme for polytopic uncertain systems with state delays," in *Proceedings of the 30th Annual Conference of IEEE Industrial Electronics Society (IECON '04)*, pp. 684–689, Busan, Republic of Korea, November 2004.

[23] H. Oya, K. Hagino, and M. Matsuoka, "Observer-based robust preview tracking control scheme for uncertain discrete-time systems," in *Proceedings of the 16th IFAC World Congress (IFAC '05)*, Mo-Ao2-TP07, Prague, Czech Republic, 2005.

5

Predictive Variable Gain Iterative Learning Control for PMSM

Huimin Xu,[1] Xuedong Zhang,[2] and Xiangjie Liu[1]

[1] *The State Key Laboratory of Alternate Electrical Power System with Renewable Energy Sources, North China Electric Power University, Beijing 102206, China*
[2] *School of Energy and Power Engineering, North China Electric Power University, Baoding 071003, China*

Correspondence should be addressed to Huimin Xu; xuhuimin@126.com

Academic Editor: Zoltan Szabo

A predictive variable gain strategy in iterative learning control (ILC) is introduced. Predictive variable gain iterative learning control is constructed to improve the performance of trajectory tracking. A scheme based on predictive variable gain iterative learning control for eliminating undesirable vibrations of PMSM system is proposed. The basic idea is that undesirable vibrations of PMSM system are eliminated from two aspects of iterative domain and time domain. The predictive method is utilized to determine the learning gain in the ILC algorithm. Compression mapping principle is used to prove the convergence of the algorithm. Simulation results demonstrate that the predictive variable gain is superior to constant gain and other variable gains.

1. Introduction

Due to their high efficiency, high power density and low noise, low loss, small size, and so forth, permanent magnet synchronous motor (PMSM) is used widely in various industrial fields. Furthermore, the applications of PMSM are expanding rapidly.

However, PMSM performance at low speed is bad because of the existence of torque ripple which deteriorates the accuracy and repeatability of PMSM and undermines potentially its suitability in precision electromechanical device. Thus, eliminating torque ripple is very important for improving PMSM performance.

Many control methods have been utilized to suppress the torque ripple. They include PID control scheme [1, 2], predictive control [3], adaptive fuzzy control [4], robust control [5], sliding-mode control [6], and so on. These control means improve the performance of PMSM system from different aspects [7], but applying conventional PID controller and modern control techniques mentioned in the above to deal with torque ripple cannot attain desired levels; moreover, some of them are too complex to employ in practice.

Because of the periodic feature of PMSM on some applications and the simplicity of iterative learning control, a large number of learning control schemes were developed to remove torque ripple in PMSM. Those learning schemes applied to permanent magnet synchronous motors can be divided into two categories according to the learning gain: the first class is fixed gain iterative learning control [8] and the other is variable gain iterative learning control. The second category has obvious advantages in instantaneous characteristics and robustness of system when compared with fixed gain iterative learning control, so designing a reasonable, objective, and effective algorithm in iterative learning controller to determine the value of learning gain at each moment is a vital important factor in solving the problem of instantaneous error growth which has aroused a strong interest of the researchers.

An iterative learning algorithm with a variable gain in iteration domain to remove measurement disturbances and guarantee that the tracking error converges to zero was developed by Zhang et al. [9], but time domain uncertainty is not considered; under this circumstance, Xu et al. [10] proposed a variable PID gain with iterative learning control scheme applied to nonlinear system to tracking the desired output; although this PID gain takes into account disturbances in both time domain and iterative domain, the choice of coefficients for PID gain is subjective.

Unlike [9, 10], we propose to use predictive control to determine the gain of iterative learning control during in both time index and iterative index. The main contributions of this method include three aspects: the first is superiority in control performance compared with constant gain and PID variable gain. The second is that this scheme can be applied not only to the linear system but also to the nonlinear system, so it has a wider application. Finally, although predictive control has been successful applied to the motor control [11–14], it is the first time, as far as the author known, that it is used to determine the gain of iterative learning control; the proposed scheme in this paper overcomes the subjectivity of the gain value choice.

2. Material and Methods

2.1. Process Description. We assume that the PMSM motion model is described as [15]

$$\begin{aligned} \dot{x}_1 &= x_2, \\ \dot{x}_2 &= \frac{1}{M}\left(F_m - \xi x_2 - f_{\text{cog}} - f_{\text{rel}} - f_{\text{fric}} + f_d\right), \\ y &= x_1; \end{aligned} \tag{1}$$

in the above, M is the load weight, ξ is a dampen coefficient, x_1, x_2 represent the position and velocity of mover, respectively, f_d is a motor load disturbance, f_{fric} is the friction, f_{cog} is a alveolar thrust ripple, f_{rel} is a reluctance thrust ripple, F_m is the reluctance motor electromagnetic force to remove part of the thrust fluctuations, and f_{fric} can be expressed as

$$f_{\text{fric}}(x_2) = \left[f_c + (f_s - f_c)\, e^{-|x_2/x_{2s}|^{\varepsilon}}\right] \operatorname{sgn}(x_2), \tag{2}$$

where f_s is the maximum static friction; f_c is a Coulomb friction, x_{2s} and ε are empirical parameters used to describe the Stribeck effect, and sgn() is a switching function; generally it is believed that f_{cog} is a periodic function which is described as

$$f_{\text{cog}}(x_1 + P) = f_{\text{cog}}(x_1). \tag{3}$$

P is polar distance. It is assumed that motor electromagnetic force $F_e = K_F(x_1)i_q$, where we assume that i_q is the q-axis mover current in the vector control mode:

$$K_F(x_1) = K_{F0} - K_{Fx}(x_1). \tag{4}$$

$K_{F0} > 0$ is an average thrust constant, K_{Fx} is reluctance thrust constant, and $F_m = K_{F0} i_q$:

$$f_{\text{rel}} = K_{Fx}(x_1)\, i_q. \tag{5}$$

f_{rel} is a periodic function to satisfy

$$K_{Fx}(x_1) = K_{Fx}(x_1 + P). \tag{6}$$

In (1), current i_q denotes the input signal, x_1, x_2 represent the states of system, and y denotes the output; generally, the exact input produces the desired output, but in practice, the torque ripple is unavoidable; as a result, the real output is not the desired.

In order to resolve above question, we develop a control scheme which merges the iterative learning and predictive control to eliminate the bad effect of the torque ripple.

The main target of this paper is to find a proper learning gain such that the output error is minimized, and arbitrary high precision output tracking is achieved.

2.2. Iterative Learning Control Law. In this section, we develop a scheme to get the appropriate learning gain by minimizing the performance function in the predictive control process.

For simplicity, we consider an open-loop iterative learning law. During the iterative index $k + 1$, the learning update is given by

$$i_{qk+1} = \varphi i_{qk} + \phi e_k, \tag{7}$$

where ϕ is the learning control gain and e_k is the output error; that is,

$$e_k = y_{qd} - y_{qk}. \tag{8}$$

y_{qd} is a realizable desired output trajectory, i_{qk+1}, i_{qk} are the system inputs in the $k + 1$st iterative and kth iterative, φ is a filter, and k is the iterative learning index.

2.3. Predictive Gain. In order to find out the appropriate learning gain by predictive control, we firstly set up the predictive model where the learning gain acts as a system input.

At sampled time t in the iterative index $k + 1$, combining (1) with (5), we can write the equation for the PMSM:

$$\begin{aligned} \dot{x}_1 &= x_2, \\ \dot{x}_2 &= \frac{1}{M}\left(\left(K_{F0} i_q - \xi x_2 - K_{Fx}\right) i_q - f_{\text{cog}} - f_{\text{fric}} + f_d\right), \\ y &= x_1. \end{aligned} \tag{9}$$

Inserting (9) into (7), we get

$$\begin{aligned} \dot{x}_{1,k} &= x_{2,k}, \\ \dot{x}_{2,k} &= \frac{1}{M}\Big(\left(K_{F0} - K_{Fx}\right)\left(\varphi i_{qk} + \phi e_k\right) - \xi x_{2,k} - f_{\text{cog}} \\ &\quad - \left[f_c + (f_s - f_c)\, e^{-|x_2/x_{2s}|^{\varepsilon}}\right] + f_d\Big), \\ y_k &= x_{1,k}. \end{aligned} \tag{10}$$

For the sake of getting an updating learning gain in next sample time $t + 1$, we make use of predictive control method. The predictive model is given by

$$\begin{aligned} \dot{x}_{1,k} &= x_{2,k}, \\ \dot{x}_{2,k} &= \frac{1}{M}\left(K_{F0}\varphi i_{qk} - \xi x_{2,k}\right) + \frac{K_{F0}}{M} e_k \phi_k, \\ y_k &= x_{1,k} \end{aligned} \tag{11}$$

ϕ_k, y_k, $x_{1,k}$, and $x_{2,k}$ denote the input, output, and states of system in the kth iterative index, respectively; the system in the above is a nonlinear system.

Secondly, we turn the nonlinear PMSM model (11) decouples into a new linear system via the input-output feedback linearization scheme. According to the exact linear theory, calculating Lie derivative of the output variable, we get the relative degree which is equal to the number of the system state variables; as a result, (11) satisfies the exact linear condition and can be linearized as

$$\dot{x_k} = Ax_k + Bv_k = \begin{bmatrix} 0 & 1 \\ 0 & 0 \end{bmatrix} x_k + \begin{bmatrix} 0 \\ 1 \end{bmatrix} v_k, \qquad Y = Cx_k = \begin{bmatrix} 1 & 0 \end{bmatrix} x_k, \tag{12}$$

where $x_k = \left[\begin{smallmatrix} x_{1k} \\ x_{2k} \end{smallmatrix}\right]$, $v_k = (1/M)(K_{F0}i_{qk} - \xi x_2 - \xi\phi_k)$, and v_k is a new control input. We can discretize (10):

$$x_k(t+1) = G_1 x_k(t) + G_2 v_k(t), \qquad y = Cx_k(t), \tag{13}$$

where $G_1 = e^{At}$, $G_2 = \int_0^T e^{At} dt$, $T = 0.1$ ms is the sampling period, and $\Delta v = v_d - v_k$, where v_d denotes desired input. We assume that the predictive horizon is H_p and the control horizon is H_v; Δv can be calculated by the following criterion:

$$V(t) = \sum_{i=0}^{H_p} \left\| y_{qd}(t+i \mid t) - y_{qk}(t+i \mid t) \right\|_L^2 + \sum_{i=0}^{H_v - 1} \left\| \Delta v(t+i \mid t) \right\|_R^2. \tag{14}$$

In the light of the least square formula, the expression of the Δv is obtained as

$$\Delta v(t) = \begin{bmatrix} S_L \Theta \\ S_R \end{bmatrix}^{-1} \begin{bmatrix} S_L \varepsilon(t) \\ 0 \end{bmatrix}, \tag{15}$$

where $\Theta = \left[\begin{smallmatrix} B & \cdots & 0 \\ AB+B & \cdots & 0 \\ \vdots & \cdots & \vdots \end{smallmatrix}\right]$, $\Psi = \left[\begin{smallmatrix} A \\ A^2 \\ \vdots \end{smallmatrix}\right]$, $Y = \left[\begin{smallmatrix} B \\ AB+B \\ \vdots \end{smallmatrix}\right]$, $C\varepsilon(t) = C(x_d - \Psi x(t) - Yv(t-1))$, S_L and S_R are square roots of eigenvalues of matrix L and R, respectively; x_d is the desired state; according to the exact linear theory, we get

$$\phi_k = \frac{v - (1/M)\left(K_{F0} i_{qk} - \xi x_2\right)}{-\xi / M}. \tag{16}$$

The open-loop learning gain $\phi_k(t)$ is given by

$$\phi_k(t) = \phi_k(t-1) - \frac{M}{\xi} \Delta v(t). \tag{17}$$

Considering (5), we have

$$i_{qk+1}(t) = \varphi i_{qk}(t) + \left(\phi(t-1) - \frac{M}{\xi} \Delta v(t)\right) e_k(t). \tag{18}$$

It can be rewritten in the form

$$i_{qk+1}(t) = \varphi i_{qk}(t) + \left(\phi(t-1) - \frac{M}{\xi} \begin{bmatrix} S_L \Theta \\ S_R \end{bmatrix}^{-1} \begin{bmatrix} S_L \varepsilon \\ 0 \end{bmatrix}\right) e_k(t). \tag{19}$$

2.4. Convergence Analysis. In this section, we give the condition under which the system output error converges to zero.

Consider (1) and it can be written as

$$\dot{x} = f(t, x(t)) + g(t) u(t), \qquad y = h(t, x(t)), \tag{20}$$

where $\dot{x} = \left[\begin{smallmatrix} \dot{x}_1 \\ \dot{x}_2 \end{smallmatrix}\right]$, $g(t) = \left[\begin{smallmatrix} 0 \\ (K_{F0} - K_{Fx})(1/M) \end{smallmatrix}\right]$, $h(t, x(t)) = x_1$, $f = \left[\begin{smallmatrix} x_2 \\ (1/M)(-\xi x_2 - f_{cog} - [f_c + (f_s - f_c) e^{-|x_2/x_{2s}|^\varepsilon}] \operatorname{sgn}(x_2) + f_d) \end{smallmatrix}\right]$, and $u(t) = i_q$.

We assume the following properties for system (20):

(1) The functions f, g, and h are assumed to satisfy the following conditions: $\forall x_1, x_2 \in R^n$

$$\begin{aligned} \left\| f(t, x_1) - f(t, x_1) \right\| &\le M_1 \left\| x_1 - x_2 \right\|, \\ \left\| g(t, x_1) - g(t, x_1) \right\| &\le M_{11} \left\| x_1 - x_2 \right\|, \\ \left\| h(t, x_1) - h(t, x_1) \right\| &\le M_{12} \left\| x_1 - x_2 \right\| \end{aligned} \tag{21}$$

for all $t \in [0, T]$ and M_1, M_{11}, and M_{12} are constants.

(2) It is assumed that, at each iterative process, the initial state error sequence $\{\delta x_k(0)\}$, $k \ge 0$, converges to zero.

(3) For any realizable output trajectory and an appropriate initial condition, there exists a unique control input generating the trajectory for the plant.

Proof. Let system satisfy assumptions (1)–(3) and (7) be applied; define the state, input and the output errors as

$$\begin{aligned} \Delta x_k(t) &= x_d(t) - x_k(t), \\ \Delta y_k(t) &= y_d(t) - y_k(t), \\ \Delta u_k(t) &= u_d(t) - u_k(t), \\ f_1(t, x_k) &= f(t, x_d) - f(t, x_d - x_k). \end{aligned} \tag{22}$$

According to (20), we get

$$\left\| f_1 \right\| \le M_1 \left\| x_k \right\|. \tag{23}$$

Then,

$$\begin{aligned} \Delta \dot{x_k}(t) &= f_1(t, \Delta x_k(t)) + g(t) \Delta u_k(t), \\ \Delta y_k(t) &= h(t, x_d(t)) - h(t, x_k(t)), \end{aligned}$$

$$h_1(t,\Delta x_k(t)) = h(t,x_d(t)) - h(t,x_k(t)),$$
$$\Delta u_{k+1}(t)$$
$$= u_d(t) - \varphi u_k(t)$$
$$-\left(\phi(t-1) - \frac{M}{\xi}\begin{bmatrix} S_L\Theta \\ S_R \end{bmatrix}^{-1}\begin{bmatrix} S_L\varepsilon \\ 0 \end{bmatrix}\right) e_k(t)$$
$$= \varphi\Delta u_k(t)$$
$$-\left(\phi(t-1) - \frac{M}{\xi}\begin{bmatrix} S_L\Theta \\ S_R \end{bmatrix}^{-1}\begin{bmatrix} S_L\varepsilon \\ 0 \end{bmatrix}\right) \Delta y_k(t)$$
$$+ (1-\varphi) u_d(t)$$
$$= \varphi\Delta u_k(t)$$
$$-\left(\phi(t-1) - \frac{M}{\xi}\begin{bmatrix} S_L\Theta \\ S_R \end{bmatrix}^{-1}\begin{bmatrix} S_L\varepsilon \\ 0 \end{bmatrix}\right) h_1$$
$$+ (1-\varphi) u_d(t). \tag{24}$$

Define the operator

$$\Upsilon_k : C_r[0,T] \longrightarrow C_r[0,T],$$
$$\Upsilon_k(u(t))$$
$$= \left(\phi(t-1) - \frac{M}{\xi}\begin{bmatrix} S_L\Theta \\ S_R \end{bmatrix}^{-1}\begin{bmatrix} S_L\varepsilon \\ 0 \end{bmatrix}\right) h_1$$
$$+ (1-\varphi) u_d(t), \tag{25}$$
$$\Delta u_{k+1}(t) = (\varphi I + \Upsilon_k)\Delta u_k(t),$$
$$\Delta u_{k+1}(t)$$
$$= (\varphi I + \Upsilon_k)(\varphi I + \Upsilon_{k-1})\cdots(\varphi I + \Upsilon_0)\Delta u_0(t).$$

I is a unit matrix; we make an estimate for Υ_k:

$$\|x(t)\| = \left\| x(0) + \int_0^t [f(s,x(s) + g(s)u(s))]\,ds \right\|,$$
$$\|x(t)\|$$
$$\leq \|x(0)\| + M_1\int_0^t \|x(s)\|\,ds + M_{11}\int_0^t \|u(s)\|\,ds,$$
$$\|x(t)\| \leq M_3\left(\|x(0)\| + \int_0^t \|u(s)\|\,ds\right), \tag{26}$$
$$\|\Upsilon_k u(t)\| \leq M_4\left(\|x(0)\| + \int_0^t \|u(s)\|\,ds\right),$$
$$\|\Upsilon_k \Delta u(t)\| \leq M_4\left(\|\Delta x(0)\| + \int_0^t \|\Delta u(s)\|\,ds\right).$$

According to lemma 2 in [16], we have the convergence conditions $0 < \varphi < 1$, $M_4 > 1$, and M_3 and M_4 are constant. □

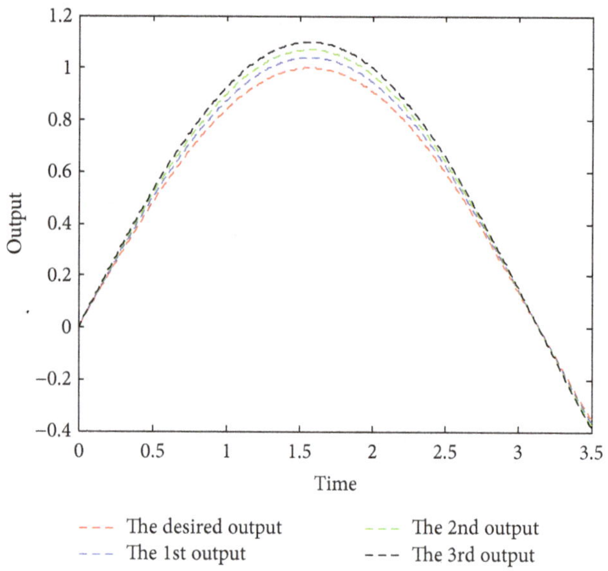

Figure 1: Desired and actual output.

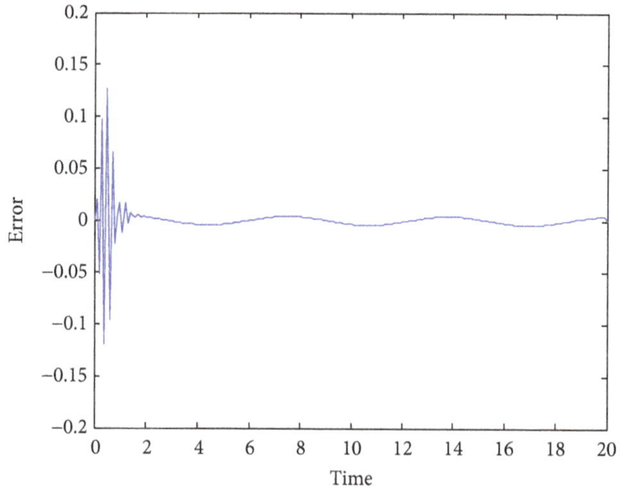

Figure 2: Error response in the 1st iterative.

2.5. Simulation. In this section, we will make a comparison between the iterative learning method based on the predicted variable gain and two kinds of efficient iterative learning control method which in recent years have been widely used for permanent magnet synchronous motors: constant gain of iterative learning control strategy and the variable PID gain iterative learning control strategy.

PMSM system preferences are as follows.

Load weight $M = 5$ kg, $l = 0.0333$ A/kg, damping coefficient is 20.99 Nm/min^{-1}, pole pitch is 60.9 mm, $F_m = (3/2)P_n\phi_f i_d$, $P_n = 2$, $\phi_f = 0.125$ wb, friction $f_s = f_c = 15$ N, control horizon $nc = 3$, and predictive horizon $np = 7$; assume that the initial state error is zero and the desired output is a sine wave and the cycle is π; amplitude is 5.

The simulation results are shown in Figures 1–7.

Figure 1 shows a comparison between the desired PMSM output and the actual PMSM output in the 1st, 2nd, and

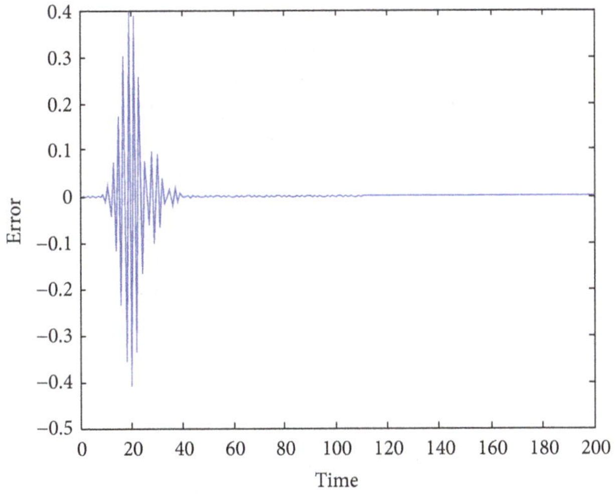

FIGURE 3: Error response in the 5th iterative.

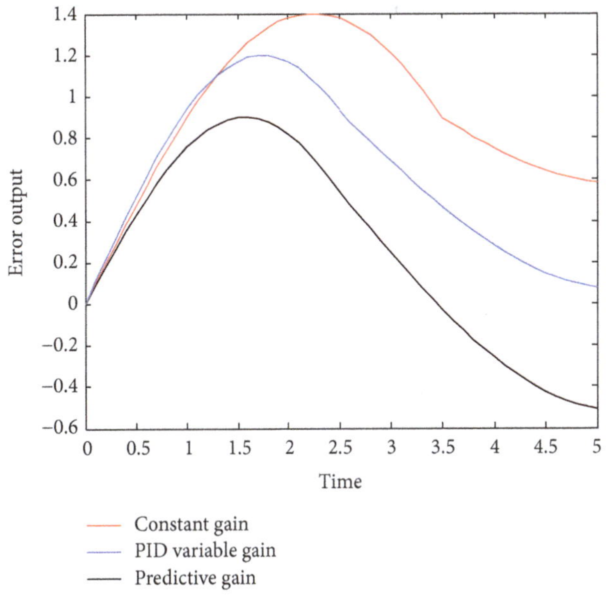

FIGURE 4: Errors of the constant gain, PID gain, and predictive gain in the 1st iterative.

3rd iteratives with the fixed gain iterative learning controller, although the actual system output will approach the desired output with the increase of the number of iterations, instantaneous vibration bandwidth is sometimes dramatic large which is in fact detrimental to system. Figures 2 and 3 give a more intuitive description about the instantaneous vibration growth which show that max vibration bandwidth increases from 0.158 mm to 1.4 mm just after four iteratives. These phenomena also give full explanation of the importance of predictive gain learning control.

Figures 4–6 show the actual system output in different iterative indices with constant gain learning controller, PID variable gain learning controller, and predictive gain learning controller, respectively. We can see from Figures 4–6 that the predictive gain is significantly better than the fixed gain

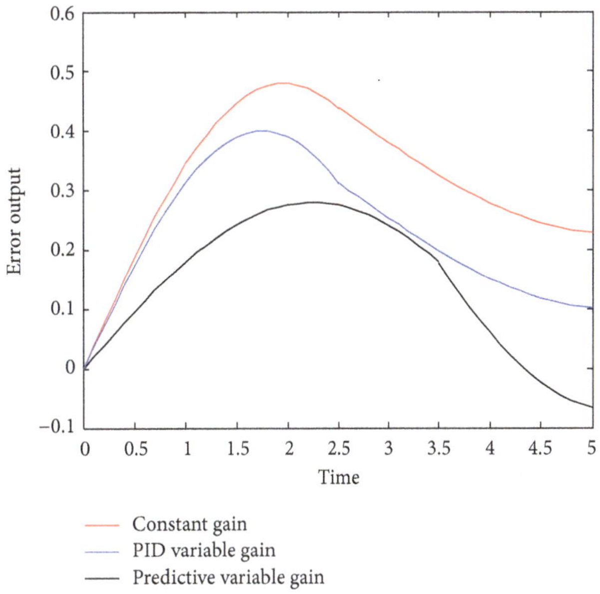

FIGURE 5: Errors of the constant gain, PID gain, and predictive gain in the 6th iterative.

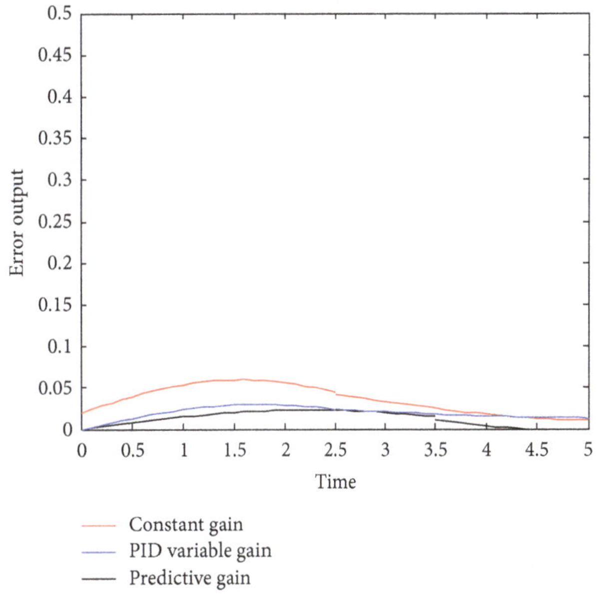

FIGURE 6: Errors of the constant gain, PID gain, and predictive gain in the 15th iterative.

and the PID gain. Firstly, the error vibration amplitude of predictive gain is less than the other two gains; secondly, error descent rate of predictive gain in time domain is also faster than fixed gain and PID gain iterative learning control.

For the sake of getting a more intuitive understanding for the advantages of the scheme proposed by this paper, Figure 7 is given. It can be seen that when the initial state error is zero, on the premise of the convergence conditions of iterative learning control, the predictive variable gain iterative learning has a higher convergent rate; meanwhile tracking accuracy also has been greatly improved.

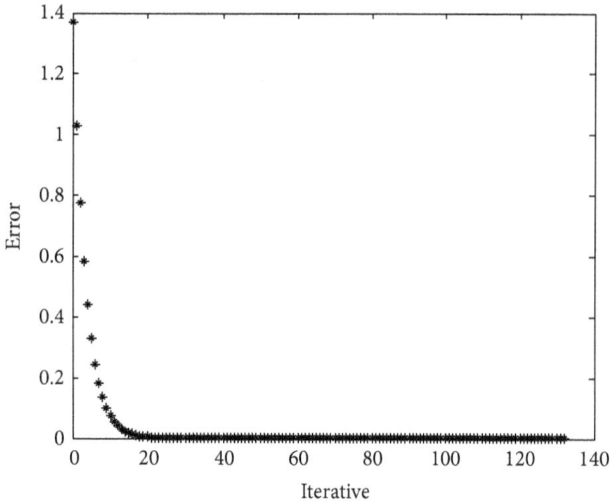

FIGURE 7: Error in iterative domain.

3. Conclusion

A variable gain iterative learning control strategy based on predictive control for PMSM has been developed. This method not only increases the iterative convergence rate, but also eliminates error in the time domain. The convergence of this technique is proved by the contractive operator theory.

Future work will consider the choosing of predictive horizon to get a better control performance.

Conflict of Interests

The authors declare that there is no conflict of interests regarding the publication of this paper.

Acknowledgment

This research was supported by the Central Universities Funds (9160215006).

References

[1] Q. Y. Xu, C. J. Zhang, L. Zhang, and C. Y. Wang, "Multiobjective optimization of PID controller of PMSM," *Journal of Control Science and Engineering*, vol. 2014, Article ID 471609, 9 pages, 2014.

[2] J. J. Xi, B. Sun, and H. Q. Zhao, "Adaptive PID controller based on single neuron for permanent magnet synchronous machine," *Electric Power Automation Equipment*, vol. 23, no. 10, pp. 59–61, 2003.

[3] A. M. Kasseml and A. A. Hassan, "Performance improvements of a permanent magnet synchronous machine via functional model predictive control," *Journal of Control Science and Engineering*, vol. 2012, Article ID 319708, 8 pages, 2012.

[4] C. Z. Zhou, D. C. Quach, N. X. Xiong et al., "An improved direct adaptive fuzzy controller of an uncertain PMSM for web-based e-service systems," *IEEE Transactions on Fuzzy Systems*, vol. 23, no. 1, pp. 58–71, 2015.

[5] M. C. Chou, C. M. Liaw, S. B. Chien, F. H. Shieh, J. R. Tsai, and H. C. Chang, "Robust current and torque controls for PMSM driven satellite reaction wheel," *IEEE Transactions on Aerospace and Electronic Systems*, vol. 47, no. 1, pp. 58–74, 2011.

[6] F. J. Lin, Y. C. Hung, and M. T. Tsai, "Fault-tolerant control for six-phase PMSM drive system via intelligent complementary sliding-mode control using TSKFNN-AMF," *IEEE Transactions on Industrial Electronics*, vol. 60, no. 12, pp. 5747–5762, 2013.

[7] S. H. Li and H. Guo, "Fuzzy adaptive internal model control schemes for PMSM speed-regulation system," *IEEE Transactions on Industrial Informatics*, vol. 8, no. 4, pp. 767–779, 2013.

[8] B. Q. Li and H. Q. Lin, "PMSM periodical torque ripple minimization using iterative learning control," *Electric Machines and Control*, vol. 15, no. 9, pp. 51–55, 2011.

[9] H. W. Zhang, F. S. Yu, X. H. Bu, and F. Z. Wang, "Robust iterative learning control for permanent magnet linear motor," *Electric Machines and Control*, vol. 16, no. 6, pp. 81–86, 2012.

[10] M. Xu, H. Lin, and Z. Liu, "Iterative learning control law with variable learning gain," *Control Theory and Applications*, vol. 24, no. 5, pp. 856–860, 2006.

[11] X. B. Kong and X. J. Liu, "Effcient nonlinear model predictive control for permanent magnet synchronous motor," *Acta Automatica Sinica*, vol. 40, no. 9, pp. 1958–1966, 2014.

[12] X. B. Kong and X. J. Liu, "Nonlinear model predictive control for DFIG-based wind power generation," *Acta Automatica Sinica*, vol. 39, no. 5, pp. 636–643, 2013.

[13] X. J. Liu, "Present situation and prospect of model predictive control application in complex power industrial process," *Proceedings of the Chinese Society for Electrical Engineering*, vol. 33, no. 5, pp. 79–85, 2012.

[14] X. J. Liu, "Continuous-time nonlinear model predictive control with input/output linearization," *Control Theory and Applications*, vol. 29, no. 2, pp. 217–224, 2014.

[15] Y. X. Song, C. H. Wang, W. S. Yin, and P. F. Jia, "Adaptive-learning control for permanent magnet linear synchronous motors," *Proceedings of the Chinese Society of Electrical Engineering*, vol. 25, no. 20, pp. 151–156, 2005.

[16] J. X. Shou, Z. J. Zhang, and D. Y. Pi, "On the convergence of open-closed-loop D-type iterative learning control for nonlinear systems," in *Proceedings of the IEEE International Symposium on Intelligent Control*, pp. 963–967, IEEE, Houston, Tex, USA, October 2003.

6

A Local Controller for Discrete-Time Large-Scale System by Using Integral Variable Structure Control

C. H. Chai and Johari H. S. Osman

Faculty of Electrical Engineering, University Technology Malaysia, 81300 Skudai, Malaysia

Correspondence should be addressed to C. H. Chai; chai8987@gmail.com

Academic Editor: Xiao He

A new local controller for discrete-time integral variable structure control of a large-scale system with matched and unmatched uncertainty is presented. The local controller is able to bring the large-scale system into stability by using only the states feedback from individual subsystem itself. A new theorem is established and proved that the controller is able to handle the effect of interconnection for the large-scale system with matched and unmatched uncertainty, and the system stability is ensured. The controller is able to control the system to achieve the quasi-sliding surface and remains on it. The results showed a fast convergence to the desired value and the attenuation of disturbance is achieved.

1. Introduction

As defined by Siljak [1], large-scale systems usually refer to systems that consist of a large number of state variables, system parametric uncertainties, a complex structure, and a strong interaction between subsystems. When the size of the dynamical system increases, the implementation of centralized control will be either impossible or uneconomical. Research in decentralized control has been motivated by the inadequacy of conventional modern control theory to deal with certain issues in large-scale system, such as the issue that it is impossible to incorporate so many feedback loops into the centralized design. Decentralized control theory has risen in response to the difficulty that there are restrictions on information transfer between certain groups of sensors or actuators [2]. In any case where there is a failure or insufficient information transfer between subsystems, local controller will play a significant role in controlling the large-scale system. Chen [3] has developed a class of local control, utilizing solely the states of each individual subsystem and the bounds of the uncertainties that assured the desired properties when the proposed requirements on a test matrix are satisfied.

In the development of decentralized control on discrete-time large-scale systems, Li et al. [4] have used decentralized control by dynamic programming method to achieve the control of three-reach river pollution problem. This paper addressed the dynamic issue of the interconnections and external disturbance of the systems. The challenge for this method is the speed of computation required for systems with fast respond. Lyou and Bien [5] used the adaptive feedback concept for stabilization of the interconnected system with uncertain system parameters and the feed-forward concept for compensation of bounded deterministic disturbances to control a large-scale interconnected discrete system. Subsequently, linear matrix inequality (LMI) technique has been used by Park and Lee [6] to derive a sufficient condition for robust stability in decentralized discrete-time large-scale systems with parametric uncertainty. Park et al. [7] applied the dynamic output feedback controller design to a discrete-time large-scale system with delay at subsystem interconnections. Lyapunov method has been combined with LMI technique to develop the dynamic output feedback controller to guarantee the cost stabilization of the systems and achieve asymptotically stable closed-loop system with adequate level of performance. Ou et al. [8] also used LMI method to achieve the stability analysis and H_∞ controller design to achieve disturbance attenuation performance by using Fuzzy Logic approach for the decentralized control of discrete-time large-scale systems.

In the early development stage of discrete-time variable structure control (VSC) theory, the basic conditions for achieving the equivalent of sliding mode as in continuous-time variable structure control have been proposed by Dote and Hoft [9], Sarpturk et al. [10], Milosavljevic [11], and Furuta [12]. Method for quasi-sliding mode design and the use of reaching law approach to develop the control law for robust control in discrete-time VSC has been proposed by Gao et al. [13]. Discrete-time integral sliding mode control for sampled data system under state regulation was reported in Abidi et al. [14]. Subsequently, Xi and Hesketh [15] demonstrated the discrete-time integral sliding mode system to deal with both matched and unmatched uncertainties focused on SISO system.

A discrete-time large-scale system in VSC has been introduced by Sheta [16] with optimum control method. His study focused on the uncertain changes in the interconnection between subsystems and these uncertainties were governed by Markov chain technique. The controller was designed off-line based on a set of expected system failure modes and switched on-line when failure was detected. Apart from this, literatures that focus on the research of discrete-time VSC for large-scale systems are rather limited.

Traditionally, the local control for a nonlinear plant by using linear systems theory is based upon linearization of the underlying nonlinear dynamics in the region of some operating points that correspond to a physical equilibrium of a plant. Thus the nonlinear system is treated as multiple time-varying linear systems that operate in a range of distinct equilibria, and gain scheduling approach was used to deal with the transition between equilibria. Other well-known methods were multiple model adaptive control and local controller network [17]. These methods usually have bigger challenge to control when the systems' uncertainty and disturbances kicked in, due to prior knowledge of the changes that are needed to generate the gain parameters. On the other hand, Markov jump large-scale system control by using local mode-dependent decentralized robust H_∞ has been considered by Zhuansun and Xiong [18] to tackle the internal uncertainties of the local subsystems and interconnection uncertainties between subsystems. Markov method requires extensive computational resources especially when the system is large; thus it is a challenge for certain application that is fast changing and requires fast respond. VSC with the advantage in rejecting the uncertainty and disturbances to local subsystems and interconnection between subsystems is a better option in this case.

In this paper, a new theorem of using integral VSC method to establish a local controller for a large-scale discrete-time system with unmatched uncertainties is proposed. The controller only utilizes the states feedback from the individual subsystem without the information from other interconnected subsystems. The simulation results have shown that the proposed controller rendered the large-scale system to be stable and able to handle the effect of the interconnections and unmatched uncertainties. This local controller is very useful for controlling large-scale system with difficulty in obtaining the feedback from other subsystems and can only utilize the states feedback information from the local subsystems.

This paper is organized into 6 sections; Section 1 is the introduction and Section 2 introduces the preliminary concept of VSC, followed by the problem statement in Section 3. The controller design and proof of the theorem are given in Section 4. Section 5 presents the simulation results of a discrete-time large-scale system under study and the conclusion is given in Section 6.

2. Preliminaries

Consider a discrete-time system in the perturbed condition:

$$x(k+1) = Ax(k) + Bu(k) + d(k), \tag{1}$$

where A represent system parameters, B represent input parameters, and $d(k)$ is the disturbances.

The VSC requires that the integral sliding surface is designed such that the system response restricted to $\sigma(k+1) = \sigma(k) = 0$ has a desired behavior such as stability or tracking. The integral VSC sliding surface equation is given by

$$\sigma(k) = Gx(k) - Gx(0) + h(k) \tag{2}$$

with $h(k)$ being iteratively computed as $h(k) = h(k-1) - Gx(k-1)$

The VSC control input portion is then given by

$$u(k) = -(GB)^{-1}\left(GAx(k) + G\hat{d}(k) - Gx(0) + h(k)\right), \tag{3}$$

where $\hat{d}(k)$ is the disturbance compensation.

3. Problem Statement

This paper considers a discrete-time large-scale system given by

$$x(k+1) = Ax(k) + Bu(k) + d(k), \tag{4}$$

where A is the system parameter, B is the input parameter, and $d(k)$ represent the system matched and unmatched uncertainty, disturbances, and interconnection between subsystems.

It is assumed that the system can be decomposed into p subsystems as follows:

$$x_i(k+1) = A_i x_i(k) + B_i u_i(k) + B_i f_{mi}(k) + f_{ui}(k) + V_i \sum_{\substack{j=1\\ j\neq i}}^{p} \emptyset_{ji} z_{ji}(k),$$

$$x_i(k+1) = A_i x_i(k) + B_i\left(u_i(k) + f_{mi}(k)\right) + V_i \sum_{\substack{j=1\\ j\neq i}}^{p} \emptyset_{ji} z_{ji}(k) + f_{ui}(k), \quad (i = 1.2, \ldots, p), \tag{5}$$

where A_i is system parameter, B_i is input parameter, $x_i(k+1) \in R^n$, $u_i(k) \in R^m$, $f_{mi}(k)$ is the matched uncertainty, $f_{ui}(k)$ is the unmatched uncertainty, V_i & $\emptyset_{ji}$ is constant matrix with appropriate dimension, $z_{ji}(k) \in R^{q_j}$ is the interconnection between subsystem j and i with

$$z_{ji}(k+1) = A_{zji} z_{ji}(k) + \psi_{ji} y_i(k), \quad (6)$$

$$y_i(k) = c_i x_i(k), \quad (7)$$

$$x_i(k) = x_{i0}, \quad (8)$$

$k = 0$, $\emptyset_{ji}$, ψ_{ji}, and A_{zji} are matrix with appropriate dimension.

It is assumed that both $f_{mi}(k)$ and $f_{ui}(k)$ are bounded; that is,

$$\begin{gathered} 0 \le f_{mi}(k) \le F_m, \\ 0 \le f_{ui}(k) \le F_u, \end{gathered} \quad (9)$$

and the bounds are known.

Assumption 1. G_iB_i is invertible (G can be arbitrarily chosen by assuming that the following conditions are met), according to Xi and Hesketh [15]:

$$\begin{gathered} \left|G_iB_i\left(f_{mi}(k) - \hat{f}_{mi}(k)\right)\right| < N, \\ 0 < N < \infty, \\ \left|G_i\left(f_{ui}(k) - \hat{f}_{ui}(k)\right)\right| < M, \\ 0 < M < \infty, \\ \left|G_iV_i\left(\sum_{\substack{j=1 \\ j\neq i}}^{p} \emptyset_{ji} z_{ji}(k) - \sum_{\substack{j=1 \\ j\neq i}}^{p} \emptyset_{ji} z_{ji}(k-1)\right)\right| < L, \\ 0 < L < \infty. \end{gathered} \quad (10)$$

It is assumed that N, M, and L are known, and $\hat{f}_{mi}(k)$ and $\hat{f}_{ui}(k)$ are the last value of disturbances.

According to Su et al. [19], the last value of a disturbance signal can be taken as estimation of its current value if the updated value is not accessible, under the assumption that the disturbance is continuous and smooth. As $\hat{f}_{mi}(k)$ represents the last value measureable from the system, it is equivalent to the previous state, $f_{mi}(k-1)$:

$$\begin{gathered} f_{mi}(k) - \hat{f}_{mi}(k) = f_{mi}(k) - f_{mi}(k-1), \\ f_{ui}(k) - \hat{f}_{ui}(k) = f_{ui}(k) - f_{ui}(k-1). \end{gathered} \quad (11)$$

The idea here is to find an approximation that is part of the disturbance to estimate the past value of disturbance for the calculation to cancel the present disturbance. The approximation is good only for a small change in the disturbance between consecutive sampling periods that the sampling rate is high as compared to the frequency composition of the disturbance [20].

The objective is then to design a decentralized controller, $u_i(k)$, such that the large-scale discrete-time system (5) and (6) can be controlled.

4. Integral Variable Structure Controller

A decentralized discrete-time controller, $u_i(k)$, based on integral VSC technique is proposed for each subsystem. The sliding surface is designed for each subsystem followed by the switching controller as presented in the following.

In order to guarantee the existence of sliding mode and reduce chattering effect, the following condition must be satisfied [10]:

(1)

$$\sigma_i(k)\left(\sigma_i(k+1) - \sigma_i(k)\right) \le 0; \quad (12)$$

(2)

$$\left|\sigma_i(k+1)\right| \le \left|\sigma_i(k)\right|. \quad (13)$$

4.1. Sliding Surface Design. A discrete-time integral sliding surface for each subsystem is designed as

$$\sigma_i(k) = G_i x_i(k) - G_i x_i(0) + h_i(k), \quad (14)$$

where $h_i(k)$ is iteratively computed as

$$\begin{aligned} h_i(k) &= h_i(k-1) \\ &\quad - \left(G_iB_iu_{i0}(k-1) + G_iA_ix_i(k-1)\right) \end{aligned} \quad (15)$$

with $\sigma_i(k) \in R^n$, $h_i(k) \in R^m$, $h_i(0) = 0$, and $G_i \in R^{1\times n}$.

It is assumed that the control law is given as

$$u_i(k) = u_{i0}(k) + u_{i1}(k), \quad (16)$$

where the first component, $u_{i0}(k) = -K_ix_i(k)$, is the equivalent control portion after system achieves the quasi-sliding mode. The gain K is to be designed later.

By taking $h_i(k+1) = h_i(k) - (G_iB_iu_{i0}(k) + G_iA_ix_i(k))$, the sliding surface dynamics can be obtained from (5), (14), and (15) as follows:

$$\begin{aligned} \sigma_i(k+1) &= G_i x_i(k+1) - G_i x_i(0) + h_i(k+1) \\ &= G_i\left(A_ix_i(k) + B_i\left(u_i(k) + f_{mi}(k)\right)\right. \\ &\quad \left. + V_i\sum_{\substack{j=1 \\ j\neq i}}^{p} \emptyset_{ji} z_{ji}(k) + f_{ui}(k)\right) - G_i x_i(0) + h_i(k) \end{aligned}$$

$$- G_i B_i u_{i0}(k) - G_i A_i x_i(k) = G_i A_i x_i(k) + G_i B_i (u_i(k) + f_{mi}(k)) + G_i f_{ui}(k) + G_i V_i \sum_{\substack{j=1 \\ j \neq i}}^{p} \emptyset_{ji} z_{ji}(k) - G_i x_i(0) + h_i(k) - G_i B_i u_0(k) - G_i A_i x_i(k). \quad (17)$$

By substituting (16) into (17), the equation of sliding surface becomes

$$\sigma_i(k+1) = G_i B_i u_{i1}(k) + G_i B_i f_{mi}(k) + G_i f_{ui}(k) + G_i V_i \sum_{\substack{j=1 \\ j \neq i}}^{p} \emptyset_{ji} z_{ji}(k) + h_i(k) - G_i x_i(0). \quad (18)$$

From (14) and (18),

$$\sigma_i(k+1) - \sigma_i(k) = G_i B_i u_{i1}(k) + G_i B_i f_{mi}(k) + G_i f_{ui}(k) + G_i V_i \sum_{\substack{j=1 \\ j \neq i}}^{p} \emptyset_{ji} z_{ji}(k) - G_i x_i(k). \quad (19)$$

4.2. Controller Design. The controller is designed according to (17) with two components, $u_i(k) = u_{i0}(k) + u_{i1}(k)$, where the second component is given as

$$u_{i1}(k) = -(G_i B_i)^{-1} \left[G_i B_i \hat{f}_{mi}(k) + G_i \hat{f}_{ui}(k) - G_i x_i(k) + \alpha_i(k)\sigma_i(k) + \beta_i(k)\operatorname{sgn}(\sigma_i(k)) \right], \quad (20)$$

where

$$\alpha_i(k) = 0, \quad \beta_i(k) = 0 \quad (21)$$

$$\text{when } \sigma_i(k) = 0$$

or

$$\alpha_i(k) = \gamma_i(k)\operatorname{sgn}(\sigma_i(k)) + \epsilon_i, \quad \text{with } \gamma_i(k) = \frac{N + M + L}{|\sigma_i(k)|}, \quad 0 < \epsilon_i < 1, \quad (22)$$

$$\beta_i(k) = \lambda_i |(1 - \epsilon_i)\sigma_i(k)|, \quad 0 < \lambda_i < 1$$

otherwise.

$u_{i1}(k)$ is the reaching mode control component that will ensure that the system is able to achieve a quasi-sliding mode.

Theorem 2. *Subject to Assumption 1 and sliding surface design of (14), the large-scale discrete-time system will achieve quasi-sliding mode and remain in it by having the control input of (16).*

Proof. The proof of the theorem is given below.

While $\sigma_i(k) \neq 0$, define a Lyapunov function:

$$J(k) = 0.5\sigma_i^2(k). \quad (23)$$

Ensuring that $J(k)$ is nonincreasing is equivalent to ensuring the condition in (12) and (13) [15].

Let

$$Q_i(k) = G_i B_i \left(f_{mi}(k) - \hat{f}_{mi}(k) \right) + G_i \left(f_{ui}(k) - \hat{f}_{ui}(k) \right) + G_i V_i \sum_{\substack{j=1 \\ j \neq i}}^{p} \emptyset_{ji} z_{ji}(k). \quad (24)$$

Substituting (23) into (20) gives

$$\sigma_i(k+1) - \sigma_i(k) = -\alpha_i(k)\sigma_i(k) - \beta_i(k)\operatorname{sgn}(\sigma_i(k)) + G_i B_i \left(f_{mi}(k) - \hat{f}_{mi}(k) \right) + G_i \left(f_{ui}(k) - \hat{f}_{ui}(k) \right) + G_i V_i \sum_{\substack{j=1 \\ j \neq i}}^{p} \emptyset_{ji} z_{ji}(k). \quad (25)$$

Multiplying both sides with $\sigma_i(k)$ gives

$$\begin{aligned} \sigma_i(k)(\sigma_i(k+1) - \sigma_i(k)) &= -\alpha_i(k)\sigma_i^2(k) - \beta_i(k)\sigma_i(k)\operatorname{sgn}(\sigma_i(k)) + Q_i(k)\sigma_i(k) \\ &= \sigma_i(k)(Q_i(k) - \alpha_i(k)\sigma_i(k)) - \beta_i(k)|\sigma_i(k)|. \end{aligned} \quad (26)$$

When $\sigma_i(k) < 0$, this implies that

$$\sigma_i(k)(\sigma_i(k+1) - \sigma_i(k)) < 0, \quad \text{provided } \alpha_i(k) > 0. \quad (27)$$

Subsequently, following the conditions stated in (10), it can be shown that

$$Q_i(k) < \|Q_i(k)\| < M + N + L = \gamma_i(k)|\sigma_i(k)|. \quad (28)$$

When $\sigma_i(k) > 0$ and since $\gamma_i(k)\sigma_i^2(k) > Q_i(k)\sigma_i(k)$, (26) becomes

$$\begin{aligned} (\sigma_i(k+1) - \sigma_i(k)) &< (\gamma_i(k) - \alpha_i(k))\sigma_i^2(k) - \beta_i(k)|\sigma_i(k)| \\ &= (\gamma_i(k)\operatorname{sgn}(\sigma_i(k)) - \alpha_i(k))\sigma_i^2(k) - \beta_i(k)|\sigma_i(k)| = -\epsilon_i\sigma_i^2(k) - \beta_i(k)|\sigma_i(k)| \leq 0. \end{aligned} \quad (29)$$

Substituting $Q_i(k)$ as defined in (24) into (25) gives

$$\sigma_i(k+1) = (\sigma_i(k) + Q_i(k) - \alpha_i(k)\sigma_i(k)) - \beta_i(k)\operatorname{sgn}(\sigma_i(k)). \quad (30)$$

Since it is defined in (20) that $\gamma_i(k) = (N+M+L)/|\sigma_i(k)|$ and $\alpha_i(k) = \gamma_i(k)\,\text{sgn}(\sigma_i(k))+\epsilon_i$, let $\varsigma_i = \gamma_i(k)|\sigma_i(k)| - Q_i(k) = M+N+L-Q_i(k)$, and it is known that $|\sigma_i(k)| = \sigma_i(k)\,\text{sgn}(\sigma_i(k))$ according to [1], substitute $Q_i(k) = \gamma_i(k)|\sigma_i(k)| - \varsigma_i$ and $\epsilon_i = \gamma_i(k)\,\text{sgn}(\sigma_i(k)) - \alpha_i(k)$ into equation below:

$$\begin{aligned}
|\sigma_i(k+1)| &= |(\sigma_i(k) + Q_i(k) - \alpha_i(k)\sigma_i(k)) \\
&\quad - \beta_i(k)\,\text{sgn}(\sigma_i(k))| = |(1-\alpha_i(k))\sigma_i(k) \\
&\quad + (\gamma_i(k)|\sigma_i(k)| - \varsigma_i) - \beta_i(k)\,\text{sgn}(\sigma_i(k))| \\
&= |(1-\alpha_i(k))\sigma_i(k) \\
&\quad + (\gamma_i(k)\,\text{sgn}(\sigma_i(k))\sigma_i(k) - \varsigma_i) \\
&\quad - \beta_i(k)\,\text{sgn}(\sigma_i(k))| \\
&= |(1-\alpha_i(k)+\gamma_i(k)\,\text{sgn}(\sigma_i(k)))\sigma_i(k) - \varsigma_i \\
&\quad - \beta_i(k)\,\text{sgn}(\sigma_i(k))| = |(1-\epsilon_i)\sigma_i(k) - \varsigma_i \\
&\quad - \beta_i(k)\,\text{sgn}(\sigma_i(k))|.
\end{aligned} \tag{31}$$

Hence,

$$\begin{aligned}
&|\sigma_i(k+1)| \\
&= |(1-\epsilon_i)\sigma_i(k) - \varsigma_i - \lambda_i|(1-\epsilon_i)\sigma_i(k)|\,\text{sgn}(\sigma_i(k))| \\
&= |(1-\lambda_i)(1-\epsilon_i)\sigma_i(k) - \varsigma_i|.
\end{aligned} \tag{32}$$

This will ensure that $|\sigma_i(k+1)| \le |\sigma_i(k)|$ is satisfied when the conditions below are met:

$$\begin{aligned}
\sigma_i(k) &< -\left[\frac{\varsigma_i}{\lambda_i + \epsilon_i - \lambda_i\epsilon_i}\right] \\
\text{or } \sigma_i(k) &> -\left[\frac{\varsigma_i}{2-(\lambda_i + \epsilon_i - \lambda_i\epsilon_i)}\right].
\end{aligned} \tag{33}$$

In the case of $\sigma_i(k) < 0$, it should be noted that, in order to ensure $\sigma_i(k)(\sigma_i(k+1) - \sigma_i(k)) < 0$, there is another condition to be met; that is, $\alpha_i(k) > 0$. Equation (20) stated that $\alpha_i(k) = \gamma_i(k)\,\text{sgn}(\sigma_i(k)) + \epsilon_i$, and $0 < \epsilon_i < 1$; therefore, in order to guarantee $\alpha_i(k) > 0$, it is necessary to have

$$\gamma_i(k) = \frac{M+N+L}{|\sigma_i(k)|} > -1. \tag{34}$$

So (34) implies that when $\sigma_i(k) < 0$, the second condition to ensure the size of $\sigma_i(k)$ decreasing is

$$\sigma_i(k) < -(M+N+L). \tag{35}$$

Since $\varsigma_i = M+N+L-Q_i(k)$ and $Q_i(k) < \|Q_i(k)\| < M+N+L$, this implies that

$$0 < \varsigma_i < 2(M+N+L). \tag{36}$$

It can be concluded that $\sigma_i(k)$ exhibits a quasi-sliding mode with lower and upper bound ($\overline{U}_l$ and $\overline{U}_u$, resp.) and the band is

$$\begin{aligned}
&U_l \le \sigma_i(k) \le U_u, \\
&U_l > \overline{U}_l = \min\left(-\frac{2(M+N+L)}{l}, -(M+N+L)\right), \\
&U_u < \overline{U}_u = \frac{2(M+N+L)}{2-l} \\
&\qquad\qquad \text{with } l = \lambda_i + \epsilon_i - \lambda_i\epsilon_i,
\end{aligned} \tag{37}$$

while $\sigma_i(k) = 0$.

By substituting $\sigma_i(k) = 0$ into (20),

$$\begin{aligned}
&u_{i1}(k) \\
&= -(G_iB_i)^{-1}\left[G_iB_i\hat{f}_{mi}(k) + G_i\hat{f}_{ui}(k) - G_ix_i(k)\right].
\end{aligned} \tag{38}$$

Since, from (14),

$$\begin{aligned}
\sigma_i(k) &= G_ix_i(k) - G_ix_i(0) - h_i(k), \\
u_{i1}(k) &= -\hat{f}_{mi}(k) \\
&\quad - (G_iB_i)^{-1}\left(G_i\hat{f}_{ui}(k) + h_i(k) - G_ix_i(0)\right),
\end{aligned} \tag{39}$$

substituting (39) into (18) gives

$$\begin{aligned}
\sigma_i(k+1) &= G_iB_i\left(f_{mi}(k) - \hat{f}_{mi}(k)\right) \\
&\quad + G_i\left(f_{ui}(k) - \hat{f}_{ui}(k)\right) \\
&\quad + G_iV_i\sum_{\substack{j=1\\ j\ne i}}^{p} \emptyset_{ji}z_{ji}(k).
\end{aligned} \tag{40}$$

This concludes the proof for Theorem 2. □

4.3. Overall System Stability. The overall system stability as the closed-loop performance while travelling along the sliding surface will be discussed in this section, that is, when $\sigma_i(k) = 0$. Substituting (16) and (20) into (5), the closed-loop dynamic is derived below:

$$\begin{aligned}
x_i(k+1) &= A_ix_i(k) + B_i(u_{i0}(k) + u_{i1}(k) + f_{mi}(k)) \\
&\quad + f_{ui}(k) + V_i\sum_{\substack{j=1\\ j\ne i}}^{p} \emptyset_{ji}z_{ji}(k) \\
&= A_ix_i(k) + B_i\left(f_{mi}(k) - \hat{f}_{mi}(k)\right) \\
&\quad + B_iu_{i0}(k) + B_i(G_iB_i)^{-1}G_ix_i(0) \\
&\quad + f_{ui}(k) - B_i(G_iB_i)^{-1}G_i\hat{f}_{ui}(k) \\
&\quad - B_i(G_iB_i)^{-1}h_i(k) + V_i\sum_{\substack{j=1\\ j\ne i}}^{p} \emptyset_{ji}z_{ji}(k).
\end{aligned} \tag{41}$$

Since $h_i(k) = \sigma_i(k) - G_i x_i(k) + G_i x_i(0)$ and $u_{i0}(k) = -K_i x_i(k)$, (41) becomes

$$\begin{aligned} x_i(k+1) &= (A_i - B_i K_i) x_i(k) \\ &+ B_i \left(f_{mi}(k) - \hat{f}_{mi}(k)\right) + f_{ui}(k) \\ &- B_i (G_i B_i)^{-1} G_i \hat{f}_{ui}(k) \\ &- B_i (G_i B_i)^{-1} (\sigma_i(k) - G_i x_i(k)) \\ &+ V_i \sum_{\substack{j=1 \\ j\neq i}}^{p} \emptyset_{ji} z_{ji}(k). \end{aligned} \tag{42}$$

Taking into account (11) and (40),

$$\begin{aligned} \sigma_i(k) &= G_i B_i (f_{mi}(k-1) - f_{mi}(k-2)) \\ &+ G_i (f_{ui}(k-1) - f_{ui}(k-2)) \\ &+ G_i V_i \sum_{\substack{j=1 \\ j\neq i}}^{p} \emptyset_{ji} z_{ji}(k-1). \end{aligned} \tag{43}$$

Thus (42) becomes

$$\begin{aligned} x_i(k+1) &= \left(A_i + B_i (G_i B_i)^{-1} G_i - B_i K_i\right) x_i(k) \\ &+ B_i [f_{mi}(k) - 2 f_{mi}(k-1) + f_{mi}(k-2)] \\ &+ \left[f_{ui}(k) - 2 B_i (G_i B_i)^{-1} G_i f_{ui}(k-1)\right. \\ &\left.+ B_i (G_i B_i)^{-1} G_i f_{ui}(k-2)\right] + \left[V_i \sum_{\substack{j=1 \\ j\neq i}}^{p} \emptyset_{ji} z_{ji}(k)\right. \\ &\left.- B_i (G_i B_i)^{-1} G_i V_i \sum_{\substack{j=1 \\ j\neq i}}^{p} \emptyset_{ji} z_{ji}(k-1)\right]. \end{aligned} \tag{44}$$

Owing to the assumption that both $f_{mi}(k)$ and $f_{ui}(k)$ are bounded, it can be assumed that

$$\begin{aligned} -W &\le B_i [f_{mi}(k) - 2 f_{mi}(k-1) + f_{mi}(k-2)] \\ &+ \left[f_{ui}(k) - 2 B_i (G_i B_i)^{-1} G_i f_{ui}(k-1)\right. \\ &\left.+ B_i (G_i B_i)^{-1} G_i f_{ui}(k-2)\right] + \left[V_i \sum_{\substack{j=1 \\ j\neq i}}^{p} \emptyset_{ji} z_{ji}(k)\right. \\ &\left.- B_i (G_i B_i)^{-1} G_i V_i \sum_{\substack{j=1 \\ j\neq i}}^{p} \emptyset_{ji} z_{ji}(k-1)\right] = w_i(k) \\ &\le W, \quad \text{with } 0 < W < \infty. \end{aligned} \tag{45}$$

Then the closed-loop system dynamic can be represented by

$$\begin{aligned} x_i(k+1) &= \left(A_i + B_i (G_i B_i)^{-1} G_i - B_i K_i\right) x_i(k) \\ &+ w_i(k). \end{aligned} \tag{46}$$

The gain, K, must be selected so that $\overline{\rho}_{\max} < 0$ with $\overline{\rho}_{\max}$ standing for the largest eigenvalue of $A_i + B_i (G_i B_i)^{-1} G_i - B_i K_i$.

Thus,

$$\begin{aligned} &x_i^T(k) [x_i(k+1) - x_i(k)] \\ &= x_i^T(k) \left(A_i + B_i (G_i B_i)^{-1} G_i - B_i K_i - I\right) x_i(k) \\ &+ x_i^T(k) w_i(k) \\ &\le \rho_{\max} x_i^T(k) x_i(k) + \left\| x_i^T(k) w_i(k) \right\| \\ &\le \rho_{\max} \left\| x_i(k) \right\|^2 + \left\| x_i^T(k) W \right\|. \end{aligned} \tag{47}$$

Since it is necessary to have $\rho_{\max} < 0$ to ensure the stability of the system,

$$\rho_{\max} < -\max\left(\frac{\left\| x_i^T(k) W \right\|}{\left\| x_i(k) \right\|^2}\right). \tag{48}$$

It can be seen from (47) that when the system is not affected by any uncertainty, that is, $W = 0$, $\rho_{\max} < 0$ is sufficient to ensure the system stability. When $W \neq 0$, the larger the uncertainty, W, is, the more negative $\rho_{\max}$ must be to guarantee the stability. This is due to the nature of discrete-time system that $x_i(k)$ will never converge to zero but stays within a band about the origin.

5. Examples and Simulation Results

In this section, a simulation of large-scale discrete-time system with three interconnected subsystems is performed to illustrate the new control strategy. This example is taken from Park and Lee [6] and the dynamic equation of the subsystems can be written as

$$\begin{aligned} x_1(k+1) &= \left(\begin{bmatrix} 0 & 1 \\ 0 & 2 \end{bmatrix} + \begin{bmatrix} 0 & 0.4\cos(k) \\ 0.2\sin(k) & 0 \end{bmatrix}\right) \\ &\cdot x_1(k) + \left(\begin{bmatrix} 0 \\ 1 \end{bmatrix} + \begin{bmatrix} 0 \\ 0.04\sin(k) \end{bmatrix}\right) u_1(k) \\ &+ \left(\begin{bmatrix} 0 & 0.1 \\ 0.1 & 0 \end{bmatrix} + \begin{bmatrix} 0 & 0 \\ 0 & 0.05\cos(k) \end{bmatrix}\right) x_2(k) \\ &+ \left(\begin{bmatrix} 0.05 & 0 \\ 0 & 0.15 \end{bmatrix} + \begin{bmatrix} 0.01\cos(k) & 0 \\ 0 & 0.01\sin(k) \end{bmatrix}\right) \\ &\cdot x_3(k), \end{aligned}$$

$$x_2(k+1) = \left(\begin{bmatrix} 0 & 1 \\ 0.5 & -0.5 \end{bmatrix} + \begin{bmatrix} 0 & 0.04\cos(k) \\ 0.04\sin(k) & 0 \end{bmatrix}\right) \cdot x_2(k) + \left(\begin{bmatrix} 1 \\ 1 \end{bmatrix} + \begin{bmatrix} 0.04\cos(k) \\ 0 \end{bmatrix}\right) u_2(k) + \left(\begin{bmatrix} 0 & 0.2 \\ 0.1 & 0 \end{bmatrix} + \begin{bmatrix} 0 & 0.04\cos(k) \\ 0 & 0 \end{bmatrix}\right) x_1(k) + \left(\begin{bmatrix} 0.15 & 0 \\ 0 & 0.1 \end{bmatrix} + \begin{bmatrix} 0.01\cos(k) & 0 \\ 0 & 0.01\sin(k) \end{bmatrix}\right) \cdot x_3(k),$$

$$x_3(k+1) = \left(\begin{bmatrix} -0.9 & 0.5 \\ 0 & 1 \end{bmatrix} + \begin{bmatrix} 0.1\cos(k) & 0 \\ 0 & 0.2\sin(k) \end{bmatrix}\right) \cdot x_3(k) + \left(\begin{bmatrix} 1 \\ 1 \end{bmatrix} + \begin{bmatrix} 0 \\ 0.05\sin(k) \end{bmatrix}\right) u_3(k) + \left(\begin{bmatrix} 0 & 0.05 \\ 0.02 & 0.1 \end{bmatrix} + \begin{bmatrix} 0 & 0.1\cos(k) \\ 0.04\sin(k) & 0 \end{bmatrix}\right) \cdot x_1(k) + \left(\begin{bmatrix} 0 & 0.05 \\ 0 & 0.1 \end{bmatrix} + \begin{bmatrix} 0 & 0.01\cos(k) \\ 0.01\sin(k) & 0 \end{bmatrix}\right) \cdot x_2(k), \quad (49)$$

where

$$x_i(k) = \begin{bmatrix} x_{i1}(k) \\ x_{i2}(k) \end{bmatrix}^T, \quad u_i(k) = \begin{bmatrix} u_{i1}(k) \\ u_{i2}(k) \end{bmatrix}^T, \quad i = 1, 2, 3. \quad (50)$$

The initial conditions for this simulation are

$$x_1(k) = [1 \;\; -0.5]^T, \quad x_2(k) = [-0.5 \;\; -1.5]^T, \quad x_3(k) = [-1 \;\; 0.5]^T. \quad (51)$$

The value of $\epsilon_i(k)$ is chosen as ϵ_1=0.1, ϵ_2=0.1, and ϵ_3=0.1. The value of $\lambda_i(k)$ is chosen as $\lambda_1 = 0.9$, $\lambda_2 = 0.2$, and $\lambda_3 = 0.3$. In this case, $G_1 = [0 \;\; 0.10]$, $G_2 = [0.10 \;\; 1]$, and $G_3 = [5 \;\; 10]$. The value of K is obtained as $K_1 = \text{place}(A_1 + B_1(G_1B_1)^{-1}G_1, B_1, [-0.9 \;\; 0.9]) = [-0.81 \;\; 3.0]$, $K_2 = \text{place}(A_2 + B_2(G_2B_2)^{-1}G_2, B_2, [0.05 \;\; 0.1]) = [0.92 \;\; -0.57]$, and $K_3 = \text{place}(A_3 + B_3(G_3B_3)^{-1}G_3, B_3, [0.1 \;\; 0.2]) = [-0.45 \;\; 1.25]$.

The sliding surfaces $\sigma_1(k)$, $\sigma_2(k)$, and $\sigma_3(k)$ are chosen to be of the same parameter. Example below shows the equation of $\sigma_1(k)$, $\sigma_2(k)$, and $\sigma_3(k)$ as implemented in the simulation:

$$\sigma_1(k) = [0 \;\; 0.10]\, x_1(k) - [0 \;\; 0.10]\, x_1(0) + h_1(k-1) - \left([0 \;\; 0.10]\begin{bmatrix} 0 \\ 1 \end{bmatrix} u_{10}(k-1) + [0 \;\; 0.10]\begin{bmatrix} 0 & 1 \\ 0 & 2 \end{bmatrix} x_1(k-1)\right),$$

$$\sigma_2(k) = [0.10 \;\; 1]\, x_2(k) - [0.10 \;\; 1]\, x_2(0) + h_2(k-1) - \left([0.10 \;\; 1]\begin{bmatrix} 1 \\ 1 \end{bmatrix} u_{20}(k-1) + [0.10 \;\; 1]\begin{bmatrix} 0 & 1 \\ 0.5 & -0.5 \end{bmatrix} x_2(k-1)\right), \quad (52)$$

$$\sigma_3(k) = [5 \;\; 10]\, x_3(k) - [5 \;\; 10]\, x_3(0) + h_2(k-1) - \left([5 \;\; 10]\begin{bmatrix} 1 \\ 1 \end{bmatrix} u_{30}(k-1) + [5 \;\; 10]\begin{bmatrix} -0.9 & 0.5 \\ 0 & 1 \end{bmatrix} x_3(k-1)\right),$$

where

$$u_{10}(k) = -[-0.81 \;\; 3.0]\, x_1(k), \quad h_1(0) = 0,$$
$$u_{20}(k) = -[0.92 \;\; -0.57]\, x_2(k), \quad h_2(0) = 0, \quad (53)$$
$$u_{30}(k) = -[-0.45 \;\; 1.25]\, x_3(k), \quad h_3(0) = 0.$$

$u_{11}(k)$ is the control signal for the first subsystem with the presence of disturbance and interconnection from the second and third subsystems:

$$u_{11}(k) = -\left([0 \;\; 0.10]\begin{bmatrix} 0 \\ 10 \end{bmatrix}\right)^{-1} \Bigg[-[0 \;\; 0.10]\, x_1(k) + [0 \;\; 0.10]\begin{bmatrix} 0 \\ 1 \end{bmatrix}\left(\begin{bmatrix} 0 \\ 0.04\sin(k) \end{bmatrix}\right) u_1(k-1) + [0 \;\; 0.10]\begin{bmatrix} 0 & 0.4\cos(k) \\ 0.2\sin(k) & 0 \end{bmatrix} x_1(k) + [0 \;\; 0.10]\begin{bmatrix} 0 & 0 \\ 0 & 0.05\cos(k) \end{bmatrix} x_2(k) + [0 \;\; 0.10]\begin{bmatrix} 0.01\cos(k) & 0 \\ 0 & 0.01\sin(k) \end{bmatrix} x_3(k) + \alpha_1(k)\,\sigma_1(k) + \beta_1(k)\,\text{sgn}(\sigma_1(k))\Bigg]. \quad (54)$$

The same conditions applied to the second subsystem with the present of disturbance and interconnection effect. The control signals, $u_{21}(k)$ and $u_{31}(k)$, are given below:

$$
\begin{aligned}
u_{21}(k) = &-\left([0.10 \quad 1]\begin{bmatrix}1\\1\end{bmatrix}\right)^{-1}\Bigg[-[0.10 \quad 1]\,x_2(k) \\
&+ [0.10 \quad 1]\begin{bmatrix}1\\1\end{bmatrix}\left(\begin{bmatrix}0.04\cos(k)\\0\end{bmatrix}\right)u_2(k-1) \\
&+ [0.10 \quad 1]\begin{bmatrix}0 & 0.04\cos(k)\\0.04\sin(k) & 0\end{bmatrix}x_2(k) \\
&+ [0.10 \quad 1]\begin{bmatrix}0 & 0.04\cos(k)\\0 & 0\end{bmatrix}x_1(k) \\
&+ [0.10 \quad 1]\begin{bmatrix}0.01\cos(k) & 0\\0 & 0.01\sin(k)\end{bmatrix}x_3(k) \\
&+ \alpha_2(k)\,\sigma_2(k) + \beta_2(k)\,\mathrm{sgn}(\sigma_2(k))\Bigg], \\
u_{31}(k) = &-\left([5 \quad 10]\begin{bmatrix}1\\1\end{bmatrix}\right)^{-1}\Bigg[-[5 \quad 10]\,x_3(k) \\
&+ [5 \quad 10]\begin{bmatrix}1\\1\end{bmatrix}\left(\begin{bmatrix}0\\0.05\sin(k)\end{bmatrix}\right)u_3(k-1) \\
&+ [5 \quad 10]\begin{bmatrix}0.1\cos(k) & 0\\0 & 0.2\sin(k)\end{bmatrix}x_3(k) \\
&+ [5 \quad 10]\begin{bmatrix}0 & 0.1\cos(k)\\0.04\sin(k) & 0\end{bmatrix}x_1(k) \\
&+ [5 \quad 10]\begin{bmatrix}0 & 0.01\cos(k)\\0.01\sin(k) & 0\end{bmatrix}x_2(k) \\
&+ \alpha_3(k)\,\sigma_3(k) + \beta_3(k)\,\mathrm{sgn}(\sigma_3(k))\Bigg].
\end{aligned}
\tag{55}
$$

The simulation has been done at a period of 50 seconds and the results are shown in Figures 1–7.

As shown in Figure 1 to Figure 3, the system trajectories of all the 3 subsystems of the discrete-time large-scale system under local control of discrete-time integral VSC are able to achieve stability and reached the desired conditions with disturbance being rejected. A comparison was made for the outputs of the systems with only the feedback control input, $u_0(k)$, without the VSC input, $u_1(k)$. It is clearly shown in Figures 4–6 that the system was unable to be controlled and unstable. Figure 7 showed that the sliding surface signals for all 3 subsystems achieved quasi-sliding mode with the discrete-time integral variable structure controller in place.

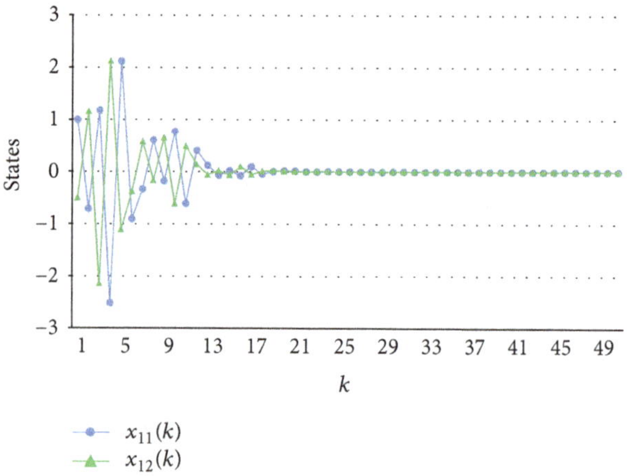

FIGURE 1: States respond of subsystem 1, $x_{11}(k)$ and $x_{12}(k)$ under discrete-time integral VSC.

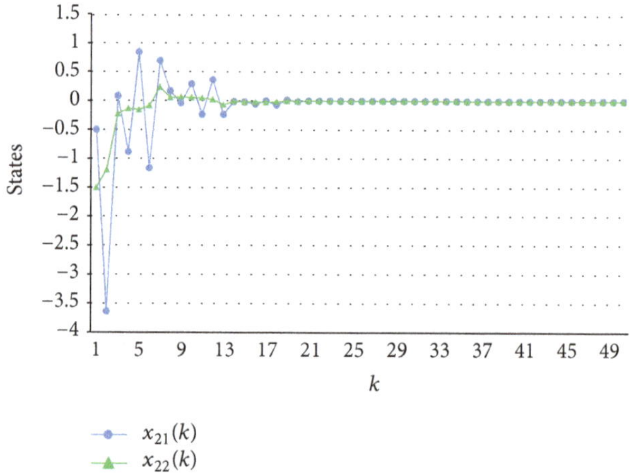

FIGURE 2: States respond of subsystem 2, $x_{21}(k)$ and $x_{22}(k)$ under discrete-time integral VSC.

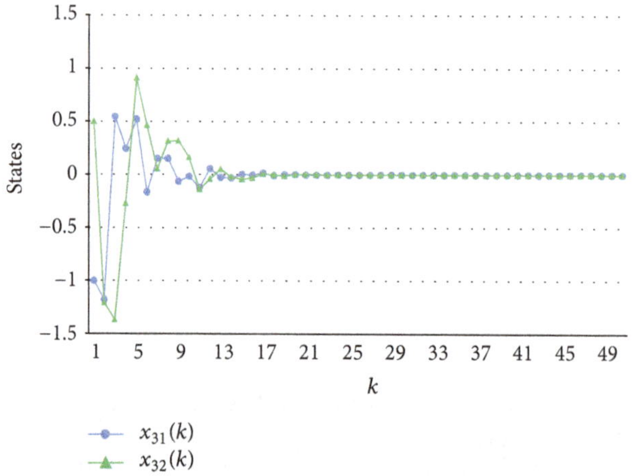

FIGURE 3: States respond of subsystem 3, $x_{31}(k)$ and $x_{32}(k)$ under discrete-time integral VSC.

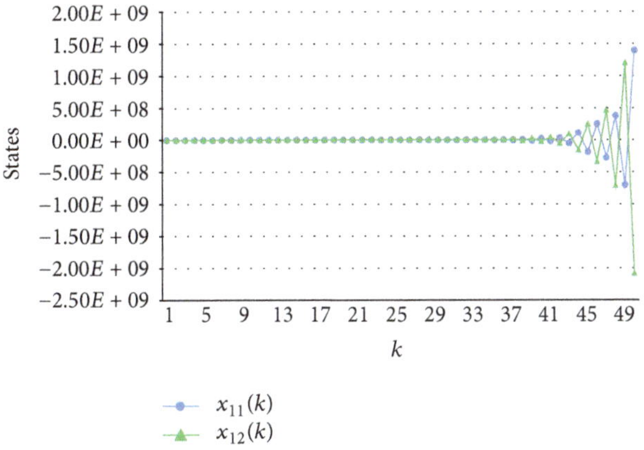

FIGURE 4: States respond of subsystem 1, $x_{11}(k)$ and $x_{12}(k)$ under feedback control, without discrete-time integral VSC.

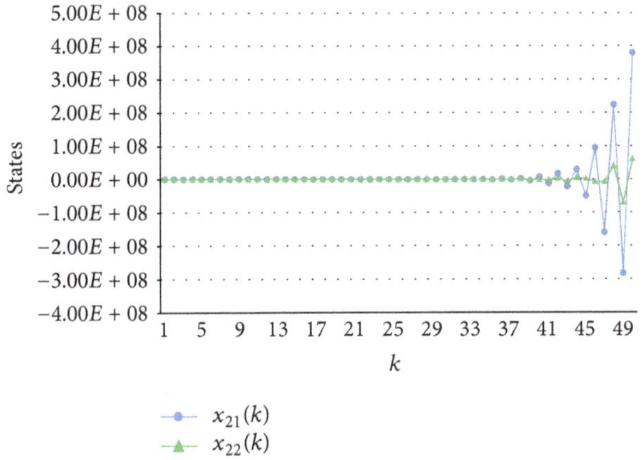

FIGURE 5: States respond of subsystem 2, $x_{21}(k)$ and $x_{22}(k)$ under feedback control, without discrete-time integral VSC.

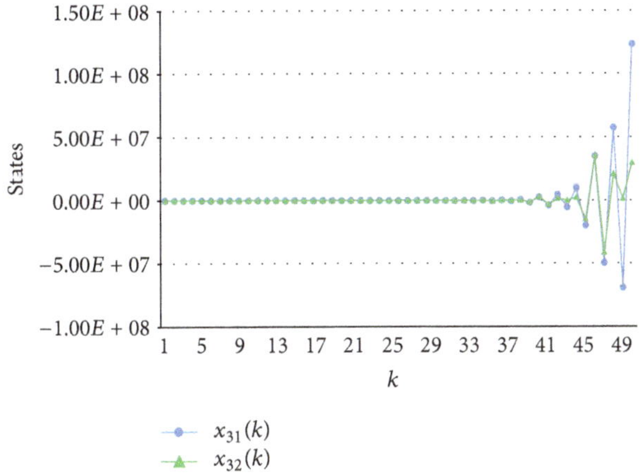

FIGURE 6: States respond of subsystem 3, $x_{31}(k)$ and $x_{32}(k)$ under feedback control, without discrete-time integral VSC.

6. Conclusion

The control of large-scale discrete-time system with unmatched uncertainty by using integral variable structure control in local control method has been proposed in this paper. A new theorem has been presented and proved that, with only the states feedback from the individual subsystem itself, the local controller is able to ensure that the system is achieving the quasi-sliding surface and remains there. The proposed local controller showed that the effect of interconnection in large-scale system is being handled well and the system stability is ensured. It is also shown that the effect of matched and unmatched uncertainty in the system is also being rejected. A discrete-time large-scale system comprised of 3 subsystems has been used to evaluate the performance of the local controller. It can be seen that the proposed controller is able to control the system to achieve the stability and desired value and also reduce the effect of disturbance as compared to the system without using VSC. As a conclusion, it can be concluded that the proposed discrete-time integral variable structure local controller has the advantage in controlling large-scale discrete-time system with matched and unmatched uncertainties.

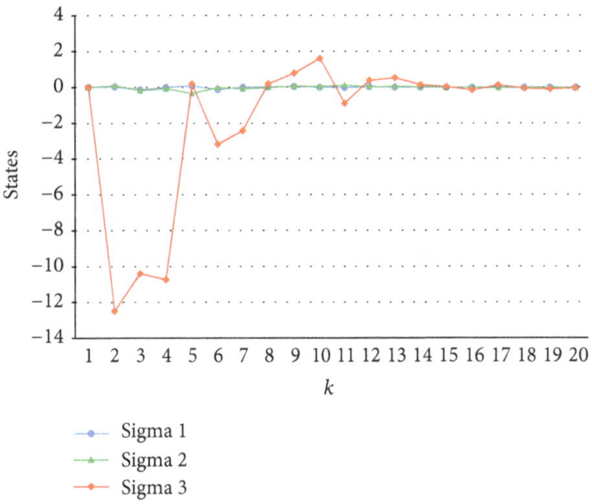

FIGURE 7: Sliding surface signal for subsystem ($\sigma_1(k)$, sigma 1), subsystem 2 ($\sigma_2(k)$, sigma 2), and subsystem 3 ($\sigma_3(k)$, sigma 3).

Conflict of Interests

The authors declare that there is no conflict of interests regarding the publication of this paper.

References

[1] D. D. Siljak, *Large-Scale Dynamic Systems: Stability and Structure*, North-Holland, Amsterdam, The Netherlands, 1978.

[2] N. R. Sandell Jr., P. Varaiya, M. Athans, and M. G. Safonov, "Survey of decentralized control methods for large scale systems," *IEEE Transactions on Automatic Control*, vol. 23, no. 2, pp. 108–128, 1978.

[3] Y. H. Chen, "Deterministic control of large-scale uncertain dynamical systems," *Journal of the Franklin Institute*, vol. 323, no. 2, pp. 135–144, 1987.

[4] G. Q. Li, K. Lee, and F. Gordon, "Decentralized control of discrete-time large-scale systems by dynamic programming," in *Proceedings of the 21st IEEE Conference on Decision and Control*, pp. 881–885, Orlando, Fla, USA, December 1982.

[5] J. Lyou and Z. Bien, "Decentralized adaptive stabilization of a class of large-scale interconnected discrete systems," *Journal of Dynamic Systems, Measurement and Control*, vol. 107, no. 1, pp. 106–109, 1985.

[6] J. H. Park and S. G. Lee, "Robust decentralized stabilization of uncertain large-scale discrete-time systems," *International Journal of Systems Science*, vol. 33, no. 8, pp. 649–654, 2002.

[7] J. H. Park, H. Y. Jung, J. I. Park, and S. G. Lee, "Decentralized dynamic output feedback controller design for guaranteed cost stabilization of large-scale discrete-delay systems," *Applied Mathematics and Computation*, vol. 156, no. 2, pp. 307–320, 2004.

[8] O. Ou, Q. Hui, and H. Zhang, "Stability analysis and $\mathbf{H}_\infty$ decentralized control for discrete-time nonlinear large-scale systems via fuzzy control approach," in *Proceedings of the 6th International Conference on Fuzzy Systems and Knowledge Discovery (FSKD '09)*, vol. 4, pp. 166–170, IEEE, Tianjin, China, August 2009.

[9] Y. Dote and R. G. Hoft, "Microprocessor based sliding mode controller for DC motor drives," in *Proceedings of the Industry Applications Society Annual Meeting*, Cincinnati, Ohio, USA, September 1980.

[10] S. Z. Sarpturk, Y. Istefanopulos, and O. Kaynak, "On the stability of discrete-time sliding mode control systems," *IEEE Transactions on Automatic Control*, vol. 32, no. 10, pp. 930–932, 1987.

[11] D. Milosavljevic, "General conditions for the existence of a quasi-sliding mode on the switching hyperplane in discrete variable structure systems," *Automation and Remote Control*, vol. 46, pp. 307–314, 1985.

[12] K. Furuta, "Sliding mode control of a discrete system," *Systems and Control Letters*, vol. 14, no. 2, pp. 145–152, 1990.

[13] W. B. Gao, Y. Wang, and A. Homaifa, "Discrete-time variable structure control systems," *IEEE Transactions on Industrial Electronics*, vol. 42, no. 2, pp. 117–122, 1995.

[14] K. Abidi, J.-X. Xu, and X. Yu,, "On the discrete-time integral sliding-mode control," *IEEE Transactions on Automatic Control*, vol. 52, no. 4, pp. 709–715, 2007.

[15] Z. Xi and T. Hesketh, "Discrete time integral sliding mode control for systems with matched and unmatched uncertainties," *IET Control Theory and Applications*, vol. 4, no. 5, pp. 889–896, 2010.

[16] A. F. Sheta, "Variable structure controller design for large-scale systems," in *Proceedings of the IEEE International Workshop on Variable Structure Systems (VSS '96)*, pp. 228–231, December 1996.

[17] K. J. Hunt and T. A. Johansen, "Design and analysis of gain-scheduled control using local controller networks," *International Journal of Control*, vol. 66, no. 5, pp. 619–651, 1997.

[18] G. Zhuansun and J. Xiong, "Local mode-dependent decentralised H∞ control of uncertain Markovian jump large-scale systems," *IET Control Theory and Applications*, vol. 7, no. 7, pp. 1029–1038, 2013.

[19] W.-C. Su, S. V. Drakunov, and Ü. Özgüner, "An $O(T^2)$ boundary layer in sliding mode for sampled-data systems," *IEEE Transactions on Automatic Control*, vol. 45, no. 3, pp. 482–485, 2000.

[20] R. Morgan and U. Ozguner, "A decentralized variable structure control algorithm for robotic manipulators," *IEEE Journal of Robotics and Automation*, vol. 1, no. 1, pp. 57–65, 1985.

7

A Speed Control Method for Underwater Vehicle under Hydraulic Flexible Traction

Yin Zhao, Ying-kai Xia, Ying Chen, and Guo-Hua Xu

School of Naval Architecture and Ocean Engineering, Huazhong University of Science and Technology, Wuhan 430074, China

Correspondence should be addressed to Guo-Hua Xu; hustxu@vip.sina.com

Academic Editor: Pedro Castillo

Underwater vehicle speed control methodology method is the focus of research in this study. Driven by a hydraulic flexible traction system, the underwater vehicle advances steadily on underwater guide rails, simulating an underwater environment for the carried device. Considering the influence of steel rope viscoelasticity and the control system traction structure feature, a mathematical model of the underwater vehicle driven by hydraulic flexible traction system is established. A speed control strategy is then proposed based on the sliding mode variable structure of fuzzy reaching law, according to nonlinearity and external variable load of the vehicle speed control system. Sliding mode variable structure control theory for the nonlinear system allows an improved control effect for movements in "sliding mode" when compared with conventional control. The fuzzy control theory is also introduced, weakening output chattering caused by the sliding mode control switchover while producing high output stability. Matlab mathematical simulation and practical test verification indicate the speed control method as effective in obtaining accurate control results, thus inferring strong practical significance for engineering applications.

1. Introduction

Oceans cover 71% of earth's surface and feature rich biological resources with vast mineral storage. Marine scientific research related to development of marine equipment and related test devices are receiving significant attention recently as resources are further evaluated for use. The underwater vehicle provides a new platform type for carried devices that incorporate the motion of the underwater craft. Underwater vehicles accommodate sensors and operation tools and utilize computers for underwater motion testing and for transferring collected data remotely through the optics network. Precise speed control of the underwater vehicles then is key to the underwater scientific research.

Commonly used speed servo-driven methods include motor-driven and hydraulic driven winches. The motor-driven winch operates with high precision and simple speed alterations yet exhibits inefficient power output, load capacity, and sealing problems in the underwater environment. The hydraulic winch, employed in this study, features higher efficiency, superior load capacity, stronger driving power, and convenient speed adjustment. Therefore, in this study we use hydraulic winch as underwater vehicle driving equipment.

Advanced control theory such as neural network, fuzzy control, and predictive control have been widely applied in the hydroserve speed control field [1]. Complexity of modern hydraulic systems due to nonlinearity, large hysteresis, time-varying parameter, and load characteristics does not allow traditional PID control to consistently meet control requirements. Newton analyzed the difference of control theory and control effects between neural network control algorithm with the traditional PID control algorithm in motor and valve control cylinder valve-controlled system [2]. Zibin et al. focused on uncertainties and designed an adaptive back-stepping neural network approaching control algorithm [3]. Azimian et al. analyzed characteristics of the electrohydraulic servomotor control and designed a neural network controller currently employed in applications [4]. Istif designed two controller types: the first is a neural network predictive controller utilizing the neural network forecasting model to predict system output and the other is a nonlinear autoregressive moving average controller converting the nonlinear system into a linear dynamic system. Characteristics of each

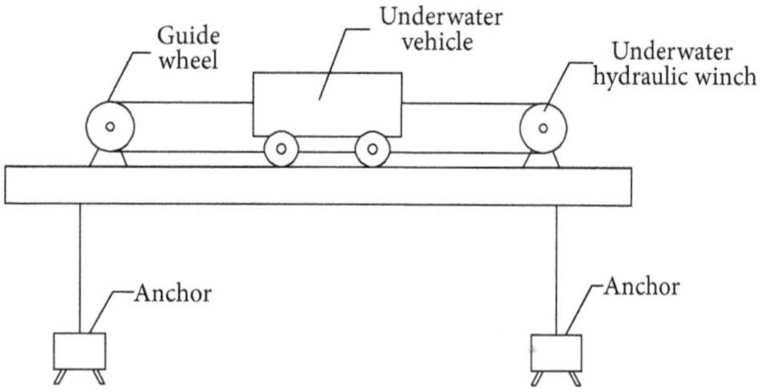

FIGURE 1: Schematic diagram of underwater vehicle propelled by hydraulic winch.

controller are analyzed through testing in the electrohydraulic servovalve control cylinder system [5].

Sliding mode control with parameter adaptability and showing good robustness for nonlinear control system has become increasingly applicable to electrohydraulic servosystems control. Mohseni et al. first introduced sliding mode control theory as applied to hydraulic cylinder servocontrol [6]. Perron et al. proposed a novel sliding mode control method for hydraulic speed control [7]. Loukianov et al., recognizing the electrohydraulic system as a typical affine nonlinear system utilizing a state feedback linear method, achieved precise control of the electrohydraulic servosystem [8]. Inherent shortcomings of the traditional sliding mode controller remain, however, restricting further development in the field of electrohydraulic servocontrol; thus application of advanced control theory to sliding mode control theory is a critical research direction for modern hydraulic control.

In this study, research was first focused on underwater vehicle control system modeling analysis. Characteristics of the hydraulic system and flexibility of the closed traction system's impact on the entire control system were examined. Sliding mode control based on reaching law was chosen as the control algorithm to control low stiffness and variable load of the system, while fuzzy control theory was applied to reduce chattering impact on the output implementation structure. The proposed underwater control method was based on fuzzy reaching law. The controlled object mathematical model was then built and digital simulation of the control method completed to verify the control algorithm and optimize parameters for practical testing.

2. Modeling of Underwater Vehicle Control System

Figure 1 displays a schematic diagram of the underwater vehicle propelled by underwater hydraulic winch. The underwater vehicle is 5 m long, 1.2 m wide, 13.5 m high, and 3000 kg, installed on the guide rail of the platform deck surface. The underwater hydraulic winch and guide wheel are located on each end of the 50 m long rail and the winch drum is a monolayer cylinder. Two traction ropes are fixed and wound on the winch drum forming the closed flexible traction structure. The upper rope end is fixed on one end of the hydraulic vehicle and the lower end fixed on the other end of the hydraulic vehicle along the guide rail across the wheel. Hydraulic vehicle traction may be achieved through the retractable rope when the hydraulic winch rotates. The hydraulic winch may then drive the underwater vehicle along the guide rail when the self-submersible platform is settled on the setting depth through flexible tension-leg.

The underwater vehicle must be controlled to accelerate smoothly from start to the setting test speed and speed stability maintained with less than 10% fluctuations for accurate motion simulations at each stage. Following completion of the related test, the speed of the underwater vehicle should gradually decelerate to zero; thus speed control system then must operate with refined precision, response, and stability. As shown in Figure 1, closed flexible traction structure allows the hydraulic winch to propel the underwater vehicle. Accurate speed control is challenged as a result of rope flexibility reducing stiffness of the traction system and change of water resistance while the underwater vehicle is moving, creating variability for the external load system.

As it is described in Figure 2, the underwater vehicle speed controller calculates the value of opening degree of valve based on the difference between the measured value and the given value of the underwater vehicle. The underwater vehicle moves with closed flexible traction structure as it is driven by the hydraulic winch and speed is determined by the open degree of the valve. It experienced variable disturbance from water resistance and flexibility of the rope during the test due to influence of water resistance and rope flexibility.

2.1. Modeling of Closed Flexible Traction Structure. The hydraulic winch drives the underwater vehicle through the closed flexible traction structure and the guide wheel attachment (Figure 3). A tension force F is applied to the rope by the hydraulic cylinder through the guide wheel to avoid speed control errors caused by fluttering of the rope. The rope is divided into three sections (L_1, L_2, L_3), according to location of the guide wheel, the underwater vehicle, and the hydraulic winch; hence, the stresses of three parts are defined as F_1, F_2, and F_3. The vehicle is subjected to three forces when propelled by the hydraulic winch, namely, the forward pulling force F_1, backward pushing force F_3, and resistance f. The guide wheel force is the sum of driving force F_2 and resistance F_1, while the hydraulic winch is subject to resistance F_2 and pulling force F_3.

The dynamics formula of the underwater vehicle may be obtained through the analysis:

$$F_1 - F_3 - f = M\frac{dv}{dt} + B_2 v_L, \tag{1}$$

where F_1 is the tension force of L_1, F_3 is the tension force of L_3, f is water resistance force applied on the vehicle, M is the mass of the underwater vehicle, B_2 is the friction coefficient between vehicle wheels and guide rail, and v_L is motion speed of the underwater vehicle.

Force analysis for steel rope is critically important for determining underwater vehicle speed control utilizing the closed flexible traction system.

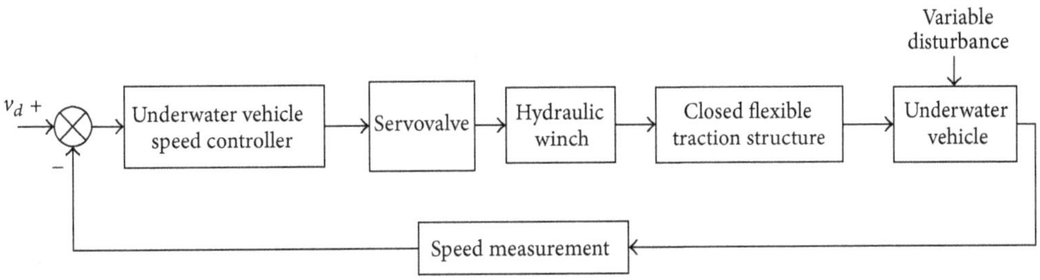

FIGURE 2: Speed control of the underwater vehicle.

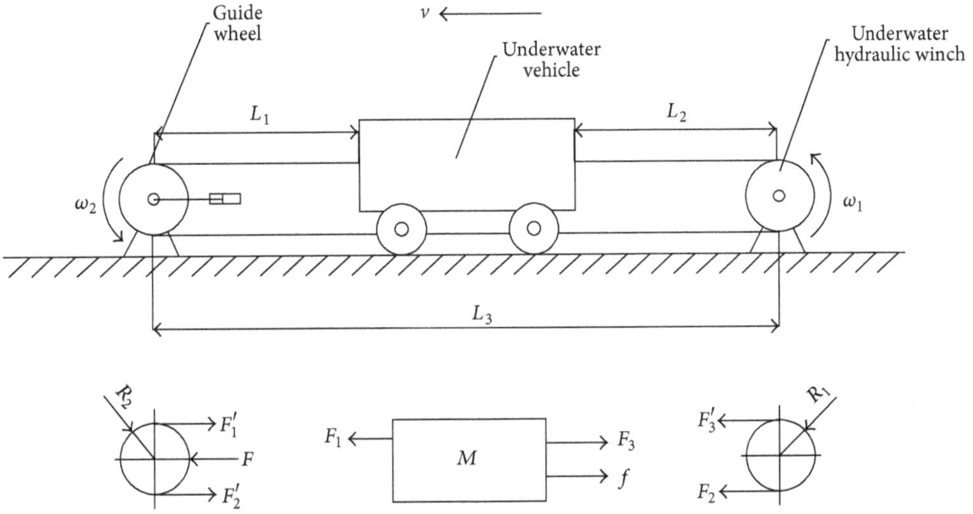

FIGURE 3: Schematic diagram of the underwater vehicle drafted by hydraulic winch.

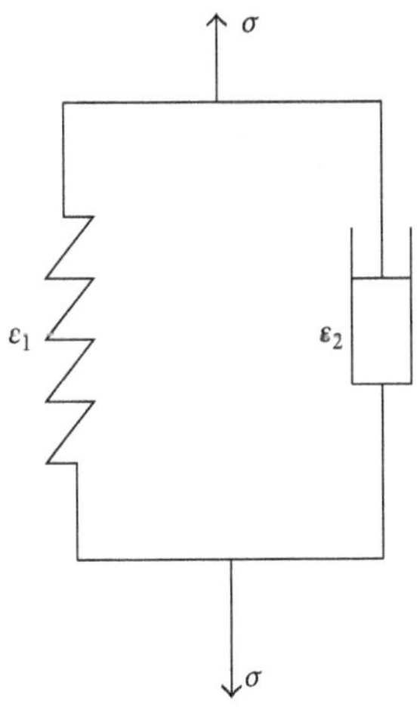

FIGURE 4: Kelvin-Voigt model.

Traditional static design assumes steel rope is a rigid body, yet the actual rope is twisted by multiple steel cores wired together, presenting more complex mechanical properties referred to as viscoelasticity when subjected to tension [9]. The Kelvin-Voigt model is employed in this study as a mechanical model of the rope (Figure 4). The model is also referred to as a nonrelaxation model as it is made with parallel springs and dampers and considers both elasticity and viciousness of the rope presented during stretching [10].

The simplified diagram of the model is a result of substituting the rope model in the closed flexible traction structure (Figure 5).

Elongation of each rope segment, referred to as static status, in the guide wheel tension force F is, respectively, denoted as x_{10}, x_{20}, and x_{30}.

The forces of each segment are as follows:

$$\begin{aligned} F_{10} &= k_1 x_{10}, \\ F_{20} &= k_2 x_{20}, \\ F_{30} &= k_3 x_{30}. \end{aligned} \tag{2}$$

When the hydraulic winch propels the vehicle, the forces for each segment are

$$\begin{aligned} F_1 &= F_{10} + \Delta F_1 = k_1 (x_{10} + \Delta x_1) + c_1 \cdot \Delta \dot{x}_1, \\ F_2 &= F_{20} + \Delta F_2 = k_2 (x_{20} + \Delta x_2) + c_2 \cdot \Delta \dot{x}_2, \\ F_3 &= F_{30} + \Delta F_3 = k_3 (x_{30} + \Delta x_3) + c_3 \cdot \Delta \dot{x}_3, \end{aligned} \tag{3}$$

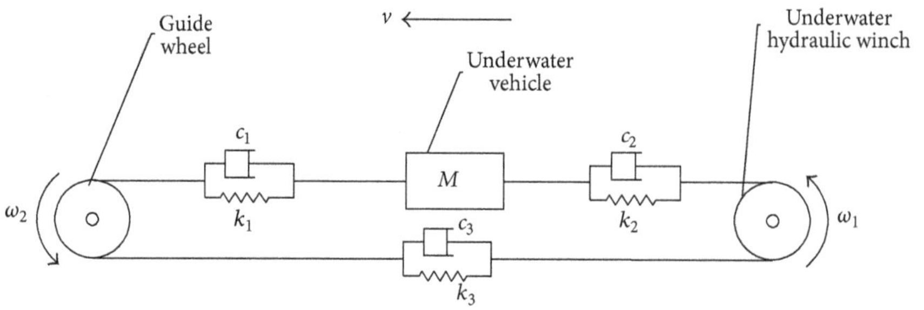

FIGURE 5: Simplified diagram of underwater vehicle propelled by hydraulic winch, where ω_1 is the angular speed of hydraulic winch rotation, R is the radium of drum, ω_2 is the guide wheel rotational speed, R represents radium, k_1 and c_1 are the elasticity and viciousness of L_1, k_2 and c_2 are the elasticity and viciousness of L_2, and k_3 and c_3 are the elasticity and viciousness of L_3.

where Δx_1, Δx_2, and Δx_3 are the elongations of each segment of rope while the vehicle moves and x_L is the distance of vehicle. Consider

$$\begin{aligned}\Delta x_1 &= \int \omega_2 dt \cdot R - x_L,\\ \Delta x_2 &= \int \omega_1 dt \cdot R - \int \omega_2 dt \cdot R,\\ \Delta x_3 &= x_L - \int \omega_1 dt \cdot R.\end{aligned} \tag{4}$$

Substitute above equations into (1) and let it be simplified into

$$\begin{aligned}F_1 - F_3 &= k_1 (x_{10} + \Delta x_1) + c_1 \cdot \Delta \dot{x}_1 - k_3 (x_{30} + \Delta x_3)\\ &\quad - c_3 \cdot \Delta \dot{x}_3\\ &= k_1 \cdot \Delta x_1 + c_1 \cdot \Delta \dot{x}_1 - k_3 \cdot \Delta x_3 - c_3 \cdot \Delta \dot{x}_3\\ &= M \cdot \dot{v}_L,\end{aligned} \tag{5}$$

where set L_1, L_3 are equally forced under pretension $F_{10} = F_{30}$.

Equation (5) can be written as:

$$\begin{aligned}&k_1 \cdot \Delta x_1 (s) + c_1 \cdot \Delta x_1 (s) s - k_3 \cdot \Delta x_3 (s) - c_3 \cdot \Delta x_3 (s) s\\ &\quad = M \cdot v_L (s) s.\end{aligned} \tag{6}$$

Substituting (4) in (6) results in:

$$\begin{aligned}&k_1 \cdot (\omega_2 (s) \cdot R - v_L (s)) + c_1 \cdot (\omega_2 (s) s \cdot R - v_L (s) s)\\ &\quad - k_3 \cdot (v_L (s) - \omega_1 (s) \cdot R) - c_3\\ &\quad \cdot (v_L (s) s - \omega_1 (s) s \cdot R) = M \cdot v_L (s) s^2.\end{aligned} \tag{7}$$

In closed flexible structure traction, the relationship between the underwater vehicle speed and hydraulic winch speed is

$$\frac{v_L (s)}{\omega (s)} = \frac{(2cs + 2k) \cdot R}{Ms^2 + 2cs + 2k}, \tag{8}$$

where set steady speed, the hydraulic winch, and guide wheel have equal rotational speed, $\omega(s) = \omega_1(s) = \omega_2(s)$. Two segments share the same properties, elasticity coefficient $k = k_1 = k_3$, viscosity coefficient $c = c_1 = c_3$, and v_L is motion speed of the underwater vehicle.

As it can be seen from (8), the relationship between the underwater vehicle speed and hydraulic winch speed showing second-order nonlinearity differs between a closed flexible traction system and a rigidly connected traction system due to viscoelastic properties of the rope. The second-order system, resulting from viscoelasticity of the rope, reduces stiffness of the traction system adding difficulty to underwater vehicle speed control.

2.2. Modeling of Hydraulic Winch System. A valve-controlled hydraulic winch is utilized as the driving mechanism of the underwater vehicle in this study.

Hydraulic motor dynamic mathematical model is constructed by the hydraulic valve flow equation, hydraulic motor continuity equation, and the motor torque balance equations [11].

(1) Hydraulic Valve Flow. Assumed valve is in ideal condition. Hydraulic valve flow equation is a typical nonlinear equation:

$$Q_L = c_d \omega x_v \sqrt{\frac{1}{\rho} (p_s - p_L \operatorname{sgn} (x_v))}, \tag{9}$$

where Q_L is load flow, c_d is the valve orifice flow coefficient, ω is the valve area gradient, p_s is oil pressure of hydraulic pump supply, p_L is the load pressure, and ρ is hydraulic oil density.

(2) Hydraulic Motor Continuity Equation. The following assumptions are made in the analysis of the hydraulic motor continuity equation: all pipes are short and thick; pipe friction loss, pipeline fluid dynamic impact, and quality may be negligible.

Provided Q_1 is the quantity flow into oil chamber of the hydraulic motor and Q_2 is the quantity flow from oil chamber of the hydraulic motor, there is

$$\begin{aligned}Q_1 &= C_{\text{im}} (p_1 - p_2) + C_{\text{em}} p_1 + D_{\text{m}} \frac{d\theta_{\text{m}}}{dt} + \frac{V_1}{\beta_e} \frac{dp_1}{dt},\\ Q_1 &= C_{\text{im}} (p_1 - p_2) - C_{\text{em}} p_2 + D_{\text{m}} \frac{d\theta_{\text{m}}}{dt} - \frac{V_2}{\beta_e} \frac{dp_2}{dt}.\end{aligned} \tag{10}$$

Working volume of the hydraulic motor may be, respectively, described as

$$V_1 = V_{01} + D_{\mathrm{m}}\theta_{\mathrm{m}},$$
$$V_2 = V_{02} + D_{\mathrm{m}}\theta_{\mathrm{m}}, \tag{11}$$

where V_{01} and V_{02} were initial volume of motor oil chamber and back oil chamber.

Since the leakage and compression resistance of the hydraulic motor, quantity flow into oil chamber Q_1, and quantity flow from oil chamber Q_2 are different, in order to simplify the analysis, define load flow Q_L as

$$\begin{aligned} Q_L &= \frac{Q_1 + Q_2}{2} \\ &= D_{\mathrm{m}}\frac{d\theta_{\mathrm{m}}}{dt} + \left(C_{\mathrm{im}} + \frac{1}{2}C_{\mathrm{em}}\right)p_L \\ &\quad + \frac{1}{2}\left(\frac{V_1}{\beta_e}\frac{dp_1}{dt} - \frac{V_2}{\beta_e}\frac{dp_2}{dt}\right) \\ &\quad + \frac{D_{\mathrm{m}}\theta_{\mathrm{m}}}{2\beta_e}\left(\frac{dp_1}{dt} + \frac{dp_2}{dt}\right), \end{aligned} \tag{12}$$

where D_{m} is the theoretical flow hydraulic motor; θ_{m} is the angle of the hydraulic motor shaft; C_{im} is the bypass leakage coefficient for internal hydraulic motor; C_{em} is the leakage coefficient for external hydraulic motors; V_1 is oil chamber volume of hydraulic motor; and V_2 is the back oil chamber volume of hydraulic motor.

Assume the inertial volume of oil chamber and back oil chamber as

$$V_{01} = V_{02} = V_0 = \frac{1}{2}V_{\mathrm{m}}, \tag{13}$$

where V_{m} is the total volume of hydraulic valve chamber, motor chamber, and connecting pipes. Because $D_{\mathrm{m}}\theta_{\mathrm{m}} \ll V_0$, $p_s = p_1 + p_2 = C$ (constant), $p_L = p_1 - p_2$, but also $dp_1/dt + dp_2/dt = 0$, formula (12) may be written as

$$Q_L = D_{\mathrm{m}}\frac{d\theta_{\mathrm{m}}}{dt} + C_{\mathrm{tm}}p_L + \frac{V_{\mathrm{m}}}{4\beta_e}\frac{dp_L}{dt}, \tag{14}$$

where C_{tm} is the total leakage coefficient of hydraulic motor $C_{\mathrm{tm}} = C_{\mathrm{im}} + (1/2)C_{\mathrm{em}}$ and β_e is elasticity modulus for the hydraulic oil.

(3) Motor Torque Balance Equations. Dynamic characteristics of the hydraulic motor power components are affected by load characteristics. Load force typically includes inertial forces, viscous damping force, elastic force, and arbitrary external load force. Disregarding the effects of nonlinear loads and oil quality of friction, according to Newton's Law, hydraulic motor torque equilibrium equation may be drawn as

$$T_s = p_L D_{\mathrm{m}} = J_{\mathrm{m}}\frac{d^2\theta_{\mathrm{m}}}{dt^2} + B_{\mathrm{m}}\frac{d\theta_{\mathrm{m}}}{dt} + G\theta_{\mathrm{m}} + T_L, \tag{15}$$

where T_s is the theoretical torque generated by hydraulic motor; J_{m} is the total inertia of the hydraulic motor and load;

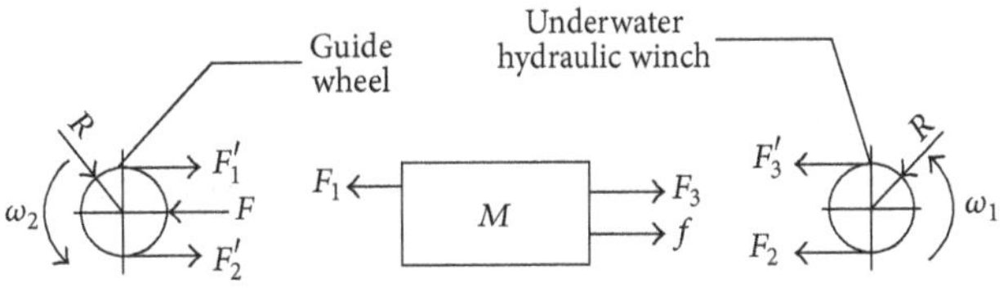

FIGURE 6: The force analysis of the closed flexible traction structure.

B_{m} is the viscous damping coefficient of load and hydraulic motor; G is the load torsion spring stiffness; and T_L is the external load torque acting on the motor shaft.

Equations (9), (14) and (15) can be written as:

$$\begin{aligned} Q_L &= K_q x_v - K_c p_L, \\ Q_L &= D_{\mathrm{m}} s\theta_{\mathrm{m}} + C_{\mathrm{tm}}p_L + \frac{V_t}{4\beta_e}sp_L, \\ T_s &= p_L D_{\mathrm{m}} = J_t s^2\theta_{\mathrm{m}} + B_{\mathrm{m}}s\theta_{\mathrm{m}} + G\theta_{\mathrm{m}} + T_L. \end{aligned} \tag{16}$$

Eliminating the middle term, the transform function of the valve opening degree and vehicle speed can be obtained [12] as follows:

$$\dot{\theta}_{\mathrm{m}} = \frac{\left(K_q/D_{\mathrm{m}}\right)x_v - \left(1/D_{\mathrm{m}}^2\right)\left(K_{ce} + \left(V_{\mathrm{m}}/4\beta_e\right)s\right)\cdot T_L}{s^2/\omega_h^2 + \left(2\zeta_h/\omega_h\right)s + 1}, \tag{17}$$

where $\omega_h = \sqrt{K_h/J} = \sqrt{4\beta_e D_{\mathrm{m}}^2/V_{\mathrm{m}}J}$ is equivalent hydraulic natural frequency; $\zeta_h = (K_{ce}/D_{\mathrm{m}})\sqrt{\beta_e J/V_{\mathrm{m}}} + (B_{\mathrm{m}}/4D_{\mathrm{m}})\sqrt{V_{\mathrm{m}}/\beta_e J}$ is damping ratio of hydraulic valve-controlled motor; and $K_{ce} = K_c + C_{\mathrm{im}} + (1/2)C_{\mathrm{em}}$ is coefficient of flow-pressure.

External load is the key factor of the underwater vehicle's speed control from the analysis above.

2.3. Modeling of External Load. The force analysis of the closed flexible traction structure is described as shown in Figure 6.

When underwater hydraulic winch drives the vehicle to move, the winch is a driving wheel and the guide wheel is a drive wheel. The force of guide wheel is the sum of driving force F_2 and resistance F_1, while the rotational speed of guide wheel is ω_2. The torque equilibrium equation of guide wheel may be obtained as

$$\left(F_2 - F_1\right)\cdot R = J_2\frac{d\omega_2}{dt} + B_1\omega_2, \tag{18}$$

where F_1 is the tension force of L_1, F_2 is the tension force of L_2, J_2 is the torque of guide wheel, B_2 is the friction coefficient between vehicle wheels and guide rail, and ω_2 is rotational speed of the guide wheel.

The dynamics formula of the underwater vehicle may be obtained utilizing Newton's Law:

$$F_1 - F_3 = M\frac{dv_L}{dt} + B_2 v_L + f, \tag{19}$$

where F_1 is the tension force of L_1, F_3 is the tension force of L_3, f is the water resistance force applied on the vehicle, M is the mass of the underwater vehicle, B_2 is the friction coefficient between vehicle wheels and the guide rail, and v_L is motion speed of the underwater vehicle.

The hydraulic winch is a driving wheel and the torque equilibrium equation may be obtained as

$$T_L = (F_2 - F_3) \cdot \frac{R}{n}, \tag{20}$$

where n is speed ratio of the hydraulic winch and R is the radium.

Substituting (18) and (19) in (20), the torque equilibrium equation of hydraulic winch may be written as

$$\begin{aligned} T_L &= [(F_2 - F_1) + (F_1 - F_3)] \cdot \frac{R}{n} \\ &= \left(\frac{J_2 (d\omega_2/dt) + B_1\omega_2}{R} + M\frac{dv_L}{dt} + B_2 v_L + f \right) \cdot \frac{R}{n} \\ &= \frac{J_2}{n}\frac{d\omega_2}{dt} + \frac{B_1\omega_2}{n} + \left(M\frac{dv_L}{dt} + B_2 v_L + f \right) \cdot \frac{R}{n}. \end{aligned} \tag{21}$$

Equation (21) can be written as:

$$\begin{aligned} T_L(s) &= [J_2 \cdot \omega_2(s)\, s + B_1 \cdot \omega_2(s) + MR \cdot v_L(s)\, s \\ &\quad + B_2 R \cdot v_L(s) + R \cdot f(s)] \cdot \frac{1}{n}. \end{aligned} \tag{22}$$

The relationship between the underwater vehicle speed and hydraulic winch speed from the analysis above is

$$\frac{v_L(s)}{\omega(s)} = \frac{(2cs + 2k) \cdot R}{Ms^2 + 2cs + 2k} \tag{23}$$

and may be written as

$$\frac{\omega(s)}{v_L(s)} = \frac{Ms^2 + 2cs + 2k}{(2cs + 2k) \cdot R}. \tag{24}$$

Substituting (24) in (22), the torque equilibrium equation of hydraulic may be obtained as

$$\begin{aligned} T_L(s) &= \left[(J_2 s + B_1) \cdot \frac{Ms^2 + 2cs + 2k}{(2cs + 2k) \cdot R} + (MRs + B_2 R) \right] \cdot \frac{v_L(s)}{n} + \frac{R}{n} \cdot f(s) \\ &= \frac{MJ_2 s^3 + \left(MB_1 + 2cJ_2 + 2cMR^2\right) s^2 + \left(2B_1 c + 2kJ_2 + 2cB_2R^2 + 2kMR^2\right) s + \left(2kB_1 + 2kB_2R^2\right) s + \left(2kB_1 + 2kB_2R^2\right)}{(2cs + 2k) \cdot nR} \\ &\quad \cdot v_L(s) + \frac{R}{n} \cdot f(s), \end{aligned} \tag{25}$$

where $T_L(s)$ is the torque equilibrium equation of hydraulic winch, $v_L(s)$ is motion speed of the underwater vehicle, and $f(s)$ is the water resistance force applied on the vehicle.

The software application, "Gambit," was applied to calculate water resistance force on the vehicle. Analyzing the shape of the underwater vehicle, drag force coefficients may be derived as

$$C_d = 0.275. \tag{26}$$

Water resistance force applied on the vehicle may be written as

$$f = \frac{1}{2} C_d \rho A v_L^2, \tag{27}$$

where C_d represents drag force coefficients of the underwater vehicle, ρ is density of water, A is horizontal projection area of underwater vehicle, and v_L is motion speed of the underwater vehicle.

The hydraulic winch external load torque in the closed flexible traction driving system (25) is affected by both the underwater vehicle speed and water resistance. The speed of the vehicle is variable during the test and the value of water resistance is related to the speed; thus the external load torque of hydraulic winch is variable.

2.4. Modeling of the Underwater Vehicle Speed Control System. Combining (17), (24), and (25), the transfer function diagram of the underwater vehicle speed control system may be obtained (Figure 7).

Compared with the conventional hydraulic winch speed servocontrol system, since steel rope is introduced as traction material, the viscoelastic properties of the steel rope present second-order system characteristics on the forward channel of the speed control system, reducing system stiffness and challenging stability for underwater vehicle speed control. The closed flexible structure and underwater environment result in certain relevance between traction load characteristics and underwater vehicle speed, while the underwater vehicle constantly changes speed through entire experiment; the load characteristic of hydraulic system is also variable and

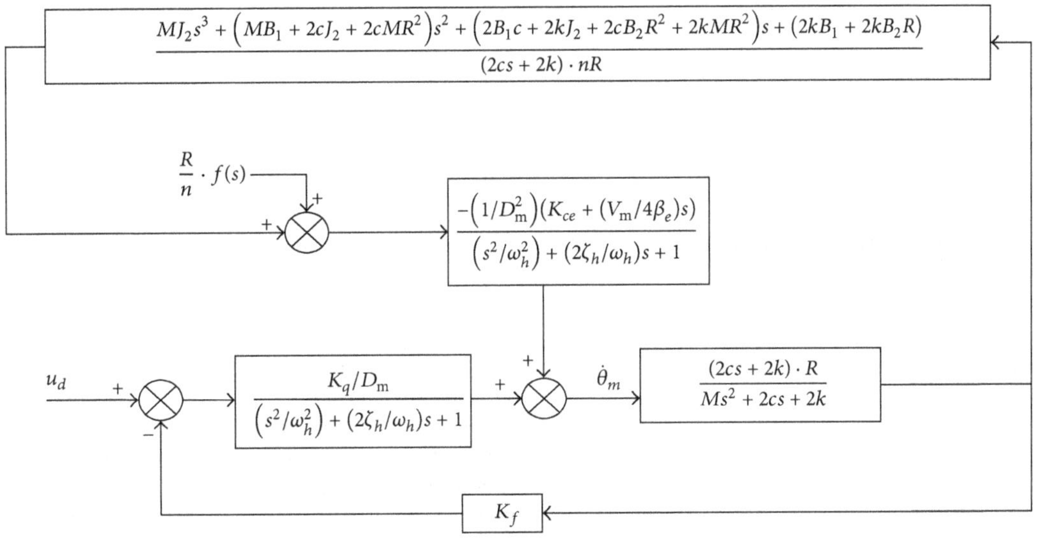

FIGURE 7: The transfer function diagram of the underwater vehicle speed control system.

consequently brings difficulty on accuracy of the underwater vehicle speed control.

3. Research on Fuzzy Sliding Mode Control

The speed control system of the underwater vehicle presents low rigidity, time-varying, and nonlinear parameters, according to the above analysis. Reliability and responsiveness for stability and accuracy of the control system then must be considered when choosing a control algorithm.

Sliding mode variable structural control theory is selected as theoretical support of the underwater vehicle speed control method in this study. Sliding mode variable structure control is essentially a special type of nonlinear control, since the discontinuity of control output. Differing from other control methods, the sliding mode control strategy features a variable system structure that evolves according to the current system state, forcing the system to move along the designed "sliding mode" state track. Sliding mode design is irrelevant to object parameters; thus it is highly responsive and insensitive to parameter variables and disturbance and shows good robustness [13].

3.1. Sliding Mode Control Based on Reaching Law. Sliding mode system motion consists of two states: first, in the normal motion state, the state space trajectory occurs outside the switching surface or across the switching surface by limited time. Second, in the sliding motion state, the system moves near the switching surface in the sliding mode. Sliding control requires high motion quality throughout the entire control process of both the normal state and the sliding mode state.

According to the sliding mode variable structure principle, the sliding reachability condition is only guaranteed when the state space reaches the switching surface from any points in limited time, but approaching trajectory is not restricted; the reaching law method can improve the dynamic quality of approaching movement. The second-order electrohydraulic system may be applied as an example to derive the sliding mode control algorithm based on exponential reaching law [14].

Consider a second-order electrohydraulic system state equation

$$\dot{x} = Ax + Bu + d, \tag{28}$$

where $x = \binom{x_1}{x_2}$ is system state variable, $A = \begin{pmatrix} a_{11} & a_{12} \\ a_{21} & a_{22} \end{pmatrix}$ and $B = \binom{b_1}{b_2}$ are system state parameters, and d is disturbance.

Then $\ddot{x} = a_{21}x_1 + a_{22}x_2 + b_2u + d_2$.

Setting a given value r, the error signal is

$$\begin{aligned} e &= r - x_1, \\ \dot{e} &= \dot{r} - x_2. \end{aligned} \tag{29}$$

Select the sliding mode control switching function as

$$S = CE = ce + \dot{e}, \tag{30}$$

where the value of the parameter c should satisfy Hurwitz conditions.

The derivation of the above equation may be obtained as

$$\begin{aligned} \dot{S} &= C\dot{E} = c\dot{e} + \ddot{e} = c(\dot{r} - x_2) + (\ddot{r} - \dot{x}_2) \\ &= c(\dot{r} - x_2) + (\ddot{r} - a_{21}x_1 - a_{22}x_2 - b_2u - d_2) \\ &= \text{rlaw}. \end{aligned} \tag{31}$$

Variable structure control law for second-order state space model is

$$u = \frac{1}{b_2}\left[\ddot{r} + c(\dot{r} - x_2) - a_{21}x_1 - a_{22}x_2 - d_2 - \text{rlaw}\right]. \tag{32}$$

As deducible from above, the conventional sliding mode control's approaching movement characteristics have not

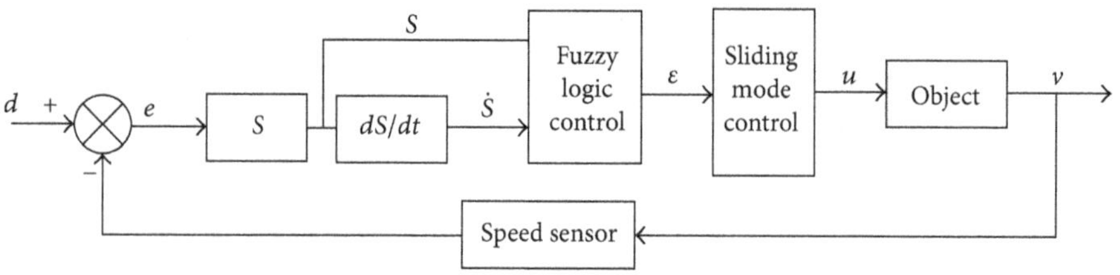

Figure 8: Sliding mode based on fuzzy reaching law control schematics.

been restricted; thus adopting reaching law is an effective method for improving motion characteristics of the approaching movement.

Here adopted exponential reaching law [15] is

$$r\text{law} = -\varepsilon \operatorname{sgn}(s) - ks, \quad \varepsilon > 0,\ k > 0. \tag{33}$$

There is

$$u = \frac{1}{b_2}\left[\ddot{r} + c\left(\dot{r} - x_2\right) - a_{21}x_1 - a_{22}x_2 - d_2 + \varepsilon \operatorname{sgn}(s) + ks\right]. \tag{34}$$

Since disturbance d_2 is unknown, the above control law cannot be achieved.

Set $d_L \le d_2(t) \le d_U$; this requires interference sector to design control law:

$$u = \frac{1}{b_2}\left[\ddot{r} + c\left(\dot{r} - x_2\right) - a_{21}x_1 - a_{22}x_2 - d_c + \varepsilon \operatorname{sgn}(s) + ks\right], \tag{35}$$

where $d_c = (d_U + d_L)/2 - ((d_U - d_L)/2)\operatorname{sgn}(s)$.

Conditions existing to satisfy sliding mode include the following:

(1) When $S > 0$, $\dot{S} = -\varepsilon - \operatorname{sgn}(s) + d_c - d$, for $\dot{S} < 0$, made $d_c = d_U$.

(2) When $S < 0$, $\dot{S} = -\varepsilon - \operatorname{sgn}(s) + d_c - d$, for $\dot{S} < 0$, made $d_c = d_L$.

Sliding mode control, from the above analysis, based on exponential reaching law, features superior response and stability; however, because of the switching function $\varepsilon \operatorname{sgn}(s)$ introduction, output characteristics vary from conventional control. $\varepsilon \operatorname{sgn}(s)$ increases robustness of the control system while increasing the value of ε and the time of system state move into the sliding mode can be reduced while $|S| < 1$, thereby enhancing the dynamic quality of sliding mode control; however, a larger value of ε will bring more significant control output chattering, as well as larger impact on implementation structure and control objects. Reasonable selection of switch function coefficients is particularly essential for sliding mode robust control based on reaching law.

3.2. Sliding Mode Control Based on Fuzzy Reaching Law. A reaching law method based on fuzzy logic is presented in this study to solve the contradiction in values of the sliding mode control switching function coefficient. Rapid response and stability of control are met as fuzzy logic control is combined with reaching law to reduce system chattering while considering slip mode stability control [16].

Figure 8 features the fuzzy controller and sliding mode controller. Fuzzy controller is a two-dimensional structure fuzzy control with double inputs and single output, switching function S and derivation $\dot{S}$ as the inputs, and is based on fuzzy rules exponential reaching law output real-time adjust parameters ε. Sliding mode variable structure controller then derives underwater speed control of test vehicle based on the changing value ε.

Exponential reaching law parameter ε may be dynamically adjusted based on the relationship between S and $\dot{S}$. When system status values are further from the switching surface, a greater amount of control is required to make the system move fast to switching surface; when $|S\dot{S}|$ reflects a small value, the system has moved to the nearby switching surface and, to prevent significant chattering phenomenon, control output must be reduced for the system to move smoothly to the surface.

4. Digital Simulation and Practical Testing

Following completion of the control algorithm design, verification and optimization of the control scheme must be conducted. The built mathematical model of the hydraulic system is utilized to realize digital simulation of the control algorithm. Practical tests then must be conducted to verify the control effect of the hydraulic system.

4.1. Model Simplification. According to design and selection of the hydraulic winch, the main nominal parameters at a given speed of the hydraulic servosystem are displayed in Table 1.

Transfer function diagram of the underwater vehicle speed control system may then be obtained (Figure 9).

Transfer function of the underwater vehicle speed control system may be written as

$$v_L(s) = \frac{2.89s + 1.534 \times 10^2}{4.218s^2 + 3.115s + 1.594 \times 10^3}u(s) - \frac{14.85s^2 + 538s + 43.68}{2.2s^2 + 1.6s + 880}v(s) - \frac{1.406s + 51.07}{625s^2 + 4.54 \times 10^2 s + 2.4 \times 10^5}f(s), \tag{36}$$

where $v_L(s)$ is motion speed of the underwater vehicle, $u(s)$ is the control input of hydraulic winch, $v(s)$ is motion speed of

TABLE 1: The main nominal parameters of hydraulic system.

Number	Parameter	Note	Unit	Nominal
1	Drum radius	R	m	0.768
2	Winch reduction ratio	N		600
3	Flow spool gain	K_q		0.003
4	Proportional valve gain	K	m/V	1.25×10^{-3}
5	Orifice flow coefficient	C_d		0.61
6	Valve area gradient	Ω	m	0.785
7	Hydraulic oil pressure	P_s	Pa	1.5×10^{7}
8	Hydraulic oil density	P	kg/m^3	850
9	Hydraulic motor displacement	D_m	m^3/rad	1.178×10^{-4}
10	Tank volume	V_t	m^3	1.47×10^{-3}
11	Bulk modulus	β_e	Pa	7.0×10^{8}
12	Total flow pressure coefficient	K_{ce}	m^5/(N·s)	1.9×10^{-11}
13	Vehicle mass	M	kg	3000
14	Horizontal projection area of underwater vehicle	A	m^2	12.99

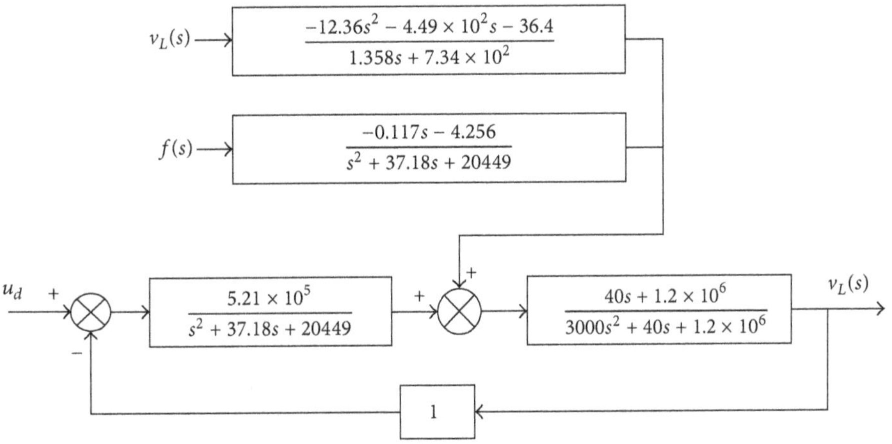

FIGURE 9: The transfer function diagram of the underwater vehicle speed control system.

the underwater vehicle on last measuring period, and $f(s)$ is water resistance force applied on the vehicle.

The function can be simplified as

$$v_L(s) = \frac{3240u(s) - (20.59s + 16.7)\,v(s) - 0.716f(s)}{2.613s^2 + 1.907s + 336.7}. \quad (37)$$

The state variables of the underwater vehicle speed control system may be chosen as $x_1 = v_L$ and $x_2 = \dot{x}_1$; then the state equation may be obtained as

$$\begin{aligned} x_1 &= v_L, \\ \dot{x}_1 &= x_2, \\ \dot{x}_2 &= -128.8x_1 - 0.729x_2 + 1239u \\ &\quad - (7.87\dot{v} + 6.39v + 0.27f), \end{aligned} \quad (38)$$

where $v(s)$ is motion speed of the underwater vehicle on last measuring period and $f(s)$ is water resistance force applied on the vehicle.

Three different algorithms have been utilized in digital simulation (Figure 10) to realize the speed control system of the underwater vehicle above.

4.2. Simulation Based on PID Control Law. Figure 11 is a digital simulation diagram utilizing PID control. A speed tracking curve is depicted on the left and a control input curve of the hydraulic winch is depicted on the right. The set speed increases to 2 m/s in 1 s and remains uniform. PID control law ultimately completes stable speed control of the underwater vehicle with relatively stable control input and a slightly longer stabilization time (approximately 5 s).

A digital simulation diagram utilizing PID control at variable speeds is displayed in Figure 12. Sinusoidal signal is selected as set speed (max speed is 1 m/s) in this simulation. Control requirements experience difficulty due to a delay as the PID controller shows a less sensitive response in tracking the sinusoidal signal.

4.3. Simulation Based on Sliding Mode Control. The sliding surface is $s = 15e + \dot{e}$.

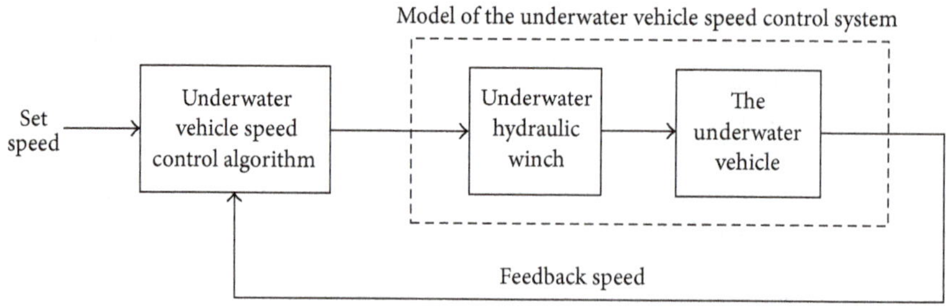

FIGURE 10: The schematic diagram of digital simulation.

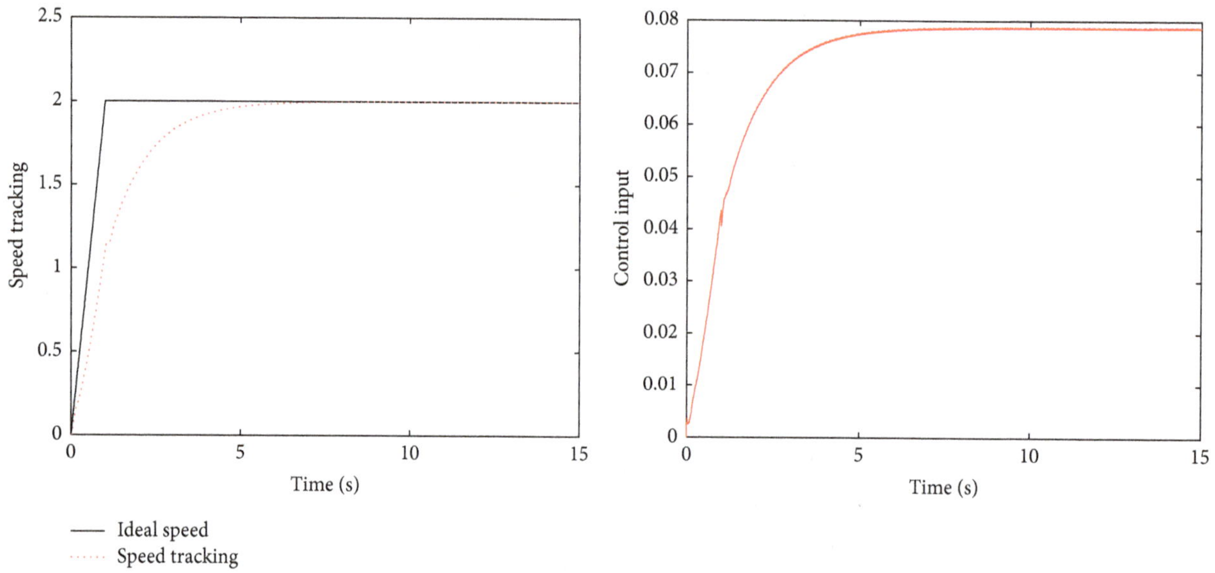

FIGURE 11: Simulation curve based on PID control law at constant speed.

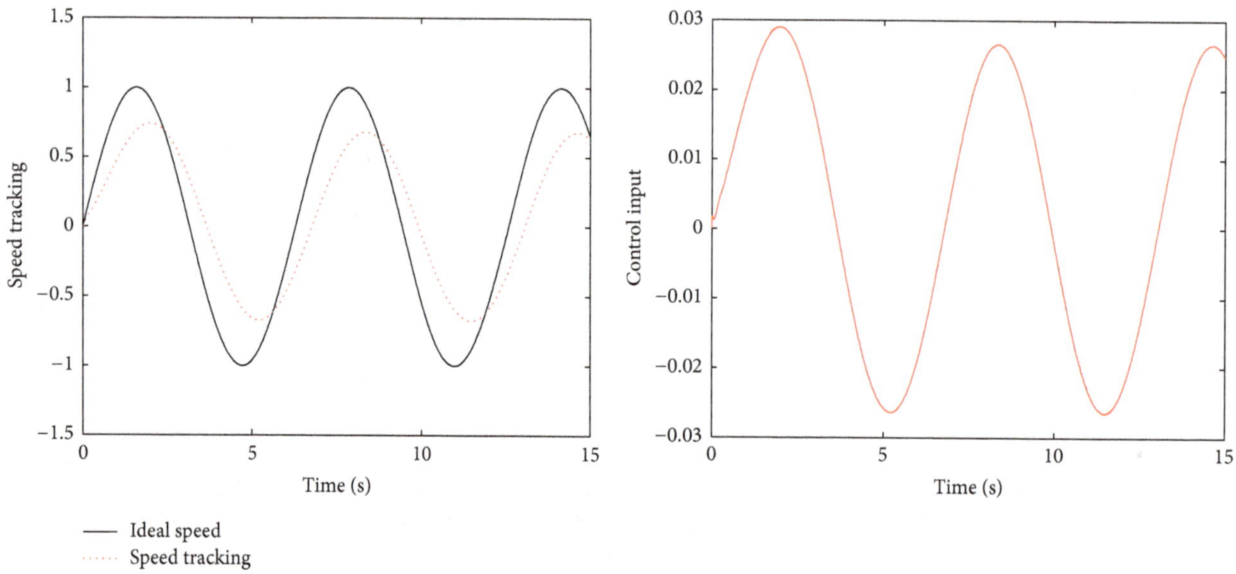

FIGURE 12: Simulation curve based on PID control law at variable speed.

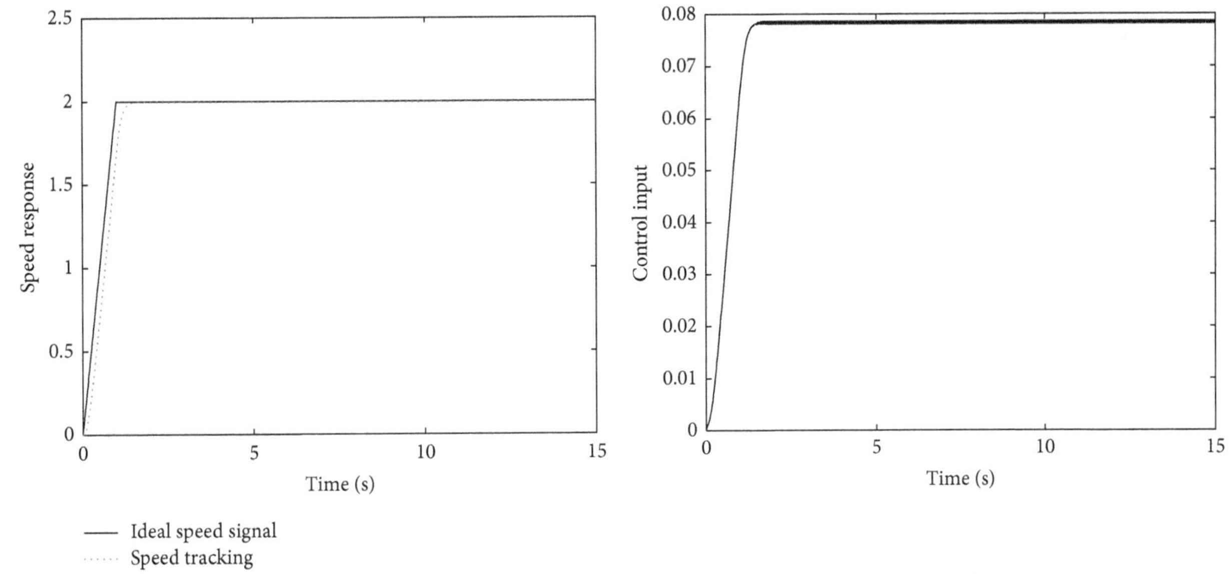

FIGURE 13: Simulation curve based on sliding mode control at constant speed.

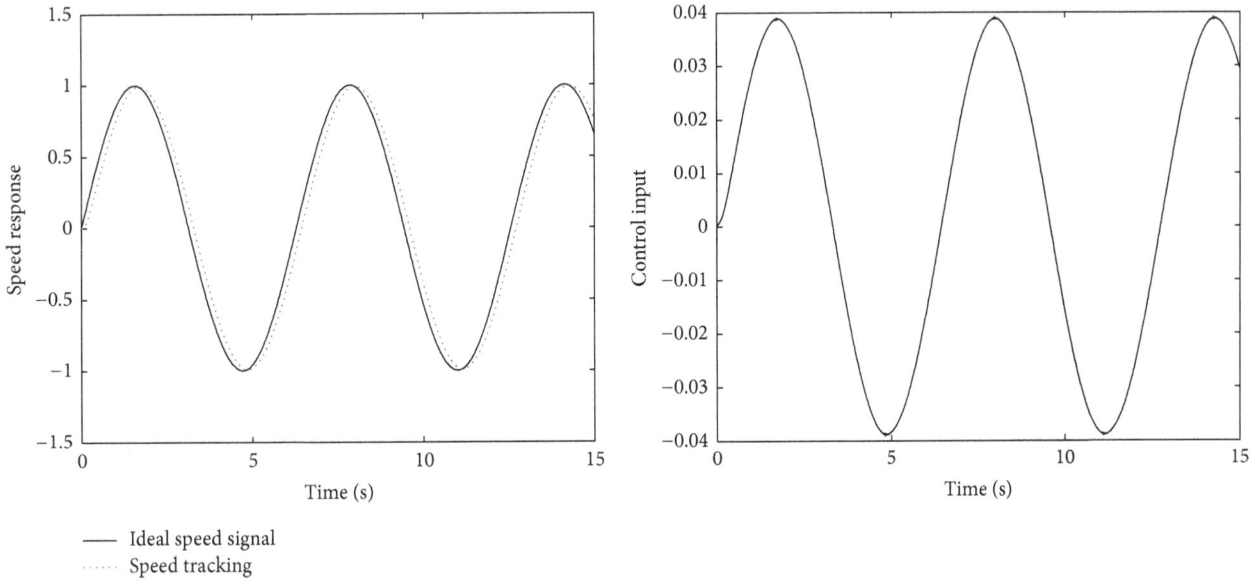

FIGURE 14: Simulation curve based on sliding mode control at variable speed.

The state equation of the underwater vehicle speed control system is

$$\begin{aligned} x_1 &= v_L, \\ \dot{x}_1 &= x_2, \\ \dot{x}_2 &= -128.8x_1 - 0.729x_2 + 1239u \\ &\quad - \left(7.87\dot{v} + 6.39v + 0.27f\right). \end{aligned} \tag{39}$$

Substituted into (35),

$$u = \frac{1}{b_2}\left[\ddot{r} + c\left(\dot{r} - x_2\right) - a_{21}x_1 - a_{22}x_2 - d_c + r\text{law}\right], \tag{40}$$

where the reaching law is $r\text{law} = -5\,\text{sgn}(s) - 10s$.

The control input of sliding mode control based on reaching law may be obtained.

Figure 13 depicts digital simulation based on sliding mode control utilizing exponential reaching law. Observations from studying the speed tracking curve show that the sliding mode variable structure control law exhibits effective control for the variable structure and the variable load of the underwater vehicle speed control system. Introduction of reaching law provides the sliding mode variable structure controller improved response quality and speed tracking effects. The control input exhibits slightly high frequency chattering in the uniform motion section, negatively impacting the control object.

Variable motion simulation based on sliding mode control utilizing exponential reaching law is pictured in Figure 14. As observed from the speed tracking curve,

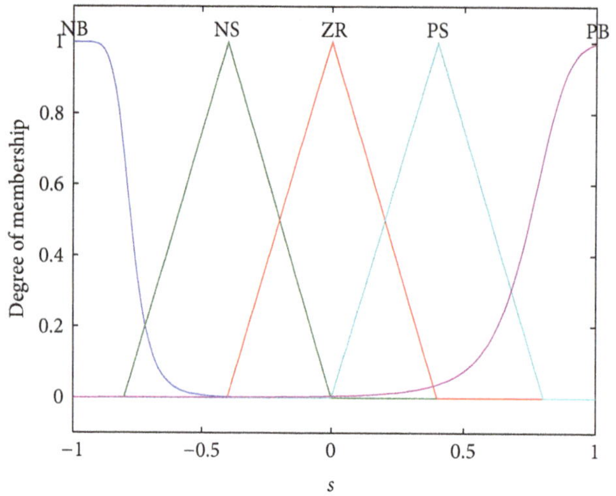

FIGURE 15: Membership function of fuzzy input s.

the algorithm exhibits satisfactory characteristics for interference, robustness, and control effect; however, from the control input curve, due to the switching function, substantial high frequency chattering of control input significantly impacts the actuator and control objects, adding difficulty to field trial applications.

4.4. Simulation Based on Fuzzy Sliding Mode Control. The sliding surface is $s = 15e + \dot{e}$.

The state equation of the underwater vehicle speed control system is

$$\begin{aligned} x_1 &= v_L, \\ \dot{x}_1 &= x_2, \\ \dot{x}_2 &= -128.8x_1 - 0.729x_2 + 1239u \\ &\quad - (7.87\dot{v} + 6.39v + 0.27f). \end{aligned} \tag{41}$$

Substituted into (35),

$$u = \frac{1}{b_2}\left[\ddot{r} + c(\dot{r} - x_2) - a_{21}x_1 - a_{22}x_2 - d_c + \text{rlaw}\right], \tag{42}$$

where the reaching law is $r\text{law} = -5 \cdot \varepsilon\,\text{sgn}(s) - 10s$, where ε is fuzzy control output.

As observed in Figures 15–17, the input and output fuzzy sets are as follows:

$$\begin{aligned} s &= \{\text{NB}, \text{NS}, \text{ZR}, \text{PS}, \text{PB}\}, \\ \dot{s} &= \{\text{NB}, \text{NS}, \text{ZR}, \text{PS}, \text{PB}\}, \\ \varepsilon &= \{\text{NB}, \text{NM}, \text{NS}, \text{ZR}, \text{PS}, \text{PM}, \text{PB}\}. \end{aligned}$$

The membership function of fuzzy inputs and outputs is shown in Figures 15, 16, and 17.

Define output fuzzy rule table as [17].

Using the fuzzy rules in Table 2 to adjust the reaching law parameter ε in real time, the digital simulation results are as shown in Figures 18 and 19.

Figures 18 and 19 are the digital simulation sliding mode control utilizing exponential reaching law based on fuzzy

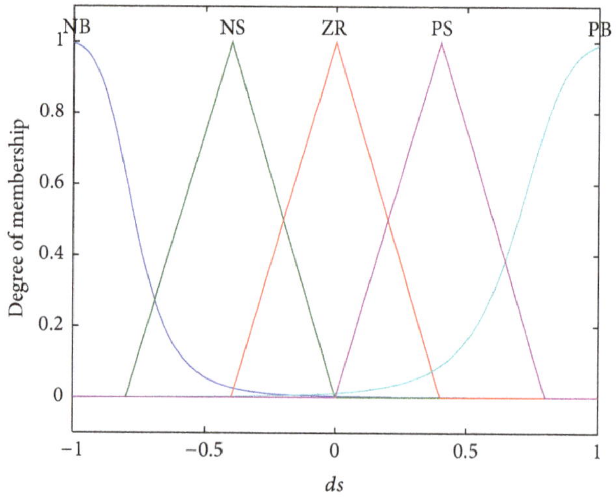

FIGURE 16: Membership function of fuzzy input $\dot{s}$.

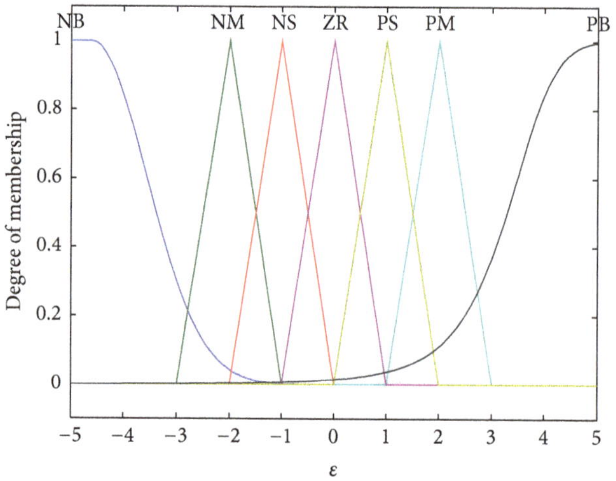

FIGURE 17: Membership function of fuzzy output ε.

logic. Figure 18 displays speed tracking curves of this control algorithm, demonstrating relatively effective control. The

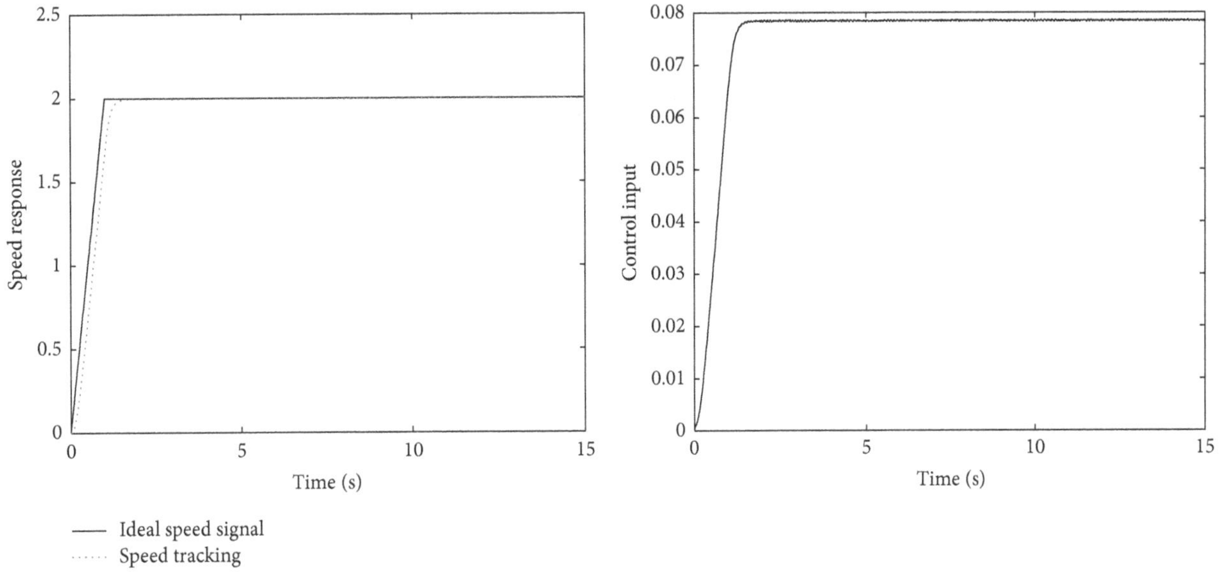

FIGURE 18: Simulation curve based on fuzzy sliding mode control at constant speed.

TABLE 2: Fuzzy rules of ε.

	S				
	NB	NS	ZR	PS	PB
$\dot{S}$					
NB	NB	NM	NS	PM	ZR
NS	NM	NS	NS	ZR	PM
ZR	NM	NS	ZR	PS	PM
PS	PS	ZR	PS	PM	PM
PB	ZR	PM	PS	PM	PB

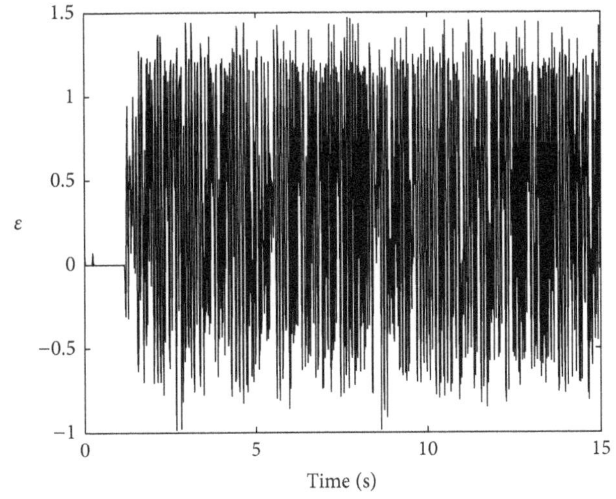

FIGURE 19: Fuzzy output ε.

control input characteristic curve is also shown to significantly improve compared with Figure 13. Introduction of fuzzy control theory (Figure 19) indicates smooth and stable control input with high reliability. Test results for the control algorithm also depict strong validation for the practical test.

Figure 20 displays the speed tracking curves and control input curve. Figure 21 displays the fuzzy control output characteristic curve. The control algorithm demonstrates effective control, especially related to variable speed motion control. The introduction of fuzzy control (Figure 21) smoothens the input of the control system and reveals sound output characteristics.

Digital simulations above reveal the following conclusions:

(1) Response of conventional PID algorithms for underwater variable speed motion is not sensitive. Difficulty is then experienced in meeting the requirements for precise control of the underwater vehicle.

(2) The sliding mode variable structure control system exhibits superior robustness over conventional continuous control systems. Output chattering adds difficulty, however, to variable structure control in application of the actual system.

(3) Sliding mode control of fuzzy adaptive reaching law exhibits sound robustness for overcoming disturbance and parameter uncertainties and the shortcomings of conventional sliding mode control. The system features greater control accuracy and superior real-time characteristics.

4.5. Practical Test. The underwater vehicle speed control unit calculates the opening degree value of valve based on the difference between the measured value and the given value of the underwater vehicle. The hydraulic winch propels the underwater vehicle with a closed flexible traction structure (Figure 22).

The underwater vehicle must be controlled to accelerate smoothly from the starting speed to the setting test speed and

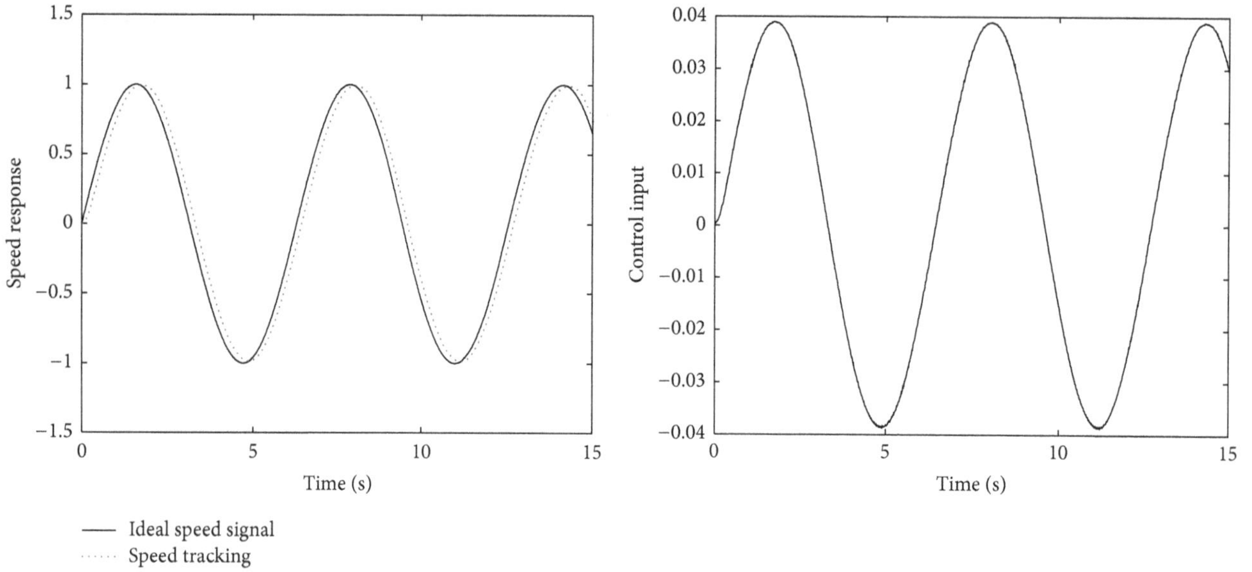

Figure 20: Simulation curve based on fuzzy sliding mode control at variable speed.

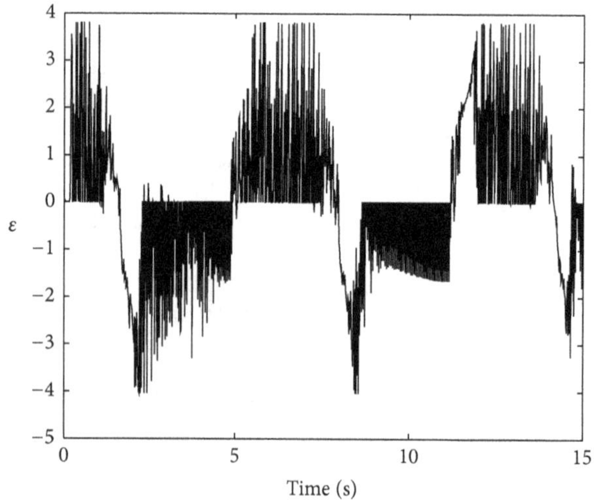

Figure 21: Fuzzy output ε.

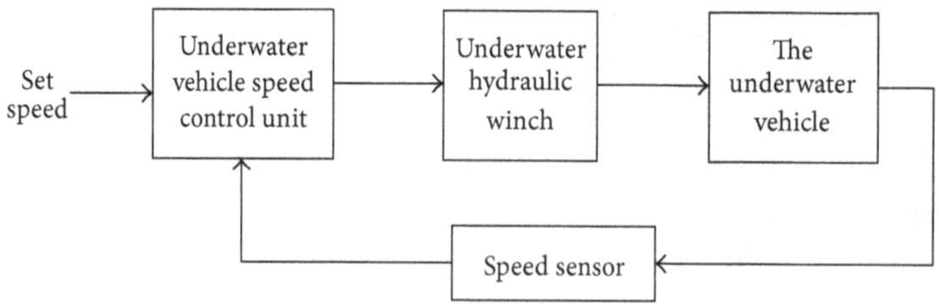

Figure 22: The schematic diagram of practical test.

to retain speed stability (less than 10% fluctuations). Speed of the underwater vehicle should decelerate smoothly to zero following completion of the related test. As it can be seen in Figure 23, t_1 is acceleration time, t_2 is test time, and t_3 is deceleration time.

The pictures illustrate practical testing: underwater hydraulic winch (Figure 24), underwater vehicle (Figure 25) and guide rails placed in self-submersible platform deck surface, and speed control unit (Figure 26) and underwater pump stations placed in the watertight pressure float on both sides of the platform. Siemens 400H redundant PLC was selected as the speed control unit processor in this study [18].

The underwater vehicle feedback speed curve is displayed at set speed 2.06 m/s (Figure 27). The curve indicates the

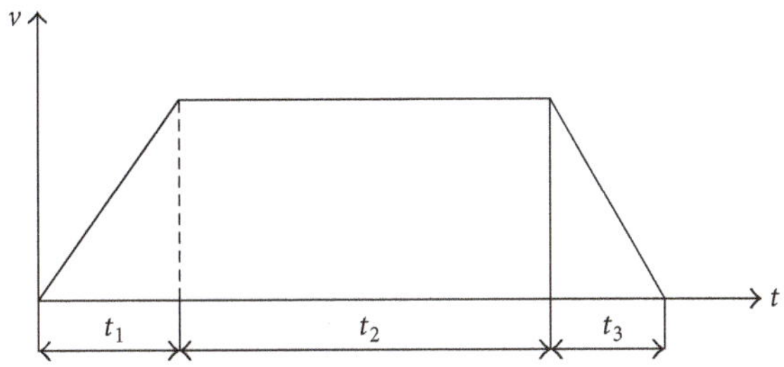

FIGURE 23: Underwater vehicle set speed curve.

FIGURE 24: Underwater hydraulic winch.

FIGURE 25: The underwater vehicle.

vehicle speed as reaching set speed in a relatively short period of time, remaining stable for a period of time to complete the test, and finally rapidly and smoothly reaching a stop. The sliding mode control based on fuzzy reaching law demonstrates effective results with the practical test.

5. Conclusion

Research is presented in this study based on speed control of the underwater vehicle as propelled by the hydraulic winch. Modeling analysis is conducted based on relevant features of the underwater vehicle as driven by the hydraulic winch, respectively, to the hydraulic driving system and enclosed flexible traction system. Flexibility of steel rope and features

FIGURE 26: Underwater vehicle speed control unit.

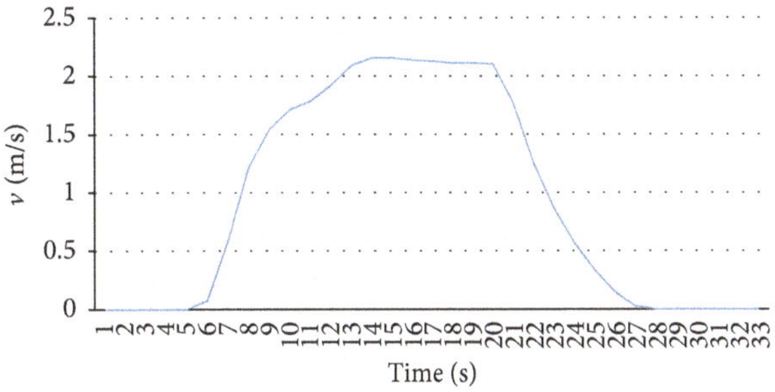

FIGURE 27: Underwater vehicle feedback speed curve.

of the hydraulic system structure reduce rigidity of the entire speed control system, increasing difficulty in speed control stability. Load of the vehicle in the movement process simultaneously presents a time-dependent nature, increasing difficulty in precise speed control. Sliding mode control algorithm based on fuzzy reaching law is then designed for the underwater vehicle speed control system with low rigidity and variable load. The sliding mode variable structure control system features superior robustness over conventional control systems. The control algorithm maintains existing features of sliding mode control, while introduction of fuzzy control law reduces output chattering caused by output switchover and stabilizing control output. Matlab mathematical simulation provided the control algorithm utilized to verify rationality and the control algorithm was realized in field tests with an ideal control effect obtained.

Conflict of Interests

The authors declare that there is no conflict of interests regarding the publication of this paper.

References

[1] B. Jin, *Study on the fuzzy sliding mode control method of electro-hydraulic position servo control system [Ph.D. thesis]*, Taiyuan University of Technology, Taiyuan, China, 2010.

[2] D. A. Newton, "Application of a neural network controller to control a rotary drive system with high power efficiency," in *Proceedings of the 7th Bath International Fluid Power Workshop, Innovations in Fluid Power*, pp. 41–54, Taunton, UK, 1995.

[3] X. Zibin, M. Jianqing, and R. Jian, "Adaptive backstepping neural network control of electro-hydraulic position servo system," in *Proceedings of the 2nd International Symposium on Systems and Control in Aerospace and Astronautics (ISSCAA '08)*, pp. 1–4, Shenzhen, China, December 2008.

[4] H. Azimian, R. Adlgostar, and M. Teshnehlab, "Velocity control of an electro hydraulic servomotor by neural networks," in *Proceedings of the International Conference on Physics and Control (PhysCon '05)*, pp. 677–682, August 2005.

[5] I. Istif, "A simulation study for the application of two different neural network control algorithms on electro-hydraulic system," in *Environmentally Conscious Manufacturing V*, vol. 5997 of *Proceedings of SPIE*, pp. 599–563, Boston, Mass, USA, November 2002.

[6] S. A. Mohseni, M. A. Shooredeli, and M. Teshnehlab, "Decoupled sliding-mode with fuzzy neural network controller for EHSS velocity control," in *Proceedings of the International Conference on Intelligent and Advanced Systems (ICIAS '07)*, pp. 7–11, IEEE, Kuala Lumpur, Malaysia, November 2007.

[7] M. Perron, J. De Lafontaine, and Y. Desjardins, "Sliding-mode control of a servomotor-pump in a position control application," in *Proceedings of the Canadian Conference on Electrical and Computer Engineering*, pp. 1287–1291, Saskatoon, Canada, May 2005.

[8] A. G. Loukianov, J. Rivera, Y. V. Orlov, and E. Y. Morales Teraoka, "Robust trajectory tracking for an electrohydraulic actuator," *IEEE Transactions on Industrial Electronics*, vol. 56, no. 9, pp. 3523–3531, 2009.

[9] R. M. Christensen, *Theory of Viscoelasticity*, Science Press, Beijing, China, 1990.

[10] T. W. Zur, "Viscoelastic properties of conveyor belts—modelling of vibration phenomena in belt conveyors during starting and stopping," *Bulk Solids Handling*, vol. 6, no. 3, pp. 553–560, 1986.

[11] H. E. Merritt, *Hydraulic Control Systems*, Wiley, New York, NY, USA, 1976.

[12] Y. Licai, *Analysis and research of the characteristic of electro-hydraulic speed servo system affected by load with large inertia and elasticity [Ph.D. thesis]*, Taiyuan University of Technology, Taiyuan, China, 2004.

[13] V. I. Utkin, "Variable structure systems with sliding modes," *IEEE Transactions on Automatic Control*, vol. 22, no. 2, pp. 212–222, 1977.

[14] K. D. Young, V. I. Utkin, and Ü. Özgüner, "A control engineer's guide to sliding mode control," *IEEE Transactions on Control Systems Technology*, vol. 7, no. 3, pp. 328–342, 1999.

[15] W. Gao, *Sliding Mode Control Theory and Design*, Science Press, Beijing, China, 1996.

[16] L. A. Zadeh, "Fuzzy sets," *Information and Computation*, vol. 8, pp. 338–353, 1965.

[17] L. A. Zadeh, "A rationale for fuzzy control," *Journal of Dynamic Systems, Measurement, and Control*, vol. 94, no. 1, pp. 3–4, 1972.

[18] C.-C. Liao, *S7-300/400 PLC Application Technology*, China Machine Press, Beijing, China, 2014.

8

MIMO Passive Control Systems Are Not Necessarily Robust

Jesús U. Liceaga-Castro,[1] Irma I. Siller-Alcalá,[1] Eduardo Liceaga-Castro,[2] and Luis A. Amézquita-Brooks[2]

[1]Departamento de Electrónica, Universidad Autónoma Metropolitana-Azcapotzalco, 02200 Ciudad de México, DF, Mexico
[2]CIIIA-FIME, Universidad Autónoma de Nuevo León, 66451 Monterrey, NL, Mexico

Correspondence should be addressed to Jesús U. Liceaga-Castro; julc@correo.azc.uam.mx

Academic Editor: Shengwei Mei

Via several cases of study it is shown that a passive multivariable linear control system, contrary to its single input single output counterpart, may not be robust. Moreover, it is shown that lack of robustness can be exposed via the multivariable structure function.

1. Introduction

The design of controllers for multivariable systems (MIMO) satisfying robustness conditions may be a long and tiresome procedure [1]. Many well-known methodologies had been proposed in order to solve this problem such as LQG, H_∞, and QFT. More recently, dissipativity which is related to the conservation, dissipation, and transport of energy is an open loop property that has been used to develop a framework for the design and analysis of control systems. In particular, passive analysis focuses on systems which dissipate energy. This open loop characteristic presents great advantages in the design of robust control systems.

For the case of a linear time invariant SISO system described by a rational transfer function $G(s) = N(s)/D(s)$ is passive if it satisfies the following [2]:

(i) $G(s)$ is stable.

(ii) $\mathbb{R}[G(j\omega)] \geq 0 \ \forall\omega \in (-\infty, \infty)$.

These conditions can be translated in terms of the classical control theory as follows: a rational SISO transfer function $G(s) = N(s)/D(s)$ is passive if

(i) $G(s)$ is stable and minimum phase,

(ii) $G(s)$ has a relative degree ≤ 1,

(iii) $G(s)$ behaves like a lead phase filter,

(iv) the Nyquist plot of $G(j\omega)$ lies in the right half plane of the complex plane $\forall\omega \in (-\infty, \infty)$.

That is, passive SISO systems are robust because they have gain and phase margins of Mg $\rightarrow \infty$ dB's and Mp $\geq 90^\circ$, respectively. Moreover, thanks to this characteristic the parallel, series, or feedback interconnection of two passive systems is also robust. In this context, in general, the strategy for passive control systems is divided in two steps: the passivation of the process and the design of a passive controller—normally a PID controller—based on the amended or passivized process. Also, in order to transform a nonpassive SISO system into a passive one, three methods are normally applied: pre- or postcompensation, feedback, and feed-forward [3, 4].

One may cast the conjecture on the existence of these frequency domain properties for linear MIMO systems. If this conjecture proves to be true, the robust design and assessment of multivariable control could be obtained in a similar manner as in SISO systems.

A square linear time invariant MIMO system can be represented by $y(s) = G(s)u(s)$, where $G(s) \in \mathbb{C}^{mxm}$ is a rational transfer function matrix; $u(t) \in \mathbb{R}^m$ is the input; and $y(t) \in \mathbb{R}^m$ the output. Assuming all the poles of $G(s)$ are in $\mathbb{R}[s] < 0$, then [2]

$$G(s) \text{ is passive} \Longleftrightarrow \lambda_{\min}\left[G(j\omega) + G^*(j\omega)\right] \geq 0 \quad \forall\omega \in (-\infty, \infty), \tag{1}$$

where $G^*(j\omega)$ is the conjugated transpose of $G(j\omega)$.

Unfortunately, as it is shown in this paper, the above conjecture is wrong. Even when the open loop MIMO control system satisfies the passivity condition, this does not necessarily result in a robust control system.

In the following sections, three cases of passivation of 2×2 linear MIMO systems are presented. The results show that in spite of passivating the systems the resulting control systems are fragile, that is, nonrobust. In addition, the cause for the lack of robustness is exposed using the multivariable structure function (MSF) [5] and the well-known Singular Values Analysis (SVA). For the sake of transparency the cases are based on 2×2 linear MIMO systems. However, the results obtained from these cases apply also to the general case of $M\times M$ MIMO systems. That is, the MSF applies to the general case of $M \times M$ linear MIMO systems [6].

2. Cases of Study

Case 1. Consider the simple 2×2 linear time invariant MIMO system of the distillation column thoroughly analysed in [7]

$$G(s) = \frac{\begin{bmatrix} 87.8 & -86.4 \\ 108.2 & -109.6 \end{bmatrix}}{(75s+1)}. \quad (2)$$

The eigenvalues of $[G(j\omega)+G^*(j\omega)]$ are shown in Figure 1.

From Figure 1 it is clear that one of the eigenvalues of $[G(j\omega)+G^*(j\omega)]$ is negative and therefore the MIMO system of (2) is nonpassive. However, it is possible not only to passivize but also to decouple the system with the static precompensator K_p given by

$$K_p = \begin{bmatrix} 0.3994 & -0.3149 \\ 0.3943 & -0.3200 \end{bmatrix}, \quad (3)$$

resulting in the passivized system $G_p(s) = G(s)K_p$ as shown in Figure 2 by the eigenvalues of $[G_p(j\omega) + G_p^*(j\omega)]$:

$$G_p(s) = \frac{\begin{bmatrix} 1 & 0 \\ 0 & 1 \end{bmatrix}}{(75s+1)}. \quad (4)$$

As mentioned above, the second step in the passive control strategy is to design a passive controller, normally a PID, based on the amended or passivized system. In this case, it is possible to design a simple decentralized PI controller $K(s)$ given by

$$K(s) = \frac{\begin{bmatrix} (s+0.01333) & 0 \\ 0 & (2s+0.02667) \end{bmatrix}}{(0.01333s)}, \quad (5)$$

resulting in the passive open loop control system $G_o(s) = G_p(s)K = G(s)K_pK(s)$,

$$G_o(s) = \frac{\begin{bmatrix} 1 & 0 \\ 0 & 2 \end{bmatrix}}{s} \quad (6)$$

with a stable closed loop matrix transfer function given by

$$G_{cl}(s) = \begin{bmatrix} \dfrac{1}{s+1} & 0 \\ 0 & \dfrac{2}{s+2} \end{bmatrix}. \quad (7)$$

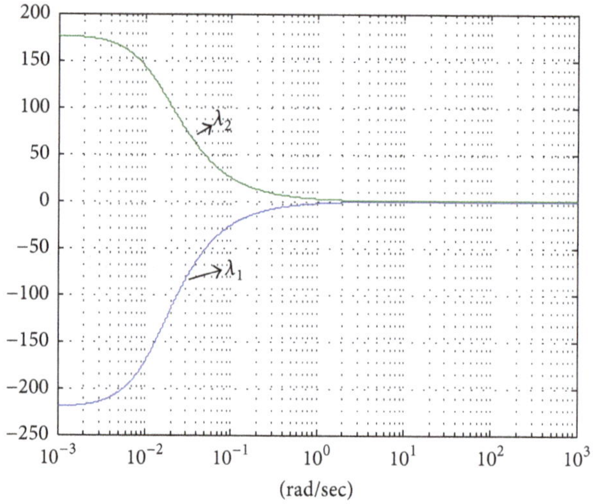

FIGURE 1: Eigenvalues of $[G(j\omega) + G^*(j\omega)]$ (Case 1).

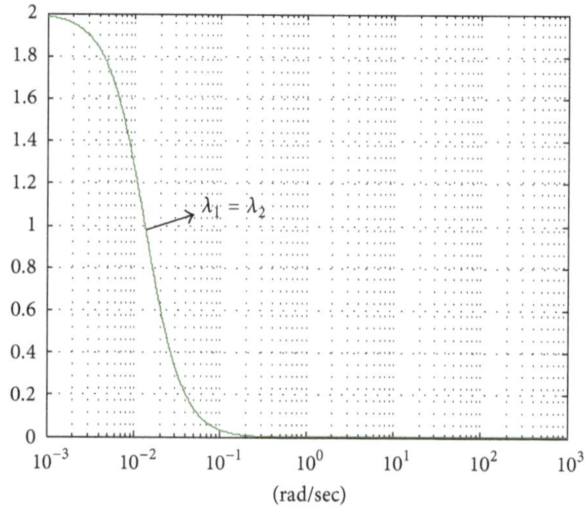

FIGURE 2: Eigenvalues of $[G_p(j\omega) + G_p^*(j\omega)]$ (Case 1).

Nonetheless, if, for instance, a small parameter perturbation or uncertainty of +4% is introduced in the gain of the individual transfer function $g_{12}(s)$ of the system in (2) the open loop control system becomes

$$G_o(s) = \frac{\begin{bmatrix} -0.3628 & 0 \\ 0 & 2 \end{bmatrix}}{s} \quad (8)$$

which is nonpassive and closed loop unstable.

Case 2. Let the well-known quadruple tank MIMO system described by the matrix transfer function [8]

$$G_2(s) = \begin{bmatrix} \dfrac{\delta_1 3.533}{(63s+1)} & \dfrac{(1-\delta_2)\,3.533}{(2457s^2+102s+1)} \\ \dfrac{(1-\delta_1)\,4.678}{(5096s^2+147s+1)} & \dfrac{\delta_2 4.678}{(91s+1)} \end{bmatrix}. \quad (9)$$

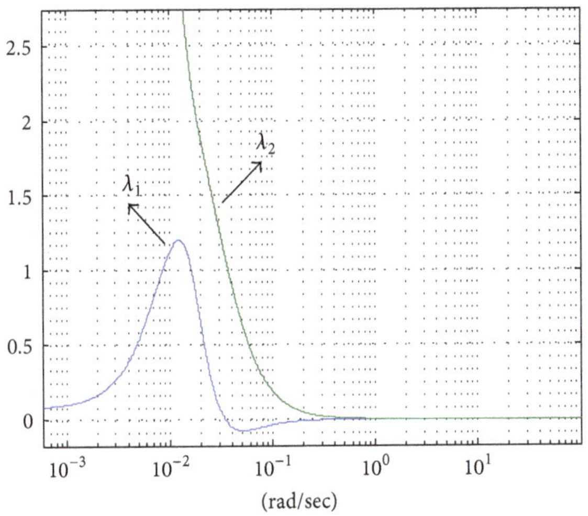

FIGURE 3: Eigenvalues of $[G_2(j\omega) + G_2^*(j\omega)]$ (Case 2).

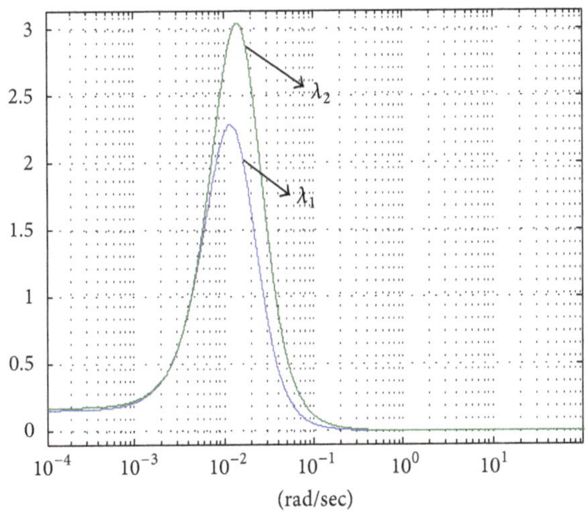

FIGURE 4: Eigenvalues of $[G_{p2}(j\omega) + G_{p2}^*(j\omega)]$ (Case 2).

If the parameters $\delta_1, \delta_2 \in (0,1)$, which depend on the positions of the system valves, are set to $\delta_1 = 0.6$ and $\delta_2 = 0.41$, this yields a system for which the eigenvalues of $[G_2(j\omega)+G_2^*(j\omega)]$ show that the quadruple tank is nonpassive (Figure 3).

Following the strategy of Case 1 the system can be decoupled by the precompensator:

$$K_{p2}(s) = \begin{bmatrix} 1 & \dfrac{-170.3s - 1.871}{9774s^2 + 281.9s + 1.918} \\ \dfrac{-131.3s - 2.084}{5208s^2 + 216.2s + 2.119} & 1 \end{bmatrix}, \quad (10)$$

resulting in the decoupled system $G_{p2}(s) = G_2(s)K_{p2}(s)$. The eigenvalues of $[G_{p2}(j\omega) + G_{p2}^*(j\omega)]$ (Figure 4) show that the precompensated system $G_{p2}(s)$ is also passive.

Therefore, it is possible to design the decentralized PI controller $K_2(s)$ to the passivized system $G_{p2}(s)$ to obtain the open loop system $G_{o2}(s) = G_{p2}(s)K_2(s) = G_2(s)K_{p2}(s)K_2(s)$ with

$$K_2(s) = \frac{\begin{bmatrix} 0.35(100s+1) & 0 \\ 0 & 0.525(100s+1) \end{bmatrix}}{s}. \quad (11)$$

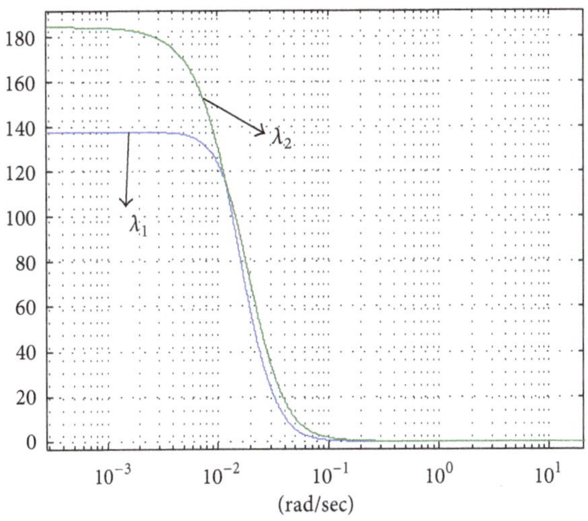

FIGURE 5: Eigenvalues of $[G_{o2}(j\omega) + G_{o2}^*(j\omega)]$ (Case 2).

The eigenvalues of $[G_{o2}(j\omega)+G_{o2}^*(j\omega)]$ (Figure 5) confirm that also the open loop system $G_{o2}(s)$ is passive. Thus, the closed loop system is stable and passive.

However, despite the passivity of the system, it can be destabilized by a small perturbation. That is, a perturbation or uncertainty of +5% in the gain of the individual transfer function $g_{12}(s)$ of the quadruple tank destabilizes the closed loop system introducing a pair of unstable poles at $\{0.00074, 0.000073\}$.

Case 3. Let 2×2 MIMO system described by the matrix transfer function $G_3(s)$

$$G_3(s) = \frac{\begin{bmatrix} \dfrac{(8.6s+8.6)}{(0.05s+1)} & \dfrac{(90s+0.9)}{(0.1215s^2+3.893s+7.3)} \end{bmatrix}}{(0.03031s^3 + 0.5304s^2 + 2.258s + 1)}. \quad (12)$$

Similar to two previous cases, the eigenvalues of $[G_3(j\omega)+G_3^*(j\omega)]$ (Figure 6) show that $G_3(s)$ is nonpassive.

Continuing with the common multivariable control strategy of decoupling via precompensation, the system $G_3(s)$ can be decoupled by the precompensator $K_{p3}(s)$ given by

$$K_{p3}(s) = \begin{bmatrix} 1 & \dfrac{(-10.47s - 0.1047)}{(s+1)} \\ \dfrac{(-0.4114s - 8.227)}{(s^2 + 32.03s + 60.06)} & 1 \end{bmatrix}. \quad (13)$$

Contrary to the two previous examples the precompensated and decoupled system $G_{p3}(s) = G_3(s)K_{p3}(s)$ remains nonpassive because the eigenvalues of $[G_{p3}(j\omega) + G_{p3}^*(j\omega)]$ are not positive at all frequencies as it is shown in Figure 7.

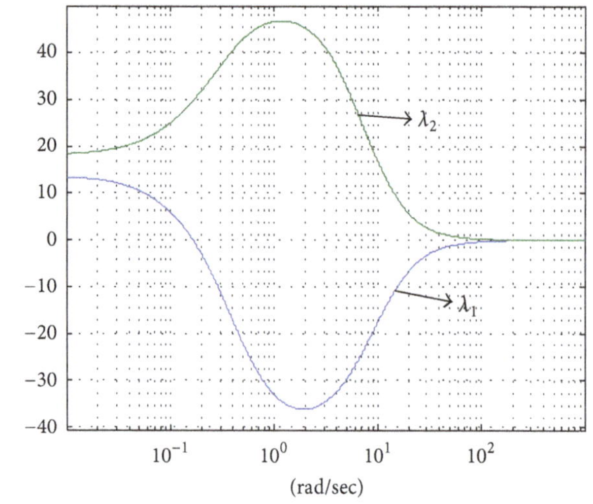

FIGURE 6: Eigenvalues of $[G_3(j\omega) + G_3^*(j\omega)]$ (Case 3).

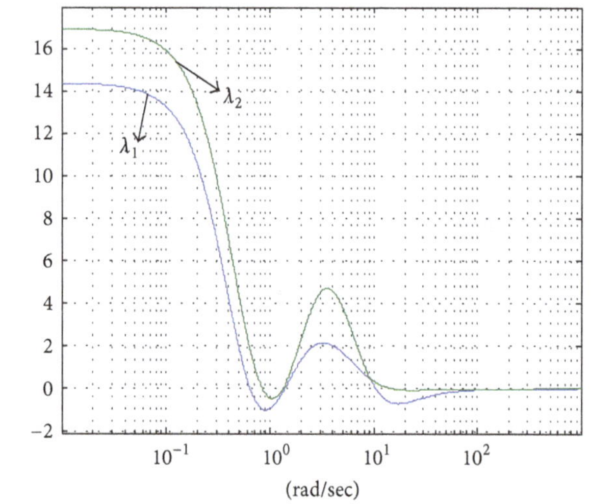

FIGURE 7: Eigenvalues of $[G_{p3}(j\omega) + G_{p3}^*(j\omega)]$ (Case 3).

Nonetheless, it is possible to both passivize and control the decoupled system $G_{3p}(s)$ by the decentralized controller $K_3(s)$ comprising two PID + lead-compensator filters:

$$K_3(s) = \begin{bmatrix} k_1(s) & 0 \\ 0 & k_2(s) \end{bmatrix}, \tag{14}$$

where

$$k_1(s) = \frac{0.079s^4 + 0.001461s^3 + 0.007979s^2 + 0.01311s + 0.00474}{s(s^2 + 0.1358s + 2.071)}, \tag{15}$$

$$k_2(s) = \frac{7.9s^4 + 0.1462s^3 + 0.7268s^2 + 0.8493s + 0.2607}{s(s^3 + 30.13s^2 + 6.097s + 62.12)}.$$

The system $G_{o3}(s) = G_{p3}(s)K_3(s) = G_3(s)K_{p3}(s)K_3(s)$ satisfies the passivity condition as the eigenvalues of $[G_{o3}(j\omega) + G_{o3}^*(j\omega)]$ (Figure 8) are positive at all frequencies. Consequently, the closed loop system is passive and stable.

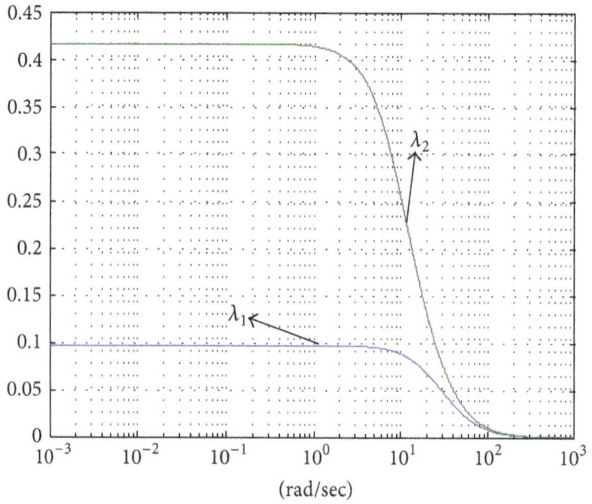

FIGURE 8: Eigenvalues of $[G_{o3}(j\omega) + G_{o3}^*(j\omega)]$ (Case 3).

Similar to the two preceding cases, regardless of the passivity condition of the open loop control system, its closed loop configuration can be easily destabilized, for instance, a small perturbation in the gain of the individual transfer function (+5%) $g_{12}(s)$ of the original system. This perturbation introduces a pair of closed loop unstable poles at $\{0.0071 \pm 1.4345j\}$.

The cases presented prove that for multivariable control systems passivity may not necessarily imply robustness. That is, a multivariable control system may remain highly sensitive and fragile to parametric perturbations despite having been passivized.

In the next section via the multivariable structure function (MSF) this situation can be easily clarified.

3. Robustness Analysis

The MSF is a key element of the framework of analysis and design for linear MIMO control systems known as *Individual Channel Analysis and Design* (ICAD) [9–11]. ICAD allows the application of the well proved SISO classical control theory in the analysis and design of linear MIMO control systems. An explanation of ICAD is out of the scope of this paper; however, a comprehensive description and applications of ICAD can be found in [12–14].

Let 2×2 MIMO system described by the matrix transfer function

$$G(s) = \begin{bmatrix} g_{11}(s) & g_{12}(s) \\ g_{21}(s) & g_{22}(s) \end{bmatrix}. \tag{16}$$

Then, the MSF $\gamma(s)$ of $G(s)$ is defined by

$$\gamma(s) = \frac{g_{12}(s)\, g_{21}(s)}{g_{11}(s)\, g_{22}(s)}. \tag{17}$$

In [5], it was proved that 2×2 MIMO system is highly sensitive to parametric uncertainties or perturbations and non-modeled dynamics if the MSF $\gamma(j\omega)$ is close to the point $(1,0)$ at some range of frequencies $\omega \in (\omega_0, \omega_1)$. This condition is

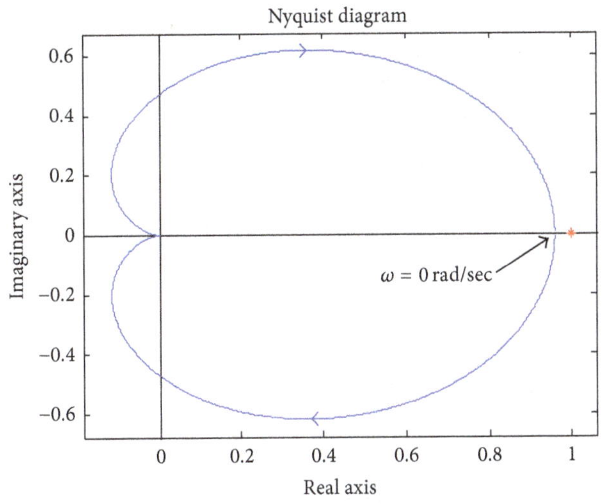

FIGURE 9: Nyquist plot of $\gamma_2(s)$ (Case 2).

known within ICAD as structurally nonrobust. This indicates that the number of minimum and nonminimum phase zeros can be modified by uncertainties or perturbations. Therefore, via the MSF, it is possible to determine the possibilities of a MIMO control system to satisfy the requirements of design and the conditions the controller must satisfy to comply with these requirements ensuring appropriate robustness.

Case 1 (robust analysis). The system $G(s)$ of (2) has no transmission zeros and its MSF $\gamma(s)$ is given by

$$\gamma(s) = \frac{g_{12}(s)\,g_{21}(s)}{g_{11}(s)\,g_{22}(s)} = 0.9715. \quad (18)$$

Hence, the system of (2) is highly sensitive at all frequencies due to the closeness of its MSF to point $(1,0)$. This condition can likewise be explained via the condition number $\kappa(j\omega)$:

$$\kappa(G(j\omega)) = \frac{\sigma_{\max}(G(j\omega))}{\sigma_{\min}(G(j\omega))}, \quad (19)$$

where $\sigma_{\max}(G(j\omega))$ and $\sigma_{\max}(G(j\omega))$ are the maximum and minimum Singular Values (SV) of $G(s)$.

The condition number $\kappa(j\omega)$ for the system of (2) is $\kappa(j\omega) = 141.732$. That is, the system is ill-conditioned at all frequencies.

Case 2 (robust analysis). The system $G_2(s)$ of (9) is minimum phase with transmission zeros at $\{-0.0431, -0.0004, -0.011, -0.159\}$. On the other hand, the Nyquist plot of the MSF $\gamma_2(s)$ of $G_2(s)$ is shown in Figure 9.

From Figure 9, it is clear that $\gamma_2(j\omega)$ is close to the point $(1,0)$ at $\omega = 0$ rad/sec. Hence, the system $G_2(s)$ is highly sensitive or ill-conditioned at $\omega = 0$ rad/sec as shown in Figure 10 by its condition number $\kappa_2(j\omega)$.

It is due to this condition that when the small perturbation of +5% was introduced in the individual transfer function $g_{12}(s)$, $G_2(s)$ became non-minimum phase with a zero at +0.000076 and consequently the closed loop control system becomes unstable.

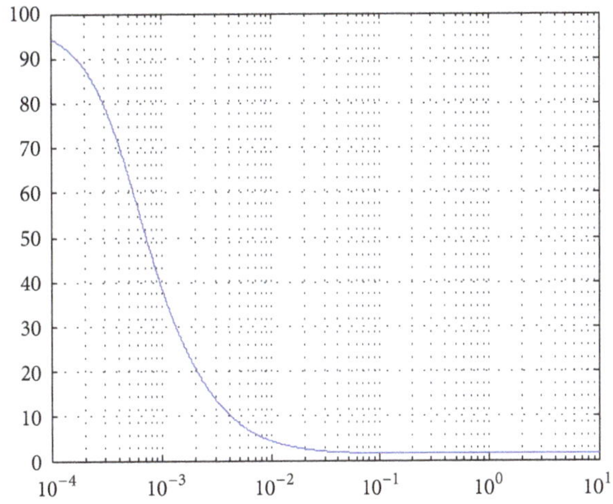

FIGURE 10: Condition number $\kappa_2(j\omega)$ (Case 2).

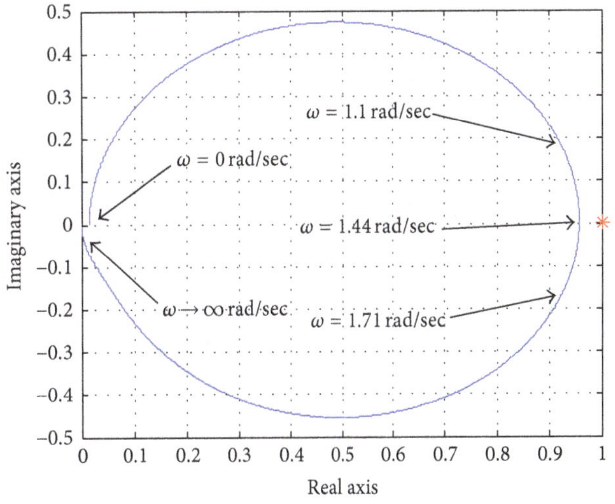

FIGURE 11: Nyquist plot of $\gamma_3(s)$ (Case 3).

Case 3 (robust analysis). System $G_3(s)$ of (12) is minimum phase with zeros at $\{-28.58, -0.0.0677 \pm 1.437j\}$. The Nyquist plot of the MSF $\gamma_3(s)$, depicted in Figure 11, indicates that $G_3(s)$ is highly sensitive at frequencies around $\omega = 1.44$ rad/sec.

This situation is confirmed by the plot of the condition number $\kappa_3(j\omega)$ (Figure 12) which clearly shows that the system is ill-conditioned in a region of frequencies around $\omega = 1.44$ rad/sec. This explains why the closed loop system was destabilized by the small perturbation of +5% in the gain of the individual transfer function $g_{12}(s)$. That is, due to this perturbation, the system became non-minimum phase with a pair of zeros at $\{0.0076 \pm 1.44j\}$ and as a result, like the previous case, closed loop unstable.

4. Conclusions

In this paper, it is shown that in the case of linear MIMO control systems passivity does not necessarily imply good robustness properties. This was exemplified by means of

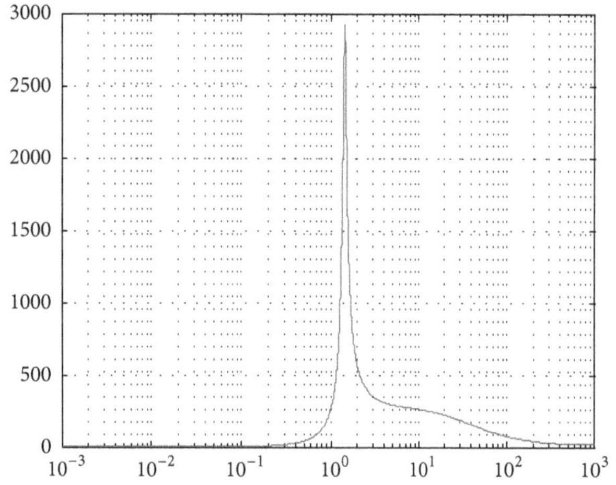

FIGURE 12: Condition number $\kappa_3(j\omega)$ (Case 3).

3 cases where despite the passivation of the open loop control systems it was possible to destabilize their closed loops configurations by small parametric perturbations. That is, the control systems are highly sensitive regardless of its passive condition. This situation was exposed via the multivariable structure function and the condition number. In particular, it was shown that the lack of structural robustness, ill-conditioning, cannot be amended via passivation. The three cases of study represent the three possible conditions in which a system is not structurally robust or ill-posed: at all frequencies, at one particular frequency, and along a region of frequencies.

It can be concluded that in order to exploit all the benefits of passivity in MIMO control systems it is necessary to consider the lack of structural robustness or ill-conditioning of the process before attempting any passivation strategy.

Finally, it is possible to extend the application of the multivariable structure function to the passive analysis of the general case of $M \times M$ MIMO systems.

Conflict of Interests

The authors declare that there is no conflict of interests related to this paper.

References

[1] E. Licéaga-Castro, J. Licéaga-Castro, C. E. Ugalde-Loo, and E. M. Navarro-López, "Efficient multivariable submarine depth-control system design," *Ocean Engineering*, vol. 35, no. 17-18, pp. 1747–1758, 2008.

[2] B. Brogliato, R. Lozano, B. Maschke, and O. Egeland, *Dissipative Systems Analysis and Control. Theory and Applications*, Communications and Control Engineering Series, Springer, London, UK, 2nd edition, 2007.

[3] A. G. Kelkar and S. M. Joshi, "Robust control of non-passive systems via passification," in *Proceedings of the American Control Conference*, vol. 5, pp. 2657–2661, IEEE, Albuquerque, NM, USA, June 1997.

[4] A. G. Kelkar and S. M. Joshi, "Robust passification and control of non-passive systems," in *Proceedings of the American Control Conference*, pp. 3133–3137, Philadelphia, Pa, USA, June 1998.

[5] J. O'Reilly and W. E. Leithead, "Multivariable control by 'individual channel design'," *International Journal of Control*, vol. 54, no. 1, pp. 1–46, 1991.

[6] W. E. Leithead and J. O'Reilly, "m-input m-output feedback control by individual channel design part 1. Structural issues," *International Journal of Control*, vol. 56, no. 6, pp. 1347–1397, 1992.

[7] S. Skogestad and I. Postlethwaite, *Multivariable Feedback Control: Analysis and Design*, Wiley, 2007.

[8] K. H. Johansson, "The quadruple-tank process: a multivariable laboratory process with an adjustable zero," *IEEE Transactions on Control Systems Technology*, vol. 8, no. 3, pp. 456–465, 2000.

[9] W. E. Leithead and J. O'Reilly, "Performance issues in the individual channel design of 2-input 2-output systems. Part 1. Structural issues," *International Journal of Control*, vol. 54, no. 1, pp. 47–82, 1991.

[10] W. E. Leithead and J. O'Reilly, "Performance issues in the individual channel design of 2-input 2-output systems. Part 2. Robustness issues," *International Journal of Control*, vol. 55, no. 1, pp. 3–47, 1992.

[11] W. E. Leithead and J. O'Reilly, "Performance issues in the individual channel design of 2-input 2-output systems, Part 3. Non-diagonal control and related issues," *International Journal of Control*, vol. 55, no. 1, pp. 265–312, 1992.

[12] E. Licéaga-Castro, J. Licéaga-Castro, and C. E. Ugalde-Loo, "Beyond the existence of diagonal controllers: from the relative gain array to the multivariable structure function," in *Proceedings of the 44th IEEE Conference on Decision and Control, and the European Control Conference (CDC-ECC '05)*, pp. 7150–7156, Sevilla, Spain, December 2005.

[13] E. Licéaga-Castro, C. E. Ugalde-Loo, J. Licéaga-Castro, and P. Ponce, "An efficient controller for SV-PWM VSI based on the multivariable structure function," in *Proceedings of the 44th IEEE Conference on Decision and Control, and the European Control Conference (CDC-ECC '05)*, pp. 4754–4759, IEEE, Sevilla, Spain, December 2005.

[14] L. Amezquita-Brooks, J. Liceaga-Castro, and E. Liceaga-Castro, "Speed and position controllers using indirect field-oriented control: a classical control approach," *IEEE Transactions on Industrial Electronics*, vol. 61, no. 4, pp. 1928–1943, 2014.

9

An Improved Optimal Slip Ratio Prediction considering Tyre Inflation Pressure Changes

Guoxing Li,[1,2] Tie Wang,[1] Ruiliang Zhang,[1] Fengshou Gu,[1,2] and Jinxian Shen[1]

[1]*Department of Vehicle Engineering, Taiyuan University of Technology, Taiyuan 030024, China*
[2]*Centre for Efficiency and Performance Engineering, University of Huddersfield, Huddersfield HD1 3DH, UK*

Correspondence should be addressed to Tie Wang; wangtie57@163.com

Academic Editor: Petko Petkov

The prediction of optimal slip ratio is crucial to vehicle control systems. Many studies have verified there is a definitive impact of tyre pressure change on the optimal slip ratio. However, the existing method of optimal slip ratio prediction has not taken into account the influence of tyre pressure changes. By introducing a second-order factor, an improved optimal slip ratio prediction considering tyre inflation pressure is proposed in this paper. In order to verify and evaluate the performance of the improved prediction, a cosimulation platform is developed by using MATLAB/Simulink and CarSim software packages, achieving a comprehensive simulation study of vehicle braking performance cooperated with an ABS controller. The simulation results show that the braking distances and braking time under different tyre pressures and initial braking speeds are effectively shortened with the improved prediction of optimal slip ratio. When the tyre pressure is slightly lower than the nominal pressure, the difference of braking performances between original optimal slip ratio and improved optimal slip ratio is the most obvious.

1. Introduction

The longitudinal motion of a vehicle is governed by the forces generated between the tyres and the road surface. Therefore, acquiring enough tyre friction is crucial to enhance vehicle dynamics. According to the friction principle, the magnitude of frictional force depends on two factors: normal pressure and friction coefficient. However, the relationship between the longitudinal friction coefficient and wheel slip ratio is complex. In the premise of constant normal pressure, when the wheel slip ratio is small, longitudinal force linearly increases with slip ratio. With further increase of the slip ratio, longitudinal force increases and then decreases nonlinearly. When the longitudinal force reaches a maximum value, the corresponding slip ratio is called optimal slip ratio.

Antilock braking system (ABS) is an automobile safety system that allows the vehicle wheels to maintain tractive contact with the road surface according to driver inputs during the braking process, preventing the wheels from both locking up and uncontrolled skidding. The logic threshold control method is widely applied in commercial ABS products [1]. As an experience based control method, in the control process, wheel slip ratio is not maintained in optimal slip ratio but fluctuated near it which cannot acquire the best braking effect. Meanwhile, if slip ratio is considered as the control target, ABS controller can maintain the practical slip ratio near the optimal slip ratio all the time during the braking process, so that vehicle controllability and stability are optimized and maximized. That is considered as the ideal braking method. The research on ABS control, aiming at optimal slip ratio, has been carried out for many years. Most researchers put the emphasis on control strategy optimization and development of control methods based on the control theory of self-turning PID, fuzzy PID, artificial neural network, and so on [2–4]. Other researchers focused on identification of road surface and optimal slip ratio [5–7].

Up to now, most studies about optimal slip ratio control are based on an optimal slip ratio estimation expression, proposed by Liu [8] and Bian [9], which gives a quantification description of influence on optimal slip ratio due to changes of road adhesion coefficient, vehicle velocity, and tyre slip angle. The optimal slip ratio expression proposed by Liu et al. is mainly developed from the tyre Magic Formula, proposed by Pacejka [10], using regression analysis. In the early time, the influence of tyre inflation pressure changes was not considered into Magic Formula based on which the influence has not been put into existing optimal slip ratio expression.

However, it is indicated in [11–13] that tyre inflation pressure changes can directly influence the relationship between tyre longitudinal force and slip ratio. That is, the value of optimal slip ratio is influenced by tyre inflation pressure changes.

Based on the existing optimal slip ratio expression and an improved Magic Formula model [14], a simulation is launched to make a study on relationship between tyre inflation pressure changes and tyre slip ratio. A second-order factor, representing the influence of tyre inflation pressure changes on the value of optimal slip ratio, is acquired. In order to verify and evaluate the improved optimal slip ratio, an ABS controller with optimal slip ratio as the control target is established. Vehicle braking processes before and after improvement are simulated and compared.

The paper is structured as follows. Section 2 gives a simple introduction of tyre dynamic principle and optimal slip ratio. Section 3 describes the proposed improved optimal slip ratio considering tyre inflation pressures. Section 4 describes the modelling of cosimulation system which consists of a fuzzy PID controller for ABS system and vehicle dynamics model in MATLAB/Simulink and CarSim software, respectively. Section 5 verifies the improved expression through comparative analyses and discussions on various scenarios. The paper is concluded in Section 6.

2. Wheel Dynamics and Optimal Slip Ratio

The dynamic differential equations for the calculation of longitudinal motion of a vehicle are described as follows:

$$m\dot{u} = -F_x, \quad (1)$$

$$J\dot{\omega} = F_x R - T_b - T_g, \quad (2)$$

$$F_{xi} = \mu_i(\lambda) F_{zi}, \quad (3)$$

where i is fl, fr, rl, and rr; m and u are a quarter of the vehicle mass and wheel velocity. F_x is the driving resistance; J is the wheel inertia; ω is the wheel rotational speed; T_b and T_g are the braking torque and the rolling resistance torque; μ_i and F_{zi} are the friction coefficient and normal force of i wheel, as shown in Figure 1.

During the braking process, the wheel speed u can be larger than its rotation speed $R\omega$ which is characterised by the wheel longitudinal slip λ:

$$\lambda = \frac{u - R\omega}{u}. \quad (4)$$

As shown in (3), the longitudinal force F_x is described as a function that depends on the longitudinal friction coefficient $\mu(\lambda)$. If the longitudinal slip λ is small, the relationship between the longitudinal force and slip is linear, but, with a further increase of the slip, the longitudinal force reaches the

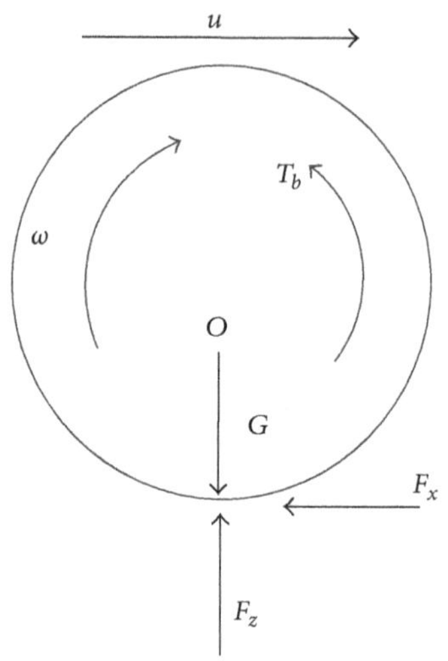

FIGURE 1: Single wheel model.

maximum at the certain value of the slip specified by tyre-road adhesion and is saturated beyond that. When the longitudinal force reaches a maximum value, the corresponding slip value is referred to as the optimal slip ratio.

In essence, the principle of ABS control is to always maintain the wheel slip ratio near the optimal slip ratio, which ensures a continuous maximum value onto tyre longitudinal force. In order to acquire the best braking effect, ABS schemes usually take the optimal slip ratio as the control target. Up to now, researchers have made numerous studies on the influence of road adhesion coefficient, vehicle velocity, and tyre slip angle changes on the optimal slip ratio [1–4].

Most studies about optimal slip ratio control focused on the real-time identification and estimation of unknown road condition, vehicle velocity, and tyre slip angle. Furthermore, based on a series of advanced control methods, ABS controller is optimized to keep the real wheel slip ratio closer to the theoretical optimal value. In principle, previous studies have been carried out mainly on the basis of the optimal slip ratio expression proposed by Liu [8] and Bian [9]:

$$\lambda_{\text{op}} = \lambda_0 + 0.165 \times \log\left(\frac{64}{u}\right) + 0.01 \times \delta^{1.5}, \quad (5)$$

where λ_{op} is the optimal slip ratio, λ_0 is the road surface friction coefficient, u is wheel velocity, and δ is the tyre slip angle. Equation (5) was developed on the basis of the tyre Magic Formula in the period from 1989 to 1994, excluding the influence of tyre inflation pressure changes. However, an increasing number of researches indicate that tyre inflation pressure change causes a direct influence on the relationship between tyre longitudinal force and slip ratio, which means that the value of optimal slip ratio could be influenced by tyre inflation pressures [6–8].

3. Optimal Slip Ratio Prediction considering the Tyre Inflation Pressure

In order to account for the influence of the tyre inflation pressure on the longitudinal friction characteristics,

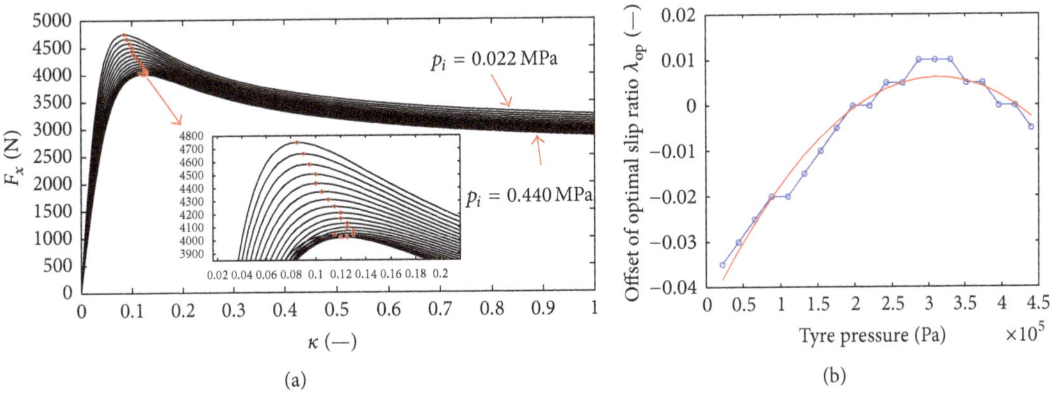

FIGURE 2: (a) F_x-λ curves under different tyre inflation pressures. (b) Relationship between optimal slip ratio and tyre inflation pressures.

Besselink et al. [14] modified and improved the longitudinal tyre characteristic formula and related parameters. The main effects of tyre inflation pressure changes on the longitudinal tyre characteristics are identified as follows [10, 11]:

(i) Changes in longitudinal slip stiffness and camber stiffness.

(ii) Changes in longitudinal peak friction coefficient.

Therefore, the Magic Formula for predicting longitudinal force can be improved to

$$F_x = (D_x \sin [C_x \cdot \arctan \{B_x\lambda_x - E_x (B_x\lambda_x - \arctan (B_x\lambda_x))\}] + S_{Vx}) G_{x\alpha}, \quad (6)$$

where λ_x is the overall longitudinal slip; B_x is the longitudinal stiffness factor; C_x is shape factor for longitudinal force; D_x is the peak value factor; F_z is the vertical force; E_x is the longitudinal curvature factor; S_{vx} is vertical shift factor; and $G_{x\alpha}$ is comprehensive factor for combined slip, which are calculated by

$$D_x = \mu_x F_z, \quad (7)$$

$$B_x = \frac{K_{x\lambda}}{C_x D_x}, \quad (8)$$

$$K_{x\lambda} = K_{x\lambda,\text{nom}} \cdot \left(1 + p_{px1} \cdot dp_i + p_{px1} \cdot dp_i^2\right), \quad (9)$$

$$\mu_x = \mu_{x,\text{nom}} \cdot \left(1 + p_{px3} \cdot dp_i + p_{px4} \cdot dp_i^2\right), \quad (10)$$

where μ, K, $\mu_{x,\text{nom}}$, and $K_{x\lambda,\text{nom}}$ are longitudinal peak friction coefficient, longitudinal slip stiffness, and their corresponding nominal values, respectively. p_i and p_{i0} are measured pressure and nominal tyre inflation pressure. And $dp_i = (p_i - p_{i0})/p_{i0}$ is dimensionless increment of tyre inflation pressure. As can be seen in both (9) and (10), there is an additional factor $(1 + p_{pxi} \cdot dp_i + p_{px(i+1)} \cdot dp_i{}^2)$ which is the product of the inflation pressure increment dp_i and Magic Formula parameter p_{pxi}, highlighting the influences of pressure changes.

In order to investigate the quantitative influence of tyre inflation pressure changes on the optimal slip ratio, with tyre inflation pressure being considered as independent variable, relation curves of longitudinal force and slip ratio under different tyre inflation pressures, F_x-λ, are presented in Figure 2 based on the improved Magic Formula of (6). The peak longitudinal friction point of each curve is crucially marked and the set of optimal slip ratio points corresponding to the tyre pressure input is acquired. In the figure, the results were placed on a standard 205/60-R15 tyre under a vertical load of 5000 N. To examine the influences of pressure changes, the tyre pressure changes from 0.1 times to 2 times of the nominal pressure, 0.22 MPa, obtaining twenty corresponding F_x-λ curves as illustrated in Figure 2(a). As shown by the maximum longitudinal friction highlighted with the circle markers, the maximum longitudinal friction increases with decreasing in tyre pressure, whereas the optimal slip ratio decreases correspondingly. When the tyre inflation pressure is higher than the nominal value, the optimal slip ratio firstly increases and then decreases slightly. Fitting results show that the quadratic fitted curve correlates to the point set of optimal slip ratio very well with a correlation coefficient of more than 0.982.

In order to take quantitative effect of inflation pressure into account, the optimal slip ratio in (5) can be modified by including two more terms as shown in the following equation:

$$\begin{aligned}\lambda_{\text{op}} &= \lambda_{\text{op,nom}} + p_{pi1} \cdot dp_i + p_{pi2} \cdot dp_i^2 \\ &= \lambda_0 + 0.165 \times \log\left(\frac{64}{u}\right) + 0.01 \times \delta^{1.5} + p_{pi1} \cdot dp_i \\ &\quad + p_{pi2} \cdot dp_i^2,\end{aligned} \quad (11)$$

where p_{pi1} is the linear influence coefficient of tyre inflation pressure changes and p_{pi2} is the quadratic influence coefficient, both of which are decided by the characteristics of the tyre itself. The impact on the optimal slip ratio given by tyre inflation pressure changes has an extremely small coherence with other impact factors. Therefore, an independent second-order factor should be essentially added into the optimal slip ratio prediction.

TABLE 1: Simulation parameters.

λ_0	0.175	Tyre type	205/60-R15	P_{px1}/P_{px2}	−0.349/0.378
v_{0i}/(km/h)	100/80/60/40	Vehicle type	Sedan/C-class	P_{px3}/P_{px4}	−0.096/0.065
δ/°	0	Engine power/kW	250	P_{pi1}	−0.213
p_i/MPa	0.132/0.176/0.22/0.264/0.308	Max. braketorque/Nm	2000	P_{pi2}	0.179

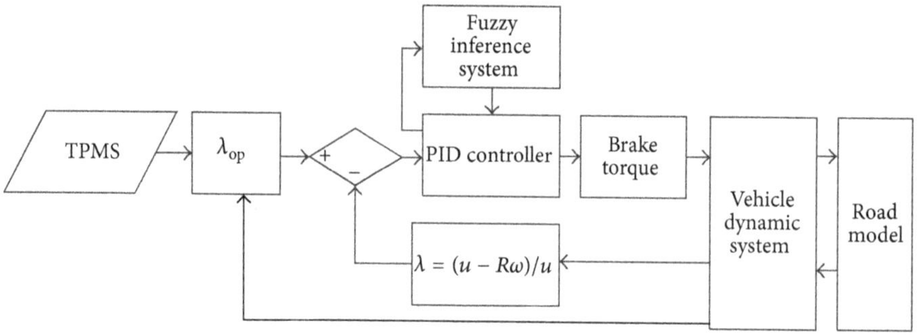

FIGURE 3: The block diagram of braking system based on the improved optimal slip ratio.

4. ABS Controller Based on the Improved Optimal Slip Ratio

The change of optimal slip ratio value can lead to a direct impact on the braking performance of ABS controller, which in turn changes both the braking distance and total braking time. In order to verify the improved expression and study the impact of tyre inflation pressure changes on braking performance, an ABS braking system based on the improved optimal slip ratio is designed, as shown in Figure 3. ABS braking system is composed of 4 parts: optimal slip ratio λ_{op} estimator, slip ratio comparator, fuzzy PID controller, and execution system. There are 3 inputs for λ_{op} estimator: tyre inflation pressure, vehicle velocity, and tyre slip angle. The road surface friction coefficient is commonly acquired by road type recognition algorithm. In order to simplify the computation, road surface friction coefficient is assigned to a constant value in this study. Real-time signal of tyre inflation pressure is acquired by direct measurement, such as commercially promoted Tyre Pressure Monitoring System (TPMS). The fuzzy PID controller is composed of a conventional PID controller and a fuzzy inference system.

Taking the error between actual slip rate and the optimal slip ratio e and error rate ec as inputs, fuzzy controller provides proportional coefficient K_p, integral coefficient K_i, and derivative coefficient K_d, its own output linguistic variables, as the input of PID controller. Design and selection of self-tuning principle, fuzzy control-rule table, membership functions, and universe range for the PID controller are referred to in [12, 13].

In order to comprehensively study the braking process and performance before and after the improvement of the optimal slip ratio, a cosimulation platform is established by combining the CarSim software and MATLAB/Simulink package based on a C-class passenger car, which then allow the simulation studies to be performed to evaluate the braking performance under different conditions. The time step is set to 0.01 s and Runge-Kutta ode45 algorithm is used. Other key simulation parameters are shown in Table 1.

5. Results and Discussion

5.1. Influence of Tyre Inflation Pressure on Wheel Velocity and Displacement. Based on the cosimulation system, the braking process and performance of the vehicle are examined for the improved optimal slip ratios by comparing them with those of original ratios. The curves of vehicle braking displacement and wheel velocity under different tyre inflation pressures and optimal slip ratios are shown in Figure 4.

If $p_i = 0.308$ MPa, for example, the wheel velocity of ABS controller aims at improved slip ratio significantly lower than the curve of the original one, rather close to the curve of the nominal pressure throughout the second-half braking process as in Figure 4(a). In order to confirm that the change in wheel velocity is due to the improvement on ABS controller, which makes the wheel velocity continuously adapted to the current optimal slip ratio, it is necessary to conduct a contrastive study of real-time slip ratios controlled by ABS controllers aiming at original and improved optimal slip ratios. Next, the real-time tyre slip ratio curves under two optimal slip ratio expressions are presented in Figure 4(b). Theoretically, according to (11) and Table 1, when tyre inflation pressure is larger than nominal pressure, the value of optimal slip ratio considering tyre pressure will slightly increase compared with the traditional slip ratio. As is shown in Figure 4(b), ABS controller aims at improved optimal slip ratio to make the *absolute* value of the wheel slip ratio continuously greater than that of a traditional one, which is consistent with the theory. Therefore, the effectiveness of improved expression and ABS controller is proved.

The differences of original velocity minus improved velocity, $u_{old} - u_{improved}$, show that a deviation occurs from the speed under different tire pressure toward the speed corresponding to the nominal pressure, as is shown in Figure 4(c).

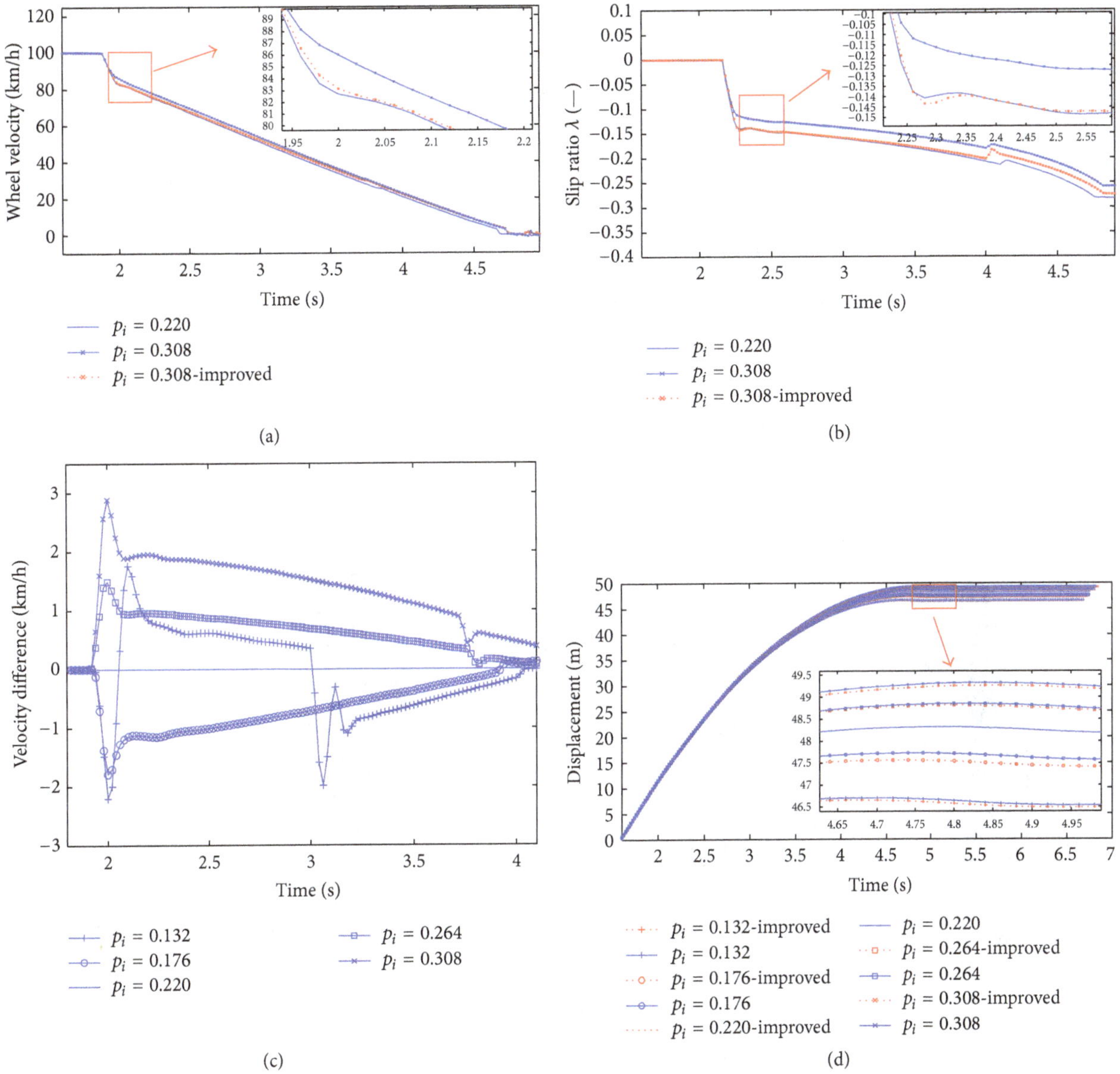

FIGURE 4: (a) Wheel velocities during the braking process. (b) Real-time tyre slip ratios during the braking process. (c) Differential value between the original velocity and the improved one. (d) Braking displacements under different tyre pressures and λ_{op} predictions.

In particular, when tyre pressure is higher than the nominal value, the wheel velocity of improved slip ratio is consistently lower than that of original one during the braking process. That is why the velocity differences of $p_i = 0.264$ MPa and $p_i = 0.308$ MPa are always positive. When the pressure is lower than the nominal value, the wheel velocity curve shows the almost opposite trend. In addition, as to the low pressure case when $p_i = 0.132$ MPa, the wheel velocity in the inception phase fluctuates strongly and remains higher than the original wheel velocity during the steady phase, which proves that ABS controller has made timely amendments to the wheel speed based on the latest improved optimal slip ratio.

Compared with the original ABS controller, braking distances and braking time under all nonnominal tyre pressures become shorter, which proves that the improved ABS controller, aiming at the improved slip ratio, has optimized the wheel velocity and thus achieved the beneficial result as shown in Figure 4(d).

5.2. Influence of Tyre Inflation Pressure on Braking Performance. In order to investigate the impact of tyre inflation pressure changes on vehicle braking performance (braking distance and braking time), simulation studies were conducted under different initial braking speeds on the basis of the original slip ratio and improved optimal slip ratio.

As shown in Figure 5(a), both original braking distance and improved braking distance increase with the initial braking speed. In addition, there is a slight rising trend with increasing tyre pressure. The braking distances before and after improvement are very close. It is hard to distinguish the influence of the improved slip ratio on the braking distance. Likewise, variation trend of the braking time is similar to that

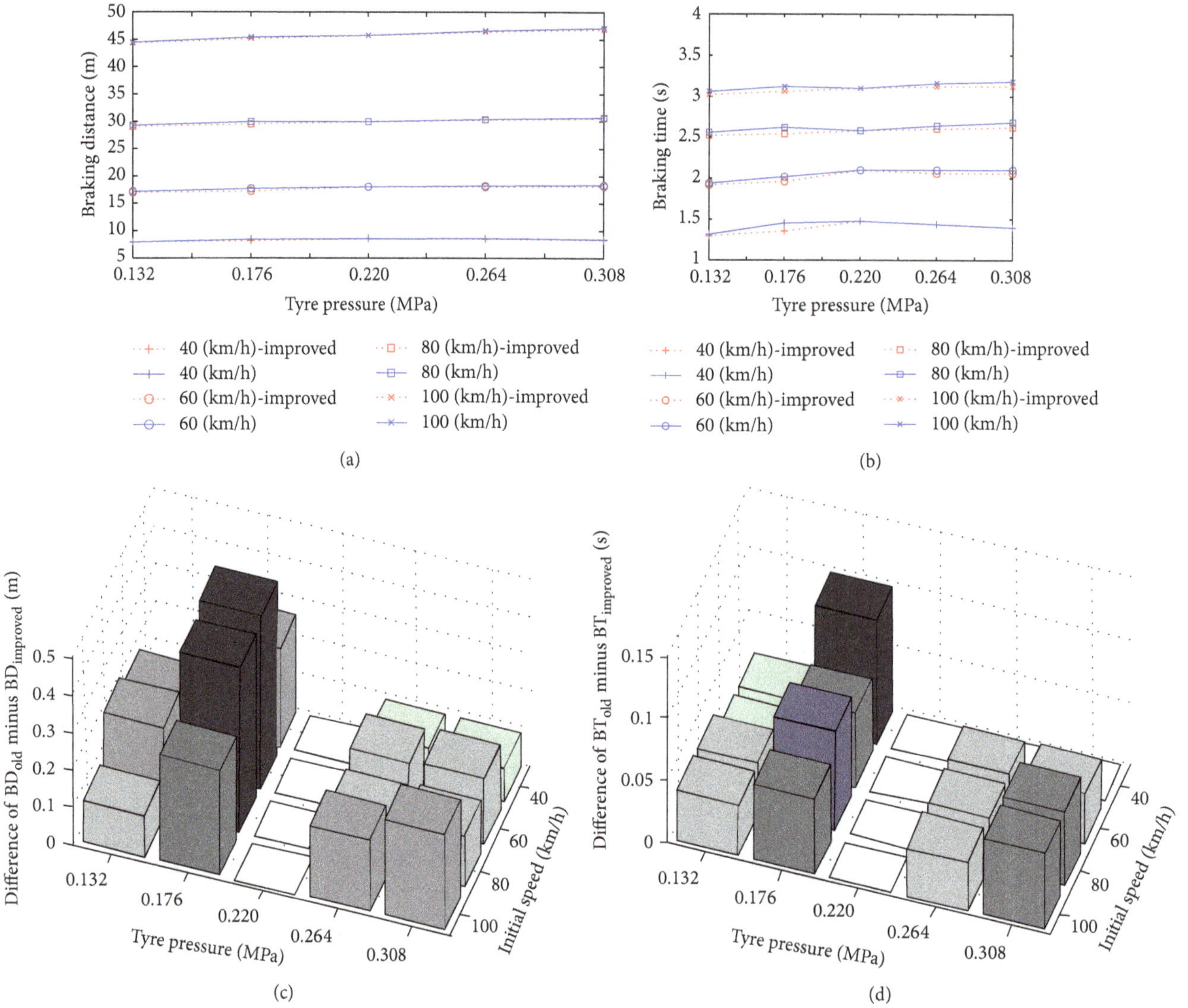

FIGURE 5: Braking performances under different initial braking speeds. (a) Comparison of braking distances before and after improvement. (b) Comparison of braking time before and after improvement. (c) Difference of original braking distance minus improved braking distance. (d) Difference of original braking time minus improved braking time.

of the braking distance, and the difference of braking time is also hard to be identified.

In order to compare and analyze the degree of reduction, the difference of braking distance and braking time before and after improvement under each condition is calculated, respectively. As shown in Figures 5(c) and 5(d), under all tyre inflation pressures and initial braking speeds, the braking distance and braking time after the improvement are all effectively shortened. According to improved optimal slip ratio expression, the value of slip ratio decreases with lowering tyre pressure, and the corresponding longitudinal force F_x increases. When tyre inflation pressure is higher than the nominal value, with the increasing pressure, the peak of F_x-κ curve began to migrate to the right, accompanied by the shape deforming. That is, when the tyre inflation pressure is different from the nominal value, the corresponding longitudinal force of the optimal slip ratio, calculated by the original prediction which does not consider the tyre pressure changes, is not the real maximum longitudinal force. The corresponding longitudinal force predicted by the improved expression considering the tyre pressure is always more closer to the maximum braking force under the real-time tyre condition. It is obvious that the greater braking force is the shorter braking time will be. Therefore, the braking performance based on the improved optimal slip ratio has been enhanced. So it is in the case of tyre pressure lower than nominal pressure.

Based on the above analysis, the essence of ABS control, aiming at the improved optimal slip ratio, is by relocating the optimal slip ratio to keep real-time wheel slip ratio near the improved value and hence the transient longitudinal force $F_{x,\text{improved}}$ always in a maximum value under current tyre pressure.

It can also be known from Figures 5(c) and 5(d) that when tyre inflation pressure is larger than nominal pressure differences of braking distance and braking time continuously increase with tyre pressure increasing. When tyre pressure is slightly lower than nominal pressure, optimization effect of the improved expression reaches the optimum. If tyre pressure continuously decreases, it tends to be less effective.

The relationship between performance differences and initial braking speed, as is shown in Figures 5(c) and 5(d), is not clearly or consistently related under the control of improved optimal slip ratio. This illustrates that the orthogonality between tyre pressure changes, wheel velocity, and other factors is good and introducing an independent second-order factor to characterize the influence of tyre inflation pressure on optimal slip ratio is reasonable and essential.

6. Conclusions

(1) Based on the ABS control aiming at improved optimal slip ratio, both vehicle braking distance and braking time, under all nonnominal tyre inflation pressures, are effectively shortened. It proves that ABS controller designed to the improved optimal slip ratio can fairly optimize the braking process under different tyre pressures.

(2) The difference of braking performances between original and improved optimal slip ratios does not change linearly with tyre inflation pressure changes. When tyre inflation pressure is slightly lower than the nominal value, the difference of braking performances is the most obvious.

(3) The influence of tyre inflation pressure changes on the prediction is less affected by initial braking speed of vehicle. Introducing an independent second-order factor to characterize the influence of tyre inflation pressure on optimal slip ratio is essential and rational.

Notations

m: A quarter of the vehicle mass
u: Wheel velocity
F_x: Longitudinal tyre force
J: Wheel inertia
ω: Wheel rotational speed
T_b: Braking torque
T_g: Rolling resistance torque
μ_i: Friction coefficient
F_{zi}: Normal force of ith wheel
R: Wheel radius
λ: Wheel longitudinal slip ratio
λ_{op}: Optimal slip ratio
δ: Tyre slip angle
λ_x: Overall longitudinal slip
B_x: Longitudinal stiffness factor
C_x: Shape factor for longitudinal force
D_x: Peak value factor
E_x: Longitudinal curvature factor
S_{vx}: Vertical shift factor
$G_{x\alpha}$: Combined slip factor
K: Longitudinal slip stiffness
p_i: Tyre inflation pressure
p_{i0}: Nominal tyre inflation pressure
dp_i: Dimensionless increment of tyre inflation pressure
p_{pxi}: Magic Formula parameter
p_{pi1}: Linear influence coefficient of tyre inflation pressure changes
p_{pi2}: Quadratic influence coefficient.

Conflict of Interests

The authors declare that there is no conflict of interests regarding the publication of this paper.

Acknowledgments

The support of the High-Tech Industrialization Projects of Shanxi Province, China (no. 2011-2368) and The Natural Science Foundation of Shanxi Province, China (no. 2012011024-2) is gratefully acknowledged. Thanks are due to Dan Song of State Grid Skill Training Centre of SEPC for her help in ABS controller modelling. The authors acknowledge the valuable comments and discussions with Hangbin Zhu of TASS International and Yonghong Wang of Shanxi Dayun Automobile Manufacturing Co., Ltd.

References

[1] G. F. Mauer, "A fuzzy logic controller for an ABS braking system," *IEEE Transactions on Fuzzy Systems*, vol. 3, no. 4, pp. 381–388, 1995.

[2] H. Wang and L. Yang, "Simulation of automotive ABS using fuzzy self-tuning PID control," *Journal of Transportation Systems Engineering and Information Technology*, vol. 12, no. 5, pp. 52–56, 2012.

[3] W. Zhu and Y. Chen, "Application and simulation of automotive ABS using fuzzy PID control," *Journal of Jiangsu University (National Science Edition)*, vol. 25, no. 4, pp. 310–314, 2004.

[4] V. Ćirović, D. Aleksendrić, and D. Mladenović, "Braking torque control using recurrent neural networks," *Proceedings of the Institution of Mechanical Engineers, Part D: Journal of Automobile Engineering*, vol. 226, no. 6, pp. 754–766, 2012.

[5] H. Guan, B. Wang, P. Lu, and L. Xu, "Identification of maximum road friction coefficient and optimal slip ratio based on road type recognition," *Chinese Journal of Mechanical Engineering*, vol. 27, no. 5, pp. 1018–1026, 2014.

[6] B. Wang, "Simulation of automobile ABS based on online tracking varying optimal slip," *Journal of Hubei Automotive Industries Institute*, no. 1, pp. 5–9, 2011.

[7] M. Mirzaei and H. Mirzaeinejad, "Optimal design of a nonlinear controller for anti-lock braking system," *Transportation Research Part C: Emerging Technologies*, vol. 24, pp. 19–35, 2012.

[8] Z. Liu, "Mathematical models of tire-longitudinal road adhesion and their use in the study of road vehicle dynamics," *Journal of Beijing Institute of Technology*, no. 2, pp. 193–204, 1996.

[9] M. Bian, "Simplified tire model for longitudinal road friction estimation," *Journal of Chongqing University of Technology (Natural Science)*, vol. 26, no. 1, pp. 1–5, 2012.

[10] H. Pacejka, *Tire and Vehicle Dynamics*, Elsevier, 2005.

[11] M. Massaro, V. Cossalter, and G. Cusimano, "The effect of the inflation pressure on the tyre properties and the motorcycle stability," *Proceedings of the Institution of Mechanical Engineers D: Journal of Automobile Engineering*, vol. 227, no. 10, pp. 1480–1488, 2013.

[12] K. Parczewski, "Effect of tyre inflation preassure on the vehicle dynamics during braking manouvre," *Eksploatacja i Niezawodnosc*, vol. 15, no. 2, pp. 134–139, 2013.

[13] H. Taghavifar and A. Mardani, "Investigating the effect of velocity, inflation pressure, and vertical load on rolling resistance of a radial ply tire," *Journal of Terramechanics*, vol. 50, no. 2, pp. 99–106, 2013.

[14] I. J. M. Besselink, A. J. C. Schmeitz, and H. B. Pacejka, "An improved Magic Formula/Swift tyre model that can handle inflation pressure changes," *Vehicle System Dynamics*, vol. 48, supplement 1, pp. 337–352, 2010.

10

Output Feedback Variable Structure Control Design for Uncertain Nonlinear Lipschitz Systems

Jeang-Lin Chang and Tsui-Chou Wu

Department of Electrical Engineering, Oriental Institute of Technology, Banciao District, New Taipei City 220, Taiwan

Correspondence should be addressed to Jeang-Lin Chang; fe035@mail.oit.edu.tw

Academic Editor: Ai-Guo Wu

This paper develops a full-order compensator-based output feedback variable structure control law for uncertain nonlinear Lipschitz systems having matched perturbations. Given that the sufficient condition is satisfied, the developed control scheme, with the observer-like technique incorporated into the design of the compensator, can achieve global exponential stabilization. An illustrative example is provided with simulation results to show the effectiveness of the proposed method.

1. Introduction

The output feedback control problem for linear and/or nonlinear systems has received considerable attention in the literature due to its importance in many practical applications. There are two major classes of output feedback control schemes: static output feedback scheme and observer-based output feedback scheme. In observer-based output feedback control, an observer is first designed to estimate unmeasurable states, and the outputs as well as the estimated states are then used to design the control law. Static output feedback control, which involves either direct output feedback control or a dynamic compensator, does not use any observer to estimate the unmeasurable states. In recent years, many static output feedback sliding mode control (OFSMC) methods [1–9] have been proposed for stabilizing multivariable plants with matched perturbations. Edwards and Spurgeon [1] have developed a synthesizing output feedback controller according to sliding mode concepts. For the plant does not satisfy the Kimura-Davison conditions, different compensator design methods have been proposed [4–9] to provide additional degree of freedom, thus obviating this constraint. However, these above-mentioned methods [1–9] cannot be implemented in nonlinear systems.

Many nonlinear systems satisfy the Lipschitz property at least locally by representing them by a linear part with Lipschitz nonlinearity around their equilibrium points. Hence, the systems with Lipschitz nonlinearities are common in practical applications. Prior studies [10–16] have developed different observer design methods for uncertain nonlinear Lipschitz systems. Nevertheless, separation principle does not generally hold for nonlinear systems. Therefore, the output feedback control problem for nonlinear Lipschitz systems is much more challenging than stabilization using full state feedback. In this paper, a solution to the output feedback variable structure control problem for uncertain nonlinear Lipschitz systems is provided. For the plant having matched perturbations and Lipschitz nonlinearity, an observer-like reference model design method is proposed and the existence condition proposed here is equivalent to the same condition of the sliding mode observer proposed by Koshkouei and Zinober [13]. Corless and Tu [17] presented a sufficient and necessary condition for the existence of this observer. Moreover, this problem can be taken as a convex optimization problem and LMI methods in prior studies [13, 18] can be employed to synthesize the parameters in the control algorithm. Note that the proposed method does not need any observer structure to estimate the system states and is a type of the dynamic compensator-based design method. Given that the condition of the controller problem is satisfied, the proposed control law can achieve global exponential stabilization.

2. Problem Formulations

Consider a class of nonlinear systems described by the following equations [11–15]:

$$\dot{\mathbf{x}}(t) = \mathbf{A}\mathbf{x}(t) + \mathbf{B}(\mathbf{u}(t) + \mathbf{d}(t)) + \mathbf{f}(\mathbf{x},\mathbf{u}), \quad \mathbf{y}(t) = \mathbf{C}\mathbf{x}(t), \tag{1}$$

where $\mathbf{x} \in \mathfrak{R}^n$, $\mathbf{u} \in \mathfrak{R}^m$, and $\mathbf{y} \in \mathfrak{R}^l$ are the system states, control inputs, and measureable outputs, respectively. The function $\mathbf{d} \in \mathfrak{R}^m$ is unknown and is a time-varying vector which represents the lump sum of matched disturbances and/or uncertainties. The function $\mathbf{f} : \mathfrak{R}^n \times \mathfrak{R}^m \rightarrow \mathfrak{R}^n$ is a nonlinear smooth vector with $\mathbf{f}(\mathbf{0},\mathbf{0}) = \mathbf{0}$ and satisfies the Lipschitz condition with respect to $\mathbf{x}$, with Lipschitz constant $\gamma > 0$; that is, for all $\mathbf{x}_1$ and $\mathbf{x}_2$

$$\|\mathbf{f}(\mathbf{x}_1,\mathbf{u}) - \mathbf{f}(\mathbf{x}_2,\mathbf{u})\| \le \gamma \|\mathbf{x}_1 - \mathbf{x}_2\|, \tag{2}$$

and $\|\mathbf{f}(\mathbf{x},\mathbf{u})\| \le \gamma\|\mathbf{x}\|$ for all $\mathbf{u} \in \mathfrak{R}^m$. In the proposed approach, the observer-like dynamic compensator is first employed to generate the reference states and the output feedback variable structure controller is then designed to stabilize system (1). Throughout this study, the following three assumptions, which were generally developed from conventional output feedback sliding mode control design methods [1–9], are made.

Assumption 1. The triple $(\mathbf{C},\mathbf{A},\mathbf{B})$ is minimum phase and $\mathrm{rank}(\mathbf{CB}) = \mathrm{rank}(\mathbf{B}) = m$.

Assumption 2. The pairs $(\mathbf{A},\mathbf{B})$ and $(\mathbf{A},\mathbf{C})$ are stabilizable and detectable, respectively.

Assumption 3. The matched perturbation is bounded by

$$\|\mathbf{d}(t)\| \le \alpha(t,\mathbf{y}), \tag{3}$$

where the scalar-valued function $\alpha(t,\mathbf{y}) > 0$ is known.

3. Output Feedback Variable Structure Controller Design

Although system (1) is imposed on the unknown disturbance and Lipschitz nonlinearity, the control objective of this section is to design the output feedback variable structure controller that can achieve globally exponential stabilization of the closed-loop system. First, the closed-loop reference model is generated by

$$\dot{\mathbf{x}}_m(t) = (\mathbf{A} - \mathbf{BK})\mathbf{x}_m(t) + \mathbf{L}(\mathbf{y}(t) - \mathbf{y}_m(t)) + \mathbf{f}(\mathbf{x}_m,\mathbf{u}), \tag{4}$$

$$\mathbf{y}_m(t) = \mathbf{C}\mathbf{x}_m(t),$$

where $\mathbf{x}_m \in \mathfrak{R}^n$ are the reference states and $\mathbf{L} \in \mathfrak{R}^{n\times l}$ is the gain designed in the latter. Let $\beta_c > 0$, $\varepsilon_1 > 0$, and $\eta_c > 0$ be the positive scalars. The gain $\mathbf{K} \in \mathfrak{R}^{m\times n}$ is decided by

$$\mathbf{K} = \frac{\beta_c}{2}\mathbf{B}^T\mathbf{P}_c, \tag{5}$$

where the positive definite matrix $\mathbf{P}_c > 0$ satisfies the following algebraic Riccati inequality:

$$\mathbf{A}^T\mathbf{P}_c + \mathbf{P}_c\mathbf{A} - \beta_c\mathbf{P}_c\mathbf{B}\mathbf{B}^T\mathbf{P}_c + \varepsilon_1\mathbf{P}_c\mathbf{P}_c + \left(\frac{\gamma^2}{\varepsilon_1} + \eta_c\right)\mathbf{I}_n < \mathbf{0}. \tag{6}$$

For the plant in (1), we design the controller as

$$\mathbf{u}(t) = -\mathbf{K}\mathbf{x}_m(t) + \mathbf{v}(t), \tag{7}$$

where the input $\mathbf{v} \in \mathfrak{R}^m$ is designed using sliding mode control. Substituting the control input (7) into system (1) yields

$$\begin{aligned}\dot{\mathbf{x}}(t) &= (\mathbf{A} - \mathbf{BK})\mathbf{x}(t) + \mathbf{B}(\mathbf{v}(t) + \mathbf{d}(t)) + \mathbf{BK}(\mathbf{x}(t) - \mathbf{x}_m(t)) + \mathbf{f}(\mathbf{x},\mathbf{u}) \\ &= (\mathbf{A} - \mathbf{BK})\mathbf{x}(t) + \mathbf{B}(\mathbf{v}(t) + \mathbf{d}(t)) + \mathbf{BK}\mathbf{e}_x(t) + \mathbf{f}(\mathbf{x},\mathbf{u}),\end{aligned} \tag{8}$$

where $\mathbf{e}_x = \mathbf{x} - \mathbf{x}_m$ and $\mathbf{e}_y = \mathbf{y} - \mathbf{y}_m = \mathbf{C}\mathbf{e}_x$ are the error states and error outputs, respectively. Then from (4) and (9), one can yield

$$\begin{aligned}\dot{\mathbf{e}}_x(t) &= (\mathbf{A} - \mathbf{BK})\mathbf{e}_x(t) + \mathbf{B}(\mathbf{v}(t) + \mathbf{d}(t)) + \mathbf{BK}\mathbf{e}_x(t) - \mathbf{LC}\mathbf{e}_x(t) + \mathbf{f}(\mathbf{x},\mathbf{u}) - \mathbf{f}(\mathbf{x}_m,\mathbf{u}) \\ &= (\mathbf{A} - \mathbf{LC})\mathbf{e}_x(t) + \mathbf{B}(\mathbf{v}(t) + \mathbf{d}(t)) + \mathbf{g}(t),\end{aligned} \tag{9}$$

where $\mathbf{g}(t) = \mathbf{f}(\mathbf{x},\mathbf{u}) - \mathbf{f}(\mathbf{x}_m,\mathbf{u})$. It follows from (2) that

$$\|\mathbf{g}(t)\| = \|\mathbf{f}(\mathbf{x},\mathbf{u}) - \mathbf{f}(\mathbf{x}_m,\mathbf{u})\| \le \gamma\|\mathbf{e}_x(t)\|. \tag{10}$$

Now we design the sliding surface as

$$\mathbf{s}(t) = \mathbf{F}\mathbf{e}_y(t), \tag{11}$$

where the matrices $\mathbf{F} \in \mathfrak{R}^{m\times l}$ are decided by

$$\mathbf{FC} = \mathbf{B}^T\mathbf{P}_o. \tag{12}$$

Furthermore, given any positive number $\varepsilon_2 > 0$, the positive definite matrix $\mathbf{P}_o > 0$ and the gain $\mathbf{L} \in \mathfrak{R}^{n\times l}$ satisfy the following algebraic Riccati inequality:

$$(\mathbf{A} - \mathbf{LC})^T\mathbf{P}_o + \mathbf{P}_o(\mathbf{A} - \mathbf{LC}) + \varepsilon_2\mathbf{P}_o\mathbf{P}_o + \left(\frac{\gamma^2}{\varepsilon_2} + \eta_o\right)\mathbf{I}_n < \mathbf{0}. \tag{13}$$

If the solution to the coupled equations (12) and (13) can be found, Koshkouei and Zinober [13] have shown that the sliding mode observer can be successfully implemented to estimate the system states. Furthermore, previous studies [17, 19] have given that there exist the matrices $\mathbf{L}$, $\mathbf{P}$, and $\mathbf{F}$ to satisfy the coupled equations (12) and (13) if and only if Assumptions 1 to 2 hold. Moreover, the way of finding the matrices **L**, **P**, and **F** that satisfy (12) and (13) has also been discussed in detail. With the help of LMI, the standard LMI forms which can be straightforwardly solved to obtain the solution of the coupled equations (12) and (13) are given in [13, 18]. To stabilize the closed-loop system and eliminate the matched perturbation, the control input $\mathbf{v}(t)$ is designed as

$$\mathbf{v}(t) = -(\alpha(t,\mathbf{y})+\rho)\frac{\mathbf{s}(t)}{\|\mathbf{s}(t)\|} = -(\alpha(t,\mathbf{y})+\rho)\frac{\mathbf{Fe}_y(t)}{\|\mathbf{Fe}_y(t)\|}, \quad (14)$$

where $\rho > 0$ is a design constant. Now consider the following Lyapunov function:

$$W(\mathbf{e}_x,\mathbf{x}_m) = \mathbf{e}_x^T(t)\mathbf{P}_o\mathbf{e}_x(t) + \varsigma\mathbf{x}_m^T(t)\mathbf{P}_c\mathbf{x}_m(t), \quad (15)$$

where $\varsigma > 0$ is a constant. Let $V(t) = \mathbf{e}_x^T(t)\mathbf{P}_o\mathbf{e}_x(t)$ and take its time derivative to obtain

$$\begin{aligned}\dot V(t) &= \mathbf{e}_x^T(t)\left((\mathbf{A}-\mathbf{LC})^T\mathbf{P}_o + \mathbf{P}_o(\mathbf{A}-\mathbf{LC})\right)\mathbf{e}_x(t) \\ &\quad + 2\mathbf{e}_x^T(t)\mathbf{P}_o\mathbf{B}(\mathbf{v}(t)+\mathbf{d}(t)) + 2\mathbf{e}_x^T(t)\mathbf{P}_o\mathbf{g}(t) \\ &\le \mathbf{e}_x^T(t)\left((\mathbf{A}-\mathbf{LC})^T\mathbf{P}_o + \mathbf{P}_o(\mathbf{A}-\mathbf{LC})\right)\mathbf{e}_x(t) \\ &\quad + 2\|\mathbf{s}(t)\|\|\mathbf{d}(t)\| - 2\|\mathbf{s}(t)\|(\alpha(t,\mathbf{y})+\rho) \\ &\quad + 2\mathbf{e}_x^T(t)\mathbf{P}_o\mathbf{g}(t) \\ &\le \mathbf{e}_x^T(t)\left((\mathbf{A}-\mathbf{LC})^T\mathbf{P}_o + \mathbf{P}_o(\mathbf{A}-\mathbf{LC})\right)\mathbf{e}_x(t) \\ &\quad - 2\rho\|\mathbf{s}(t)\| + 2\|\mathbf{P}_o\mathbf{e}_x(t)\|\|\mathbf{g}(t)\|.\end{aligned} \quad (16)$$

Since (2) holds, one can yield

$$\begin{aligned}2\|\mathbf{P}_o\mathbf{e}_x(t)\|\|\mathbf{g}(t)\| &\le \varepsilon_2\mathbf{e}_x^T(t)\mathbf{P}_o\mathbf{P}_o\mathbf{e}_x(t) + \frac{1}{\varepsilon_2}\mathbf{g}^T(t)\mathbf{g}(t) \\ &\le \varepsilon_2\mathbf{e}_x^T(t)\mathbf{P}_o\mathbf{P}_o\mathbf{e}_x(t) + \frac{\gamma^2}{\varepsilon_2}\mathbf{e}_x^T(t)\mathbf{e}_x(t).\end{aligned} \quad (17)$$

Substituting this inequality into (16) and applying the inequality in (16) to it yield

$$\begin{aligned}\dot V(t) &\le \mathbf{e}_x^T(t) \cdot \left((\mathbf{A}-\mathbf{LC})^T\mathbf{P}_o + \mathbf{P}_o(\mathbf{A}-\mathbf{LC}) + \varepsilon_2\mathbf{P}_o\mathbf{P}_o + \frac{\gamma^2}{\varepsilon_2}\mathbf{I}_n\right) \\ &\quad \cdot \mathbf{e}_x(t) - 2\rho\|\mathbf{s}(t)\| \le -\eta_o\|\mathbf{e}_x(t)\|^2 - 2\rho\|\mathbf{s}(t)\| \\ &\le -\eta_o\|\mathbf{e}_x(t)\|^2.\end{aligned} \quad (18)$$

Then taking the time derivative of (15) and applying the above inequality to it give

$$\begin{aligned}\dot W(\mathbf{e}_x,\mathbf{x}_m) &\le -\eta_o\|\mathbf{e}_x(t)\|^2 \\ &\quad + \varsigma\left(\mathbf{x}_m^T(t)\left(\mathbf{A}^T\mathbf{P}_c + \mathbf{P}_c\mathbf{A} - \beta\mathbf{P}_c\mathbf{BB}^T\mathbf{P}_c\right)\mathbf{x}_m(t)\right. \\ &\quad \left.+ 2\mathbf{x}_m^T(t)\mathbf{P}_c\mathbf{f}(t,\mathbf{x}_m,\mathbf{u}) + 2\mathbf{x}_m^T(t)\mathbf{P}_c\mathbf{LCe}_x(t)\right) \\ &\le -\eta_o\|\mathbf{e}_x(t)\|^2 \\ &\quad + \varsigma\left(\mathbf{x}_m^T(t)\left(\mathbf{A}^T\mathbf{P}_c + \mathbf{P}_c\mathbf{A} - \beta\mathbf{P}_c\mathbf{BB}^T\mathbf{P}_c\right)\mathbf{x}_m(t)\right. \\ &\quad \left.+ 2\gamma\|\mathbf{P}_c\mathbf{x}_m(t)\|\|\mathbf{x}_m(t)\| + 2\mathbf{x}_m^T(t)\mathbf{P}_c\mathbf{LCe}_x(t)\right).\end{aligned} \quad (19)$$

From

$$\begin{aligned}2\gamma\|\mathbf{P}_c\mathbf{x}_m(t)\|\|\mathbf{x}_m(t)\| \le \varepsilon_1\mathbf{x}_m^T(t)\mathbf{P}_c\mathbf{P}_c\mathbf{x}_m(t) + \frac{\gamma^2}{\varepsilon_1}\mathbf{x}_m^T(t)\mathbf{x}_m(t),\end{aligned} \quad (20)$$

it follows that

$$\begin{aligned}\dot W(\mathbf{e}_x,\mathbf{x}_m) &\le \varsigma\Bigg(\mathbf{x}_m^T(t) \cdot \left(\mathbf{A}^T\mathbf{P}_c + \mathbf{P}_c\mathbf{A} - \beta\mathbf{P}_c\mathbf{BB}^T\mathbf{P}_c + \varepsilon_1\mathbf{P}_c\mathbf{P}_c + \frac{\gamma^2}{\varepsilon_1}\mathbf{I}_n\right) \\ &\quad \cdot \mathbf{x}_m(t) + 2\kappa\|\mathbf{e}_x(t)\|\|\mathbf{x}_m(t)\|\Bigg) - \eta_o\|\mathbf{e}_x(t)\|^2 \\ &\le -\eta_o\|\mathbf{e}_x(t)\|^2 - \eta_c\varsigma\|\mathbf{x}_m(t)\|^2 + 2\varsigma\kappa\|\mathbf{e}_x(t)\| \\ &\quad \cdot \|\mathbf{x}_m(t)\|,\end{aligned} \quad (21)$$

where $\kappa = \|\mathbf{P}_c\|\|\mathbf{LC}\|$. Choosing $\varsigma = \eta_c\eta_o/\kappa^2$ results in

$$\dot W(\mathbf{e}_x,\mathbf{x}_m) \le -\frac{1}{2}\eta_o\|\mathbf{e}_x(t)\|^2 - \frac{1}{2}\eta_c\varsigma\|\mathbf{x}_m(t)\|^2. \quad (22)$$

Therefore, $\mathbf{e}_x$ and $\mathbf{x}_m$ converge exponentially to zero. Since $\mathbf{e}_x = \mathbf{x} - \mathbf{x}_m$, then

$$\mathbf{x}(t) \longrightarrow \mathbf{0} \quad \text{as } t \longrightarrow \infty. \quad (23)$$

Remark 4. To obtain the similar effect of the Luenberger observer, in this paper the term $\mathbf{LCe}_x = \mathbf{Le}_y$ is introduced into the reference model. Lavretsky and Wise [20] have first proposed this technique and applied it in adaptive control. Hence, the reference states can be considered equivalent to the system states.

Remark 5. From (4), (10), and (14), we can rewrite the sliding surface and the controller as

$$\begin{aligned}\mathbf{s}(t) &= \mathbf{Fy}(t) - \mathbf{FCx}_m(t), \\ \mathbf{u}(t) &= -\mathbf{Kx}_m(t) - (\alpha(t,\mathbf{y})+\rho)\frac{\mathbf{s}(t)}{\|\mathbf{s}(t)\|}, \\ \dot{\mathbf{x}}_m(t) &= (\mathbf{A}-\mathbf{BK})\mathbf{x}_m(t) + \mathbf{L}(\mathbf{y}(t) - \mathbf{y}_m(t)) \\ &\quad + \mathbf{f}(\mathbf{x}_m,\mathbf{u}).\end{aligned} \quad (24)$$

Hence, the proposed control algorithm does not use any observer structure and is a type of the full-order compensator-based output feedback controller.

Remark 6. When system (1) does not involve Lipschitz nonlinearity, different compensator design methods for output feedback sliding mode control have been proposed in the literature [5–9]. The design problem of the sliding surface is thus formulated as a new static output feedback problem for an augmented system. Since it is well known that the general static output feedback problem is not convex, no complete solution to this problem is known. In the proposed method, the existing well-built LMI-tools can be employed to straightforwardly solve the coupled equations (12) and (13). As a result, the advantage of the proposed method is that an analytic solution has been obtained which obviates the numerical complexities of the above-mentioned approaches [5–9].

Remark 7. The main problem of implementing controller (24) is that a phenomenon called chattering is generated due to the discontinuous function $\mathbf{s}(t)/\|\mathbf{s}(t)\|$. It can be considered the undesired chattering effect produced by the high switching action of the control input. As a result, the chattering becomes the main implementation problem of sliding mode control. Numerous techniques have been proposed to eliminate this phenomenon. One of the most common solutions to reduce the chattering is the continuous approximation techniques. However, the continuous approximation techniques have the trade-off relation between the control performance and the chattering migration. Another drawback with applying the continuous approximation techniques is the reduction of the control accuracy.

4. Numerical Example

Consider the train system, proposed by Aldeen and Sharma [15], in which the state-space form of the model including both matched disturbance and Lipschitz nonlinearity is given by

$$\dot{\mathbf{x}}(t)=\begin{bmatrix}0 & 1 & 0 & 0\\ \dfrac{-k}{M_E} & -\alpha g & \dfrac{k}{M_E} & 0\\ 0 & 0 & 0 & 1\\ \dfrac{k}{M_C} & 0 & \dfrac{-k}{M_C} & -\alpha g\end{bmatrix}\mathbf{x}(t)+\begin{bmatrix}0\\ \dfrac{1}{M_E}\\ 0\\ 0\end{bmatrix}(u(t)+d(t))+\begin{bmatrix}0\\ \dfrac{\rho}{M_E}(x_2+x_4)^2\\ 0\\ \dfrac{\rho}{M_E}(x_2+x_4)^2\end{bmatrix},$$

$$\mathbf{y}(t)=\begin{bmatrix}1 & 0 & 0 & 0\\ 0 & 1 & 0 & 0\end{bmatrix}\mathbf{x}(t), \tag{25}$$

where $\mathbf{x}=[x_1\ x_2\ x_3\ x_4]^T$ and $d(t)=\cos(t)+\sin(\pi t)$ is set as the unknown disturbance. For simulation purpose, the parameter values, being the same as those reported in [15], are $M_E=10\,\text{Kg}$, $M_C=10\,\text{Kg}$, $k=4.87\,\text{N/s}$, $\alpha=0.5\,\text{s/m}$, $\rho=1\,\text{N/s}$, and $g=9.8\,\text{m/s}^2$. It is easy to obtain this train system which does not satisfy the Kimura-Davison conditions and contains stable invariant zeros at $\{-4.6924, -0.2076\}$. Applying the developed control algorithm, we can obtain the following matrices:

$$\begin{gathered}\mathbf{K}=[37.4080\ \ 22.6673\ \ 14.3300\ \ 3.1712],\\ \mathbf{L}=\begin{bmatrix}10 & 0\\ 0 & 0.6212\\ 1.4169 & 0\\ 1.4169 & 0\end{bmatrix},\\ \mathbf{F}=[0\ \ 0.1],\\ \mathbf{P}_c=\begin{bmatrix}1.4408 & 0.1496 & 0.5065 & 0.1240\\ 0.1496 & 0.0907 & 0.0573 & 0.0127\\ 0.5065 & 0.0573 & 2.6441 & 0.4962\\ 0.1240 & 0.0127 & 0.4962 & 0.2028\end{bmatrix},\\ \mathbf{P}_o=\begin{bmatrix}0.05 & 0 & -0.0072 & -0.0010\\ 0 & 1 & 0 & 0\\ -0.0072 & 0 & 2.7804 & -0.5102\\ -0.0010 & 0 & -0.5102 & 0.2035\end{bmatrix}.\end{gathered} \tag{26}$$

Hence, the reference model is designed as

$$\dot{\mathbf{x}}_m=\begin{bmatrix}0 & 1 & 0 & 0\\ -4.2278 & -7.1667 & -0.9460 & -0.3171\\ 0 & 0 & 0 & 1\\ 0.9740 & 0 & -0.9740 & -4.9000\end{bmatrix}\mathbf{x}_m+\begin{bmatrix}10 & 0\\ 0 & 0.6212\\ 1.4169 & 0\\ 1.4169 & 0\end{bmatrix}(\mathbf{y}-\mathbf{y}_m)+\begin{bmatrix}0\\ 0.1(x_{m2}+x_{m4})^2\\ 0\\ 0.2(x_{m2}+x_{m4})^2\end{bmatrix},$$

$$\mathbf{y}_m=\begin{bmatrix}1 & 0 & 0 & 0\\ 0 & 1 & 0 & 0\end{bmatrix}\mathbf{x}_m \tag{27}$$

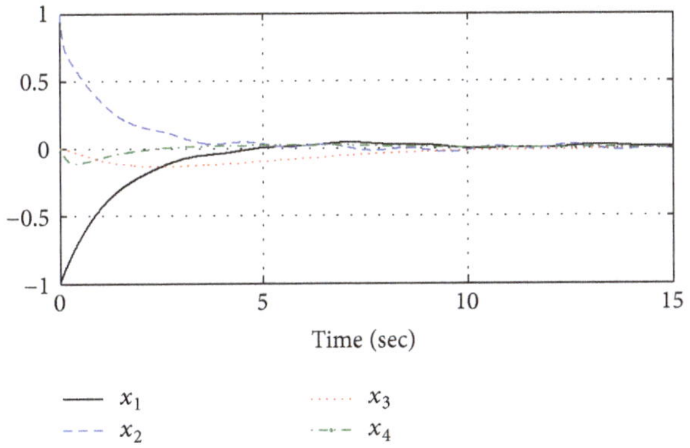

FIGURE 1: System states of the proposed method.

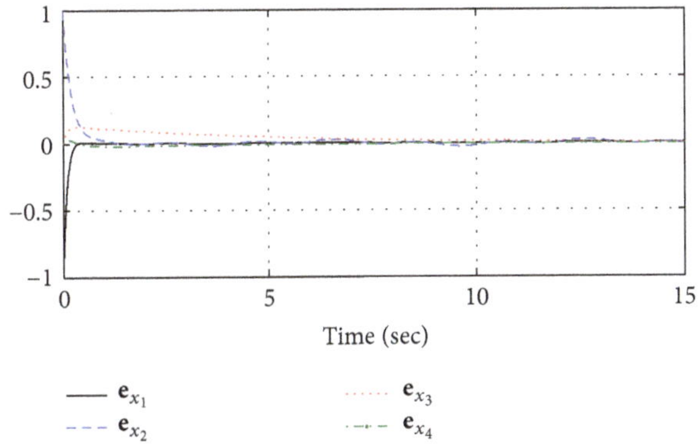

FIGURE 2: Error states of the proposed method.

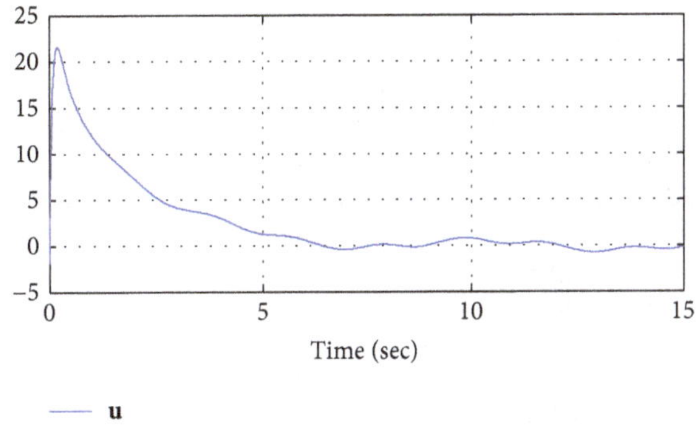

FIGURE 3: Response of the control input.

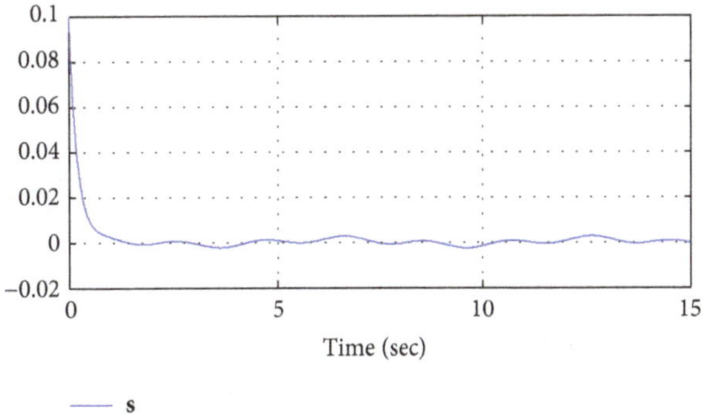

FIGURE 4: Response of the sliding surface.

and the sliding surface is chosen as

$$\mathbf{s}(t) = \begin{bmatrix} 0 & 0.1 \end{bmatrix} (\mathbf{y}(t) - \mathbf{y}_m(t)). \tag{28}$$

To avoid the chattering problem, we apply the continuous approximation in the controller and then design the control inputs as

$$\mathbf{u}(t) = -\begin{bmatrix} 37.4080 & 22.6673 & 14.3300 & 3.1712 \end{bmatrix} \mathbf{x}_m(t) - 2\frac{\mathbf{s}(t)}{\|\mathbf{s}(t)\| + 0.01}. \tag{29}$$

Figures 1–4 illustrate the simulation results obtained using the initial conditions $\mathbf{x}(0) = \begin{bmatrix} -1 & 1 & 0 & 0 \end{bmatrix}^T$. Figures 1 and 2 plot the evolution of the system states and error states, respectively. Figure 3 depicts the control input. The response of the sliding surface is given in Figure 4. As seen in Figure 1, although the nominal system has both matched disturbance and Lipschitz nonlinearity, the proposed control law can successfully restrain the effect of matched disturbance and achieve the desired performance. Moreover, robust stability of the closed-loop system can be guaranteed.

5. Conclusions

This study examines the output feedback variable structure control problem for a class of Lipschitz nonlinear systems with matched perturbations. The proposed method, which is simple by nature involving only its original system matrices and does not need any observer structure, can obtain exponential stabilization of the closed-loop system. The existence condition, which is equivalent to that used in sliding mode observer, is developed for the design of the proposed dynamic compensator-based controller. The numerical example demonstrates that the proposed algorithm can be successfully implemented.

Conflict of Interests

The authors declare that there is no conflict of interests regarding the publication of this paper.

References

[1] C. Edwards and S. K. Spurgeon, *Sliding Mode Control Theory and Application*, Taylor & Francis, London, UK, 1998.

[2] R. El-Khazali and R. Decarlo, "Output feedback variable structure control design," *Automatica*, vol. 31, no. 6, pp. 805–816, 1995.

[3] H. H. Choi, "Variable structure output feedback control design for a class of uncertain dynamic systems," *Automatica*, vol. 38, no. 2, pp. 335–341, 2002.

[4] J.-L. Chang, "Dynamic output integral sliding-mode control with disturbance attenuation," *IEEE Transactions on Automatic Control*, vol. 54, no. 11, pp. 2653–2658, 2009.

[5] S. K. Bag, S. K. Spurgeon, and C. Edwards, "Output feedback sliding mode design for linear uncertain systems," *IEE*

Proceedings—Control Theory and Applications, vol. 144, no. 3, pp. 209–216, 1997.

[6] C. Edwards and S. K. Spurgeon, "Compensator based output feedback sliding mode controller design," *International Journal of Control*, vol. 71, no. 4, pp. 601–614, 1998.

[7] C. Edwards and S. K. Spurgeon, "Linear matrix inequality methods for designing output feedback sliding mode controllers," *IEE Proceedings—Control Theory and Applications*, vol. 203, pp. 539–545, 2003.

[8] C. Edwards, S. K. Spurgeon, and R. G. Hebden, "On the design of sliding mode output feedback controller," *International Journal of Control*, vol. 76, no. 9-10, pp. 893–905, 2003.

[9] H. H. Choi, "Sliding-mode output feedback control design," *IEEE Transactions on Industrial Electronics*, vol. 55, no. 11, pp. 4047–4054, 2008.

[10] R. Rajamani, "Observers for Lipschitz nonlinear systems," *IEEE Transactions on Automatic Control*, vol. 43, no. 3, pp. 397–401, 1998.

[11] A. M. Pertew, H. J. Marquez, and Q. Zhao, "H_∞ synthesis of unknown input observers for non-linear Lipschitz systems," *International Journal of Control*, vol. 78, no. 15, pp. 1155–1165, 2005.

[12] Q. P. Ha and H. Trinh, "State and input simultaneous estimation for a class of nonlinear systems," *Automatica*, vol. 40, no. 10, pp. 1779–1785, 2004.

[13] A. J. Koshkouei and A. S. I. Zinober, "Sliding mode state observation for non-linear systems," *International Journal of Control*, vol. 77, no. 2, pp. 118–127, 2004.

[14] P. R. Pagilla and Y. Zhu, "Controller and observer design for Lipschitz nonlinear systems," in *Proceedings of the American Control Conference (AAC '04)*, pp. 2379–2384, IEEE, Boston, Mass, USA, July 2004.

[15] M. Aldeen and R. Sharma, "Estimation of states, faults and unknown disturbances in non-linear systems," *International Journal of Control*, vol. 81, no. 8, pp. 1195–1201, 2008.

[16] H.-L. Choi and J.-T. Lim, "Output feedback stabilization for a class of Lipschitz nonlinear systems," *IEICE Transactions on Fundamentals of Electronics, Communications and Computer Sciences*, vol. 88, no. 2, pp. 602–605, 2005.

[17] M. Corless and J. Tu, "State and input estimation for a class of uncertain systems," *Automatica*, vol. 34, no. 6, pp. 757–764, 1998.

[18] C. P. Tan and C. Edwards, "An LMI approach for designing sliding mode observers," *International Journal of Control*, vol. 74, no. 16, pp. 1559–1568, 2001.

[19] C. Edwards, X.-G. Yan, and S. K. Spurgeon, "On the solvability of the constrained Lyapunov problem," *IEEE Transactions on Automatic Control*, vol. 52, no. 10, pp. 1982–1987, 2007.

[20] E. Lavretsky and K. Wise, *Robust and Adpative Control: With Aerospace Applications*, Springer, Berlin, Germany, 2012.

11

Fuzzy Weight Cluster-Based Routing Algorithm for Wireless Sensor Networks

Teng Gao,[1,2] Jin-Yan Song,[3] Jin-Hua Ding,[1] and De-Quan Wang[1]

[1]*School of Mechanical Engineering & Automation, Dalian Polytechnic University, Dalian 116034, China*
[2]*School of Control Science and Engineering, Dalian University of Technology, Dalian 116024, China*
[3]*School of Information Engineering, Dalian Ocean University, Dalian 116023, China*

Correspondence should be addressed to Teng Gao; gaoteng@dlpu.edu.cn

Academic Editor: Onur Toker

Cluster-based protocol is a kind of important routing in wireless sensor networks. However, due to the uneven distribution of cluster heads in classical clustering algorithm, some nodes may run out of energy too early, which is not suitable for large-scale wireless sensor networks. In this paper, a distributed clustering algorithm based on fuzzy weighted attributes is put forward to ensure both energy efficiency and extensibility. On the premise of a comprehensive consideration of all attributes, the corresponding weight of each parameter is assigned by using the direct method of fuzzy engineering theory. Then, each node works out property value. These property values will be mapped to the time axis and be triggered by a timer to broadcast cluster headers. At the same time, the radio coverage method is adopted, in order to avoid collisions and to ensure the symmetrical distribution of cluster heads. The aggregated data are forwarded to the sink node in the form of multihop. The simulation results demonstrate that clustering algorithm based on fuzzy weighted attributes has a longer life expectancy and better extensibility than LEACH-like algorithms.

1. Introduction

In recent years, applications of sensor networks have evolved in many areas due to their large applicability and development possibilities, especially in the wireless sensor networks (WSNs) area [1]. A wireless sensor network consists of a large number of light-weight sensor nodes having limited battery life, computational capabilities, storage, and bandwidth [2]. The potential applications of sensor networks are highly varied, such as environmental monitoring, target tracking, battle field surveillance, monitoring the enemy territory, detection of attacks, and security etiquette [3]. An important aspect of such networks is that the nodes are unattended, resource-constrained, their energy cannot be replenished, and network topology is unknown [4]. The node which lost energy may cause the malfunction of the entire network. Therefore, the research on WSNs has mainly been focused on saving the limited energy and extending the life time of wireless sensor networks. Researchers have developed many theories to save energy from almost every aspect, but we have our sight on routing protocol.

An efficient routing protocol is the one which consumes minimum energy and provides large coverage area [5]. Based on the logical structure, the routing protocol is divided into two categories. The first category is flat routing, in which all nodes in the network are coequal and there are no special nodes. The advantage of this type of protocols is their robustness. The other category is hierarchical-based routing. One of the most classical paradigms of hierarchical-based routing is the clustering, in which cluster is an infrastructure and nodes play different roles. In a clustering architecture, cluster head nodes can be used to process and send the information to the sink node while member nodes can be used to perform the sensing in the proximity of the target and transmit the information to corresponding cluster head. Clustering provides an efficient way of saving energy within a cluster and outside cluster and inside a wireless sensor network. The cluster head acts as a bridge between other sensor nodes and sink node and sometimes between one cluster head and other cluster head in multihop cases [6]. This means that creation of clusters and assigning special tasks to cluster heads can greatly contribute to overall system scalability, lifetime, and energy efficiency [7]. Thus, cluster-based routing takes great advantage over the plat-based one at above performances. However, the disadvantage of cluster-based routing is that the cluster head is so vital that it becomes the bottleneck of the entire network. Therefore, the selection of the cluster head will influence the performance of the entire network. The existing clustering algorithms differ on the criterium for the selection of the cluster heads. According to the current research findings, the selection method of the cluster head can be divided into several categories below.

The classical routing protocols that are based on k-means clustering such as LEACH [8], TEEN [9], and APTEEN [10] select cluster head based on a random acquired value. If this value is less than a certain threshold, the nodes will be the cluster head. Whereas because of the randomicity during the selection, the selected cluster head is prone to be distributed improperly and unevenly, this could cause the uneven distribution of the traffic flow in different cluster head nodes. One of the direct consequences is that some cluster heads exhaust energy; at the mean time the performance of the entire network is affected.

Some distributed routing protocols based on a certain attribute are proposed in DCHS [11], HCDA (the Highest-Connectivity Degree Algorithm) [12], and ACMWN [13]. The attributes that can determine cluster head selection include residual energy, neighbors number, the cost that communicate in intracluster, and the distance between the node and the sink node and ID. Because only one attribute is taken into account in these protocols, the selected cluster head cannot be the most suitable node. Although the rationality of the cluster head selection is improved to a certain extent, certain problems such as the unevenly distributed cluster head and the imbalance load remain unsolved.

Multiattribute cluster head selection protocol such as HEED [14] and WCA [15], which use several attribute to determine the cluster head, are greatly favored due to the consideration of various factors. The advantage of multiattribute cluster head selection is that a better partition of cluster can be obtained. The two protocols both adopt successive screening method to determine the cluster head, by which the finite iteration must be implemented. The major drawback of the former is that distributed algorithm makes each node unaware of global information so that some nodes may not join any clusters, while the latter need to iterate many times if many attributes are used to gain a better performance, which will increase time complexity and consume more energy.

WCA-LEACH [16], MWBC [17], WCA-GSEN [18], and AOW-LEACH [19] combine multifactors such as residual energy, communication cost, and neighbor nodes number in order to avoid the randomicity in the cluster head selection of LEACH. However, all the algorithms above determine weight of each factor using trial and error method, which will influence the performance of the whole protocol.

From the analysis and comparison mentioned above, multiattribute cluster head selection can obtain the unparalleled rationality in partition of cluster; therefore, we consider the residual energy, neighbors number, the cost that communicate in intracluster, and the distance between the node and the sink node as the attribute to propose a new clustering routing algorithm using fuzzy weight multiattribute (CFWA) to determine cluster head selection, by which the energy can be saved and the lifetime of the whole network will be extended.

2. Conform the Weight of Attribute

To save energy and balance load, the residual energy is the most crucial factor of the attributes during the process of cluster head selection. The cost that is used in the communication in intracluster and the neighbors number also influence cluster head selection. Nevertheless, the distance between the node and the sink node will not be considered due to the fact that uniform distribution of the cluster head is required by energy-efficient cluster-based algorithm. In this paper, the direct method [20] based on the abutting object relative membership degree in engineering fuzzy theory and intelligence decision-making is adopted in order to confirm the proportion of each attribute that is hold during cluster head selection.

Definition 1. Compare the member O_k with another member O_l on duality about weightiness in the object set O. When O_k is more important than O_l,

$$0.5 < \beta_{kl} \leqslant 1; \tag{1}$$

when O_l is more important than O_k,

$$\begin{gathered} 0 \leqslant \beta_{kl} < 0.5, \\ \beta_{kl} = 1 - \beta_{lk}; \end{gathered} \tag{2}$$

when O_k has the same importance as O_l,

$$\begin{gathered} \beta_{kl} = 0.5, \\ \text{especially, } \beta_{kk} = 0.5, \end{gathered} \tag{3}$$

where β_{kl} is named relatively weightiness fuzzy value between the object O_l and O_k. Particularly, if the object sequencing about weightiness is $O_1 \prec O_2 \prec \cdots \prec O_m$, $\beta_{k1,k1+1}$ $(k_1 = 1, 2, \ldots, m-1)$ is defined as the abutting object relatively weightiness fuzzy value.

Assumption 2. In the available attributes, it is assumed that residual energy (E_r) is more significant than the cost that communicate in intracluster (Cost), and the latter is more important than neighbors number (Deg).

That is, residual energy has the unexampled importance than the cost that communicate in intracluster while the latter is more important than neighbors number ratherish. The relevant fuzzy value that $\beta_{Er,\mathrm{Cost}}$ is 1 and $\beta_{\mathrm{Cost,Deg}}$ is 0.55 can be found out based on Table 1 [21].

Based on the assumption about relative significance fuzzy scale value, provided that the object E_r is more important than Cost, $\beta_{Er,\mathrm{Cost}}$ is the corresponding significance degree when just comparing objects E_r and Cost, of which benchmark is E_r, the more important one between these two objects. Because $\beta_{Er,Er}$, which is the fuzzy scale value that

TABLE 1: Relationships between mood operator and fuzzy value.

Mood operator	Fuzzy scale value	Memberships value
Similar	0.5	1
Ratherish	0.55	0.905
Slightly	0.6	0.667
Relatively	0.65	0.538
Obviously	0.7	0.429
Markedly	0.75	0.333
Quite	0.8	0.25
Very	0.85	0.176
Extremely	0.9	0.111
Violently	0.95	0.053
Incomparable	1	0

the object E_r compares to itself, is 0.5, if the only two objects E_r and Cost are still compared, the degree that E_r belongs to significance is $\beta_{Er} = 1$, and the one of Cost is $\beta_{\text{Cost}} = 1.5 - \beta_{Er,\text{Cost}}$. Therefore, the relationship of the significance degree between E_r and Cost is

$$\frac{\beta_{Er}}{\beta_{\text{Cost}}} = \frac{1}{1.5 - \beta_{Er,\text{Cost}}}. \quad (4)$$

The nonnormalization weight may be figured out:

$$\begin{aligned} \omega'_{Er} &= 1, \\ \omega'_{\text{Cost}} &= \omega'_{Er}(1.5 - 1) = 0.5, \\ \omega'_{\text{Deg}} &= \omega'_{\text{Cost}}(1.5 - 0.55) = 0.475. \end{aligned} \quad (5)$$

The object weight vector that obtained after normalization and reverting suffix is

$$\omega = \left(\omega_{Er}, \omega_{\text{Cost}}, \omega_{\text{Deg}}\right) = (0.5063, 0.2532, 0.2405). \quad (6)$$

On account of the diffidence in the unit of each attribute, normalization procedure will be implemented.

The normalization expression of residual energy is $E_r/E_{\max}$, in which $E_{\max}$ is the original energy of each node. A proportional function relationship exists between residual energy and cluster head selection; that is, the node whose residual energy is higher has more chances to be a cluster head.

The costs that communicate in intracluster, which is obtained by calculating received signal strength information (RSSI), is normalized as $\text{RSSI}_{\text{ave}}/\text{TSSI}$, where TSSI is the transmission signal strength value, which will be the same in the broadcasting phase of each node. RSSI_{ave} denotes the average strength value of all the wireless signals that have been received. The bigger the value is, the lower the cost is. RSSI_{ave} is also proportional to the probability that the node can be selected as the cluster head.

According to the conclusion that is drawn by Heinzelman et al. [22], the relationship of cluster number and energy consumption in the scene of 100 nodes is showed in Figure 1, from which the optimal nodes number in a cluster can be deduced; in other words, the optimal neighbor number can be confirmed. The normalization function $F(\text{Deg})$ that denotes the neighbor number and the energy expenditure relationship is fit based on Figure 1:

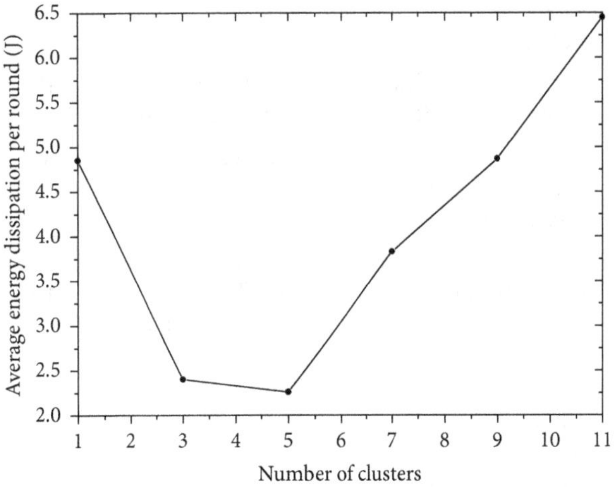

FIGURE 1: The relationship of cluster number and energy consumption in the scene of 100 nodes in LEACH.

$$F(\text{Deg}) = \frac{6.479 \times x^2 - 226.6x + 3001}{n \times E_{\max} \times (x^2 - 10.17x + 243.4)}. \quad (7)$$

Here, n is the total number of nodes.

The absolute attribute degree value of each node based on OWA operator can be calculated out by the following object function:

$$F_1 = \omega_{Er} \cdot \left(\frac{E_r}{E_{\max}}\right) + \omega_{\text{Cost}} \cdot \left(\frac{\text{RSSI}_{\text{ave}}}{\text{TSSI}}\right) + \omega_{\text{Deg}} \cdot F(\text{Deg}). \quad (8)$$

In the same manner, the function

$$F_2 = \left(\frac{E_r}{E_{\max}}\right)^{\omega_{Er}} + \left(\frac{\text{RSSI}_{\text{ave}}}{\text{TSSI}}\right)^{\omega_{\text{Cost}}} + F(\text{Deg})^{\omega_{\text{Deg}}} \quad (9)$$

is the absolute attribute degree value based on GOWA plus operator while the expression

$$F_3 = \left(\frac{E_r}{E_{\max}}\right)^{\omega_{Er}} \cdot \left(\frac{\text{RSSI}_{\text{ave}}}{\text{TSSI}}\right)^{\omega_{\text{Cost}}} \cdot F(\text{Deg})^{\omega_{\text{Deg}}} \quad (10)$$

calculates the absolute attribute degree value based on GOWA multiplication operator

$$\text{Obviously, } \omega_{Er} + \omega_{\text{Cost}} + \omega_{\text{Deg}} = 1. \quad (11)$$

3. System Module

3.1. Network Module

(a) All sensor nodes cannot move after being deployed, and each node has a unique ID.

(b) There is the only one sink node which lies outside the network.

(c) All sensor nodes are homogeneous, with no GPS equipment on it. All nodes are time synchrony.

(d) Each node has the ability to aggregate data; as a result several data packages can be compressed as one package.

(e) If the node knows the transmission power, it can calculate out the approximate distance between the transmitter and receiver based on the RSSI

$$\text{RSSI} = A - 10n \log 10 (d), \tag{12}$$

where d represents the distance; A is the RSSI value when transmitter and receiver are 1 m apart; n is the environmental factor.

(f) The battery that cannot be supplied is the main energy supply of the node. However, the node is able to adjust transmission power freely to save energy based on the distance from the receiver.

(g) The energy of the sink node is infinite.

(h) The bidirectional channel is defined through the whole network.

3.2. Wireless Channel Module. The same wireless channel module is put to use in LEACH [8] and this paper, which is composed of free space module and two-ray ground module. The boundary distance d_0 is used to differentiate the service conditions, when communication distance between transmitter and receiver is less than d_0 and the free space module will be adopted. Otherwise, if the communication distance is beyond d_0, two-ray ground module will be used, in which the energy that is consumed in transmitter sending data is in proportion to the biquadrate of the communication distance. Therefore, the trait of the module mentioned above is that the transmitter automatically uses different wireless channel module to work out the energy amount required in sending data in terms of communication distance.

Energy efficiency is the pivotal issue of WSNs, which requires free space module to be used at best in the communication between the transmitter and receiver, for which the communication distance between nodes should keep within the distance d_0. In a clustering structure network, the distance between cluster head usually is longer than that between cluster head and its corresponding member node, which need communication radius to be less than the distance $d_0/2$ in intracluster if the distance that is less than d_0 is anticipant in intercluster. By limiting communication distance, the energy is saved at last.

According to the wireless channel module defined above, the energy module below is available.

3.3. Energy Module. The energy consumption that the transmitter sends k bits data to the receiver with the distance d is

$$E_t(k,d) = \begin{cases} kE_{\text{elec}} + k\varepsilon_{fs}d^2 & d < d_0 \\ kE_{\text{elec}} + k\varepsilon_{mp}d^4 & d \geq d_0. \end{cases} \tag{13}$$

The node received k bits data, which consumes energy as follows:

$$E_r(k) = kE_{\text{elec}}. \tag{14}$$

If a node spends E_{fusion} energy to aggregate one bit, then the energy used in aggregating m data packages to a single package is

$$E_f(m,k) = mkE_{\text{fusion}}. \tag{15}$$

4. CFWA Algorithm Description

The first goal of our work is to tackle the problem of the cluster head maldistribution which will result in unbalanced load in the whole network and premature death of some nodes. The resolvent is to use the fuzzy weight attribute degree algorithm to establish a cluster-based routing in network layer.

CFWA algorithm is composed of two phases, initialization and operation. There are several time slices in initialization phase used to receive the signal from the sink node and implement flooding to obtain the grads level. Operation phase contains setup phase and steady-state phase. In the setup phase, there are 4 subperiods, including node broadcasting, cluster head broadcasting, member joining cluster, and TDMA schedule broadcasting. The steady-state phase contains a few rounds, and these rounds consist of several frames. The time structure is shown in Figure 2.

4.1. Initialization. The sink node broadcasts a beacon at a certain power; the sensor node who received the beacon signal should limit within a region at the radius of ($d_{\max} - d_0/8$).

After a period of delay, the sink node broadcasts another beacon at the maximum power of the sensor node, by which the radio wave covers a circle region at the radius of $d_{\max}$. The node that has received this beacon evaluates the distance between the sink node and itself based on the RSSI, as well as gaining the grads level 0.

After evaluating distance, the node who received either beacon turns off transceiver and goes into dormancy. The node who only received the second beacon wakes up and starts broadcasting its own grads level at the radius of $d_0/8$ at a random time in the certain interval during which all nodes that hold the same grads level will complete broadcasting their own grads levels and then goes to sleep again. And the nodes, which have never received any signals before, receive this message and set its grads level as 1 (received message plus 1), from which the distance between the sink node and itself is considered as ($d_{\max} + d_0/8$). When a node has received any message about grads level, it goes into dormancy immediately. After the broadcasting that is implemented by the nodes whose grads level is 0 and which has only received the second beacon is ended, the receivers broadcast their grads levels at the same radius of $d_0/8$ at a random time before going into dormancy. The node who receives this message sets their grads level to 2 and goes to sleep until the timeout of the nodes who broadcast the message "1." The rest may be deduced by analogy until each node in the network has a grads level, as shown in Figure 3.

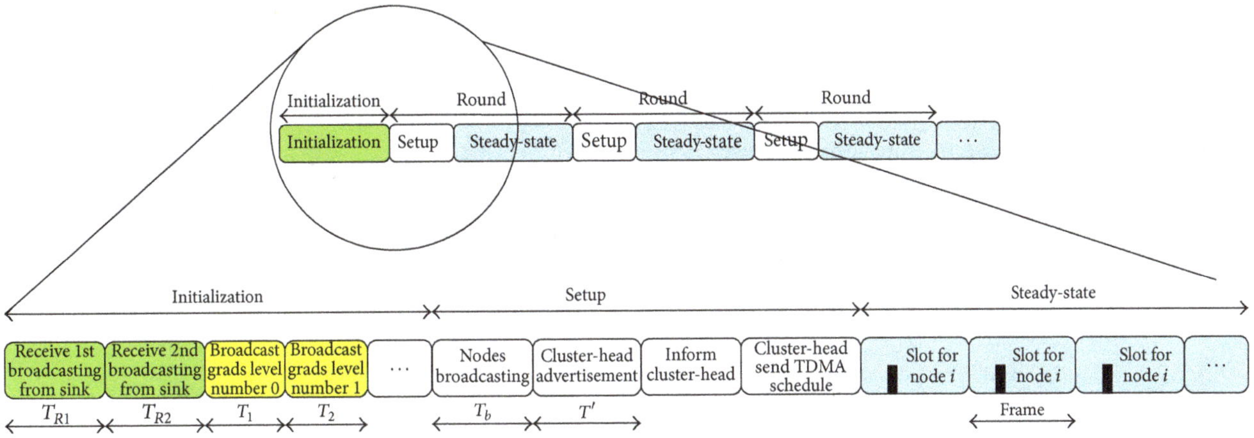

FIGURE 2: CFWA algorithm time structure diagram.

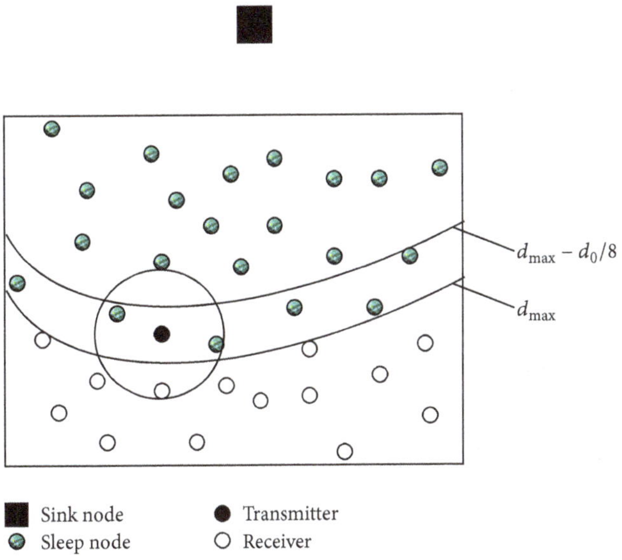

FIGURE 3: Flooding in initialization phase.

Algorithm 1 is the pseudocode of the initialization.

4.2. Clustering. Each node broadcasts a message $\langle \text{ID}, E_r \rangle$ at a certain power in a period of time T_b, which covers a region at the radius of $d_0/2$. Each node receives the messages from neighbors and stores the information into memory after the end of broadcasting, based on which each node calculates out neighbor number Deg, the average residual energy, and the cost in intracluster communication RSSI_{ave}. Thus the node can obtain all attributes it wants.

A calculation will be implemented in terms of formula (8) or (9) or (10) to obtain the absolute attribute degree F by each node. The node whose absolute attribute degree F is bigger has higher probability of being the cluster head than the smaller one, because the former has great advantage over the latter in the energy efficiency. Then the absolute attribute degree F is mapped onto the time axis before the cluster head broadcasts by means of the timer triggering, from which the node whose absolute attribute degree is bigger broadcasts cluster head information earlier.

A timer T_i, whose time span is determined by the absolute attribute degree F_i, is set for each node. For the reason that the node whose absolute attribute degree is bigger broadcasts cluster head information at earlier time, the following equation is available:

$$T_i'' = (1 - F_i) \times T'. \quad (16)$$

Here T' is the total time in which all cluster heads broadcast information. However, data packages collision is inevitable if the nodes that hold the same absolute attribute degree value in the network implement the simultaneous cluster head broadcasting. To avoid this, a random number between 0 and 1 is introduced to generate disturbance. A constant λ is set to be 0.9, by which the relationship between T_i and the absolute attribute degree F_i would not be affected. Thus the improved equation is described as follows:

$$T_i = \{1 - [\lambda \times F_i + (1 - \lambda) \times \text{rand}(0, 1)]\} \times T'. \quad (17)$$

Equation (17) makes the absolute attribute degree value map onto the time axis, based on which the node whose absolute attribute degree F is bigger will have a timeout earlier. When time is up, a $\langle \text{ID} \rangle$ package is broadcasted at the radius of d_0 by the cluster head. The nodes who can receive and parse the package correctly will lost the chance of being the cluster heads if the sender is lying in the neighbor list that is stored in memory, which is the radio coverage method, which makes the cluster head distribute evenly in the whole network.

The cluster head broadcasting phase is finished when time T' is up.

4.3. The Establishment of Routing in Intercluster and in Intracluster. After receiving the broadcasting of cluster heads, the member nodes select the nearest cluster head based on the RSSI and send the join information to it. The distance between the member node and the corresponding cluster head is evaluated as well.

```
void time_snychronization( );
int grads = 0;
if (receive(hellol)) {flag = 1;}
else if (receive(hello2))
{           /*estimate distance between sink and itself. d_0 is the
             demarcation point between free space model and
      two ray ground model. RSSI(SK) and RSSI(ID)
      represent the received signal strength at sink node
      and this node, respectively.*/
     double dtosk = pow(10.0, (RSSI(SK) − RSSI(ID) + x_i(sigma))/(10 * lambda))/d_0;
     if (flag != 1) broadcast(grads);
}
/*X and Y are the bandaries of deployment area*/
while (d_max + grads * d_0/8 < d_max + Y)
{
         If (receive(grads))
         {
            grads++;
            //d_max is the farthest distance to receive the sink's signal
            dtosk = d_max + grads * d_0/8;
            broadcast(grads);
         }
}
```

ALGORITHM 1: The pseudocode implementation of the initialization.

If the distance between the member node and the cluster head is more than its distance to the sink node, the member node will communicate with the sink node directly at a fixed time slice regardless of cluster head while going to dormancy at the rest time to save energy.

Algorithm 2 shows the pseudocode of the setup phase. As for a cluster head, the nearest cluster head will be selected to join into based on the RSSI if the distance between the sink node and the selected cluster head is shorter than the distance between the source cluster heads and the sink node. The distance between the relational cluster heads is evaluated in the same way. If the distance between cluster heads is more than the distance between the cluster head and the sink node or there is no cluster head nearer the sink node than itself, this cluster head communicates directly with the sink node.

The cluster head assigns a time slot for each member after receiving all join information, by which a TDMA schedule is schemed. The cluster heads who communicate directly with the sink node promulgate the schedule firstly, and the other cluster heads, who cannot communicate directly with the sink node, promulgate the schedule only when it received the schedule from its cluster head of upper level.

As one of the members, the cluster head communicate with its cluster head of upper level at the appointed time slot, when the routing in intercluster is established.

The routing is simpler in intracluster. The member of nodes, who go to dormancy at the rest time to save energy, communicates directly with the cluster head at the appointed time slot.

Similar to LEACH, the usage of a TDMA/CDMA MAC will reduce inter- and intracluster collisions in CFWA family algorithms.

4.4. Data Transmission. The interval used in data transmission is much longer than the time of setup phase so that the energy dissipation can be reduced further. Compared with the LEACH-like algorithms, CFWA family algorithms have longer time in data transmission.

At data transmission phase, the member nodes send information to the cluster head according to the schedule and then go into dormancy, while the cluster head must keep under working state to receive the information coming from its members and send the aggregated data to the next hop at the time slot that is assigned by the cluster head of upper level. The cluster head who communicates directly with the sink node implements data fusion after a frame and then sends the aggregated data to the sink node.

5. Simulation and Analysis

5.1. The Selection of Simulation Platform. NS2 is adopted as the simulation platform in this paper. As a discrete event simulator, NS2, in which the object-oriented design technique is introduced and plenty of function modules are furnished, can simulate and analyze various network protocols and draw very intuitionistic conclusions about the performance analysis of the system.

LEACH-like algorithms such as LEACH, AOW-LEACH, and DCHS are simulated and compared with CFWA family algorithms in the same scene, as the parameters are set in Table 2.

5.2. Description Comparison. From Figure 4 it is clear that CFWA achieved more well-proportioned cluster description among the algorithms. Due to not many limitations on the

```
broadcast(ID, E_r);
calculate(Deg); //Obtain the neighbor number
calculate(RSSI_ave); //Obtain the average RSSI of my neighbors
//normalization procedure of Deg
F(Deg) = (6.479 * Deg^2 - 226.6Deg + 3001)/(n * E_max * (x^2 - 10.17x + 243.4));
if (CFWA_1)
{F = w1 * (E_r/E_max) + w2 * (RSSI_ave/TSSI) + w3 * F(Deg); }
elseif (CFWA_2)
{F = pow(E_r/E_max, w1) + pow(RSSI_ave/TSSI, w2) + pow(F(Deg), w3); }
elseif (CFWA_3)
{F = pow(E_r/E_max, w1) * pow(RSSI_ave/TSSI, w2) * pow(F(Deg), w3); }
T_i = (0.9 * (1 - F) + 0.1 * rand(0, 1)) * T';
if (!receive(cluster head) && T_i == now)//now is the current time, not receive//any advertisement
{
  broadcast(cluster head); //declare itself as cluster head
  headFlag = 1; //cluster head mark is set
}
if (T' == now)//if has received some message
{
    *p = receivedCHList[ ]; //load in the list
    if (length(receivedCHList) != 0)
    {
      currentCH = selectCH(*p); //select cluster head
       join(currentCH); //send join information
    }
    else
    {
            sleep( );
    }
}
if (HeadFlag == 1)//cluster head
{
  creatTDMA( ); //generate TDMA schedule
  if (currentCH == Sink) broadcast(TDMA);
  if (receive(TDMA)) broadcast(TDMA);
}
else{
    receive(TDMA);
}
```

ALGORITHM 2: The pseudocode implementation of the setup phase.

TABLE 2: Parameters in simulation.

Parameter	Value
Initial energy (J)	2
Data packet size (Bytes)	500
Threshold distance (d_0) (m)	86
Packet header size (Bytes)	25
E_{elec} (nJ/bit)	50
E_{fusion} (nJ/bit/signal)	5
ε_{fs} (pJ/bit/m^2)	10
ε_{mp} (pJ/bit/m^4)	0.0013

radius of the clusters and cluster heads selection, so the cluster distribution is casual in LEACH. DCHS only limited the cluster heads selection on energy; hence the cluster description is not even. AOW-LEACH took the cluster heads selection into account, in which some parameters play roles for the even distribution of cluster. However, the scale of the cluster was not restricted, so that the cluster description was not very homogeneous. CFWA adopted multiple approaches such as limiting cluster radius and selecting cluster heads according to several parameters, to ensure the uniform cluster description.

CFWA family algorithms are composed of CFWA_1, CFWA_2, and CFWA_3, which is developed in terms of the different absolute attribute degree value F from (8), (9), and (10), respectively. The simulation firstly takes place in the network with 100 nodes, in which the deployment area is $100\,\text{m} \times 100\,\text{m}$ and the sink node is located at (50, 175). The simulation results are described in Figures 5, 6, and 7.

5.3. Performance Analysis. Figure 5 denotes the relation between nodes number alive and runtime, from which it is obvious that CFWA family algorithms enhance 30%

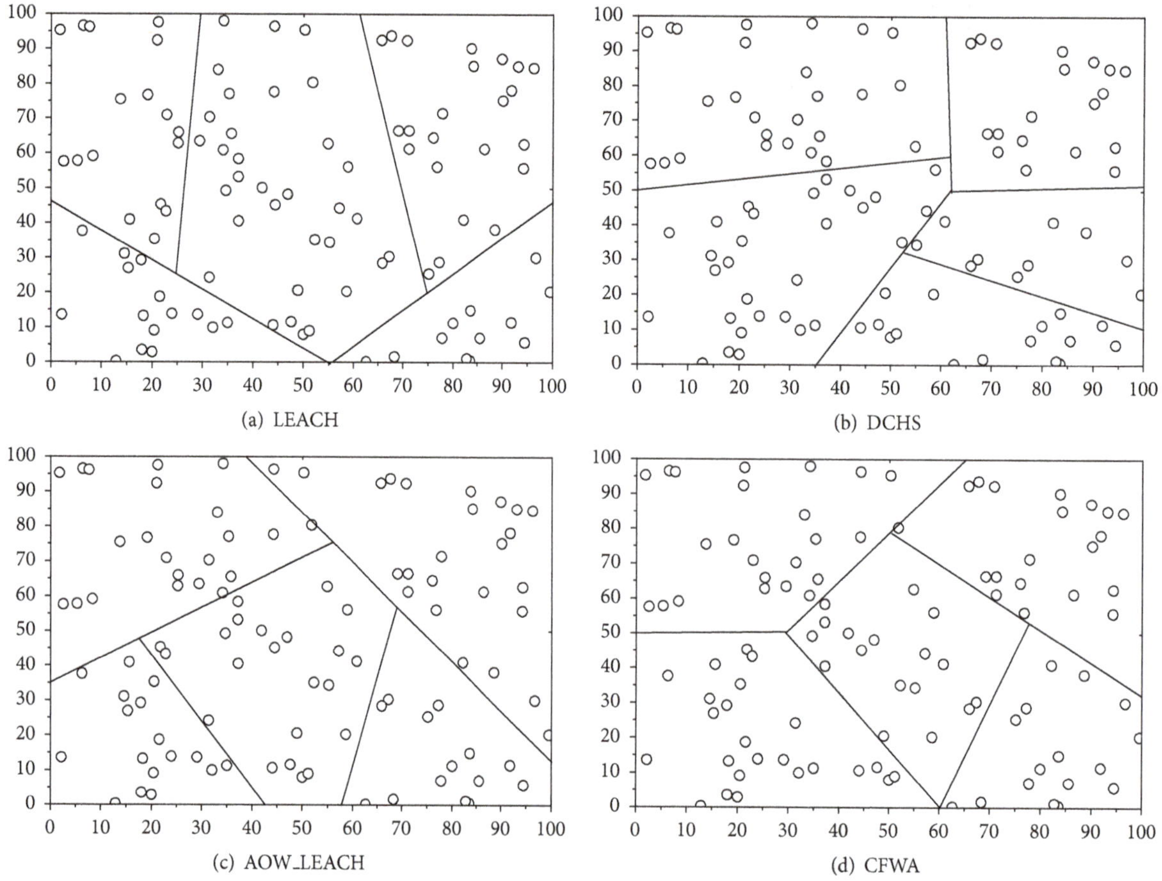

Figure 4: Cluster description comparison in a certain round.

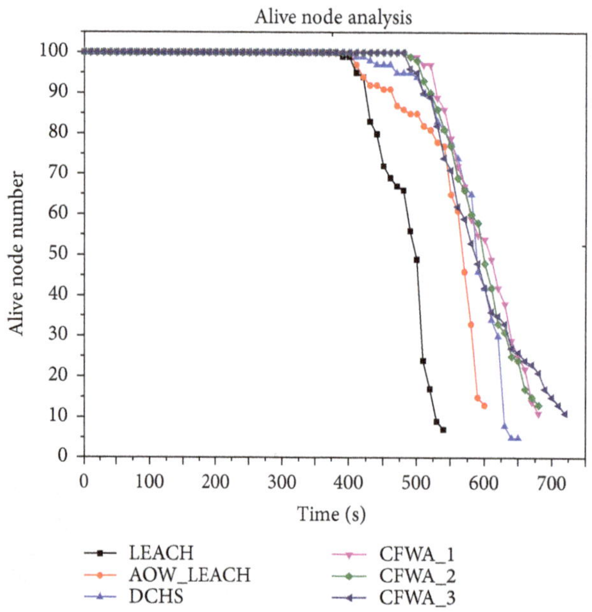

Figure 5: Alive nodes comparison diagram between CFWA family algorithms and LEACH-like algorithms in 100 nodes.

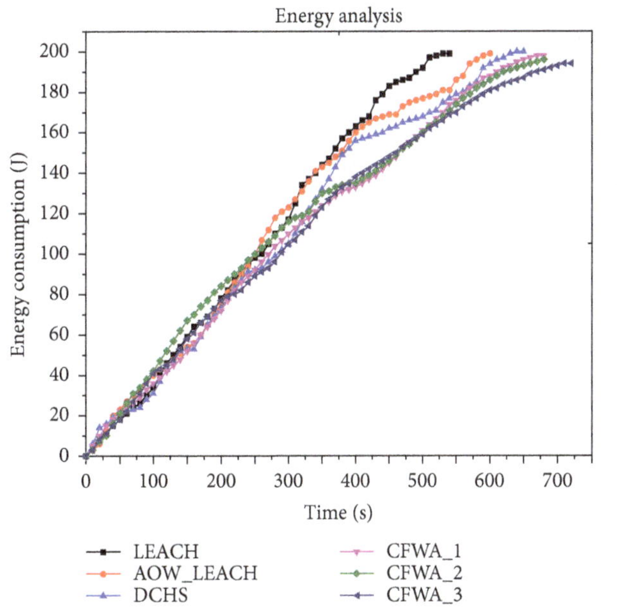

FIGURE 6: Energy analysis chart between CFWA family algorithms and LEACH-like algorithms in 100 nodes.

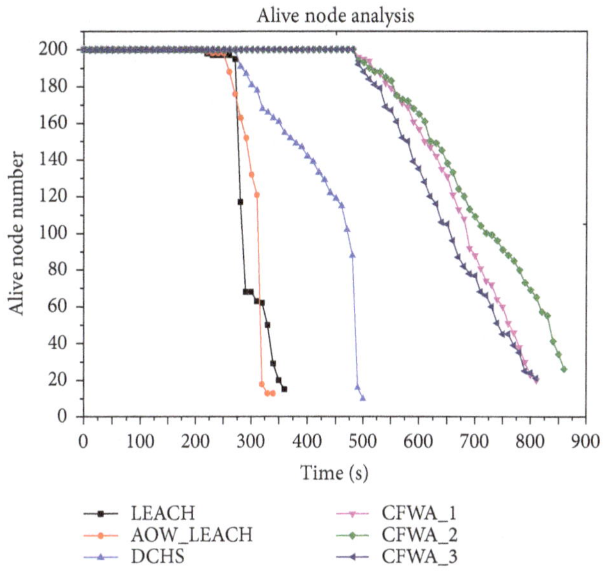

FIGURE 8: Alive nodes comparison diagram between CFWA family algorithms and LEACH-like algorithms in 200 nodes.

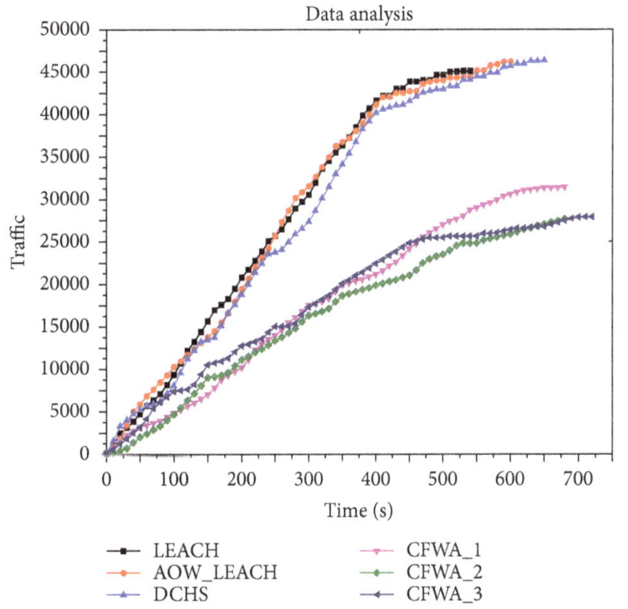

FIGURE 7: Traffic comparison diagram between CFWA family algorithms and LEACH-like algorithms in 100 nodes.

approximately more than LEACH on the total runtime of the entire network, as well as 5–10% more than AOW_LEACH and DCHS. The time of the first dead node is at 480th second in CFWA family algorithms while 400th second is available in LEACH-like algorithms, which is a great improvement. The main reason for this result is that the power that can cover the circle region at the radius of $d_0/2$ is used to broadcast information to neighbors, in addition to the multihop routing in intercluster, and the longer interval of data transmission is adopted. These measures reduce and balance the energy consumption of the whole network. From the point of view of the individual, the energy of each node is saved and used efficiently, so that the lifetime of the network is extended.

Energy analysis demonstrates that CFWA family algorithms have consumed similar energy at each round during the network operation while they have the different energy dissipation at each round in LEACH-like algorithms especially after the first node death, as described in Figure 6. The reason for this phenomenon is that the radio coverage method is carried out to ensure the uniform distribution of the clusters, which is conducive to balancing the energy depletion of the entire network. Furthermore, CFWA family algorithms consume less energy than LEACH-like algorithms at each round. This is because the multihop routing is used to forward data to the sink, which makes cluster heads avoid sending the data to the sink directly.

The traffic that is received by the sink node is shown in Figure 7, from which it is indicated that the traffic of CFWA family algorithms is much less than that of LEACH-like algorithms. The reason for the great differences of traffic in the operation of two kinds of algorithms is that cluster head node only implements data aggregation once before data is sent to the sink node in LEACH-like algorithms while multiple data aggregations are run during the process of data being transmitted to the sink node.

The performance of the network with 200 nodes is also evaluated through the simulations. The same parameters as that in 100 nodes scene are used to create the simulation model, and the results are demonstrated in Figures 8, 9, and 10, respectively. From the charts we can clearly see that the phenomenons emerged from the simulations in the scene with 200 nodes which is more obvious than that in the scene of 100 nodes. This is due to the increased cluster number. When nodes quantity increases, the cluster number

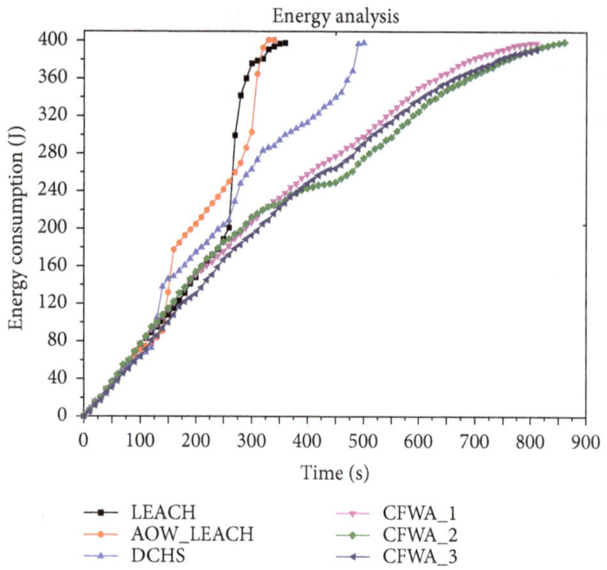

FIGURE 9: Energy analysis chart between CFWA family algorithms and LEACH-like algorithms in 200 nodes.

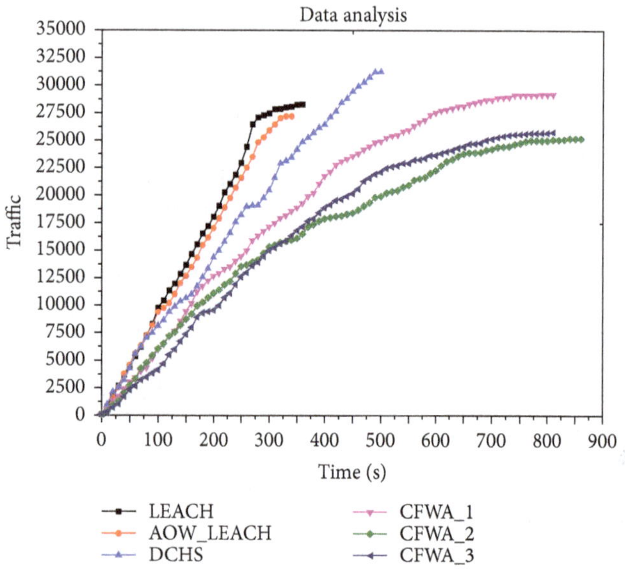

FIGURE 10: Traffic comparison diagram between CFWA family algorithms and LEACH-like algorithms in 200 nodes.

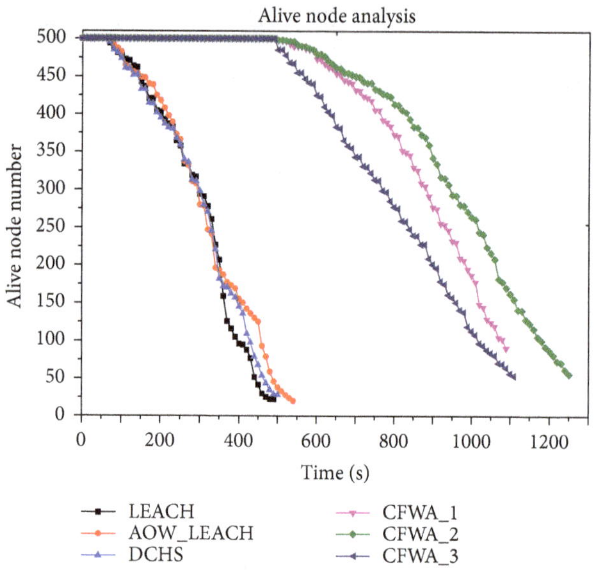

FIGURE 11: Alive nodes comparison diagram between CFWA family algorithms and LEACH-like algorithms in 500 nodes.

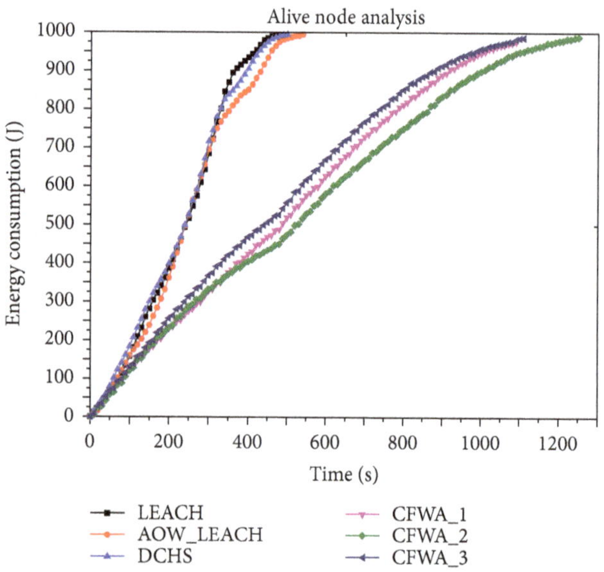

FIGURE 12: Energy analysis chart between CFWA family algorithms and LEACH-like algorithms in 500 nodes.

is also increased. In this case, multiple cluster heads transmit data to the sink node at random time in the interval of a round, which will result in the severe collisions, bakeoffs, and retransmission of data if there is the lack of time management. When the total number of nodes in the whole network is 200, the actions mentioned above will consume lots of energy and shorten the lifetime of the network. Oppositely, the TDMA mechanism is adopted in CFWA family algorithms to avoid data collision so that the death time of the first node and the lifespan of the whole network of CFWA family algorithms are longer than that of LEACH-like algorithms.

Similar situation happens in the network of 500 nodes, in which the deployment area is 200 m × 200 m and the sink node is located at (100, 275), as described in Figures 11, 12, and 13.

The performance of CFWA_1, CFWA_2, and CFWA_3 is similar in the three scenes, which denotes that the reliability, the stability, and the scalability of CFWA family algorithms are especially excellent.

5.4. Parameters

5.4.1. Node Broadcasting Time T_b. Each node broadcasts information to the neighbors at the radius of $d_0/2$ at the beginning of each round, the time span of which is the pivotal

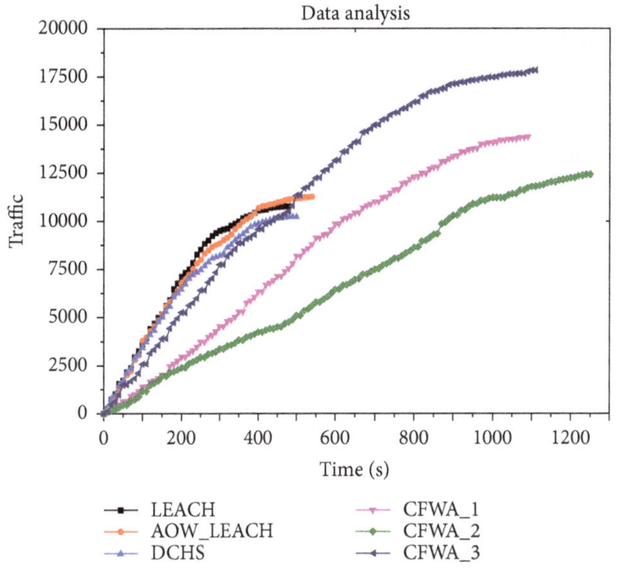

Figure 13: Traffic comparison diagram between CFWA family algorithms and LEACH-like algorithms in 500 nodes.

factor that may influence the usage of the energy. If T_b is too large, each node will increase waiting time so as to consume unwanted energy of idle state. However, the parameter is connected with the network size. If the network size is too large, T_b must be enlarged in order to avoid the collision that happened on account of broadcasting in the limited time.

5.4.2. Cluster Head Broadcasting Time T'. Cluster head broadcasting is transmitted by radio in turn based on the time order that is mapped by the absolute attribute degree of its own. T', which is the total time span in the process of the broadcasting of cluster heads, is also a significant factor that influences energy efficiency. If T' is too small, the radio coverage method will not be implemented. That is, there is delay during the propagation of radio wave. If a node has broadcasted the cluster head advertisement, the information needs a short period of time to transfer. Just in this period, a certain neighbor node may declare itself to be the cluster head because the timer has been triggered and no information is received. In this case, several cluster heads maybe lie in the adjacent regions or the same cluster. The large T' will result in energy consumption in the waiting time of idle state, which makes the lifetime of the whole network shorten.

5.5. Complexity Analysis. It is assumed that there are n nodes in the network, and the nodes broadcast $n\langle \mathrm{ID}, E_r\rangle$ messages during the cluster head selection, followed by k cluster head broadcasting if k cluster heads are selected all over the network. Even if only one cluster head can communicate directly with the sink node, $n-1$ join messages will be broadcasted by all nodes. Furthermore, k cluster heads will broadcast at most k TDMA schedule subsequently. Thus, the total message spending in the phase of cluster forming is $n+k+n-1+k=2n$ in the whole network, which denotes that the message complexity of CFWA family algorithms in the setup phase is $O(n)$.

All nodes finish broadcasting within T_b, while the timer of each node will stop when cluster head broadcasting interval T' is over. Likewise, the process of nodes' joining clusters and cluster head broadcasting TDMA schedule is also accomplished in fixed interval. Therefore, the time complexity of the algorithm CFWA in setup phase is $O(1)$.

5.6. Network Scalability Analysis. The direct communication with the sink node is adopted in LEACH-like algorithms, which will limit the network size to a great extent. This is mostly because some cluster heads are far away from the sink node and cannot communicate with it even if the largest power is used, which results in the waste of the energy of some cluster. The multihop relay is used to forward data in CFWA, with the distributed algorithm that only needs local information in cluster forming algorithm. Therefore, the routing, which is established by CFWA algorithm, is suitable to large scale wireless sensor networks.

6. Conclusions

Clustering routing algorithm is an important research issue, which will also influence the operational efficiency of network. On the basis of analysis and comparison of some classical algorithm, a novel clustering routing algorithm CFWA is proposed. The fuzzy weight absolute degree is introduced to make the most factors that can influence energy efficiency become an organic whole to determine the selection of the cluster head, which is the main innovation and improvement of the classical algorithms. Moreover, CFWA supports data fusion both in intercluster and in intracluster, which can eliminate the redundant data effectively so as to reduce the traffic and save the energy. In addition, CFWA selects the nearest path to forward the aggregated data to the sink node at the type of multihop by comparing the distance between node and the corresponding cluster head and that between node and the sink node. The simulation results show that the lifetime and energy efficiency of CFWA family algorithms is better than the classical algorithm.

Although improvements are made in some performance, there are some limits such as time synchrony and fault tolerant in using this algorithm yet. CFWA belongs to the table-driven routing algorithm so that this protocol is most appropriate when constant monitoring by the sensor network is needed.

Another disadvantage of CFWA algorithm is that delay generating in the data transmission process from a node to the sink node is too long. This is because data fusion is implemented at each cluster head in the path toward the sink node.

Furthermore, the ant colony optimization technique should be introduced into the direct methodology in order to achieve a better cluster head distribution all over the network, and spare cluster head and path should be used to promote the robustness in further work.

Conflict of Interests

The authors declare that there is no conflict of interests regarding the publication of this paper.

Acknowledgments

This work was funded in part by a grant from National Natural Science Foundation of China no. 5307012. This work was also partly supported by the fund of the general program of Liaoning Provincial Department of Education Science Research, no. L2013210, and Dalian Polytechnic University Youth Grants, no. QNJJ201307.

References

[1] Z. Zinonos, C. Chrysostomou, and V. Vassiliou, "Wireless sensor networks mobility management using fuzzy logic," *Ad Hoc Networks*, vol. 16, pp. 70–87, 2014.

[2] K. W. Sha, J. Gehlot, and R. Greve, "Multipath routing techniques in wireless sensor networks: a survey," *Wireless Personal Communications*, vol. 70, no. 2, pp. 807–829, 2013.

[3] B. A. Said, E. Abdellah, and M. Ahmed, "Gateway and cluster head election using fuzzy logic in heterogeneous wireless sensor networks," in *Proceedings of the International Conference on Multimedia Computing and Systems (ICMCS '12)*, pp. 761–766, Tangier, Morocco, May 2012.

[4] L. Barolli, Q. Wang, E. Kulla, B. Kamo, F. Xhafa, and M. Younas, "A fuzzy-based simulation system for cluster-head selection and sensor speed control in wireless sensor networks," in *Proceedings of the 3rd International Conference on Emerging Intelligent Data and Web Technologies (EIDWT '12)*, pp. 16–22, Bucharest, Romania, September 2012.

[5] S. Maurya and A. K. Daniel, "Hybrid routing approach for heterogeneous wireless sensor networks using fuzzy logic technique," in *Proceedings of the 4th International Conference on Advanced Computing and Communication Technologies (ACCT '14)*, pp. 202–207, IEEE, Rohtak, India, February 2014.

[6] P. Kumari, M. P. Singh, and P. Kumar, "Survey of clustering algorithms using fuzzy logic in wireless sensor network," in *Proceedings of the International Conference on Energy Efficient Technologies for Sustainability (ICEETS '13)*, pp. 924–928, April 2013.

[7] J. N. Al-Karaki and A. E. Kamal, "A taxonomy of routing techniques in wireless sensor networks," in *Handbook of Sensor Networks: Compact Wireless and Wired Sensing Systems*, M. Ilyas and I. Mahgoub, Eds., pp. 116–139, CRC Press, 2005.

[8] W. R. Heinzelman, A. P. Chandrakasan, and H. Balakrishnan, "Energy-efficient communication protocol for wireless microsensor networks," in *Proceedings of the 33rd Annual Hawaii International Conference on System Siences (HICSS '00)*, pp. 3005–3014, January 2000.

[9] A. Manjeshwar and D. P. Agarwal, "TEEN: a routing protocol for enhanced efficiency in wireless sensor networks," in *Proceedings of the 15th International Parallel and Distributed Processing Symposium*, pp. 2009–2015, IEEE, San Francisco, Calif, USA, April 2000.

[10] A. Manjeshwar and D. P. Agarwal, "APTEEN: a hybrid protocol for efficient routing and comprehensive information retrieval in wireless sensor networks," in *Proceedings of the 2nd International Workshop on Parallel and Distributed Computing Issues in Wireless Networks and Mobile Computing (IPDPS '02)*, pp. 195–202, Lauderdale, Fla, USA, April 2002.

[11] M. J. Handy, M. Haase, and D. Timmermann, "Low energy adaptive clustering hierarchy with deterministic cluster-head selection," in *Proceedings of the 4th IEEE Conference on Mobile and Wireless Communications Networks*, pp. 368–372, Stockholm, Sweden, 2002.

[12] A. K. Parekh, "Selecting routers in ad-hoc wireless networks," in *Proceedings of the the SBT/IEEE International Telecommunications Symposium (ITS '94)*, pp. 420–424, Rio de Janeiro, Brazil, August 1994.

[13] C. R. Lin and M. Gerla, "Adaptive clustering for mobile wireless networks," *IEEE Journal on Selected Areas in Communications*, vol. 15, no. 7, pp. 1265–1275, 1997.

[14] O. Younis and S. Fahmy, "Heed: a hybrid, energy-efficient, distributed clustering approach for ad hoc sensor networks," *IEEE Transactions on Mobile Computing*, vol. 3, no. 4, pp. 366–379, 2004.

[15] M. Chatterjee, S. K. Das, and D. Turgut, "WCA: a weighted clustering algorithm for mobile ad hoc networks," *Journal of Cluster Computing*, vol. 5, no. 2, pp. 193–204, 2002.

[16] J.-W. Zhang, Y.-Y. Ji, J.-J. Zhang, and C.-L. Yu, "A weighted clustering algorithm based routing protocol in wireless sensor networks," in *Proceedings of the ISECS International Colloquium on Computing, Communication, Control, and Management (CCCM '08)*, pp. 599–602, IEEE, Guangzhou, China, August 2008.

[17] H.-Q. Huang, D.-Y. Yao, J. Shen, K. Ma, and H.-T. Liu, "Multi-weight based clustering algorithm for wireless sensor networks," *Journal of Electronics & Information Technology*, vol. 30, no. 6, pp. 1489–1492, 2008.

[18] Y. Y. Ji, J. W. Zhang, and C. L. Yu, "An improvement routing protocol by weighted clustering algorithm in WSN," *Journal of Hangzhou Dianzi University*, vol. 28, no. 6, pp. 29–32, 2008.

[19] B. Cai and X. D. Chen, "Clustering algorithm based on automatic on-demand weighted for sensor networks," *Microelectronics & Computer*, vol. 25, no. 11, pp. 129–132, 2008.

[20] X. C. Huang, *A study on theories and methodologies for fuzzy multi-objective decision makings with their applications [Ph.D. dissertation]*, Dalian University of Technology, Dalian, China, 2003.

[21] S. Y. Chen, *Engineering Fussy Theories and Application*, Press of Dalian University of Technology, Dalian, China, 1998.

[22] W. B. Heinzelman, A. P. Chandrakasan, and H. Balakrishnan, "An application-specific protocol architecture for wireless microsensor networks," *IEEE Transactions on Wireless Communications*, vol. 1, no. 4, pp. 660–670, 2002.

12

High-Order Sliding Mode-Based Synchronous Control of a Novel Stair-Climbing Wheelchair Robot

Juanxiu Liu, Yifei Wu, Jian Guo, and Qingwei Chen

School of Automation, Nanjing University of Science and Technology, Nanjing 210094, China

Correspondence should be addressed to Yifei Wu; wuyifei0911@163.com

Academic Editor: Hung-Yuan Chung

For the attitude control of a novel stair-climbing wheelchair with inertial uncertainties and external disturbance torques, a new synchronous control method is proposed via combing high-order sliding mode control techniques with cross-coupling techniques. For this purpose, a proper controller is designed, which can improve the performance of the system under conditions of uncertainties and torque perturbations and also can guarantee the synchronization of the system. Firstly, a robust high-order sliding mode control law is designed to track the desired position trajectories effectively. Secondly, considering the coordination of the multiple joints, a high-order sliding mode synchronization controller is designed to reduce the synchronization errors and tracking errors based on the controller designed previously. Stability of the closed-loop system is proved by Lyapunov theory. The simulation is performed by MATLAB to verify the effectiveness of the proposed controller. By comparing the simulation results of two controllers, it is obvious that the proposed scheme has better performance and stronger robustness.

1. Introduction

With the rapid increase of the elderly population over the age of 60, population aging will be an outstanding performance of global population trends in the 21st century. Aging society brings a lot of problems, such as nursing for the elderly and medical problems. At the same time, thousands of people lost the ability to walk each year caused by a variety of accidents, natural disasters, and diseases. With the development of the society and the improvement of human civilization, the people with disabilities need to use modern high-tech to improve their freedom and quality of life. Hence, wheelchair robot used to help the disabled or elderly people walking has become a hot research area in recent years.

Although the barrier-free accessibility has been disseminated in recent years, stairs and other architectural barriers still exist in many cities and buildings. Since the standard wheelchair has no capability of crossing barriers, a number of stair-climbing wheelchairs which can help the disabled or elderly people overcoming obstacles have been researched. The common stair-climbing mechanisms used in the stair-climbing wheelchairs are tracks, wheels, and hybrid structures. Tracked stair-climbing wheelchair can guarantee the stability of the users in the process of ascending and descending stairs. Lawn et al. [1] designed a tracked wheelchair capable of negotiating large number of twisting and irregular stairs. In [2], a wheelchair using cluster wheels was developed. The wheelchair seat was kept stable during the stairs ascending process and the user needed not to face down the stairs. In [3], Wheelchair.q using triple cluster wheels was designed. The cluster wheels systems usually have complex mechanisms and the stability is lower than crawler systems. A stair-climbing robot with legs and wheels was designed in [4]. Chen and Pham [5] designed a prototype which was comprised of a pair of rotational multilimbed structures. There are some other design schemes adopting hybrid structures in [6–8].

Since the stair-climbing wheelchair is used in complex terrain, the first thing that should be considered is high precious position control for wheelchair system. In [2], PID control was used to provide appropriate torque during climbing process. A fuzzy controller was applied to correct the errors in direction and position misalignment, so that the final posture of the tracked mobile robot was corrected in [9]. In [10], an active tension control law combined with the computed torque method was obtained for wheelchair robot during the stair-climbing process, which can track the reference input curve of homonymic constraint force when tracking reference input curve of each joint. Although the control strategies mentioned above can make some efforts for the control of stair-climbing wheelchair, various system uncertainties and torque perturbations were not taken into consideration.

Since the sliding mode control has strong robustness for system disturbances and unmodeled dynamics, it has been widely used in robot control. Conventional sliding mode control can achieve the first-order sliding motion of the system states, which means the relative degree of the sliding variable s is 1. When the system states slide along the preset manifold, only the sliding variable is guaranteed to converge to zero, and its derivative is nonsmooth, so the chattering phenomenon always comes up in the sliding mode control system. In order to solve the chattering which exists in conventional sliding mode control, a variety of methods were proposed such as state-dependent gain method, observer-based chattering suppression, and the high-order sliding model control (HOSMC) in [11, 12]. In [13], with the HOSMC used in attitude control of large-scale spacecraft, the robustness of the system with respect to uncertainties and external disturbances was improved, and the chattering phenomenon was attenuated. In [14], a HOSMC was designed for a flexible link space robotic arm with payload, which exploited the robustness properties of SMC, while also increasing accuracy by reducing chattering effects.

Another important problem that should be considered is the coordination of the stair-climbing wheelchair system due to its characteristic of multiple joints. If the stair-climbing wheelchair works in a noncoordination manner, the assembly task will be failed, and a more serious consequence is that the users will be injured. In [15], a new control approach to position synchronization of multiple motion axes was developed, by incorporating cross-coupling technology into adaptive control architecture. A novel robust adaptive terminal sliding mode position synchronized control approach was proposed for the operation of multiple motion axes system and the convergence of position errors and synchronization errors could be guaranteed in [16]. In [17], the synchronous control of a dual linear motor servosystem was developed by a cross-coupled intelligent complementary sliding mode control system; a better control performance and robustness with regard to uncertainties can be achieved.

The motivation of the presented work here is to find a proper controller which has strong rejection capacity against external disturbances and robustness to deal with uncertainties and also can guarantee the synchronization of the system. Based on a novel stair-climbing wheelchair robot which was designed in [18], a new control method is proposed in this paper via combining the HOSMC techniques and cross-coupling techniques. The mechanical structure of the stair-climbing robot is described at first, and then its dynamics with uncertainties and perturbations is analyzed. After that, single joint position tracking controller adopting HOSMC techniques is designed to assure the high accuracy tracking under conditions of uncertainties and torque perturbations, and then a synchronous controller based on HOSMC techniques

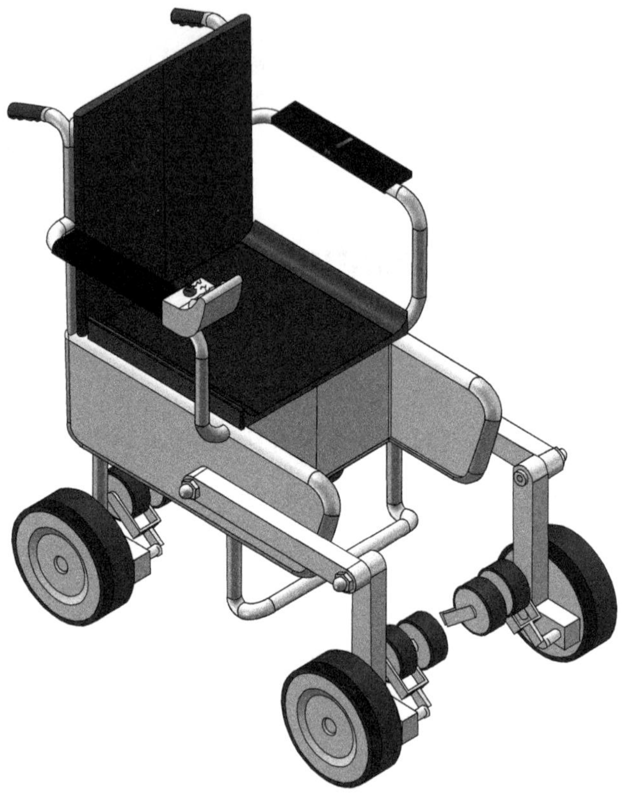

FIGURE 1: Virtual prototype of wheelchair.

and cross-coupling techniques is developed to reduce the synchronization errors and position errors at the same time. The stability of the closed-loop system is proved by Lyapunov theory. In the last section, simulations are completed under the same condition by using the HOSM controller and the HOSM synchronous controller, respectively. The two results are compared to validate the effectiveness of the proposed method.

This paper is organized as follows. Section 2 describes mechanical structure of the stair-climbing wheelchair and its dynamic model with uncertainties and perturbations. In Section 3, a HOSM position controller and a HOSM synchronous controller are proposed and the stability of the closed-loop system is analyzed. Section 4 presents the simulation results. Finally, the conclusion is given in Section 5.

2. System Description and Modeling

2.1. System Description. It seems to be a quite complex problem to design a staircase climbing wheelchair which is adaptable to various terrains. However, this problem can be solved by simply splitting the process of staircase climbing into two different problems: (a) climbing a single step of variable height; (b) providing stability for the entire mechanism while the wheelchair is on the stair [19]. Based on this design idea, a stair-climbing wheelchair is proposed. Figure 1 shows the virtual prototype of the stair-climbing wheelchair. Figure 2 presents the structure of one side of the wheelchair. The parts of climbing mechanism and two-link

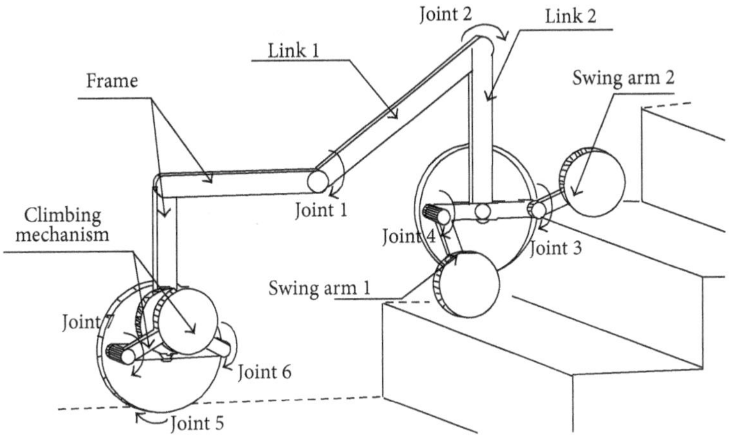

FIGURE 2: Mechanical structure of wheelchair.

mechanism are shown in Figure 2. In this system, all joints are revolute joints driven by motors. The climbing mechanisms, the wheels, and the two-link mechanisms are independent of each other.

The climbing mechanism which contains two swing arms (Figure 2) is designed to surmount a single step. The system has four such climbing mechanisms: two front and two rear. The front climbing mechanisms are joined to two-link mechanism, and the rear mechanisms are fixed on the frame. When the wheelchair reaches a step, the swing arms which are driven by motors rotate at a constant speed. When the swing arms touch the top of the step, the wheelchair's weight is supported by them. The front wheels can now be lifted to surmount the step. When the front wheels touch the top of the step, the weight is now transferred to them. The swing arms continue to rotate to their original positions. This process ends, and the system is now ready to climb the next step.

The two-link mechanism is designed to ensure the wheelchair seat always stays at the upright position. Because the wheelchair seat is joined to the frame, when the front wheels rise, the frame will rotate. This should be adjusted by link mechanism. Similarly, the rear climbing mechanisms move when the top of staircase is reached, and the link mechanism should accommodate the seat.

The task of the sensorial system is to measure the distances between the wheels and the steps and the information about the steps. Proposed placement of stair sensors is shown in Figure 3. Two ultrasound sensors are placed on front wheels in horizontal position to detect stair edge. To measure the distances to next step, two ultrasound sensors are placed on the link mechanism in horizontal position. The width of the step can be calculated by comparing the data of two sets of sensors. There are fourteen rotary transformers (one per joint) to ensure each joint position is measurable. When the positions of the joints are known, the height of swing arms relative to the ground can be calculated, which is defined as h_1. Then, two ultrasound sensors are placed on front climbing mechanisms in vertical position to measure the height of swing arms relative to the step, which can be defined as h_2. So the step height is the height difference between h_1 and h_2. Finally, two ultrasound sensors are placed on the frame to measure the positions of rear wheels with respect to the stair edge.

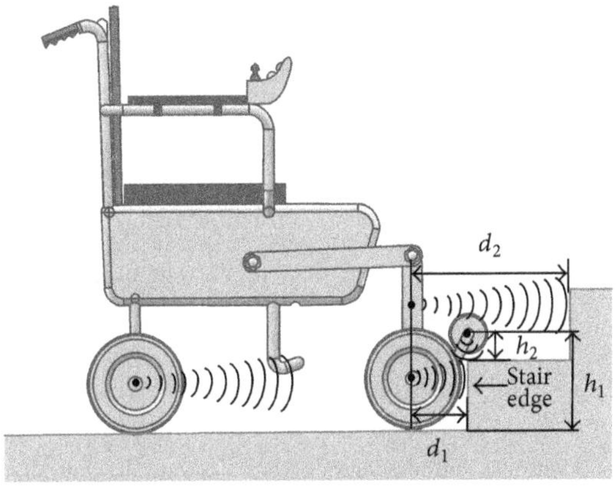

FIGURE 3: Proposed stair sensor placement.

The process of climbing stairs is shown in Figure 4 and achieved by the following steps.

(1) When the front wheels sense a step, the front climbing mechanisms rotate up at a speed defined by the program. At the same time, the two-link mechanisms also rotate for adjusting the attitude of the seat. The front wheels are raised to surmount the step. The wheelchair moves forward until next step is sensed.

(2) Repeat step (1) until the wheelchair surmounts the second step.

(3) The front and rear climbing mechanisms climb synchronously.

(4) When the top of the staircase is reached, the front climbing mechanisms detect no step edge and remain motionless. The rear mechanisms continue to move to the top of the stair.

2.2. System Kinematics and Dynamics. In order to analyze the model of wheelchair, some assumptions are made as follows.

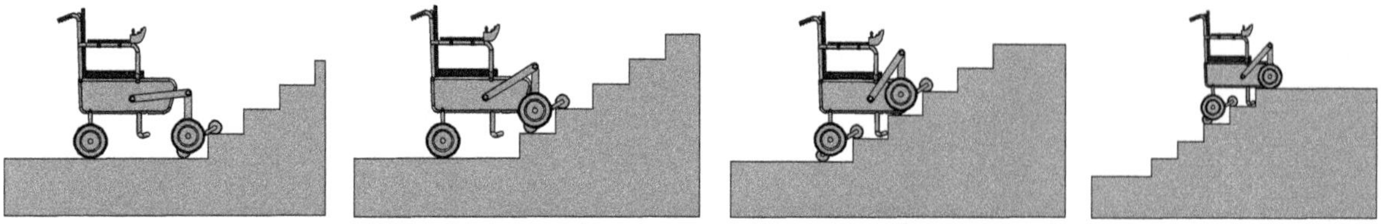

FIGURE 4: Steps of stairs ascending process.

Assumption 1. The wheelchair is a rigid body system. The tires are point contact with the ground, and the contact points have no relative sliding. The tires have no elastic deformation.

Assumption 2. The structure of the robot is symmetrical, so the dynamic analysis is based on the structure which is simplified to planar mechanism.

Assumption 3. The position and attitude of wheelchair are not affected by active control. Each joint has one rotational degree-of-freedom, and the joints are affected by active control.

Figure 5 shows the kinematic scheme of stair-climbing wheelchair. l_i $(i = 1, 2, \ldots, 7)$ is the length of wheelchair's each part. R is the radius of wheels, and r is the radius of small wheels. h is the step height. d is the width of the step. d_1 is the distance between the front wheels and the step edge. q_i $(i = 1, 2, \ldots, 7)$ is the angle of joint i. δ is the original angle of the swing arm. q_0, $\dot{q}_0$, and $\ddot{q}_0$ denote the angle of the seat with respect to ground, velocity, and acceleration, respectively. p_g is the wheelchair's center of gravity. m_i and j_i are the mass and moment of inertia of wheelchair's each part, respectively.

The desired trajectories can be determined by the inverse kinematic model according to the kinematic parameters above. Define the q_{di} $(i = 0, 1, \ldots, 7)$ as desired trajectory of the joint i. For the safety of user, the trajectories must satisfy $q_{d0} = \dot{q}_{d0} = \ddot{q}_{d0} = 0$, making sure that the wheelchair seat always stays at the upright position. For simplicity, the inverse kinematic model is given here, and the derivation process is presented in [20]. When the wheelchair climbs the first and second steps, the inverse kinematic model is the following:

$$
\begin{aligned}
q_{d0} &= 0, \\
q_{d1} &= \arcsin\left[\frac{(h + r + l_4 + l_6 \sin(-\delta - q_{d3}) - R - l_1 - l_2 \sin(q_{d0}))}{l_3}\right], \\
q_{d2} &= \frac{\pi}{2} - q_{d1}, \\
q_{d3} &= -\delta + vt, \\
q_{d4} &= -\pi + \delta + vt, \\
q_{d5} &= \frac{[d + R - d_1 + l_1 \sin(q_{d0}) + l_2 (1 - \cos(q_{d0})) + l_3 (1 - \cos(q_{d0} + q_{d1}))]}{l_3}, \\
q_{d6} &= -\delta, \\
q_{d7} &= -\pi + \delta,
\end{aligned}
\tag{1}
$$

where v is the set rotational speed of climbing mechanisms. When h, d, and d_1 are accurately measured by sensorial system, the desired trajectories can be generated by (1).

According to the previous works in [20], the dynamics can be well approximated by the following equation:

$$M(q)\ddot{q} + C(q, \dot{q})\dot{q} + G(q) = \tau, \tag{2}$$

where q is a vector of generalized coordinates, $q = [q_0, q_1, \ldots, q_n]^T \in R^8$, and $\dot{q}$ and $\ddot{q}$ are velocity vector and acceleration vector, respectively. $M(q) \in R^{8\times 8}$ is the positive definite symmetric inertial matrix. $C(q, \dot{q}) \in R^{8\times 8}$ is the matrix containing Coriolis force and centripetal force. $G(q) \in R^8$ is the gravity vector and $\tau = [\tau_0, \tau_1, \ldots, \tau_7] \in R^8$ is torque input vector. The detailed expressions can be found in the Appendix.

In practical applications, there are a lot of uncertain factors such as friction and disturbing torque. Considering the uncertainties and the modeling errors, the dynamic equation (2) can be rewritten as follows:

$$M_0(q)\ddot{q} + C_0(q, \dot{q})\dot{q} + G_0(q) = \tau + \rho(t), \tag{3}$$

$$\rho(t) = d(t) - \Delta M(q)\ddot{q} - \Delta C(q, \dot{q})\dot{q} - \Delta G(q), \tag{4}$$

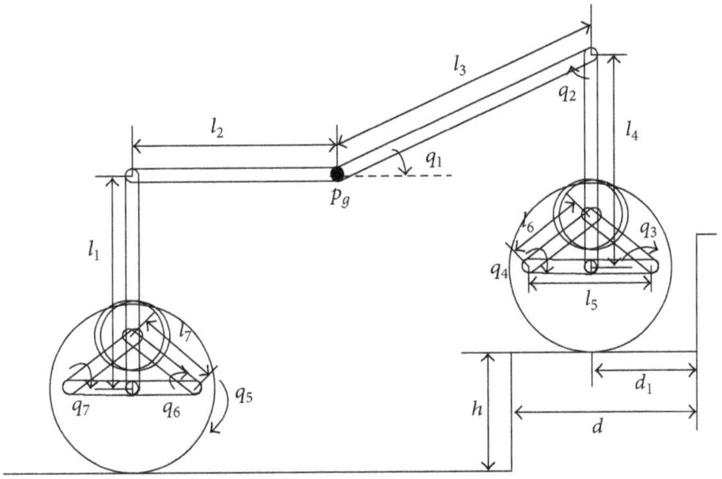

FIGURE 5: Kinematic scheme of stair-climbing wheelchair.

where $M_0(q), C_0(q,\dot{q})$, and $G_0(q)$ are nominal values of $M(q)$, $C(q,\dot{q})$, and $G(q)$, respectively. $\Delta M(q)$, $\Delta C(q,\dot{q})$, and $\Delta G(q)$ are uncertainties of $M(q), C(q,\dot{q})$, and $G(q)$, respectively. $d(t)$ is torque perturbations vector, and $\rho(t) \in R^8$ denotes the total uncertainties of the dynamics caused by parameter variations and perturbations.

Some assumptions for the system parameters and variables are made as follows.

Assumption 4. $M_0(q)$ is a positively definite matrix, and there are known positive constants $M_{\max}$ and $M'_{\max}$ such that the following inequalities hold: $\|M_0^{-1}(q)\| \leq M_{\max}$ and $\|\dot{M}_0^{-1}(q)\| \leq M'_{\max}$.

Assumption 5. There are known positive constants $\rho_{\max}$ and $\rho'_{\max}$ satisfying the following inequalities: $\|\rho(t)\| \leq \rho_{\max}$ and $\|\dot{\rho}(t)\| \leq \rho'_{\max}$.

3. Control System Design and Stability Analysis

The wheelchair kinematics and dynamics have been given in the previous section, which may make the controller design complicated. A robust control law with respect to uncertainties and perturbations is needed in the high accuracy position tracking of the stair-climbing wheelchair system. In this paper, single joint position tracking controller and a synchronous controller are developed, respectively. Stability of the closed-loop system is analyzed by Lyapunov theory. Here is the design process in detail.

3.1. HOSM Controller Design. This section considers the position tracking problem for wheelchair joints, the kinematic and dynamic parameters are known, and the desired trajectory is denoted as q_{di}. The measureable states are joint position q_i and the joint velocity $\dot{q}_i$. The control objective is to design robust chattering-free control signal τ, which ensures that q_i always tracks the desired trajectory q_{di}, in spite of the inertial uncertainties and external disturbance torques.

The position tracking error of a single joint is defined as

$$e_i = q_i - q_{di}. \tag{5}$$

Set $s = [e_1, e_2, \ldots, e_n]^T \in R^n$, in order to achieve HOSM control of sliding variable s, we define σ as

$$\sigma = \ddot{s} + c_1\dot{s} + c_0 s, \tag{6}$$

where c_1 and c_0 are both positive constants. The HOSM control law for sliding variable s is as follows:

$$\tau = u_{eq} + M_0(q)\,u_{vss}, \tag{7}$$

$$\begin{aligned} u_{eq} &= M_0(q)\,\ddot{q}_d + C_0(q,\dot{q})\,\dot{q} + G_0(q) \\ &\quad + M_0(q)\left(-c_1\dot{s} - c_0 s\right), \end{aligned} \tag{8}$$

$$\dot{u}_{vss} = -k_1\sigma - k_2\frac{\sigma}{\|\sigma\|}, \tag{9}$$

where k_1 and k_2 are positive constants, satisfying $k_1 > 0$ and $k_2 > \|\dot{M}_0^{-1}(q)\|\|\rho(t)\| + \|M_0^{-1}(q)\|\|\dot{\rho}(t)\|$.

Theorem 1. *Considering system (3), the sliding surface defined in (6) is chosen. When the parameters satisfy $k_1 > 0$ and $k_2 > \|\dot{M}_0^{-1}(q)\|\|\rho(t)\|+\|M_0^{-1}(q)\|\|\dot{\rho}(t)\|$, the position tracking error fulfills the condition that $\lim_{t\to\infty} s = 0$, under the control effort of the control law given in (7).*

Proof. Choose the following Lyapunov function:

$$V = \frac{1}{2}\sigma^T\sigma. \tag{10}$$

Taking the time derivative of V, it can be written as

$$\dot{V} = \sigma^T\dot{\sigma} = \sigma^T\left(\dddot{s} + c_1\ddot{s} + c_0\dot{s}\right). \tag{11}$$

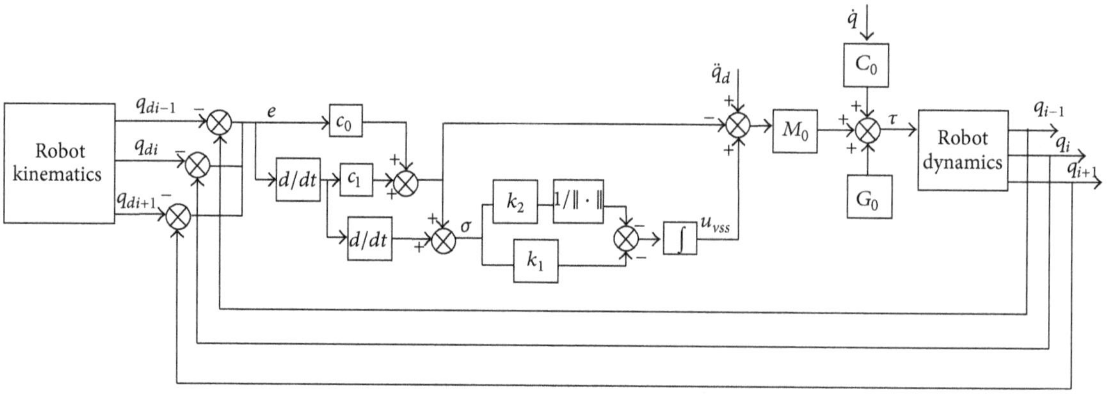

FIGURE 6: Structure of the HOSM controller.

Differentiating (5) and taking (3) yield

$$\begin{aligned}\ddot{s} &= \ddot{q} - \ddot{q}_d \\ &= M_0^{-1}(q)(\tau + \rho(t) - G_0(q) - C_0(q,\dot{q})\dot{q}) - \ddot{q}_d.\end{aligned} \tag{12}$$

Substituting (7), (8), and (9) into (12), we have

$$\begin{aligned}\ddot{s} &= M_0^{-1}(q)[M_0(q)\ddot{q}_d + \rho(t) + M_0(q)(-c_1\dot{s} - c_0 s) \\ &\quad + M_0(q)u_{vss}] - \ddot{q}_d = M_0^{-1}(q)\rho(t) - c_1\dot{s} - c_0 s \\ &\quad + u_{vss}.\end{aligned} \tag{13}$$

Differentiating (13), and considering (9), it can be obtained as follows:

$$\begin{aligned}\dddot{s} &= \dot{M}_0^{-1}(q)\rho(t) + M_0^{-1}(q)\dot{\rho}(t) - c_1\ddot{s} - c_0\dot{s} + \dot{u}_{vss} \\ &= \dot{M}_0^{-1}(q)\rho(t) + M_0^{-1}(q)\dot{\rho}(t) - c_1\ddot{s} - c_0\dot{s} - k_1\sigma \\ &\quad - k_2\frac{\sigma}{\|\sigma\|}.\end{aligned} \tag{14}$$

Substituting (14) into (11), we get

$$\begin{aligned}\dot{V} &= \sigma^T\Big(\dot{M}_0^{-1}(q)\rho(t) + M_0^{-1}(q)\dot{\rho}(t) - c_1\ddot{s} - c_0\dot{s} \\ &\quad - k_1\sigma - k_2\frac{\sigma}{\|\sigma\|} + c_1\ddot{s} + c_0\dot{s}\Big) = \sigma^T\dot{M}_0^{-1}(q)\rho(t) \\ &\quad + \sigma^T M_0^{-1}(q)\dot{\rho}(t) - \sigma^T k_1\sigma - k_2\|\sigma\| \le \|\sigma^T\| \\ &\quad \cdot \|\dot{M}_0^{-1}(q)\|\|\rho(t)\| + \|\sigma^T\|\|M_0^{-1}(q)\|\|\dot{\rho}(t)\| \\ &\quad - \sigma^T k_1\sigma - k_2\|\sigma\| = \|\sigma^T\|\Big(\|\dot{M}_0^{-1}(q)\|\|\rho(t)\| \\ &\quad + \|M_0^{-1}(q)\|\|\dot{\rho}(t)\| - k_2\Big) - \sigma^T k_1\sigma < -\sigma^T k_1\sigma.\end{aligned} \tag{15}$$

Therefore, $\sigma^T\dot{\sigma} < 0$, when $\|\sigma\| \ne 0$, the system is asymptotically stable, and $\lim_{t\to\infty} s = 0$ also can be assured. □

From Theorem 1, it can be obtained that the tracking error can converge to zero in spite of uncertainties and perturbation. Figure 6 presents the structure of the HOSM position tracking controller without coordinating the motion of each joint. With the information provided by sensorial system, the reference trajectory q_{di} is obtained by using the inverse kinematic model. The tracking error e_i can be obtained by the feedback signal, and then the controller calculates the joint driving moment τ which is applied to the dynamic model for joint control.

3.2. HOSM Synchronous Controller Design. Because the position and attitude of the wheelchair are affected by each joint's motion, the coordination among the joints is needed to be considered besides the position tracking error of each joint. If the motions of climbing mechanisms are asynchronous, the seat will lean in the process of climbing stairs, and it will result in a series of problems such as the system damage and injury accidents. This section considers the synchronization problem of multiple joints system. The control objective is to design robust chattering-free control signal τ', which ensures that the position tracking error and synchronization error converge to zero, when q_{di}, q_i, and $\dot{q}_i$ are known.

For a multiple joints system, the position tracking errors must satisfy the following condition in [15]:

$$\lim_{t\to\infty} e_1 = \lim_{t\to\infty} e_2 = \cdots = \lim_{t\to\infty} e_n = 0. \tag{16}$$

Considering (17), the synchronization errors are defined as

$$\begin{aligned}\varepsilon_1 &= e_1 - e_2 + e_1, \\ \varepsilon_2 &= e_2 - e_3 + e_2 - e_1, \\ &\vdots \\ \varepsilon_i &= e_i - e_{i+1} + e_i - e_{i-1}, \\ &\vdots \\ \varepsilon_n &= e_n + e_n - e_{n-1}.\end{aligned} \tag{17}$$

The expression means that the synchronization errors of the multiple joints system ε_i are defined as differential position errors among multiple joints system. Setting $\varepsilon = [\varepsilon_1, \varepsilon_2, \ldots, \varepsilon_n]^T \in R^n$, function (17) can be written as

$$\begin{bmatrix} \varepsilon_1 \\ \varepsilon_2 \\ \vdots \\ \varepsilon_{n-1} \\ \varepsilon_n \end{bmatrix} = \begin{bmatrix} 2 & -1 & 0 & \cdots & 0 \\ -1 & 2 & -1 & \cdots & 0 \\ \vdots & \ddots & \ddots & \ddots & \vdots \\ 0 & \cdots & -1 & 2 & -1 \\ 0 & 0 & \cdots & -1 & 2 \end{bmatrix} \begin{bmatrix} e_1 \\ e_2 \\ \vdots \\ e_{n-1} \\ e_n \end{bmatrix} = Te, \quad (18)$$

where T is the positive definite matrix.

Setting $s_2 = \varepsilon$, in order to achieve HOSM control of sliding variable s_2, define σ_s as

$$\sigma_s = \ddot{s}_2 + c_1 \dot{s}_2 + c_0 s_2. \quad (19)$$

Substituting (18), there is

$$\sigma_s = T\ddot{s} + c_1 T\dot{s} + c_0 Ts. \quad (20)$$

The HOSM control law for sliding variable s_2 is designed as

$$\dot{u}'_{vss} = -k_3 \sigma_s - k_4 \frac{\sigma_s}{\|\sigma_s\|}, \quad (21)$$

where $k_3 > 0$ and $k_4 > \|\dot{M}_0^{-1}(q)\|\|\rho(t)\| + \|M_0^{-1}(q)\|\|\dot{\rho}(t)\|$.

The HOSM synchronous control law which is the overall control law is developed as follows:

$$\tau' = u_{eq} + M_0(q) u_{vss} + M_0(q) u'_{vss}. \quad (22)$$

Theorem 2. *Considering system (3), the sliding surface defined in (19) is chosen. When the parameters satisfy* $k_1 > 0$, $k_2 > \|\dot{M}_0^{-1}(q)\|\|\rho(t)\| + \|M_0^{-1}(q)\|\|\dot{\rho}(t)\|$, $k_4 > \|\dot{M}_0^{-1}(q)\|\|\rho(t)\| + \|M_0^{-1}(q)\|\|\dot{\rho}(t)\|$, *and* $k_3 > 0$, *the position tracking error e and synchronization error* ε *can converge to zero simultaneously, under the control effort of the control law given in (22).*

Proof. Choose the following Lyapunov function:

$$V = \frac{1}{2}\sigma^T \sigma + \frac{1}{2}\sigma_s^T T^{-1} \sigma_s. \quad (23)$$

Differentiating V with respect to time yields

$$\dot{V} = \sigma^T \dot{\sigma} + \sigma_s^T T^{-1} \dot{\sigma}_s. \quad (24)$$

Considering (14), and differentiating (6) and (20), we get

$$\begin{aligned} \dot{\sigma} &= \dot{u}_{vss} + \dot{u}'_{vss} + \dot{M}_0^{-1}(q)\rho(t) + M_0^{-1}(q)\dot{\rho}(t), \\ \dot{\sigma}_s &= T\dot{u}_{vss} + T\dot{u}'_{vss} + T\dot{M}_0^{-1}(q)\rho(t) \\ &\quad + TM_0^{-1}(q)\dot{\rho}(t). \end{aligned} \quad (25)$$

Substituting (25) into (24), we have

$$\begin{aligned} \sigma^T\dot{\sigma} &= \sigma^T\left(\dot{M}_0^{-1}(q)\rho(t) + M_0^{-1}(q)\dot{\rho}(t) - k_1\sigma - k_2\frac{\sigma}{\|\sigma\|} - k_3\sigma_s - k_4\frac{\sigma_s}{\|\sigma_s\|}\right) = \sigma^T\dot{M}_0^{-1}(q)\rho(t) \\ &\quad + \sigma^T M_0^{-1}(q)\dot{\rho}(t) - \sigma^T k_1\sigma - k_2\|\sigma\| - \sigma^T k_3\sigma_s \\ &\quad - \sigma^T k_4\frac{\sigma_s}{\|\sigma_s\|} \le \|\sigma^T\|\left\|\dot{M}_0^{-1}(q)\right\|\|\rho(t)\| + \|\sigma^T\| \\ &\quad \cdot \left\|M_0^{-1}(q)\right\|\|\dot{\rho}(t)\| - \sigma^T k_1\sigma - k_2\|\sigma\| - \sigma^T k_3 T\sigma \\ &\quad - k_4\frac{\sigma^T T\sigma}{\|\sigma_s\|} = \|\sigma^T\|\left(\left\|\dot{M}_0^{-1}(q)\right\|\|\rho(t)\| \right. \\ &\quad \left. + \left\|M_0^{-1}(q)\right\|\|\dot{\rho}(t)\| - k_2\right) - \sigma^T k_1\sigma - \sigma^T k_3 T\sigma \\ &\quad - k_4\frac{\sigma^T T\sigma}{\|\sigma_s\|} < -\sigma^T k_1\sigma - \sigma^T k_3 T\sigma - k_4\frac{\sigma^T T\sigma}{\|\sigma_s\|}, \end{aligned} \quad (26)$$

$$\begin{aligned} \sigma_s^T T^{-1}\dot{\sigma}_s &= \sigma_s^T T^{-1} T\left(\dot{M}_0^{-1}(q)\rho(t) + M_0^{-1}(q)\dot{\rho}(t) - k_1\sigma - k_2\frac{\sigma}{\|\sigma\|} - k_3\sigma_s - k_4\frac{\sigma_s}{\|\sigma_s\|}\right) = \sigma_s^T\dot{M}_0^{-1}(q) \\ &\quad \cdot \rho(t) + \sigma_s^T M_0^{-1}(q)\dot{\rho}(t) - \sigma^T Tk_1\sigma - \frac{\sigma^T Tk_2\sigma}{\|\sigma\|} \\ &\quad - \sigma_s^T k_3\sigma_s - k_4\sigma_s \le \|\sigma_s^T\|\left(\left\|\dot{M}_0^{-1}(q)\right\|\|\rho(t)\| \right. \\ &\quad \left. + \left\|M_0^{-1}(q)\right\|\|\dot{\rho}(t)\| - k_4\right) - \sigma^T Tk_1\sigma - \frac{\sigma^T Tk_2\sigma}{\|\sigma\|} \\ &\quad - \sigma_s^T k_3\sigma_s < -\sigma^T Tk_1\sigma - \frac{\sigma^T Tk_2\sigma}{\|\sigma\|} - \sigma_s^T k_3\sigma_s. \end{aligned}$$

□

From (18), T is positive definite matrix, so it can be obtained that $\sigma^T\dot{\sigma} < 0$ and $\sigma_s^T T^{-1}\dot{\sigma}_s < 0$. It is proved that the system is stable simultaneously, and the convergence of the position tracking error e and synchronization error ε is assured; that is, $\lim_{t\to\infty} s = 0$ and $\lim_{t\to\infty}\varepsilon = 0$. Figure 7 presents the structure of the HOSM synchronous controller. It can be found that the synchronous control signal is added to the HOSMC control signal.

4. Simulation Results

To demonstrate the performance of the proposed approach in Section 3, simulations are performed by MATLAB. In this section, the HOSM controller and HOSM synchronous

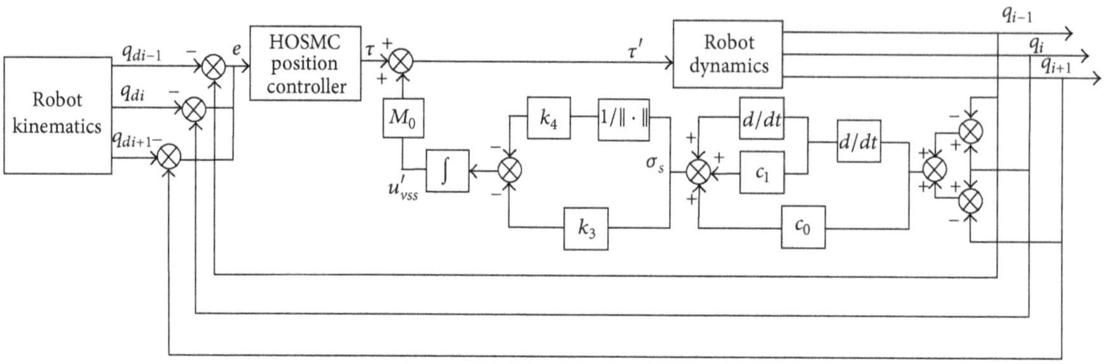

FIGURE 7: Structure of the HOSM synchronous controller.

controller are used to control the dynamic model of the wheelchair, respectively.

The values of kinematic and dynamic parameters are assigned as $l_1 = 0.3$ m, $l_2 = 0.3$ m, $l_3 = 0.405$ m, $l_4 = 0.3$ m, $l_5 = 0.18$ m, $l_6 = 0.12$ m, and $l_7 = 0.12$ m. The radius of wheel is $R = 0.12$, and the radius of small wheel is $r = 0.05$ m. The step parameters are $h = 0.17$ m and $d = 0.25$ m. Moreover, $\delta = 0.66$ rad, $m_1 = 5$ kg, $m_2 = 4$ kg, $m_3 = 3$ kg, $m_4 = 3$ kg, $m_5 = 6$ kg, $m_6 = 3$ kg, and $m_7 = 3$ kg, and $j_1 = 0.6\,\text{kg·m}^2$, $j_2 = 0.7\,\text{kg·m}^2$, $j_3 = 0.03\,\text{kg·m}^2$, $j_4 = 0.03\,\text{kg·m}^2$, $j_5 = 0.04\,\text{kg·m}^2$, $j_6 = 0.03\,\text{kg·m}^2$, and $j_7 = 0.03\,\text{kg·m}^2$. The rotational speed of climbing mechanisms $v = 1.675$ rad/s. The dynamics of the wheelchair is obtained in Section 2 by using these parameters. Choosing the process of the wheelchair climbing two steps, the desired position trajectories are designed according to analyzing the kinematic model in Section 2.

In this section, the position responses and tracking errors of Joint 1 and Joint 2 achieved by the two controllers are shown and compared. For comparison purpose, the simulation is implemented under the same condition by using the HOSM controller and the HOSM synchronous controller, respectively. The parameters of the two controllers are $c_0 = 61$, $c_1 = 70$, $k_1 = 80$, $k_2 = 120$, $k_3 = 100$, and $k_4 = 120$. They are properly chosen to assure that the system can get similar performance using different controllers, and they keep constant during the whole simulation.

First is the moment of inertia uncertainties simulation. The inertia of Joint 1 keeps constant during the whole simulation. The inertia of Joint 2 has been changed at 4 s, 8 s, and 12 s. The value of the step amplitude is $0.7\,\text{kg·m}^2$; that means the inertia of Joint 2 increases or decreases by 100% every time.

Figures 8(a) and 8(c) show the position responses achieved by HOSM controller. The results of HOSM synchronous controller are shown in Figures 8(b) and 8(d). It can be seen that the position responses of the two joints are always tracking the desired trajectories effectively by using the two control schemes. There is no obvious overshoot at the startup stage, and the tracking curves of the synchronous controller are smoother at the startup stage.

TABLE 1: Simulation results in case 1.

Item	HOSMC maximum value (rad)	HOSM synchronous controller maximum value (rad)
Position error of Joint 1	2×10^{-4}	1.4×10^{-4}
Position error of Joint 2	4×10^{-4}	2.2×10^{-4}
Synchronization error	8.1×10^{-4}	3.75×10^{-4}

The tracking errors and synchronization errors are shown in Figure 9. When the inertia of Joint 2 changes abruptly, the maximum position error of Joint 1 reaches 2×10^{-4} rad (Figure 9(a)) using the HOSMC strategy, and the maximum position error of Joint 2 reaches 4×10^{-4} rad (Figure 9(c)) at the same time. When the HOSM synchronous controller is applied, the maximum position errors of the two joints are 1.4×10^{-4} rad (Figure 9(b)) and 2.2×10^{-4} rad (Figure 9(d)), respectively.

The simulation results are listed in Table 1. From the comparison of the data above, the system's tracking errors and synchronous error are reduced over 30% and 50%, respectively, by adopting the HOSM synchronous scheme. It is clear that the proposed scheme has a better control performance with respect to inertia uncertainties.

Then, the torque disturbance simulation is carried out. The system has torque disturbance and parameter uncertainties in this case. Torque disturbance is a square wave with a period of 10 s, and the amplitude is 50 N·m.

The parameters of the two controllers are the same as case 1. Figures 10(a)–10(d) show the position responses of the two joints. It can be found in Figure 11 that both controllers give response to the disturbance and uncertainties, but the proposed controller gives smaller error curves. After the disturbance is added, the maximum error of Joint 1

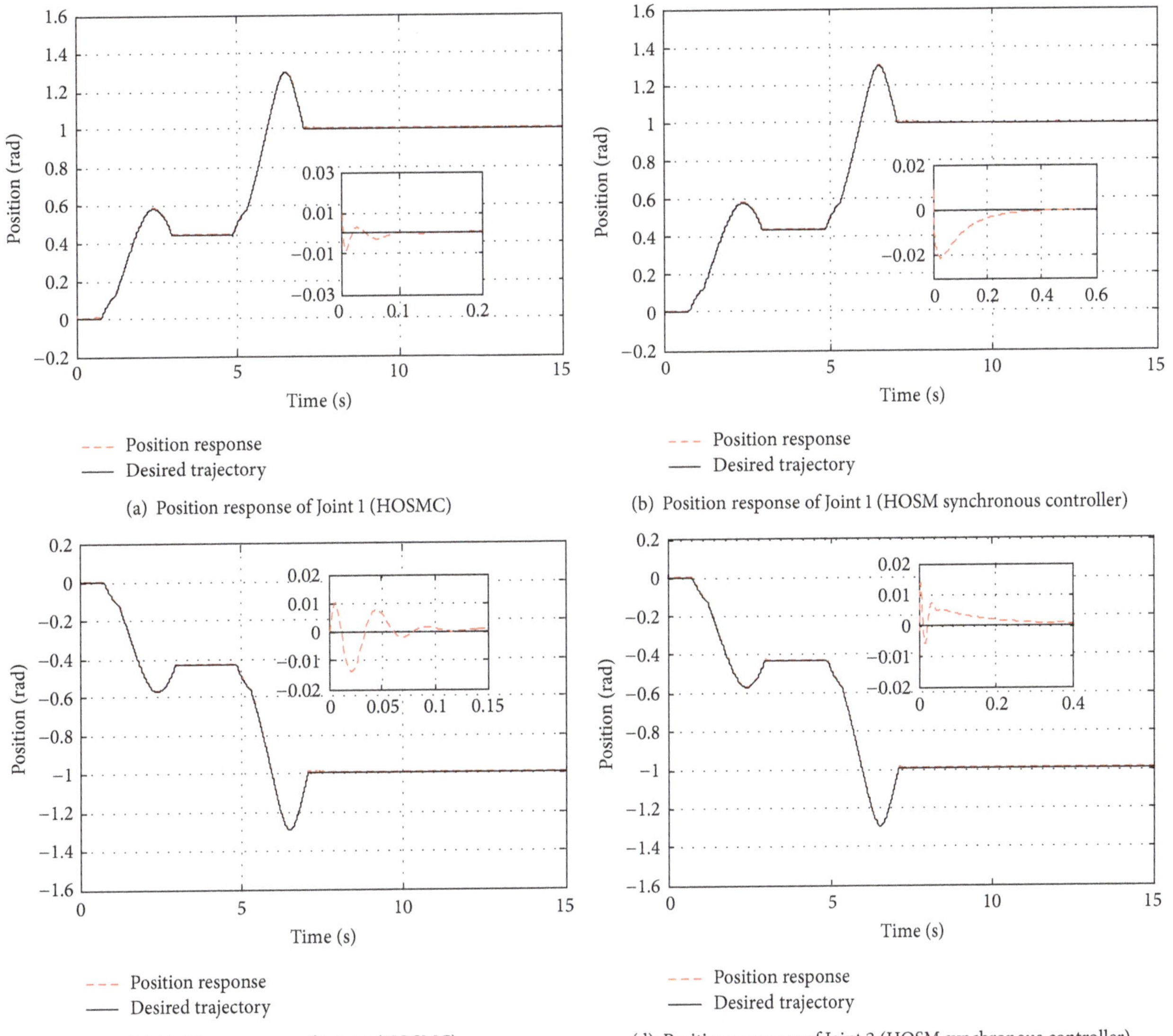

(a) Position response of Joint 1 (HOSMC)

(b) Position response of Joint 1 (HOSM synchronous controller)

(c) Position response of Joint 2 (HOSMC)

(d) Position response of Joint 2 (HOSM synchronous controller)

FIGURE 8: Position responses under inertia uncertainties.

TABLE 2: Simulation results in case 2.

Item	HOSMC maximum value (rad)	HOSM synchronous controller maximum value (rad)
Position error of Joint 1	3.1×10^{-4}	2×10^{-4}
Position error of Joint 2	4×10^{-4}	2.1×10^{-4}
Synchronization error	9.5×10^{-4}	5.5×10^{-4}

reaches 3.1×10^{-4} rad (Figure 11(a)) using the HOSMC strategy, and the maximum error of Joint 2 reaches 4×10^{-4} rad (Figure 11(c)) at the same time. When the HOSM synchronous controller is applied, the maximum errors of the two joints are 2×10^{-4} rad (Figure 11(b)) and 2.1×10^{-4} rad (Figure 11(d)), respectively. Figures 11(e) and 11(f) show the synchronization error curves.

The simulation results in case 2 are listed in Table 2. From the comparison of the data above, the system's position errors and synchronous error are reduced over 36% and 40% by adopting the HOSM synchronous scheme. It is clear that the HOSM synchronous controller has stronger robustness and higher performance. It also validates the correctness of the proposed scheme in this paper.

5. Conclusions

Considering a novel stair-climbing wheelchair with inertia uncertainties and torque disturbances, a HOSM controller is established, and closed-loop stability of the system is proved. Considering the synchronization of the system, a synchronous controller which combines

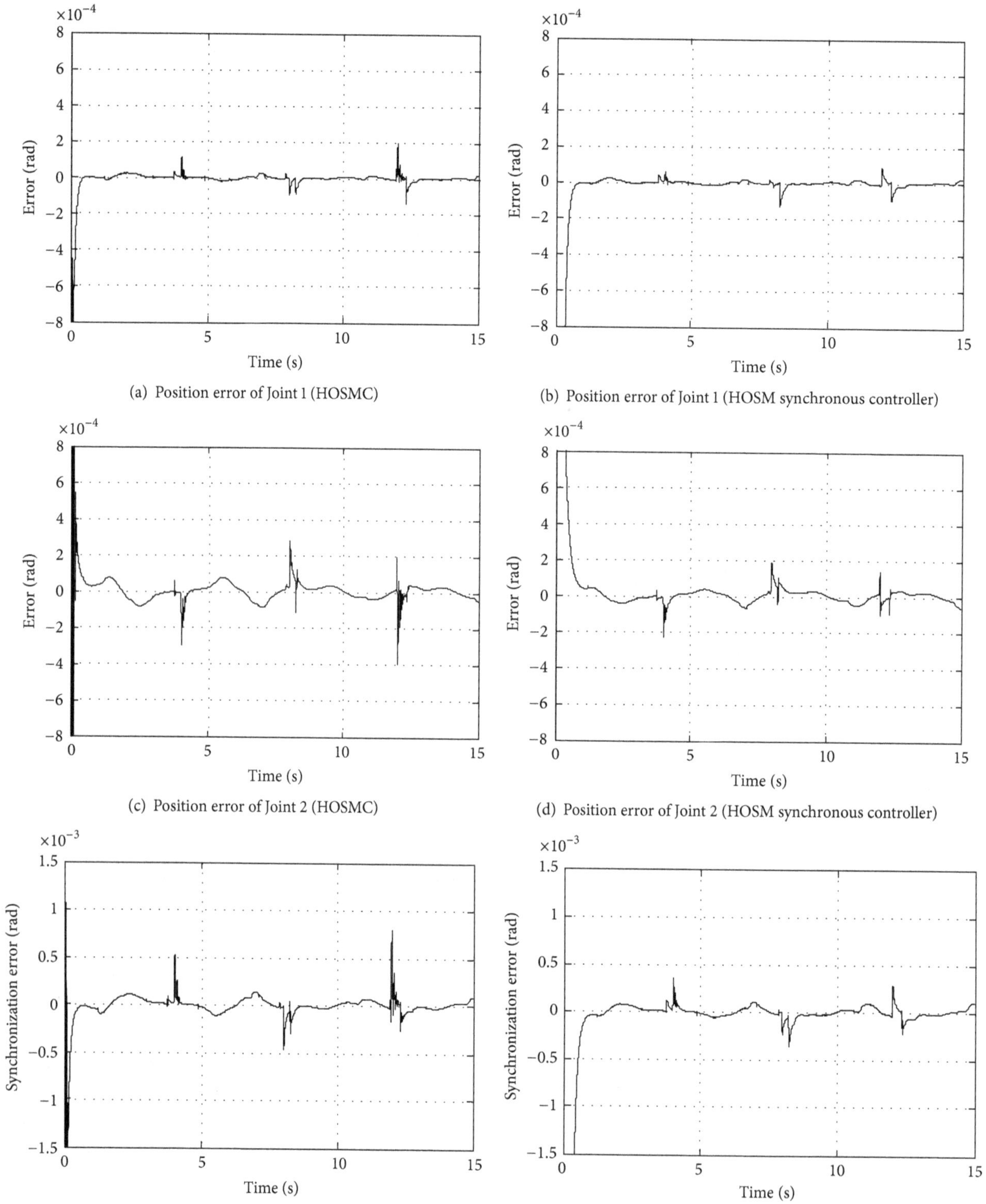

(a) Position error of Joint 1 (HOSMC)

(b) Position error of Joint 1 (HOSM synchronous controller)

(c) Position error of Joint 2 (HOSMC)

(d) Position error of Joint 2 (HOSM synchronous controller)

(e) Synchronization error (HOSMC)

(f) Synchronization error (HOSM synchronous controller)

FIGURE 9: Simulation results with inertia uncertainties.

the HOSMC techniques and cross-coupling techniques is proposed, and closed-loop stability of the system is proved, too. The simulation is performed in MATLAB to demonstrate the effectiveness of the proposed controller. The simulation results show that the proposed scheme can give better tracking performance with inertia uncertainties and torque disturbance and better synchronization.

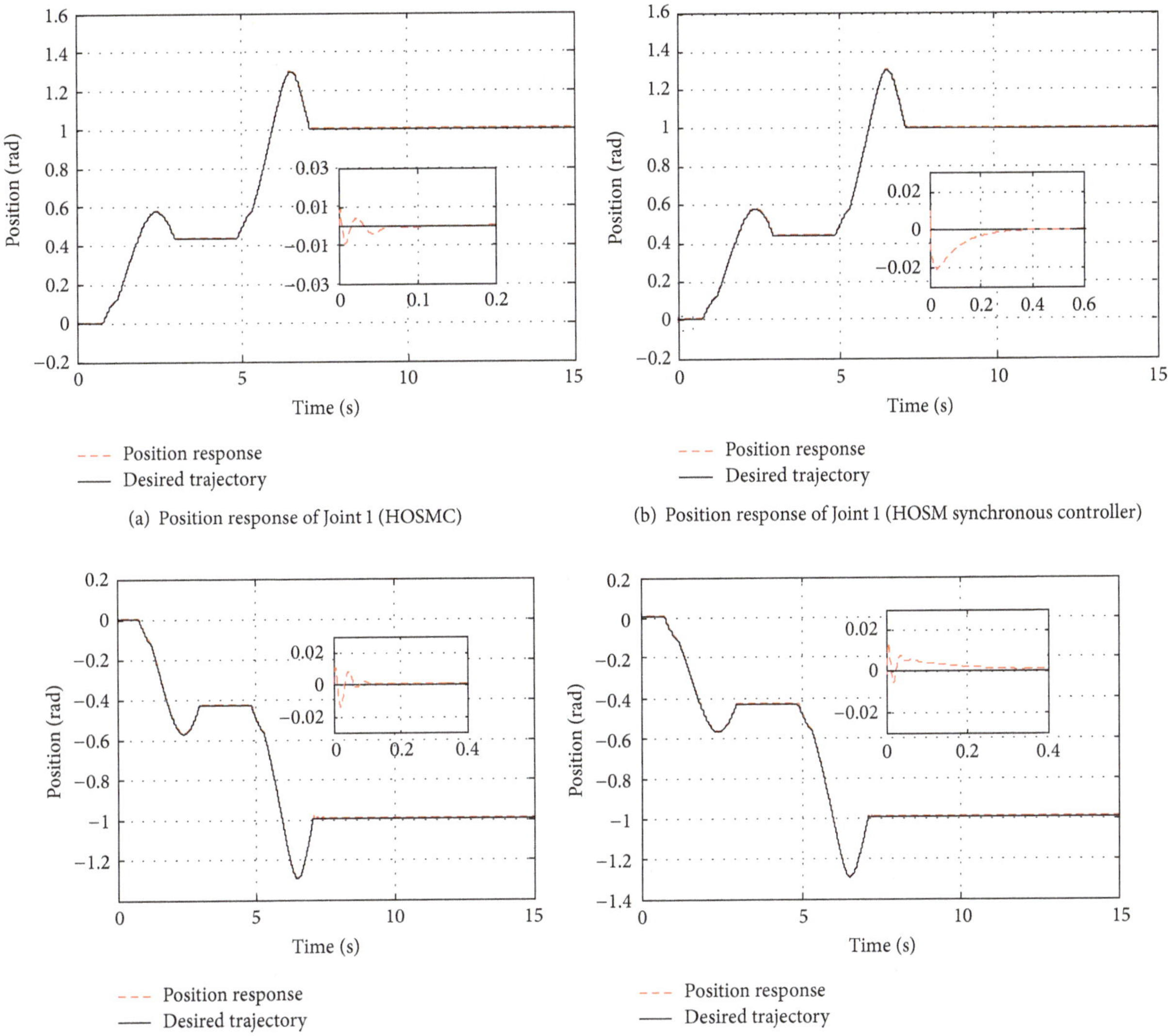

(a) Position response of Joint 1 (HOSMC)

(b) Position response of Joint 1 (HOSM synchronous controller)

(c) Position response of Joint 2 (HOSMC)

(d) Position response of Joint 2 (HOSM synchronous controller)

FIGURE 10: Position responses under inertia uncertainties and torque disturbance.

Appendix

The dynamic model in detail is

$$\begin{bmatrix} M_{11} & M_{12} & \cdots & M_{18} \\ M_{21} & M_{22} & & \\ \vdots & & \ddots & \vdots \\ M_{81} & & \cdots & M_{88} \end{bmatrix} \begin{bmatrix} \ddot{q}_0 \\ \ddot{q}_1 \\ \vdots \\ \ddot{q}_7 \end{bmatrix} + \begin{bmatrix} C_{11} & C_{12} & \cdots & C_{18} \\ C_{21} & C_{22} & & \\ \vdots & & \ddots & \vdots \\ C_{81} & & \cdots & C_{88} \end{bmatrix} \begin{bmatrix} \dot{q}_0 \\ \dot{q}_1 \\ \vdots \\ \dot{q}_7 \end{bmatrix} + \begin{bmatrix} G_1 \\ G_2 \\ \vdots \\ G_8 \end{bmatrix} g = \begin{bmatrix} \tau_0 \\ \tau_1 \\ \vdots \\ \tau_n \end{bmatrix}, \quad \text{(A.1)}$$

where g is the gravitational acceleration. M_{ij}, C_{ij}, and G_i are as follows:

$$M_{11} = j_0 + j_1 + j_2 + j_3 + j_4 + j_5 + j_6 + j_7 + (m_3 + m_4 + m_6 + m_7)(l_6 + r)^2 + (m_5 + m_6 + m_7) l_2^2 + (m_1 + m_2 + m_3 + m_4) l_3^2 + (m_2 + m_3 + m_4$$

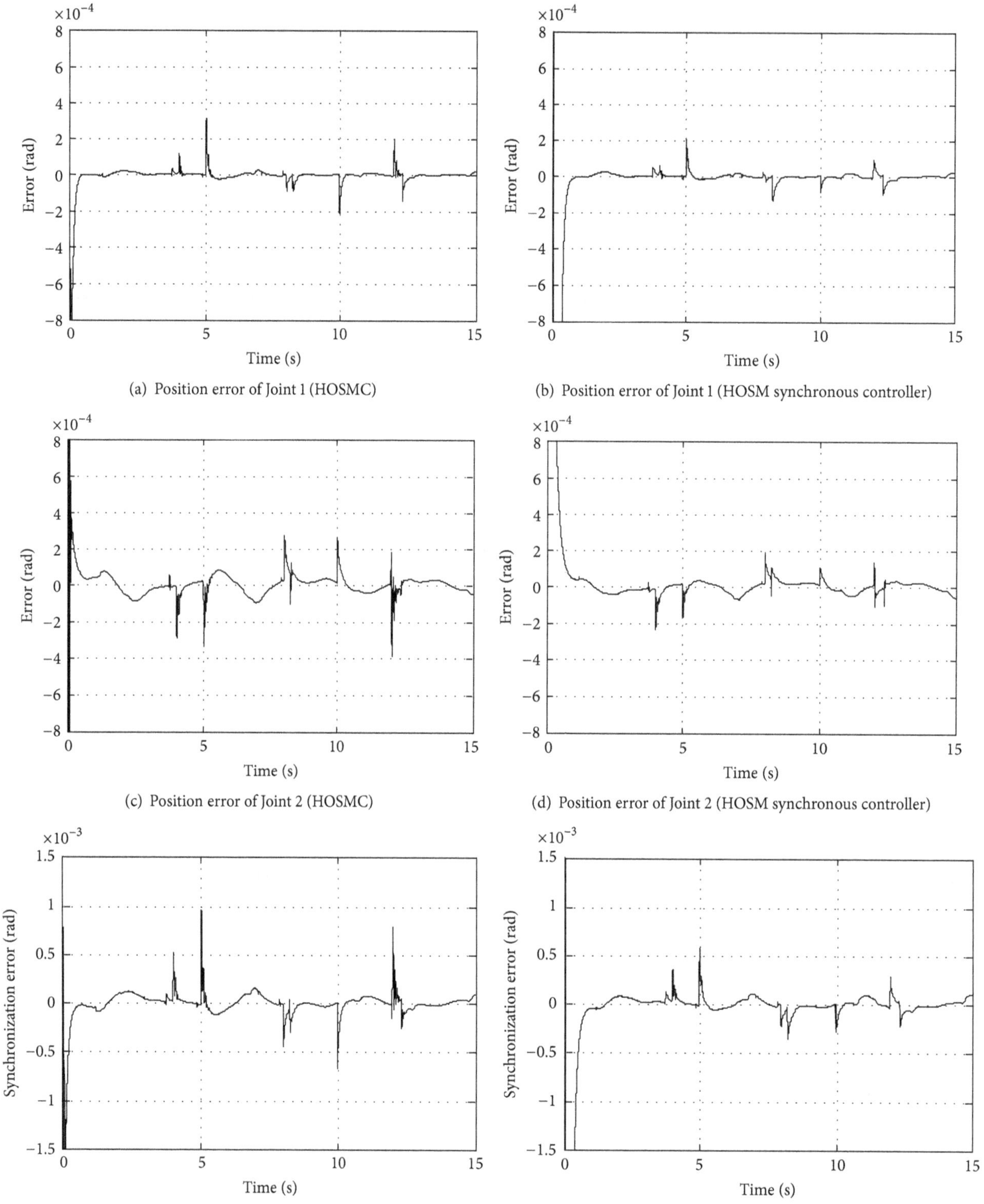

(a) Position error of Joint 1 (HOSMC)

(b) Position error of Joint 1 (HOSM synchronous controller)

(c) Position error of Joint 2 (HOSMC)

(d) Position error of Joint 2 (HOSM synchronous controller)

(e) Synchronization error (HOSMC)

(f) Synchronization error (HOSM synchronous controller)

FIGURE 11: Simulation results with inertia uncertainties and torque disturbance.

$$+ m_5 + m_6 + m_7) l_4^{\,2} + (m_3 + m_4 + m_6 + m_7) l_5^{\,2}$$

$$+ (m_2 + m_3 + m_4) l_3^{\,2} \cos(q_2) - (m_3 - m_4) l_3 l_5 (1$$

$$+ \cos(q_2)) - (m_3 \cos(q_3) - m_4 \cos(q_4)) l_3 (l_6$$

$$+ r) - (m_3 \sin(q_3) - m_4 \sin(q_4)) l_4 (l_6 + r)$$

$$+ (m_3 \cos(q_3) + m_4 \cos(q_4)) l_5 (l_6 + r) - (m_2$$

$$+ m_3 + m_4) l_3 l_4 \sin(q_2) - (m_3 \cos(q_3 - q_1) - m_4$$

$$\cdot \cos(q_4 - q_1)) l_3 (l_6 + r) + (m_6 \cos(q_6 - q_1) - m_7 \cos(q_7 - q_1)) l_2 (l_6 + r) + (m_6 \cos(q_6 - q_5) + m_7 \cos(q_7 - q_5)) l_5 (l_6 + r) - (m_6 \sin(q_6 - q_1) - m_7 \sin(q_7 - q_1)) l_4 (l_6 + r) + (m_6 - m_7) \cdot (\cos(q_1 - q_5) l_2 - \sin(q_5 - q_1) l_4) l_5,$$

$$M_{12} = j_1 + j_2 + j_3 + j_4 + m_3 (l_6 + r)^2 + (m_1 + m_2 + m_3 + m_4) l_3^2 + (m_2 + m_3 + m_4) (l_4^2 + l_3^2 \cdot \cos(q_2) - l_3 l_4 \sin(q_2)) + (m_3 + m_4) l_5^2 - (m_3 - m_4) l_3 l_5 (1 + \cos(q_2)) - (m_3 \cos(q_3) - m_4 \cdot \cos(q_4)) l_3 (l_6 + r) + (m_3 \cos(q_3) + m_4 \cos(q_4)) \cdot l_5 (l_6 + r) - (m_3 \sin(q_3) - m_4 \sin(q_4)) l_4 (l_6 + r) - (m_3 \cos(q_1 - q_3) - m_4 \cos(q_1 - q_4)) l_3 (l_6 + r),$$

$$M_{13} = j_2 + j_3 + j_4 + (m_3 + m_4) (l_6 + r)^2 + (m_2 + m_3 + m_4) (l_4^2 - l_3 l_4 \sin(q_2)) + (m_3 + m_4) l_5^2 - (m_3 - m_4) l_3 l_5 (1 + \cos(q_2)) - (m_3 \cos(q_3) - m_4 \cos(q_4)) l_3 (l_6 + r) + (m_3 \cos(q_3) + m_4 \cdot \cos(q_4)) l_5 (l_6 + r) - (m_3 \sin(q_3) - m_4 \sin(q_4)) \cdot l_4 (l_6 + r) - (m_3 \cos(q_1 - q_3) - m_4 \cdot \cos(q_1 - q_4)) l_3 (l_6 + r),$$

$$M_{14} = j_3 + m_3 (l_6 + r)^2 - m_3 (\cos(q_3) (l_3 - l_5) + \sin(q_3) l_4 + \cos(q_1 \quad q_3) l_3) (l_6 + r),$$

$$M_{15} = j_4 + m_4 (l_6 + r)^2 + m_4 (\cos(q_4) (l_3 + l_5) + \sin(q_4) l_4 + \cos(q_1 - q_4) l_3) (l_6 + r),$$

$$M_{16} = j_5 + m_5 (\cos(q_1 - q_5) l_2 - \sin(q_5 - q_1) l_4) R,$$

$$M_{17} = j_6 + m_6 (l_6 + r)^2 + m_6 (\cos(q_1 - q_6) l_2 + \cos(q_5 - q_6) l_5 - \sin(q_6 - q_1) l_4) (l_6 + r),$$

$$M_{18} = j_7 + m_7 (l_6 + r)^2 - m_7 (\cos(q_1 - q_7) l_2 - \cos(q_5 - q_7) l_5 - \sin(q_7 - q_1) l_4) (l_6 + r);$$

$$M_{21} = M_{12},$$

$$M_{22} = j_1 + j_2 + j_3 + j_4 + m_3 (l_6 + r)^2 + (m_2 + m_3 + m_4) (l_3^2 + \cos(q_2) l_3^2 - \sin(q_2) l_3 l_4 + l_4^2) + (m_3 + m_4) (l_5^2 - l_3 l_5) - m_3 \cos(q_3) (l_3 - l_5) (l_6 + r) - m_3 \sin(q_3) l_4 (l_6 + r) - m_3 \cos(q_2) l_3 l_5 + m_4 \cos(q_2) l_3 l_5 - m_3 \cos(q_1 - q_3) l_3 (l_6 + r),$$

$$M_{23} = j_2 + j_3 + j_4 + m_3 (l_6 + r)^2 + (m_2 + m_3 + m_4) \cdot (l_4^2 - l_3 l_4 \sin(q_2)) + (m_3 + m_4) l_5^2 - (m_3 - m_4) l_3 l_5 (1 + \cos(q_2)) - (m_3 \cos(q_3) - m_4 \cdot \cos(q_4)) l_3 (l_6 + r) + (m_3 \cos(q_3) + m_4 \cos(q_4)) \cdot l_5 (l_6 + r) - (m_3 \sin(q_3) - m_4 \sin(q_4)) l_4 (l_6 + r) - (m_3 \cos(q_1 - q_3) - m_4 \cos(q_1 - q_4)) l_3 (l_6 + r),$$

$$M_{24} = j_3 + m_3 (l_6 + r)^2 - m_3 ((l_3 - l_5) \cos(q_3) + l_4 \cdot \sin(q_3) + l_3 \cos(q_1 - q_3)) (l_6 + r),$$

$$M_{25} = j_4 + m_4 ((l_3 + l_5) \cos(q_4) + l_4 \sin(q_4) + l_3 \cdot \cos(q_1 - q_4)) (l_6 + r),$$

$$M_{26} = M_{27} = M_{28} = 0;$$

$$M_{31} = M_{13},$$

$$M_{32} = M_{23},$$

$$M_{33} = j_2 + j_3 + j_4 + (m_3 + m_4) (l_6 + r)^2 + (m_2 + m_3 + m_4) (l_4^2) + (m_3 + m_4) l_5^2 + (m_3 \cos(q_3) + m_4 \cos(q_4)) l_5 (l_6 + r) - (m_3 \sin(q_3) - m_4 \cdot \sin(q_4)) l_4 (l_6 + r),$$

$$M_{34} = j_3 + m_3 (l_6 + r)^2 + m_3 (\cos(q_3) l_5 - \sin(q_3) \cdot l_4) (l_6 + r),$$

$$M_{35} = j_4 + m_4 (l_6 + r)^2 + m_4 (\cos(q_4) l_5 + \sin(q_4) \cdot l_4) (l_6 + r),$$

$$M_{36} = M_{37} = M_{38} = 0;$$

$$M_{41} = M_{14},$$

$$M_{42} = M_{24},$$

$$M_{43} = M_{34},$$

$$M_{44} = j_3 + m_3 (l_6 + r)^2,$$

$$M_{45} = M_{46} = M_{47} = M_{48} = 0;$$

$$M_{51} = M_{15},$$

$$M_{52} = M_{25},$$

$M_{53} = M_{35}$,

$M_{54} = 0$,

$M_{55} = j_4 + m_4(l_6 + r)^2$,

$M_{56} = M_{57} = M_{58} = 0$;

$M_{61} = M_{16}$,

$M_{62} = M_{63} = M_{64} = M_{65} = 0$,

$M_{66} = j_5 + m_5 R^2$,

$M_{67} = M_{68} = 0$;

$M_{71} = M_{17}$,

$M_{72} = M_{73} = M_{74} = M_{75} = M_{76} = 0$,

$M_{77} = j_6 + m_6(l_6 + r)^2$,

$M_{78} = 0$;

$M_{81} = M_{18}$,

$M_{82} = M_{83} = M_{84} = M_{85} = M_{86} = M_{87} = 0$,

$M_{88} = j_7 + m_7(l_6 + r)^2$,

$$C_{11} = ((m_2 l_3 + m_3(l_3 - l_5) + m_4(l_3 + l_5)) l_3 \sin(q_1) - m_5(\sin(q_5 - q_1) l_2 + \cos(q_5 - q_1) l_4) R - m_7 \cdot \sin(q_7 - q_1) l_2 (l_6 + r)) \dot{q}_0,$$

$$C_{12} = \left((m_2 + m_3 + m_4) l_3{}^2 - (m_3 - m_4) l_3 l_5\right) \sin(q_1) \cdot (\dot{q}_1 + \dot{q}_0),$$

$$C_{13} = -(m_2 + m_3 + m_4) l_3{}^2 \dot{q}_2 - ((m_2 + m_3 + m_4) l_4 \cdot \cos(q_2) + (m_3 - m_4) l_5 \sin(q_1)) l_3 (\dot{q}_0 + \dot{q}_1 + \dot{q}_2) - (m_2 + m_3 + m_4) l_3 l_4 (\dot{q}_0 + \dot{q}_1) + (m_3 \cdot \sin(q_3) + m_3 \sin(q_3 - q_1) - m_4 \sin(q_4) - m_4 \cdot \sin(q_4 - q_1)) l_3 (l_6 + r)(\dot{q}_0 + \dot{q}_1 + \dot{q}_2),$$

$$C_{14} = m_3(l_3 \sin(q_3) - l_4 \cos(q_3) - l_5 \sin(q_3) + l_3 \cdot \sin(q_3 - q_1))(l_6 + r)(\dot{q}_0 + \dot{q}_1 + \dot{q}_2 + \dot{q}_3),$$

$$C_{15} = m_4(l_4 \cos(q_4) - l_5 \sin(q_4) - l_3 \sin(q_4) - l_3 \cdot \sin(q_4 - q_1))(l_6 + r)(\dot{q}_0 + \dot{q}_1 + \dot{q}_2 + \dot{q}_4),$$

$$C_{16} = -m_5 l_2 R \sin(q_5 - q_1)(\dot{q}_0 + \dot{q}_5) - m_5 l_4 R \cos(q_5 - q_1)(\dot{q}_0 + \dot{q}_5),$$

$$C_{17} = -m_6(l_4 \cos(q_6 - q_1) + l_5 \sin(q_6 - q_5) + l_2 \cdot \sin(q_6 - q_1))(l_6 + r)(\dot{q}_0 + \dot{q}_6),$$

$$C_{18} = m_7(l_4 \cos(q_7 - q_1) - l_5 \sin(q_7 - q_5) - l_2 \cdot \sin(q_7 - q_1))(l_6 + r)(\dot{q}_0 + \dot{q}_7);$$

$$C_{21} = ((m_2 l_3 + m_3(l_3 - l_5) + m_4(l_3 + l_5)) l_3 \sin(q_1) + m_4(l_4 \cos(q_4) - (l_3 + l_5)\sin(q_4) - l_3 \sin(q_4 - q_1))(l_6 + r)) \dot{q}_0,$$

$$C_{22} = (((m_2 + m_3 + m_4) l_3 - m_3 l_5 + m_4 l_5) l_3 \sin(q_1) + m_4(l_4 \cos(q_4) - (l_3 + l_5)\sin(q_4) - l_3 \sin(q_4 - q_1))(l_6 + r))(\dot{q}_0 + \dot{q}_1),$$

$$C_{23} = -((m_2 + m_3 + m_4) l_3 l_4 (1 + \cos(q_2)) + (m_3 - m_4) l_3 l_5 \sin(q_1) - (m_3 l_3(\sin(q_3) + \sin(q_3 - q_1)) - m_4(l_3 + l_5)\sin(q_4) + m_4 l_4 \cos(q_4) - m_4 l_3 \sin(q_4 - q_1))(l_6 + r))(\dot{q}_0 + \dot{q}_1 + \dot{q}_2),$$

$$C_{24} = m_3(l_3 \sin(q_3) - l_4 \cos(q_3) - l_5 \sin(q_3) + l_3 \cdot \sin(q_3 - q_1))(l_6 + r)(\dot{q}_0 + \dot{q}_1 + \dot{q}_2 + \dot{q}_3),$$

$$C_{25} = m_4(l_4 \cos(q_4) - l_5 \sin(q_4) - l_3 \sin(q_4) - l_3 \cdot \sin(q_4 - q_1))(l_6 + r)(\dot{q}_0 + \dot{q}_1 + \dot{q}_2 + \dot{q}_4),$$

$C_{26} = C_{27} = C_{28} = 0$;

$$C_{31} = (m_2 + m_3 + m_4) l_3 l_4 (1 + \cos(q_2)) \dot{q}_0 - (m_3 \cdot \sin(q_3) + m_4 \sin(q_4) + m_3 \sin(q_3 - q_1) - m_4 \cdot \sin(q_4 - q_1)) l_3 (l_6 + r) \dot{q}_0,$$

$$C_{32} = (m_2 + m_3 + m_4) l_3 l_4 (1 + \cos(q_2))(\dot{q}_0 + \dot{q}_1) - (m_3 \sin(q_3) - m_4 \sin(q_4) + m_3 \sin(q_3 - q_1) - m_4 \sin(q_4 - q_1)) l_3 (l_6 + r)(\dot{q}_0 + \dot{q}_1),$$

$C_{33} = 0$,

$$C_{34} = -m_3(l_4 \cos(q_3) + l_5 \sin(q_3))(l_6 + r)(\dot{q}_0 + \dot{q}_1 + \dot{q}_2 + \dot{q}_3),$$

$$C_{35} = m_4(l_4 \cos(q_4) - l_5 \sin(q_4))(l_6 + r)(\dot{q}_0 + \dot{q}_1 + \dot{q}_2 + \dot{q}_3),$$

$C_{36} = C_{37} = C_{38} = 0$;

$$C_{41} = m_3(l_4 \cos(q_3) - (l_3 - l_5)\sin(q_3) - l_3 \cdot \sin(q_3 - q_1))(l_6 + r)\dot{q}_0,$$

$$C_{42} = m_3(l_4 \cos(q_3) - (l_3 - l_5)\sin(q_3) - l_3 \cdot \sin(q_3 - q_1))(l_6 + r)(\dot{q}_0 + \dot{q}_1),$$

$$C_{43} = m_3 (l_4 \cos(q_3) + l_5 \sin(q_3)) (l_6 + r) (\dot{q}_0 + \dot{q}_1 + \dot{q}_2),$$

$$C_{44} = C_{45} = C_{46} = C_{47} = C_{48} = 0;$$

$$C_{51} = m_4 ((l_3 + l_5) \sin(q_4) - l_4 \cos(q_4) + l_3 \cdot \sin(q_4 - q_1)) (l_6 + r) \dot{q}_0,$$

$$C_{52} = m_4 ((l_3 + l_5) \sin(q_4) - l_4 \cos(q_4) + l_3 \cdot \sin(q_4 - q_1)) (l_6 + r) (\dot{q}_0 + \dot{q}_1),$$

$$C_{53} = m_4 (l_5 \sin(q_4) - l_4 \cos(q_4)) (l_6 + r) (\dot{q}_0 + \dot{q}_1 + \dot{q}_2),$$

$$C_{54} = C_{55} = C_{56} = C_{57} = C_{58} = 0;$$

$$C_{61} = m_5 (l_2 \sin(q_5 - q_1) + l_4 \cos(q_5 - q_1)) R\dot{q}_0,$$

$$C_{62} = C_{63} = C_{64} = C_{65} = C_{66} = C_{67} = C_{68} = 0;$$

$$C_{71} = m_6 (l_2 \sin(q_6 - q_1) + l_4 \cos(q_6 - q_1) + l_5 \cdot \sin(q_6 - q_5)) (l_6 + r) \dot{q}_0,$$

$$C_{72} = C_{73} = C_{74} = C_{75} = C_{76} = C_{77} = C_{78} = 0;$$

$$C_{81} = m_7 (-l_2 \sin(q_7 - q_1) - l_4 \cos(q_7 - q_1) + l_5 \cdot \sin(q_7 - q_5)) (l_6 + r) \dot{q}_0,$$

$$C_{82} = C_{83} = C_{84} = C_{85} = C_{86} = C_{87} = C_{88} = 0,$$

$$G_1 = (m_2 + m_3 + m_4) l_3 - (m_3 - m_4) l_5 + (m_1 + m_2 + m_3 + m_4) l_3 \cos(q_1) - (m_5 + m_6 + m_7) (l_2 \cdot \cos(q_1) + l_4 \sin(q_1)) - (m_6 - m_7) l_5 \cos(q_5) - (m_3 \cos(q_3) + m_6 \cos(q_6) + m_4 \sin(q_4) + m_7 \cdot \sin(q_7)) (l_6 + r),$$

$$G_2 = (m_2 + m_3 + m_4) l_3 - (m_3 - m_4) l_5 + (m_1 + m_2 + m_3 + m_4) l_3 \cos(q_1) - m_3 (l_6 + r) \cos(q_3),$$

$$G_3 = -(m_3 - m_4) l_5 - (m_3 \cos(q_3) + m_4 \sin(q_4)) \cdot (l_6 + r),$$

$$G_4 = -m_3 \cos(q_3) (l_6 + r),$$

$$G_5 = -m_4 \sin(q_4) (l_6 + r),$$

$$G_6 = -m_5 \cos(q_5) R,$$

$$G_7 = -m_6 \cos(q_6) (l_6 + r),$$

$$G_8 = -m_7 \sin(q_7) (l_6 + r). \tag{A.2}$$

Conflict of Interests

The authors declare that there is no conflict of interests regarding the publication of this paper.

Acknowledgment

This work was supported by the National Natural Science Foundation of China under Grant no. 61074023.

References

[1] M. J. Lawn, T. Sakai, M. Kuroiwa, and T. Ishimatsu, "Development and practical application of a stairclimbing wheelchair in Nagasaki," *Journal of HWRS-ERC*, vol. 2, no. 2, pp. 33–39, 2001.

[2] N. M. A. Ghani, M. O. Tokhi, A. N. K. Nasir, and S. Ahmad, "Control of a stair climbing wheelchair," *IAES International Journal of Robotics and Automation*, vol. 1, no. 4, pp. 203–213, 2012.

[3] G. Quaglia, W. Franco, and R. Oderio, "Wheelchair.q, a mechanical concept for a stair climbing wheelchair," in *Proceedings of the IEEE International Conference on Robotics and Biomimetics (ROBIO '09)*, pp. 800–805, IEEE, Guilin, China, December 2009.

[4] T. Mabuchi, T. Nagasawa, K. Awa, K. Shiraki, and T. Yamada, "Development of a stair-climbing mobile robot with legs and wheels," *Artificial Life and Robotics*, vol. 2, no. 4, pp. 184–188, 1998.

[5] C.-T. Chen and H.-V. Pham, "Design and fabrication of a statically stable stair-climbing robotic wheelchair," *Industrial Robot*, vol. 36, no. 6, pp. 562–569, 2009.

[6] M. J. Lawn and T. Ishimatsu, "Modeling of a stair-climbing wheelchair mechanism with high single-step capability," *IEEE Transactions on Neural Systems and Rehabilitation Engineering*, vol. 11, no. 3, pp. 323–332, 2003.

[7] Y. Sugahara, N. Yonezawa, and K. Kosuge, "A novel stair-climbing wheelchair with transformable wheeled four-bar linkages," in *Proceedings of the 23rd IEEE/RSJ International Conference on Intelligent Robots and Systems (IROS '10)*, pp. 3333–3339, IEEE, Taipei, Taiwan, October 2010.

[8] D. Davies and S. Hirose, "Continuous high-speed climbing control and leg mechanism for an eight-legged stair-climbing vehicle," in *Proceedings of the IEEE/ASME International Conference on Advanced Intelligent Mechatronics (AIM '09)*, pp. 1606–1612, IEEE, Singapore, July 2009.

[9] E. Mihankhah, A. Kalantari, E. Aboosaeedan, H. D. Taghirad, and S. A. A. Moosavian, "Autonomous staircase detection and stair climbing for a tracked mobile robot using fuzzy controller," in *Proceedings of the IEEE International Conference on Robotics and Biomimetics (ROBIO '08)*, pp. 1980–1985, IEEE, Bangkok, Thailand, February 2009.

[10] J. Wang, T. Wang, C. Yao, X. Li, and C. Wu, "Dynamic modeling and control for WT wheelchair robot during the stair-climbing process," *Journal of Mechanical Engineering*, vol. 50, no. 13, pp. 22–34, 2014.

[11] H. Lee and V. I. Utkin, "Chattering suppression methods in sliding mode control systems," *Annual Reviews in Control*, vol. 31, no. 2, pp. 179–188, 2007.

[12] G. Bartolini, A. Ferrara, and E. Usai, "Chattering avoidance by second-order sliding mode control," *IEEE Transactions on Automatic Control*, vol. 43, no. 2, pp. 241–246, 1998.

[13] K.-M. Ma, "Design of higher order sliding mode attitude control laws for large-scale spacecraft," *Control and Decision*, vol. 28, no. 2, pp. 201–210, 2013.

[14] A. Arisoy, M. K. Bayrakceken, S. Basturk, M. Gokasan, and O. S. Bogosyan, "High order sliding mode control of a space robot manipulator," in *Proceedings of the 5th International Conference on Recent Advances in Space Technologies (RAST '11)*, pp. 833–838, IEEE, Istanbul, Turkey, June 2011.

[15] D. Sun, X. Shao, and G. Feng, "A model-free cross-coupled control for position synchronization of multi-axis motions: theory and experiments," *IEEE Transactions on Control Systems Technology*, vol. 15, no. 2, pp. 306–314, 2007.

[16] D. Zhao, S. Li, F. Gao, and Q. Zhu, "Robust adaptive terminal sliding mode-based synchronised position control for multiple motion axes systems," *IET Control Theory & Applications*, vol. 3, no. 1, pp. 136–150, 2009.

[17] F.-J. Lin, P.-H. Chou, C.-S. Chen, and Y.-S. Lin, "DSP-based cross-coupled synchronous control for dual linear motors via intelligent complementary sliding mode control," *IEEE Transactions on Industrial Electronics*, vol. 59, no. 2, pp. 1061–1073, 2012.

[18] J. X. Liu and Q. W. Chen, "Dynamics and control study of a stair-climbing walking aid robot," in *Proceedings of the International Conference on Mechanic Automation and Control Engineering (MACE '10)*, pp. 6190–6194, IEEE, Wuhan, China, June 2010.

[19] R. Morales, A. Gonzalez, V. Feliu, and P. Pintado, "Environment adaptation of a new staircase-climbing wheelchair," *Autonomous Robots*, vol. 23, no. 4, pp. 275–292, 2007.

[20] J. X. Liu, Y. F. Wu, J. Guo, R. Li, and Q. W. Chen, "Modeling and simulation analysis of a walking assistant robot for both plane and stair," *Chinese Journal of Engineering Design*, vol. 22, no. 4, pp. 344–350, 2015.

13

A Robust Control Method for *Q-S* Synchronization between Different Dimensional Integer-Order and Fractional-Order Chaotic Systems

Adel Ouannas[1] and Raghib Abu-Saris[2]

[1]*Department of Mathematics and Computer Science, University of Tebessa, 12002 Tebessa, Algeria*
[2]*Department of Health Informatics, College of Public Health and Health Informatics, King Saud Bin Abdulaziz University for Health Science, Riyadh 11481, Saudi Arabia*

Correspondence should be addressed to Adel Ouannas; ouannas_adel@yahoo.fr

Academic Editor: Xiao He

A robust control approach is presented to study the problem of *Q-S* synchronization between Integer-order and fractional-order chaotic systems with different dimensions. Based on Laplace transformation and stability theory of linear integer-order dynamical systems, a new control law is proposed to guarantee the *Q-S* synchronization between n-dimensional integer-order master system and m-dimensional fractional-order slave system. This paper provides further contribution to the topic of *Q-S* chaos synchronization between integer-order and fractional-order systems and introduces a general control scheme that can be applied to wide classes of chaotic and hyperchaotic systems. Illustrative example and numerical simulations are used to show the effectiveness of the proposed method.

1. Introduction

Fractional derivatives provide an excellent instrument for the description of memory and hereditary properties of various materials and processes. The advantages or the real objects of the fractional-order systems are that we have more degrees of freedom in the model and that a "memory" is included in the model [1–6].

Recently, more and more attentions were paid to synchronization of integer-order chaotic systems and fractional-order chaotic systems. Many control methods have been proposed and different synchronization types have been studied between integer-order and fractional-order chaotic system. For example, general control schemes have been described in [7, 8]. A sliding mode method has been designed in [9–11]. A synchronization method of a class of hyperchaotic systems is given in [12]. In [13], a nonlinear feedback control method has been introduced and some robust observer techniques have been used in [14, 15]. Also, complete synchronization and antisynchronization have been observed, for example, in [16–18], and function projective synchronization has been studied in [19]. Until now, a variety of control schemes have been proposed to study the problem of chaos synchronization between different dimensional systems such as modified function projective synchronization [20], generalized matrix projective synchronization [21], generalized synchronization [22–24], inverse generalized synchronization [25], full state hybrid projective synchronization [26], *Q-S* synchronization [27], increased order synchronization [28, 29], and reduced order generalized synchronization [30]. Amongst all kinds of synchronization, *Q-S* synchronization has been extensively considered [31–39], due to its universality and its great potential applications in applied sciences and engineering.

Motivated by the above discussions, the main aim of this paper is to present constructive scheme to investigate *Q-S* synchronization between n-dimensional integer-order master system and m-dimensional fractional-order slave system in n-D. By using Laplace transformation, stability property of integer-order linear systems, and suitable fractional-order control law, a new criterion is derived to achieve the *Q-S* chaos synchronization between n-dimensional integer-order master system and m-dimensional fractional-order slave system in n-D. The proposed control method is simple, efficient, and easy to implement in applications. The outline of the rest of this paper is organized as follows. First, Section 2 provides some basic concepts of fractional derivative. In Section 3, the problem of Q-S synchronization between integer-order master system and fractional-order slave system is formulated. Our main result is presented in Section 4. In Section 5, numerical example is used to verify the effectiveness and feasibility of the proposed control scheme. Finally, Section 6 is the brief conclusion.

2. Basic Concepts of Fractional Derivative

There are several definitions of a fractional derivative of order $\alpha > 0$ [40–42]. The two most commonly used definitions are the Riemann-Liouville and Caputo definitions. Each definition uses Riemann-Liouville fractional integration and derivatives of whole order. The difference between the two definitions is in the order of evaluation. The Riemann-Liouville fractional integral operator of order $\alpha \geq 0$ of the function $f(t)$ is defined as

$$J^{\alpha} f(t) = \frac{1}{\Gamma(\alpha)} \int_0^t (t-\tau)^{\alpha-1} f(\tau)\, d\tau, \quad \alpha > 0,\ t > 0. \tag{1}$$

Some properties of the operator J^{α} can be found, for example, in [43, 44]. We recall only the following, for $\mu \geq -1, \alpha, \beta \geq 0$, and $\gamma > -1$; we have

$$\begin{aligned} J^{\alpha} J^{\beta} f(t) &= J^{\alpha+\beta} f(t), \\ J^{\alpha} t^{\gamma} &= \frac{\Gamma(\gamma+1)}{\Gamma(\alpha+\gamma+1)} t^{\alpha+\gamma}. \end{aligned} \tag{2}$$

In this study, Caputo definition is used and the fractional derivative of $f(t)$ is defined as

$$\begin{aligned} {}^{C}D_t^{\alpha} f(t) &= J^{m-\alpha}\left(\frac{d^m}{dt^m} f(t)\right) \\ &= \frac{1}{\Gamma(m-\alpha)} \int_0^t \frac{f^{(m)}(\tau)}{(t-\tau)^{\alpha-m+1}}\, d\tau, \end{aligned} \tag{3}$$

for $m-1 < \alpha \leq m, m \in \mathbb{N}, t > 0$. The fractional differential operator ${}^{C}D_t^{\alpha}$ is left-inverse (and not right-inverse) to the fractional integral operator J^{α}; that is, ${}^{C}D_t^{\alpha} J^{\alpha} = I$, where I is the identity operator. The Laplace transform of the Caputo fractional derivative rule reads

$$\begin{gathered} L\left\{{}^{C}D_t^{\alpha} f(t)\right\} = s^{\alpha} F(s) - \sum_{k=0}^{n-1} s^{\alpha-k-1} f^{(k)}(0), \\ (\alpha > 0,\ n-1 < \alpha \leq n). \end{gathered} \tag{4}$$

Particularly, when $\alpha \in (0,1]$, we have $L\{{}^{C}D_t^{\alpha} f(t)\} = s^{\alpha}F(s) - s^{\alpha-1}f(0)$. The Laplace transform of the Riemann-Liouville fractional integral rule satisfies

$$L\{J^{\alpha} f(t)\} = s^{-\alpha} F(s), \quad (\alpha > 0). \tag{5}$$

Caputo fractional derivative appears more suitable to be treated by the Laplace transform technique in that it requires the knowledge of the (bounded) initial values of the function and of its integer derivatives of order $k = 1, 2, \ldots, m-1$, in analogy with the case when $\alpha = n$.

3. Q-S Synchronization Problem for Integer-Order and Fractional-Order Chaotic System

We assume that the master chaotic systems can be considered in the following form:

$$\dot{X}(t) = AX(t) + F(X(t)), \tag{6}$$

where $X(t) \in \mathbb{R}^n$ is the state vector of the master system (6), $A \in \mathbb{R}^{n\times n}$ is a constant matrix, and $F : \mathbb{R}^n \to \mathbb{R}^n$ is a nonlinear vector function. Also, consider the slave chaotic system as

$${}^{C}D_t^{p} Y(t) = G(Y(t)) + U, \tag{7}$$

where $Y(t) \in \mathbb{R}^m$ is the state vector of the slave system (7), $G : \mathbb{R}^m \to \mathbb{R}^m$, p is a rational number between 0 and 1, ${}^{C}D_t^{p}$ is the Caputo fractional derivative of order p, and $U = (u_i)_{1\leq i\leq m}$ is a vector controller. The problem of Q-S synchronization for the master system (6) and the slave system (7) in dimension n is to find the controller U such that the synchronization error,

$$e(t) = Q(Y(t)) - S(X(t)), \tag{8}$$

satisfies that

$$\lim_{t\to+\infty} \|e(t)\| = 0, \tag{9}$$

where $Q : \mathbb{R}^m \to \mathbb{R}^n$, $S : \mathbb{R}^n \to \mathbb{R}^n$ are continuously differentiable functions. We assume that $n < m$.

4. General Control Scheme

The integer-order derivative of the error system (8) can be derived as

$$\dot{e}(t) = \mathbf{D}Q(Y(t)) \times \dot{Y}(t) - \mathbf{D}S(X(t)) \times \dot{X}(t), \tag{10}$$

where $\mathbf{D}Q(Y(t)) \in \mathbb{R}^{n\times m}$, $\mathbf{D}S(X(t)) \in \mathbb{R}^{n\times n}$ are the Jacobian matrices of the functions Q and S, respectively,

$$\mathbf{D}Q(Y(t)) = \begin{pmatrix} \frac{\partial Q_1}{\partial y_1} & \frac{\partial Q_1}{\partial y_2} & \cdots & \frac{\partial Q_1}{\partial y_m} \\ \frac{\partial Q_2}{\partial y_1} & \frac{\partial Q_2}{\partial y_2} & \cdots & \frac{\partial Q_2}{\partial y_m} \\ \vdots & \vdots & \ddots & \vdots \\ \frac{\partial Q_n}{\partial y_1} & \frac{\partial Q_n}{\partial y_2} & \cdots & \frac{\partial Q_n}{\partial y_m} \end{pmatrix},$$

$$\mathbf{D}S(X(t)) = \begin{pmatrix} \frac{\partial S_1}{\partial x_1} & \frac{\partial S_1}{\partial x_2} & \cdots & \frac{\partial S_1}{\partial x_n} \\ \frac{\partial S_2}{\partial x_1} & \frac{\partial S_2}{\partial x_2} & \cdots & \frac{\partial S_2}{\partial x_n} \\ \vdots & \vdots & \ddots & \vdots \\ \frac{\partial S_n}{\partial x_1} & \frac{\partial S_n}{\partial x_2} & \cdots & \frac{\partial S_n}{\partial x_n} \end{pmatrix}. \quad (11)$$

Hence, we have the following result.

Theorem 1. *There exists a suitable feedback gain matrix $C \in \mathbb{R}^{n\times n}$ to realize the Q-S synchronization between the master system (6) and the slave system (7) in n-D under the following controller:*

$$U = -G(Y(t)) + J^{1-p}(G(Y(t)) + V), \quad (12)$$

where the vector quantity $V = (V_i)_{1\le i\le m}$ is defined as

$$(V_1, V_2, \ldots, V_n)^T = -M^{-1}R, \quad (13)$$

$$(V_{n+1}, V_{n+2}, \ldots, V_m)^T = 0, \quad (14)$$

where M^{-1} is the inverse of matrix M,

$$M = \begin{pmatrix} \frac{\partial Q_1}{\partial y_1} & \frac{\partial Q_1}{\partial y_2} & \cdots & \frac{\partial Q_1}{\partial y_n} \\ \frac{\partial Q_2}{\partial y_1} & \frac{\partial Q_2}{\partial y_2} & \cdots & \frac{\partial Q_2}{\partial y_n} \\ \vdots & \vdots & \ddots & \vdots \\ \frac{\partial Q_n}{\partial y_1} & \frac{\partial Q_n}{\partial y_2} & \cdots & \frac{\partial Q_n}{\partial y_n} \end{pmatrix}, \quad (15)$$

$$R = (C - A)e(t) + \mathbf{D}Q(Y(t)) \times G(Y(t)) - \mathbf{D}S(X(t)) \times (AX(t) + F(X(t))). \quad (16)$$

Proof. By inserting the control law described by (12) into (7), we can rewrite the slave system as follows:

$${}^{C}D_t^{p}Y(t) = J^{1-p}(G(Y(t)) + V). \quad (17)$$

Applying the Laplace transform to (17) and letting $\mathbf{F}(s) = \mathbf{L}(Y(t))$, we obtain

$$s^p\mathbf{F}(s) - s^{p-1}Y(0) = s^{p-1}\mathbf{L}(G(Y(t)) + V); \quad (18)$$

multiplying both the left-hand and right-hand sides of (18) by s^{1-p} and applying the inverse Laplace transform to the result, we obtain a new equation for the slave system

$$\dot{Y}(t) = G(Y(t)) + V. \quad (19)$$

Now, the error system (10) can be described as

$$\begin{aligned}\dot{e}(t) = (A - C)e(t) &+ \mathbf{D}Q(Y(t)) \times G(Y(t)) \\ &+ (C - A)(Q(Y(t)) - S(X(t))) - \mathbf{D}S(X(t)) \\ &\times (AX(t) + F(X(t))) + \mathbf{D}Q(Y(t)) \times V,\end{aligned} \quad (20)$$

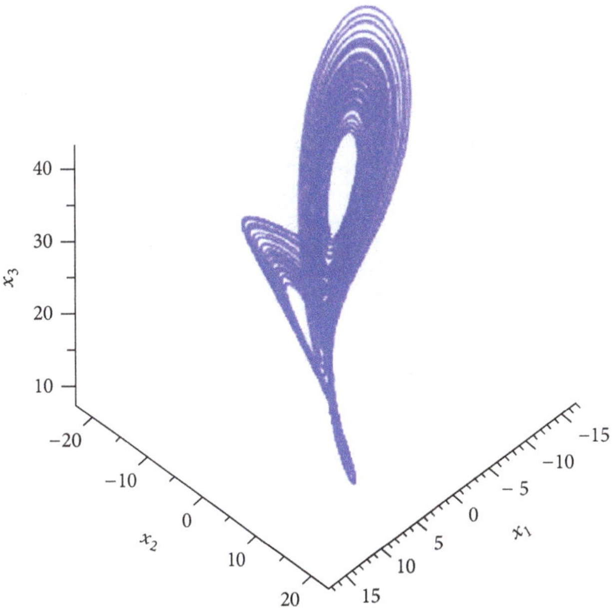

FIGURE 1: The chaotic attractor of Lorenz system when $(\alpha, \beta, \gamma) = (10, 28, -8/3)$.

where $C \in \mathbb{R}^{n\times n}$ is a feedback gain matrix to be chosen. The error system (20) can be simplified as follows:

$$\dot{e}(t) = (A - C)e(t) + R + \mathbf{D}Q(Y(t)) \times V, \quad (21)$$

where R was defined by (16). By using (14), the error system (21) can be written as

$$\dot{e}(t) = (A - C)e(t) + R + M \times (V_1, V_2, \ldots, V_n)^T. \quad (22)$$

Hence, by substituting (13) into (22), we get

$$\dot{e}(t) = (A - C)e(t). \quad (23)$$

With respect to the asymptotic stability property of linear continuous-time dynamical systems, if the feedback gain matrix C is selected such that all eigenvalues of $A - C$ are strictly negative, it is immediate that all solutions of the error system (23) go to zero as $t \to \infty$. Therefore, systems (6) and (7) are globally Q-S synchronized in n-D. □

5. Illustrative Example

In this section, to validate the Q-S synchronization method proposed in the previous section, we consider the Lorenz system as a master system and the fractional-order hyperchaotic Lorenz system as a slave system. The Lorenz system can be described as

$$\begin{aligned}\dot{x}_1 &= \alpha(x_2 - x_1), \\ \dot{x}_2 &= \beta x_1 - x_2 - x_1x_3, \\ \dot{x}_3 &= \gamma x_3 + x_1x_2,\end{aligned} \quad (24)$$

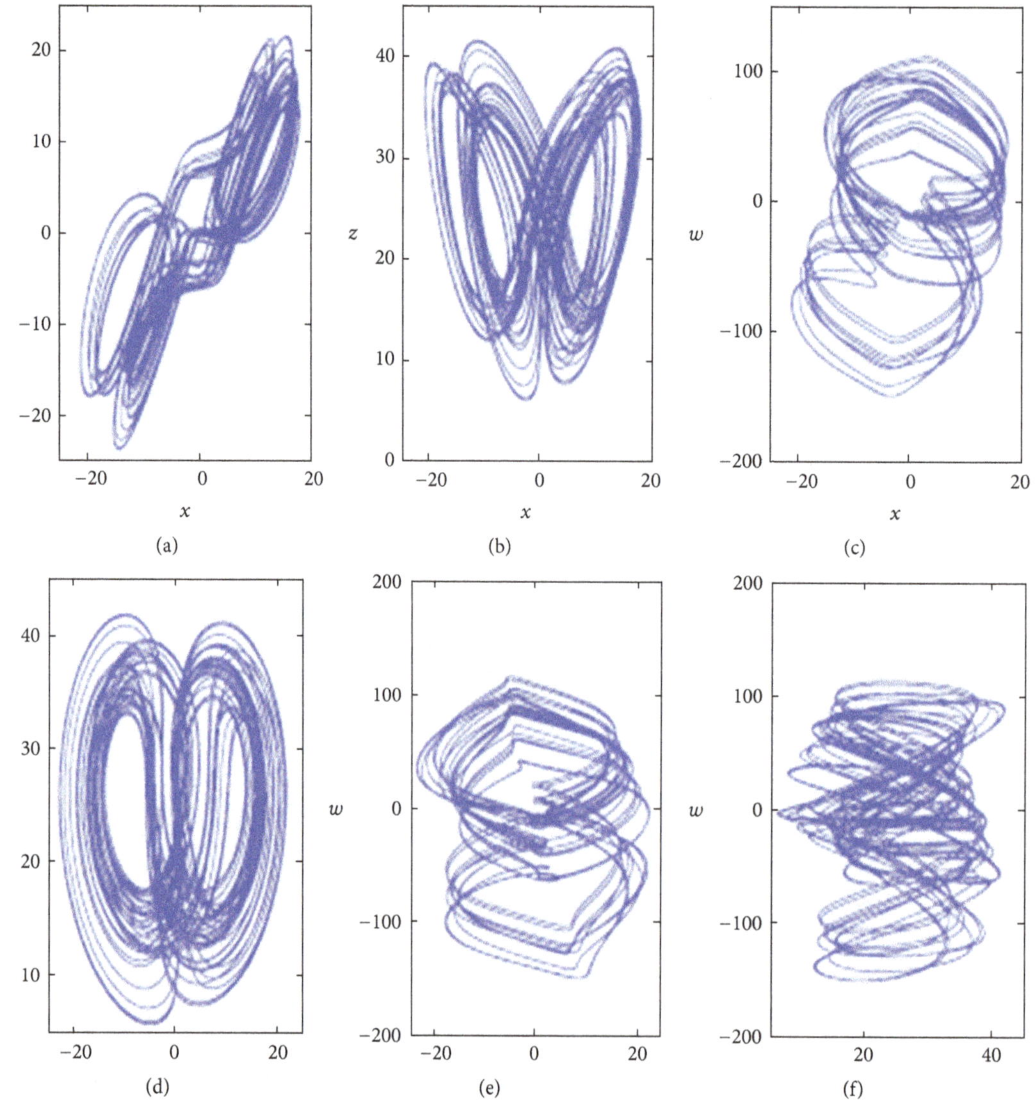

FIGURE 2: Chaotic attractor projections of fractional-order hyperchaotic Lorenz system when $p = 0.98$ and $(a, b, c, r) = (10, 8/3, 28, -1)$.

which has a chaotic attractor, for example, when $(\alpha, \beta, \gamma) = (10, 28, -8/3)$ [45]. Then the linear part $A = (a_{ij})_{3\times3}$ and the nonlinear part F of the Lorenz system (24) are given by

$$A = \begin{pmatrix} -\alpha & \alpha & 0 \\ \beta & -1 & 0 \\ 0 & 0 & \gamma \end{pmatrix}, \qquad F = \begin{pmatrix} 0 \\ -x_1 x_3 \\ x_1 x_2 \end{pmatrix}. \tag{25}$$

The Lorenz chaotic attractor is shown in Figure 1.

The controlled fractional-order hyperchaotic Lorenz system can be described by the following nonlinear fractional-order ODE:

$$\begin{aligned} D^p y_1 &= a(y_2 - y_1) + y_4 + u_1, \\ D^p y_2 &= c y_1 - y_2 - y_1 y_3 + u_2, \\ D^p y_3 &= -b y_3 + y_1 y_2 + u_3, \\ D^p y_4 &= r y_4 - y_2 y_3 + u_4, \end{aligned} \tag{26}$$

where u_i, $i = 1, 2, 3, 4$, are the controllers. This system, as shown in [46], exhibits hyperchaotic behaviors when $p = 0.98$ and $(a, b, c, r) = (10, 8/3, 28, -1)$. Chaotic attractor projections of the fractional-order hyperchaotic Lorenz system are shown in Figure 2.

According to Q-S synchronization control technique proposed in the previous section, the functions $Q : \mathbb{R}^4 \to \mathbb{R}^3$ and $S : \mathbb{R}^3 \to \mathbb{R}^3$ are selected as follows:

$$\begin{aligned} Q(y_1, y_2, y_3, y_4) &= \left(y_1 + \frac{1}{3} y_1^3 + y_4, y_2 + 2y_4, y_3 + 3y_4\right)^T, \\ S(x_1, x_2, x_3) &= (x_1, x_2, x_3 + x_1)^T. \end{aligned} \tag{27}$$

Then,

$$\mathbf{DQ}(y_1,y_2,y_3,y_4)=\begin{pmatrix}1+y_1^2 & 0 & 0 & 1\\ 0 & 1 & 0 & 2\\ 0 & 0 & 1 & 3\end{pmatrix},$$
$$\mathbf{DS}(x_1,x_2,x_3)=\begin{pmatrix}1 & 0 & 0\\ 0 & 1 & 0\\ 1 & 0 & 1\end{pmatrix}. \quad (28)$$

It is easy to show that if we choose the feedback gain matrix C as

$$C=\begin{pmatrix}0 & 10 & 0\\ 28 & 0 & 0\\ 0 & 0 & 0\end{pmatrix}, \quad (29)$$

then the eigenvalues of the matrix $A-C$ are strictly negative. Then the controllers u_1, u_2, u_3, and u_4 can be designed as follows:

$$\begin{aligned}u_1 &= J^{0.02}\left[-\alpha e_1(t) - \frac{1}{1+y_1^2}\left(ry_4 - y_2y_3 - \alpha\left(x_2-x_1\right)\right)\right] - a\left(y_2 - y_1\right) - y_4,\\ u_2 &= -cy_1 + y_2 + y_1y_3 + J^{0.02}\left[-e_2(t) - 2ry_4 + 2y_2y_3 + \beta x_1 - x_2 - x_1x_3\right],\\ u_3 &= by_3 - y_1y_2 + J^{0.02}\left[-\gamma e_3(t) - 3ry_4 + 3y_2y_3 + \alpha\left(x_2-x_1\right) + \gamma x_3 + x_1x_2\right],\\ u_4 &= -ry_4 + y_2y_3 + J^{0.02}\left[ry_4 - y_2y_3\right],\end{aligned} \quad (30)$$

and the error system can be written as

$$\begin{aligned}\dot{e}_1 &= -10e_1,\\ \dot{e}_2 &= -e_2,\\ \dot{e}_3 &= -\frac{8}{3}e_3.\end{aligned} \quad (31)$$

Therefore, according to Theorem 1, systems (24) and (26) are globally Q-S synchronized in 3D. The numerical simulations of the error functions evolution are shown in Figure 3.

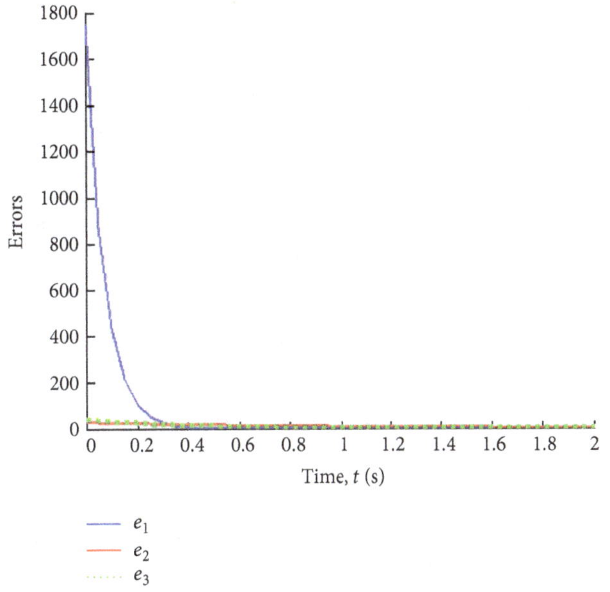

FIGURE 3: Time evolution of Q-S synchronization errors between systems (24) and (26).

6. Conclusion

In this paper, new control scheme, based on fractional control law, Laplace transformation, and stability theory of linear integer-order dynamical system, for Q-S synchronization was presented between n-dimensional integer-order master system and m-dimensional fractional-order slave system. To observe Q-S synchronization with respect to dimension n, the synchronization criterion was obtained via controlling the linear part of the master system. Numerical example and simulations result were used to verify the effectiveness of the proposed control method.

Conflict of Interests

The authors declare that there is no conflict of interests regarding the publication of this paper.

References

[1] I. Petráš, "A note on the fractional-order Chua's system," *Chaos, Solitons and Fractals*, vol. 38, no. 1, pp. 140–147, 2008.

[2] D. Cafagna, "Fractional calculus: a mathematical tool from the past for present engineers," *IEEE Industrial Electronics Magazine*, vol. 1, no. 2, pp. 35–40, 2007.

[3] P. J. Torvik and R. L. Bagley, "On the appearance of the fractional derivative in the behavior of real materials," *Journal of Applied Mechanics*, vol. 51, no. 2, pp. 294–298, 1984.

[4] B. J. West, M. Bologna, and P. Grigolini, *Physics of Fractal Operators*, Springer, New York, NY, USA, 2002.

[5] S. Westerlund, *Dead Matter Has Memory!*, Causal Consulting, Kalmar, Sweden, 2002.

[6] A. Oustaloup, *La Derivation Non Entiere: Theorie, Synthese et Applications*, Hermès, Paris, France, 1995.

[7] G.-Q. Si, Z.-Y. Sun, and Y.-B. Zhang, "A general method for synchronizing an integer-order chaotic system and a fractional-order chaotic system," *Chinese Physics B*, vol. 20, no. 8, Article ID 080505, 2011.

[8] Z. Ping, Y.-M. Cheng, and K. Fei, "Synchronization between fractional-order chaotic systems and integer orders chaotic

systems (fractional-order chaotic systems)," *Chinese Physics B*, vol. 19, no. 9, Article ID 090503, 2010.

[9] D.-Y. Chen, R.-F. Zhang, X.-Y. Ma, and J. Wang, "Synchronization between a novel class of fractional-order and integer-order chaotic systems via a sliding mode controller," *Chinese Physics B*, vol. 21, no. 12, Article ID 120507, 2012.

[10] D. Chen, R. Zhang, J. C. Sprott, H. Chen, and X. Ma, "Synchronization between integer-order chaotic systems and a class of fractional-order chaotic systems via sliding mode control," *Chaos*, vol. 22, Article ID 023130, 2012.

[11] Y.-P. Wu and G.-D. Wang, "Synchronization between fractional-order and integer-order hyperchaotic systems via sliding mode controller," *Journal of Applied Mathematics*, vol. 2013, Article ID 151025, 5 pages, 2013.

[12] Y. Wu and G. Wang, "Synchronization of a class of fractional-order and integer order hyperchaotic systems," *Journal of Vibration and Control*, vol. 20, no. 10, pp. 1584–1588, 2013.

[13] L.-X. Jia, H. Dai, and M. Hui, "Nonlinear feedback synchronisation control between fractional-order and integer-order chaotic systems," *Chinese Physics B*, vol. 19, no. 11, Article ID 110509, 2010.

[14] I. El Gammoudi and M. Feki, "Synchronization of integer order and fractional order Chua's systems using robust observer," *Communications in Nonlinear Science and Numerical Simulation*, vol. 18, no. 3, pp. 625–638, 2013.

[15] D. Chen, C. Wu, H. H. Iu, and X. Ma, "Circuit simulation for synchronization of a fractional-order and integer-order chaotic system," *Nonlinear Dynamics*, vol. 73, no. 3, pp. 1671–1686, 2013.

[16] A. Khan and P. Tripathi, "Synchronization between a fractional order chaotic system and an integer order chaotic system," *Nonlinear Dynamics and Systems Theory*, vol. 14, no. 4, 2013.

[17] L.-X. Yang, W.-S. He, and X.-J. Liu, "Synchronization between a fractional-order system and an integer order system," *Computers & Mathematics with Applications*, vol. 62, no. 12, pp. 4708–4716, 2011.

[18] Y. Wu and G. Wang, "Synchronization and anti-synchronization between a class of fractional-order and integer-order chaotic systems with only one controller term," *Journal of Theoretical & Applied Information Technology*, vol. 48, no. 1, pp. 145–151, 2013.

[19] Z. Ping and Y.-X. Cao, "Function projective synchronization between fractional-order chaotic systems and integer-order chaotic systems," *Chinese Physics B*, vol. 19, no. 10, Article ID 100507, 2010.

[20] H.-J. Liu, Z.-L. Zhu, H. Yu, and Q. Zhu, "Modified function projective synchronization of fractional order chaotic systems with different dimensions," *Discrete Dynamics in Nature and Society*, vol. 2013, Article ID 763564, 7 pages, 2013.

[21] Z. Wu, X. Xu, G. Chen, and X. Fu, "Generalized matrix projective synchronization of general colored networks with different-dimensional node dynamics," *Journal of the Franklin Institute*, vol. 351, no. 9, pp. 4584–4591, 2014.

[22] G. Zhang, Z. Liu, and Z. Ma, "Generalized synchronization of different dimensional chaotic dynamical systems," *Chaos, Solitons & Fractals*, vol. 32, no. 2, pp. 773–779, 2007.

[23] Y. Yu, H.-X. Li, and J. Yu, "Generalized synchronization of different dimensional chaotic systems based on parameter identification," *Modern Physics Letters B*, vol. 23, no. 22, pp. 2593–2606, 2009.

[24] X. He, C. Li, J. Huang, and L. Xiao, "Generalized synchronization of arbitrary-dimensional chaotic systems," *Optik*, vol. 126, pp. 454–461, 2015.

[25] A. Ouannas and Z. Odibat, "Generalized synchronization of different dimensional chaotic dynamical systems in discrete time," *Nonlinear Dynamics*, vol. 81, no. 1-2, pp. 765–771, 2015.

[26] M. Hu, Z. Xu, R. Zhang, and A. Hu, "Adaptive full state hybrid projective synchronization of chaotic systems with the same and different order," *Physics Letters A*, vol. 365, no. 4, pp. 315–327, 2007.

[27] A. M. El-Sayed, H. M. Nour, A. Elsaid, A. E. Matouk, and A. Elsonbaty, "Circuit realization, bifurcations, chaos and hyperchaos in a new 4D system," *Applied Mathematics and Computation*, vol. 239, pp. 333–345, 2014.

[28] S. Ogunjo, "Increased and reduced order synchronization of 2D and 3D dynamical systems," *International Journal of Nonlinear Science*, vol. 16, no. 2, pp. 105–112, 2013.

[29] M. M. Al-Sawalha and M. S. M. Noorani, "Adaptive increasing-order synchronization and anti-synchronization of chaotic systems with uncertain parameters," *Chinese Physics Letters*, vol. 28, no. 11, Article ID 110507, 2011.

[30] K. S. Ojo, S. T. Ogunjo, A. N. Njah, and I. A. Fuwape, "Increased-order generalized synchronization of chaotic and hyperchaotic systems," *Pramana*, vol. 84, no. 1, pp. 33–45, 2015.

[31] H. Manfeng and X. Zhenyuan, "A general scheme for Q-S synchronization of chaotic systems," *Nonlinear Analysis. Theory, Methods & Applications*, vol. 69, no. 4, pp. 1091–1099, 2008.

[32] Q. Wang and Y. Chen, "Generalized Q–S (lag, anticipated and complete) synchronization in modified Chua's circuit and Hindmarsh–rose systems," *Applied Mathematics and Computation*, vol. 181, no. 1, pp. 48–56, 2006.

[33] Z. L. Wang and X.-R. Shi, "Adaptive *Q-S* synchronization of non-identical chaotic systems with unknowns parameters," *Nonlinear Dynamics*, vol. 59, no. 4, pp. 559–567, 2010.

[34] Z. Yan, "Q-S (lag or anticipated) synchronization backstepping scheme in a class of continuous-time hyperchaotic systems—a symbolic-numeric computation approach," *Chaos*, vol. 15, no. 2, Article ID 023902, 2005.

[35] Y. Yang and Y. Chen, "The generalized Q-S synchronization between the generalized Lorenz canonical form and the Rössler system," *Chaos, Solitons and Fractals*, vol. 39, no. 5, pp. 2378–2385, 2009.

[36] J. Zhao and T. Ren, "Q–S synchronization between chaotic systems with double scaling functions," *Nonlinear Dynamics*, vol. 62, no. 3, pp. 665–672, 2010.

[37] J. Zhao and K. Zhang, "A general scheme for Q-S synchronization of chaotic systems with unknown parameters and scaling functions," *Applied Mathematics and Computation*, vol. 216, no. 7, pp. 2050–2057, 2010.

[38] Z. Yan, "Q-S synchronization in 3D Hénon-like map and generalized Hénon map via a scalar controller," *Physics Letters A*, vol. 342, no. 4, pp. 309–317, 2005.

[39] L.-X. Yang and W.-S. He, "Adaptive *Q-S* synchronization of fractional-order chaotic systems with nonidentical structures," *Abstract and Applied Analysis*, vol. 2013, Article ID 367506, 8 pages, 2013.

[40] K. B. Oldham and J. Spanier, *The Fractional Calculus*, Academic Press, New York, NY, USA, 1974.

[41] K. S. Miller and B. Ross, *An Introduction to the Fractional Calculus and Fractional Differential Equations*, Wiley, New York, NY, USA, 1993.

[42] M. Caputo, "Linear models of dissipation whose Q is almost frequency independent. II," *Geophysical Journal International*, vol. 13, no. 5, pp. 529–539, 1967.

[43] I. Podlubny, *Fractional Differential Equations*, Academic Press, New York, NY, USA, 1999.

[44] R. Goren and F. Mainardi, "Fractional calculus: integral and differential equations of fractional order," in *Fractals and Fractional Calculus in Continuum Mechanics*, A. Carpinteri and F. Mainardi, Eds., Springer, New York, NY, USA, 1997.

[45] E. N. Lorenz, "Deterministic nonperiodic flow," *Journal of the Atmospheric Sciences*, vol. 20, no. 2, pp. 130–141, 1963.

[46] Z. Wang, X. Huang, Y.-X. Li, and X.-N. Song, "A new image encryption algorithm based on the fractional-order hyperchaotic Lorenz system," *Chinese Physics B*, vol. 22, no. 1, Article ID 010504, 2013.

Hybrid Particle Swarm and Differential Evolution Algorithm for Solving Multimode Resource-Constrained Project Scheduling Problem

Lieping Zhang,[1,2] Yingxiong Luo,[3] and Yu Zhang[1,2]

[1] *Guangxi Key Laboratory of New Energy and Building Energy Saving, Guilin University of Technology, Guilin 541004, China*
[2] *College of Mechanical and Control Engineering, Guilin University of Technology, Guilin 541004, China*
[3] *College of Information Science and Engineering, Guilin University of Technology, Guilin 541004, China*

Correspondence should be addressed to Lieping Zhang; zlp_gx_gl@163.com

Academic Editor: Petko Petkov

In order to find a feasible solution for the multimode resource-constrained project scheduling problem (MRCPSP), a hybrid of particle swarm optimization (PSO) and differential evolution (DE) algorithm is proposed in this paper. The proposed algorithm uses a two-level coding structure. The upper-level structure is coded for scheduling sequence, which is optimized by PSO algorithm. The lower-level structure is coded for project execution mode, and DE algorithm is used to solve the optimal scheduling model. The effectiveness and advantages of the proposed algorithm are illustrated by using the test function of project scheduling problem library (PSPLIB) and comparing with other scheduling methods. The results show that the proposed algorithm can well solve MRCPSP.

1. Introduction

Resource-constrained project scheduling problem (RCPSP) is a project scheduling problem of how to arrange the task start-up time to minimize the total project time reasonably in the conditions of resource constraints and project timing constraints [1]. Multimode resource-constrained project scheduling problem (MRCPSP) is an extension of RCPSP, which is a kind of Nondeterministic Polynomial (NP) problem. On the basis of traditional RCPSP, MRCPSP considers a variety of optional modes of the task and the dependent relationships of resources. It can select task model and attribute dynamically. Comparing with the RCPSP, MRCPSP is closer to the real problem and is of more theoretical as well as practical significance [2]. The optimization solution methods of MRCPSP can be divided into precise algorithm, heuristic algorithm, and intelligent optimization algorithm [3]. Precise algorithm adopts the branch-and-bound method as the main solving method, and it can get the optimal solution. But it is not suitable for solving large-scale scheduling problem. Heuristic algorithm has a strong ability to solve large-scale scheduling problem and the characteristic of the computing speed. But it cannot guarantee obtaining the optimal solution. Intelligent optimization algorithm can get the optimization solution or suboptimal solution by using the algorithm optimization mechanism and evaluation mechanism in a limited set of feasible solutions. Intelligent optimization algorithm is an effective solving method for MRCPSP. Jarboui et al. proposed a combinatorial particle swarm optimization (PSO) algorithm to solve MRCPSP [4]. Van Peteghem and Vanhoucke proposed a scatter search algorithm for MRCPSP, which is executed with different improvement methods [5]. Tseng and Chen presented a two-phase genetic local search algorithm that combines the genetic algorithm and the local search method to solve MRCPSP [6]. Damak et al. proposed a differential evolution (DE) algorithm to solve MRCPSP with multiple execution modes for each activity and minimization of the makespan [7]. Wang and Fang proposed an estimation of distribution algorithm to solve MRCPSP [8]. Li and Zhang presented an ant colony optimization-based methodology for solving the MRCPSP considering both renewable and nonrenewable resources [9]. Liu et al. presented a memetic algorithm to solve MRCPSP, in which a new fitness function and two very effective local search procedures are used in the proposed algorithm [10].

The methods mentioned above have their different merits and shortcomings. But if we take the combination of several methods into account properly, a better balance between quality and efficiency of solving may be achieved. In this case, we can get the optimum solution and obtain the ultimate solution for different actual demands. Liu et al. proposed a novel hybrid algorithm named PSO-DE, which integrates PSO with DE algorithm to solve constrained numerical and engineering optimization problems [11]. Zhang and Kang proposed a hybrid of ant colony and particle swarm optimization algorithms for solving MRCPSP, which can well solve MRCPSP [12]. Chen and Sandnes proposed a two-PSO algorithm to solve MRCPSP, in which constriction PSO is proposed for the activity priority determination while discrete PSO is employed for mode assignment [13].

A hybrid of particle swarm and differential evolution algorithm is proposed for solving MRCPSP. In that algorithm, the scheduling orders are determined by the upper-level algorithm with PSO algorithm, and the task execution modes are determined by the lower-level algorithm with DE algorithm, which can get faster convergence of the algorithm and avoids falling into local optimum. Based on the verification of many MRCPSP's instances in PSPLIB, the simulation results show that the proposed algorithm can effectively solve the MRCPSP.

2. Descriptions of MRCPSP

The MRCPSP can be described as follows. A project has a series of tasks, and there is a certain logic sequence between them as the technical process and many other reasons. Meanwhile, among several patterns, individual task can choose one of them to complete, and each pattern corresponds to a set of known time limits for project and resource requirements. Solution of the problem is to generate a scheduling scheme that enables one or some of the goals to achieve the optimization, which can meet the work of the precedence constraints relations and resource constraints condition.

In this paper, what we study is to minimize the time limit for a project in MRCPSP. The upper bound of the time limit for a project is defined as $\overline{D}$. The task j ($j = 1, 2, \ldots, J$) must select one from execution models M_j. Furthermore, in the process of implementation, the execution model cannot be interrupted or changed. In m ($1 \leqslant m \leqslant M_j$) kinds of mode to perform the work j requires k kinds of renewable resources, denoted by r^{ρ}_{jmk}, and n kinds of nonrenewable resources, denoted by r^{ν}_{jmn}, and the execution time is denoted by d_{jm}. There is only one execution mode for virtual task 1 and J, which does not consume resources, and the time limit for the project is zero. Among them, renewable resources are denoted by ρ and nonrenewable resources are denoted by ν. In the whole time of the project and each stage of the project, the kth kind of renewable resources can amount to a constant, denoted by R^{ρ}_k ($k = 1, 2, \ldots, K$), and the nth kind of nonrenewable total resources, denoted by R^{ν}_n ($n = 1, 2, \ldots, N$). If the task J selects the mth kind of modes to execute and complete in the stage of t, then $\chi_{jmt} = 1$; otherwise, $\chi_{jmt} = 0$, wherein χ_{jmt} is a decision variable. The mathematical model for MRCPSP can be described as follows [14]:

$$\min \quad \sum_{t=EF_j}^{LF_j} t \cdot x_{jmt} \tag{1}$$

$$\text{s.t.} \quad \sum_{m=1}^{M_j} \sum_{t=EF_j}^{LF_j} x_{jmt} = 1, \quad j = 1, 2, \ldots, J \tag{2}$$

$$\sum_{m=1}^{M_i} \sum_{t=EF_i}^{LF_i} t \cdot x_{imt} \leq \sum_{m=1}^{M_j} \sum_{t=EF_j}^{LF_j} t \cdot x_{jmt} - d_{jm}, \quad j = 2, \ldots, J, \; i \in P_j \tag{3}$$

$$\sum_{j=1}^{J} \sum_{m=1}^{M_j} r^{\rho}_{jmk} \sum_{q=\max\{t, EF_j\}}^{\min\{t+d_m-1, LF_j\}} x_{jmq} \leq R^{\rho}_k, \quad k = 1, \ldots, K, \; t = 1, \ldots, \overline{D} \tag{4}$$

$$\sum_{j=1}^{J} \sum_{m=1}^{M_j} r^{\nu}_{jmn} \sum_{t=EF_j}^{LF_j} x_{jmt} \leq R^{\nu}_n, \quad n = 1, \ldots, N. \tag{5}$$

Among them, $x_{jmt} \in \{0, 1\}$, and $j = 1, 2, \ldots, J$, and $m = 1, 2 \ldots, M_j$, and $t = EF_j, \ldots, LF_j$, EF_j, LF_j, respectively, corresponds to the earliest completion time and finish time at the latest of the task j, and P_j is the direct predecessor task set of j. Equation (1) is the objective function which means the shortest time limit for the total project. Equation (2) means that a task can only be completed once in an execution mode. Equation (3) represents the precedence constraints. According to the constructed network model and network planning methods, the end time of each task is equal to the sum of its start time and the time limit for a task, and the tight task must be carried out at the end of the work. Equation (4) ensures the amount of renewable resources in every stage will not be larger than the amount available. Equation (5) ensures that the amount of nonrenewable resources consumed in the entire project will not be larger than the amount available.

3. Algorithm Design

The upper-level individual coding of the designed algorithm controls the task scheduling orders, and the lower-level individual coding manages the task execution modes. After determining the task scheduling orders, it can find the optimal execution mode in the scheduling order. Due to the fact that the range of execution mode's values is just an integer in the range [1, 3], it is a very small range and it will make the PSO algorithm easily fall into local optimum. Hence, the DE algorithm was introduced into the proposed algorithm to solve the optimal execution mode, which has stronger global convergence ability and robustness, and then optimize the task scheduling sequence with the PSO algorithm.

3.1. Coding Design. The upper-level individual of algorithm uses the particle coding based on priority rules. The numbers of tasks are represented by the search space dimension of particle swarm, a total of J. At the same time, all dimension values x_{ij} of particle x_i are random real numbers in the range $[0,1]$. The value represents the priority order of particle j kinds of dimension. The higher the value, the greater the priority. While the virtual start task of the project's first task is the first task to be executed, the last task is the virtual end task of the project, which is the final project task. All the priority values that can set virtual start task are $x_{ij} = 1$, the priority values of the virtual end task are $x_{ij} = 0$, and the other priority values of task are in the range $(0,1)$. Moreover, the task priority values are different.

In order to let the priority rules satisfy the task's precedence constraints, we need to adjust the task priority value. Specific adjustment method is as follows. Compare the priority values of current task and preceding task; the value will not be adjusted if it is lower than the latter. Otherwise, exchange both of the priority values. And then, consider the preceding task as the current task and compare it with its preceding task. As long as they are exchanged, repeat this step until no values need to be exchanged or there is no preceding task. After this adjustment, we can get the priority rules that met the requirements of the task precedence constraints. The low level of algorithm controls the execution mode coding. Assuming that the execution modes of the task j are M_j kinds, the execution mode of J is in the range $[1, M_j]$; note that the value of the execution mode is integer.

3.2. A Hybrid of PSO and DE Algorithm

3.2.1. PSO Algorithm Based on Inertia Weight. In an n-dimensional search space, population $X = \{x_1, \dots, x_2, \dots, x_m\}$ is composed of m particles, and the particle position is $x_i = (x_{i1}, x_{i2}, \dots, x_{im})^T$, and the particle velocity is $V_i = (v_{i1}, v_{i2}, \dots, v_{in})^T$. The corresponding individual extreme value is $P_i = (p_{i1}, p_{i2}, \dots, p_{im})^T$, and the global extreme value of the entire population is $P_g = (p_{g1}, p_{g2}, \dots, p_{gm})^T$. Taking into account the premature convergence and poor global convergence to the basic PSO algorithm in the practical application, we adopted the PSO algorithm based on inertia weight to promote the global searching ability and the local search ability [15]. According to the principle of following the current optimal particle, x_i particle, which introduced the inertia weight w, will change its speed and position according to

$$v_{id}^{(t+1)} = wv_{id}^{(t)} + c_1 r_1 \left(p_{id}^{(t)} - x_{id}^{(t)}\right) + c_2 r_2 \left(p_{gd}^{(t)} - x_{id}^{(t)}\right) \quad (6)$$

$$x_{id}^{(t+1)} = x_{id}^{(t)} + v_{id}^{(t+1)}, \quad (7)$$

where $d = 1, 2, \dots, n$, which means the dth dimension of the particle, and $i = 1, 2, \dots, m$, which means the particle i. m is the population size. t means the current evolution times. r_1 and r_2 are random numbers distributed in the range $[0,1]$. c_1 and c_2, respectively, correspond to accelerated constant of the individual particle and the group particle. $v_{id}^{(t+1)}$, $v_{id}^{(t)}$, $x_{id}^{(t+1)}$, and $x_{id}^{(t)}$, respectively, represent the velocity and the position of the ith particle in the $(t+1)$th generation and the tth generation.

The value of inertia weight w can affect the algorithm's global search ability and local search ability. Related research results show that when w decline linearly from 0.9 to 0.4, the algorithm can converge quickly to the optimal solution [16]. Therefore, this paper adopts the dynamic inertia weight, and the value of w changes in line in the searching process of the particle swarm algorithm, as shown in

$$w = 0.9 - \left(\frac{\text{nc}}{\text{ncmax}}\right) * 0.5, \quad (8)$$

where nc is the current number of cycles and ncmax is the maximum number of cycles.

3.2.2. DE Algorithm. DE algorithm is a simple and efficient parallel search algorithm, which has better robustness and faster convergence capability than other evolution algorithms. And it can be used to solve the MRCPSP. The low-level individual of the designed algorithm uses the DE algorithm to find the optimal execution mode in the scheduling sequence given.

For the individual $X_{r_1,G}$, a new individual $V_{i,G+1}$ can be obtained by

$$V_{i,G+1} = X_{r_1,G} + F \times \left(X_{r_2,G} - X_{r_3,G}\right), \quad (9)$$

where $i = 1, 2, \dots, N$. r_1 and r_2 as well as r_3 are different random integers in the range $[1, N]$, and they are different from the subscript index i. The variation factor F is a real constant in the range $[0,2]$, the main function of which is to control the degree of amplification of differential vector $(X_{r_2,G} - X_{r_3,G})$.

The interlace operation is as shown below:

$$U_{i,G+1} = \left(u_{1i,G+1}, u_{2i,G+1}, \dots, u_{Di,G+1}\right). \quad (10)$$

In the equation

$$u_{ji,G+1} = \begin{cases} V_{ji,G+1}, & \text{if } (\text{rand}b(j) \le \text{CR}) \text{ or } (j = mbr(i)) \\ X_{ji,G+1}, & \text{if } (\text{rand}b(j) > \text{CR}),\ (j \ne mbr(i)), \end{cases} \quad (11)$$

$$j = 1, 2, \dots, D,$$

where rand$b(j)$ is a uniform distribution probability in the range $[0,1]$, CR is the crossover probability predefined by the user, CR $\in (0,1)$, and $mbr(i)$ is a random number generated in the range $[1, D]$.

In order to determine whether vector $U_{i,G+1}$ can be $(G+1)$th generation of population individuals, compare $U_{i,G+1}$ with $X_{i,G}$; if the former fitness value is better than the latter, replace $X_{i,G}$ with $U_{i,G+1}$ in $(G+1)$th generation; otherwise, reserve $X_{i,G}$.

3.3. Scheduling Generation Strategy. Scheduling generation strategy based on priority rules includes serial scheduling scheme and parallel scheduling scheme. In this paper, we use serial scheduling scheme to solve the problem. For an upper individual x_i, we can get x_i' according to sorting the order of x_i from big to small. And the serial number of original individual x_i is stored in the order of x_i' to y_i. For example, if the original individuals are $[1, 0.3, 0.5, 0.2, 0.4, 0.7, 0]$, then the sorted individuals are $[1, 0.7, 0.5, 0.4, 0.3, 0.2, 0]$, and the value of y_i is $[1, 6, 3, 5, 2, 4, 7]$. Select an element y_{ij} in order from y_i according to the value of y_{ij} to determine the start time of scheduling task x_i and y_{ij}. Choose a larger one from the current time and all the preceding task finish time of the current task as the current task start time, and the end time of the current task is determined by the lower layer determined time limit for project plus the current task start time. Update the current time in accordance with the use of resources. When the available resources run out, update the current time. And when the end time of a task is greater than the current time, release its consumed resources in turn. Meanwhile, make the current time equal to the end time of the task that releases resources, until available resources meet the requirements of the current task. The end time of virtual end of task is the project completion time, namely, the value of algorithm fitness function.

3.4. Algorithm Implementation Processes. The major steps of the proposed algorithm can be outlined as follows.

Step 1. Initialize the upper-level populations to generate m individuals position (priority rules) $(x_1, x_2, \ldots, x_m)$ and velocity $(v_1, v_2, \ldots, v_m)$.

Step 2. Adjust the priority rules to meet the task precedence constraints represented in (3).

Step 3. Initialize the lower-level populations to generate n individuals (the task execution modes).

Step 4. The lower-level populations, respectively, execute mutation and interlace operation according to (9) and (10).

Step 5. The lower-level populations execute selection operation and optimize the execution mode.

Step 6. If the lower-level populations reached the maximum numbers of iterations, then go to Step 7, or go to Step 4 to continue to optimize the execution mode.

Step 7. Respectively, update the velocity and position of the upper populations according to (6) and (7).

Step 8. If the upper-level populations reached the maximum numbers of iterations, the algorithm ends, or go to Step 2.

4. Simulation Analyses

In order to test the effectiveness of the algorithm, the standard MRCPSP instances of PSLIB are used to validate the effectiveness of the algorithm. The selected task is J18. The task contains 2 kinds of renewable resources and 2 kinds of nonrenewable resources, in which the renewable resources are $R_1 = 26$, $R_2 = 27$, and the nonrenewable resources are $N_1 = 48$, $N_2 = 51$. We compare the proposed algorithm with different iterations. The upper-level PSO algorithm parameters of the proposed algorithm are described as follows: the individual particle acceleration constant is selected as $c_1 = 2$ and the group of particle acceleration constants is selected as $c_2 = 2$, the number of populations is selected as 10, and the number of iterations is taken as 10. The parameters of lower-level DE algorithm are described as follows [17]: the interlace probability is selected as CR = 0.4, the variation factor is selected as $F = 0.7$, and both of the numbers of populations and iterations are selected as 10. A Gantt chart of optimal solution is shown in Figure 1.

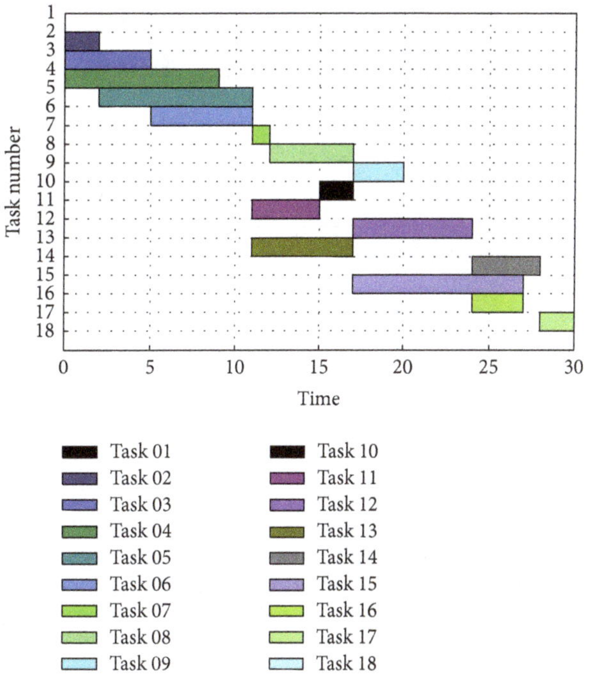

FIGURE 1: Gantt chart of the optimal solution.

TABLE 1: Computing results vary from different number of iterations in the upper level.

Upper-level number of iterations	Average deviation (%)	Optimal solution ratio (%)
10	0.78	83.33
15	0.44	93.33

When the upper-level iteration numbers are, respectively, 10 and 15, the solved results of the proposed algorithm are shown in Table 1. In accordance with Table 1, we can know that the proposed algorithm is effective for MRCPSP, and, with the increase of the number of iterations, the probability of obtaining the optimal solution increased.

In order to test the feasibility of the proposed algorithm, we compare the proposed algorithm with two layers' PSO algorithm based on inertia weight. The parameters of two

Table 2: Results comparison of PSODE algorithm and PSO algorithm.

Algorithm	Average deviation (%)	Optimum ratio (%)
PSODE	0.78	83.33
PSO	5.67	40

Table 3: Results comparison of the proposed algorithm and other methods.

Project	Population size	J10		J20	
		Average deviation	Optimal solution rate	Average deviation	Optimal solution rate
This paper	10-10	0%	100%	1.82%	76.67%
Reference [12]	10-10	0%	100%	1.28%	71.5%

layers' PSO algorithm based on inertia weight are described as follows: the individual particle acceleration constants c_1 and c_2 are selected as 2, the numbers of populations are selected as 10, and the numbers of iterations are selected as 10. The parameters of the proposed algorithm are described as before, in which the number of iterations of PSO algorithm is selected as 10. The corresponding results are shown in Table 2. In accordance with Table 2, we can know that the average deviation and optimum ratio of PSODE algorithm are higher than those of PSO algorithm.

In order to further test the feasibility and the effectiveness of the proposed algorithm, we compare it with the solving results of [12]. Projects J10 and J20 have 536 and 554 optimal solutions, respectively, and projects J10 and J20 have two kinds of renewable resources and two kinds of nonrenewable resources, respectively, and every instance has three modes. Project J10 has a total of 12 tasks, and project J20 has a total of 22 tasks, and both the first task and the last task are virtual tasks, which do not consume the time and the resources. The parameters of the proposed algorithm are described as before. The results comparison of the proposed algorithm and the method of [12] is shown in Table 3. According to the results in Table 3, the proposed algorithm is better than the method of [12] in average deviation and optimal solution rate.

5. Conclusions

A new algorithm for MRCPSP combined PSO algorithm and DE algorithm is proposed in this paper. The optimal project execution modes are obtained by DE algorithm after the determination of the task scheduling sequence; then the task scheduling sequences are optimized by PSO algorithm. In order to confirm the validity of the algorithm, we take an experiment based on the standard MRCPSP instances of PSLIB and other scheduling methods. The experimental and comparison results show that the proposed algorithm has a better performance in average deviation and optimal solution rate.

Conflict of Interests

The authors declare that there is no conflict of interests regarding the publication of this paper.

Acknowledgments

This research was supported by Guangxi Natural Science Foundation (no. 2014GXNSFAA118371) and the research fund of Guangxi Key Laboratory of New Energy and Building Energy Saving (no. 12-03-21-3).

References

[1] F. Ballestín and R. Blanco, "Theoretical and practical fundamentals for multi-objective optimisation in resource-constrained project scheduling problems," *Computers and Operations Research*, vol. 38, no. 1, pp. 51–62, 2011.

[2] P. Ghoddousi, E. Eshtehardian, S. Jooybanpour, and A. Javanmardi, "Multi-mode resource-constrained discrete time-cost-resource optimization in project scheduling using non-dominated sorting genetic algorithm," *Automation in Construction*, vol. 30, pp. 216–227, 2013.

[3] P. Brucker, A. Drexl, R. Möhring, K. Neumann, and E. Pesch, "Resource-constrained project scheduling: notation, classification, models, and methods," *European Journal of Operational Research*, vol. 112, no. 1, pp. 3–41, 1999.

[4] B. Jarboui, N. Damak, P. Siarry, and A. Rebai, "A combinatorial particle swarm optimization for solving multi-mode resource-constrained project scheduling problems," *Applied Mathematics and Computation*, vol. 195, no. 1, pp. 299–308, 2008.

[5] V. Van Peteghem and M. Vanhoucke, "Using resource scarceness characteristics to solve the multi-mode resource-constrained project scheduling problem," *Journal of Heuristics*, vol. 17, no. 6, pp. 705–728, 2011.

[6] L.-Y. Tseng and S.-C. Chen, "Two-phase genetic local search algorithm for the multimode resource-constrained project scheduling problem," *IEEE Transactions on Evolutionary Computation*, vol. 13, no. 4, pp. 848–857, 2009.

[7] N. Damak, B. Jarboui, P. Siarry, and T. Loukil, "Differential evolution for solving multi-mode resource-constrained project scheduling problems," *Computers and Operations Research*, vol. 36, no. 9, pp. 2653–2659, 2009.

[8] L. Wang and C. Fang, "An effective estimation of distribution algorithm for the multi-mode resource-constrained project scheduling problem," *Computers & Operations Research*, vol. 39, no. 2, pp. 449–460, 2012.

[9] H. Li and H. Zhang, "Ant colony optimization-based multi-mode scheduling under renewable and nonrenewable resource constraints," *Automation in Construction*, vol. 35, pp. 431–438, 2013.

[10] S. Liu, D. Chen, and Y. Wang, "Memetic algorithm for multimode resource-constrained project scheduling problems," *Journal of Systems Engineering and Electronics*, vol. 25, no. 4, pp. 609–617, 2014.

[11] H. Liu, Z. Cai, and Y. Wang, "Hybridizing particle swarm optimization with differential evolution for constrained numerical and engineering optimization," *Applied Soft Computing Journal*, vol. 10, no. 2, pp. 629–640, 2010.

[12] W. Zhang and K. Kang, "Ant colony and particle swarm optimization algorithm-based solution to multi-mode resource-constrained project scheduling problem," *Computer Engineering and Applications*, vol. 43, no. 34, pp. 213–216, 2007.

[13] R.-M. Chen and F. E. Sandnes, "An efficient particle swarm optimizer with application to man-day project scheduling problems," *Mathematical Problems in Engineering*, vol. 2014, Article ID 519414, 9 pages, 2014.

[14] R. Kolisch and A. Sprecher, "PSPLIB—a project scheduling problem library," *European Journal of Operational Research*, vol. 96, no. 1, pp. 205–216, 1997.

[15] A. El-Gallad, M. El-Hawary, A. Sallam, and A. Kalas, "Enhancing the particle swarm optimizer via proper parameters selection," in *Proceedings of the IEEE Canadian Conference on Electrical & Computer Engineering*, pp. 792–797, Winnipeg, Canada, May 2002.

[16] Y. H. Shi and R. C. Eberhart, "Empirical study of particle swarm optimization," in *Proceedings of the Congress on Evolutionary Computation (CEC '99)*, pp. 1945–1950, Piscataway, NJ, USA, July 1999.

[17] Z. Huang and Y. Chen, "An improved differential evolution algorithm based on adaptive parameter," *Journal of Control Science and Engineering*, vol. 2013, Article ID 462706, 5 pages, 2013.

15

The Stabilization of Continuous-Time Networked Control Systems with Data Drift

Qixin Zhu,[1,2] Kaihong Lu,[2] and Yonghong Zhu[3]

[1]*School of Mechanical Engineering, Suzhou University of Science and Technology, Suzhou 215009, China*
[2]*School of Electrical and Electronic Engineering, East China Jiaotong University, Nanchang 330013, China*
[3]*School of Mechanical and Electronic Engineering, Jingdezhen Ceramic Institute, Jingdezhen 333001, China*

Correspondence should be addressed to Qixin Zhu; bob21cn@163.com

Academic Editor: Kalyana C. Veluvolu

By data drift, we mean the data received by the controller may be different from that sent by the sensor, or the data received by actuator may be different from that sent by the controller. The issues of guaranteed cost control for a class of continuous-time networked control systems with data drift are investigated. Firstly, with the consideration of data drift between sensor and controller, a closed-loop model of networked control systems including network factors such as time-delay and data-dropouts is established. And then, selecting an appropriate Lyapunov function, a guaranteed cost controller in terms of linear matrix inequality (LMI) is designed to asymptotically stabilize the networked control system with data drift. Finally, simulations are included to demonstrate the theoretical results.

1. Introduction

Networked control systems (NCSs) are often encountered in practice for widespread fields of applications because of their suitable and flexible structure [1–3]. However, in practical NCS, it inevitably causes time-delay and data-dropouts because of the introduction of the communication network [4, 5], which could cause negative impact on the system, including performance decline and instability. Thus, the issues about time-delay and data-dropouts have attracted considerable attention in the control field [1, 3, 6–14]. An augmented state vector method is proposed in [8] to control a linear system over a periodic delay network. Queuing mechanisms are developed in [9, 10], which utilize some deterministic or probabilistic information of NCSs for control purpose. Random delays are discussed in [11] via an optimal stochastic control methodology. Packet dropouts and network-induced delays were lumped into one item in Xiong and Lam [12] to study stabilization of discrete-time NCSs. By combing packet dropouts and network-induced delays into one item, Liu and Fridman [13] and Meng et al. [14] studied the stability and stabilization of continuous-time NCSs.

Recently, guaranteed cost control is widely used in NCS to keep the stability of the system and to make it meet a certain performance indicator [15–21]. Guaranteed cost control of multi-input and multioutput (MIMO) networked control systems (NCSs) with multichannel packet disordering are discussed by Li et al. [15]. An observer-based guaranteed cost control problem in networked control systems with random data packet dropouts is proposed in [17], in which sensor-to-controller and controller-to-actuator packet dropouts are both modeled by two mutually independent stochastic variables satisfying the Bernoulli binary distribution. Xie et al. [19] are concerned with the state-feedback guaranteed cost controller design for a class of networked control systems (NCSs) with state-delay. In [21], Li and Wu investigated the issue of integrity against actuator faults for NCS under variable-period sampling, in which the existence conditions of guaranteed cost faults-tolerant control law are testified in terms of Lyapunov stability theory.

In NCSs, the occurrence of uncertain network factors such as quantization errors and network noises could induce the phenomenon that data received by the controller may be different from the data sent by the sensor, or data received by

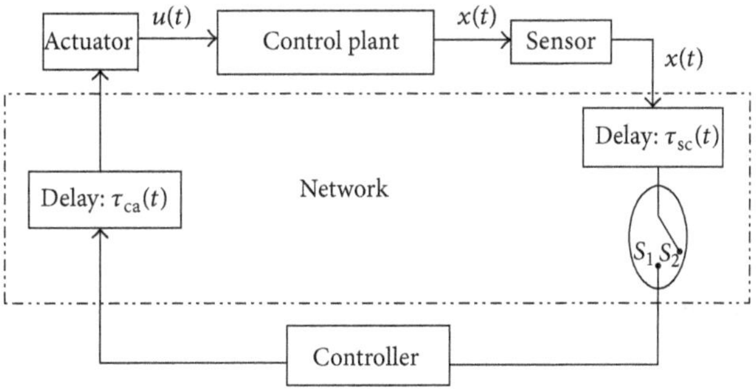

FIGURE 1: The structure of NCS.

actuator may be different from the data sent by the controller, which is called data drift. Data drift could cause negative impact on the system such as performance decline, even leading to instability. However, data drift in an NCS has not been taken into account in the literature above. In [22], Wang and Han first introduce a class of channel utilization-based switched controllers with controller-to-actuator data drift considered, and its results are established under the condition that data drift exists between controller and actuator, which may not be usable under the condition that data drift exists between sensor and controller. So, it is necessary to establish the results for NCS with sensor-to-controller data drift. Moreover, in [22], the guaranteed cost problem has not been considered, which motivates us to do this study.

This paper aims at investigating the problem of guaranteed cost control for a class of continuous-time networked control systems with data drift. First of all, the sensor-to-controller data drift is modeled as time-varying parameters. And a closed-loop model of networked control systems with time-varying parameters is established by lumping the network-induced delay and data-dropouts to a synthetically time-varying delay. Then, by using an appropriate Lyapunov functional, a guaranteed cost controller in terms of linear matrix inequality (LMI) is designed to cope with the effect of data drift and enhance the NCSs' performance.

The paper is organized in 5 sections including the Introduction. Section 2 presents problem formulations and modeling of NCS with data drift. Section 3 presents guaranteed cost controller design for NCS with data drift. There are some examples to illustrate the results in Section 4. Section 5 summarized this paper.

Notations. R^n denotes the n-dimensional Euclidean space. The superscript "T" stands for matrix transposition. The notation $X > 0$ means that the matrix X is a real positive definite matrix. I is the identity matrix of appropriate dimensions. $\left[\begin{smallmatrix} X & Z \\ * & Y \end{smallmatrix}\right]$ denotes a symmetric matrix, where $*$ denotes the entries implied by symmetry.

2. Modeling of NCS with Data Drift

Consider the linear control plant of NCS as follows:

$$\dot{x}(t) = Ax(t) + Bu(t), \tag{1}$$

where $x(t) \in R^n$ and $u(t) \in R^m$ represent state value and input and output separately; A and B are matrices with appropriate dimensions.

The typical structure of NCS is shown in Figure 1. Transmission delays induced by the network are sensor-to-controller delay τ_{sc} and controller-to-actuator delay τ_{ca}. In fact, these two delays can be lumped together as $\tau = \tau_{sc} + \tau_{ca}$ when the feedback controller is static. It assumes the state of the system is completely measurable. A piecewise static continuous feedback controller, which is realized by a zero-order-hold (ZOH), is employed:

$$u(t) = Kx(t-\tau), \quad t \in [t_k, t_{k+1}), \; k = 1, 2, \ldots, \tag{2}$$

where K is the static state feedback gain matrix to be designed and t_k is the sampling instant.

Because the bandwidth is limited, data packet dropouts also happen in NCS. Considering that data packet dropouts may occur, the network is modeled as a switch. When the switch is located in position of S_1, the data packet containing $x(t_k)$ is transmitted, and the controller utilizes the updated data; but when it is located in position S_2, the data packet dropouts occur, and the controller uses the old data. Here only sensor-to-controller dropouts are considered. For a fixed sampling period h, the dynamics of the switch can be expressed as follows:

The NCS with no packet dropout at time t_k:

$$u(t) = Kx(t_k - \tau). \tag{3}$$

The NCS with one packet dropout at time t_k:

$$u(t) = Kx(t_k - \tau - h) \\ \vdots \tag{4}$$

The NCS with $d_k \in Z^+$ packet dropout at time t_k: (2) can be rewritten as

$$u(t) = Kx(t_k - \tau - d_k h). \tag{5}$$

When the control inputs are transmitted through network medium, data drift is unavoidable. In what follows, we take the sensor-to-controller data drift into account; it means the data received by controller is different from that sent by sensor. Considering the effect of data draft, we denote the ratio of data received by ith controller to the corresponding data sent by the ith controller by $D_i(t)$ for any $i = 1, 2, \ldots, n$. And we define $D(t) = \operatorname{diag}(D_1(t), D_2(t), \ldots, D_n(t))$. Under the consideration of controller-to-actuator data drift, the control law in (5) is converted into

$$u(t) = KD(t)\, x(t_k - \tau - d_k h). \tag{6}$$

Let $\eta(t) = t - t_k + \tau_{ca} + \tau_{sc} + d_k h$; $t \in [t_k, t_{k+1})$; (6) can be expressed as follows:

$$u(t) = KD(t)\, x(t - \eta(t)). \tag{7}$$

We assume it satisfies

$$\eta_m \le \eta(t) = t - t_k + \tau_{ca} + \tau_{sc} + d_k h \le \eta_M \quad t \in [t_k, t_{k+1}). \tag{8}$$

The upper bound of variable $D(t)$ is defined as $D_u = \operatorname{diag}(D_{u1}, D_{u2}, \ldots, D_{un})$, $1 \ge D_{ui} > 0$, while the lower bound of variable $D(t)$ is defined as $D_l = \operatorname{diag}(D_{l1}, D_{l2}, \ldots, D_{ln})$, $1 > D_{li} \ge 0$. That is to say, $D(t) \in [D_l, D_u]$, which is time-varying. The average value of these two constant matrices can be obtained as

$$D_0 = \operatorname{diag}(D_{01}, D_{02}, \ldots, D_{0n}), \quad D_{0i} = \frac{D_{ui} + D_{li}}{2}. \tag{9}$$

Furthermore, the following time-varying matrix is introduced:

$$F(t) = \operatorname{diag}(f_1(t), f_2(t), \ldots, f_n(t)), \quad f_i(t) = \frac{D_i(t) - D_{0i}}{D_{0i}}. \tag{10}$$

Obviously, we have

$$-1 \le \frac{D_{li} - D_{0i}}{D_{0i}} \le f_i(t) = \frac{D_i(t) - D_{0i}}{D_{0i}} \le \frac{D_{ui} - D_{0i}}{D_{0i}} = \frac{D_{ui} - D_{li}}{D_{ui} + D_{li}} \le 1. \tag{11}$$

Based on (11), we have $-I_{n\times n} \le F(t) \le I_{n\times n}$. Based on (10), it knows $D_i = D_{0i}(1 + f_i(t))$, $i = 1, 2, \ldots, n$. Naturally, we have $D = D_0(I + F(t))$. Submitting this into (7), we have

$$u(t) = KD_0(I + F(t))\, x(t - \eta(t)). \tag{12}$$

Submitting (12) into (1), the following follows:

$$\dot{x}(t) = Ax(t) + BKD_0(I + F(t))\, x(t - \eta(t)). \tag{13}$$

Remark 1. The networked control systems with sensor-to-controller data drift are modeled as system (13). From (6), we know that if $D(t) = I$, which means $D_i(t) = 1$ for any $i = 1, 2, \ldots, n$, sensor-to-controller data drift does not happen. In model (13), the uncertain matrix $D(t)$ denoting data drift is transformed to a bounded matrix $F(t)$ by introducing an upper bound and a lower bound for data drift. From (11), it is obvious that $F^T F = F^2 \le I$.

Remark 2. Different from the models in literatures [18–21], the static feedback control gain matrix here is located between input matrix and the uncertain matrix induced by sensor-to-controller data drift, which makes it more difficult to achieve the static control gain that can cope with the time-varying data drift.

3. Guaranteed Cost Controller Design of NCS with Data Drift

For system model (13) established in Section 2, the cost function is given as follows:

$$J = \int_0^{\infty} x^T(t)\, Sx(t)\, dt, \tag{14}$$

where S is a symmetric positive definite matrix.

Definition 3. For system (13) and its cost function (14), if there exist a control gain matrix K and a constant J_0, the cost function satisfies $J \le J_0$. It is called matrix K and is the guaranteed cost control gain of NCS.

To analyze the stability of the system expediently, the following lemmas are introduced.

Lemma 4 (see [23]). *For any matrices M, N, and $F(t)$ with $F^T F \le I$ and any scalar $\varepsilon > 0$, the inequality holds as*

$$MF(t)N + N^T F^T(t) M^T \le \varepsilon M M^T + \varepsilon^{-1} N^T N. \tag{15}$$

The fundamental preliminary result is presented in the following theorem.

Theorem 5. *Given symmetric positive definite matrices S and matrix K, if there exist a set of symmetric positive definite matrices Q_1, Q_2, R_1, and R_2 and matrix $P > 0$, as well as matrices M_1, M_2, M_3, M_4, and M_5 and a constant $\varepsilon > 0$, satisfying the matrix inequality as*

$$\begin{bmatrix} \Pi & \Xi & \hat{I} \\ * & \varepsilon^{-1} I & 0 \\ * & * & \varepsilon I \end{bmatrix} < 0, \tag{16}$$

where

$$\Pi = \begin{bmatrix} Q_1 + M_1A + A^T M_1{}^T - R_1 + S & A^T M_3{}^T + R_1 & A^T M_4{}^T & M_1BKD_0 + A^T M_2{}^T & A^T M_5{}^T - M_1 + P \\ * & Q_2 - Q_1 - R_1 - \frac{1}{\eta_M - \eta_m} R_2 & \frac{1}{\eta_M - \eta_m} R_2 & M_3BKD_0 & -M_3 \\ * & * & -Q_2 - \frac{1}{\eta_M - \eta_m} R_2 & M_4BKD_0 & -M_4 \\ * & * & * & M_2BKD_0 + (BKD_0)^T M_2{}^T & (BKD_0)^T M_5{}^T - M_2 \\ * & * & * & * & \Omega - M_5 + M_5{}^T \end{bmatrix},$$

$$\Omega = \eta_m^2 R_1 + (\eta_M - \eta_m) R_2, \quad \Xi = \begin{bmatrix} M_1BKD_0 \\ M_3BKD_0 \\ M_4BKD_0 \\ M_2BKD_0 \\ M_5BKD_0 \end{bmatrix}, \quad \hat{I} = \begin{bmatrix} 0 \\ 0 \\ 0 \\ I \\ 0 \end{bmatrix} \tag{17}$$

then system (13) is asymptotically stable. And the cost function J satisfies

$$J \le x^T(0) P x(0) + \int_{-\eta_m}^{0} x^T(\alpha) Q_1 x(\alpha) d\alpha + \int_{-\eta_M}^{-\eta_m} x^T(\alpha) Q_2 x(\alpha) d\alpha + \eta_m \int_{-\eta_m}^{0} \int_{\theta}^{t} \dot{x}^T(\delta) R_1 \dot{x}(\delta) d_\theta d_\delta + \int_{-\eta_M}^{-\eta_m} \int_{\tau}^{t} \dot{x}^T(\omega) R_2 \dot{x}(\omega) d_\tau d_\omega. \tag{18}$$

Proof. First of all, we consider the Lyapunov-Krasovskii function as follows:

$$V(t) = V_1(t) + V_2(t) + V_3(t), \tag{19}$$

where

$$V_1(t) = x^T(t) P x(t),$$
$$V_2(t) = \int_{t-\eta_m}^{t} x^T(\alpha) Q_1 x(\alpha) d\alpha + \int_{t-\eta_M}^{t-\eta_m} x^T(\alpha) Q_2 x(\alpha) d\alpha,$$
$$V_3(t) = \eta_m \int_{t-\eta_m}^{t} \int_{\theta}^{t} \dot{x}^T(\delta) R_1 \dot{x}(\delta) d_\theta d_\delta + \int_{t-\eta_M}^{t-\eta_m} \int_{\tau}^{t} \dot{x}^T(\omega) R_2 \dot{x}(\omega) d_\tau d_\omega, \tag{20}$$

with $P > 0$, $Q_1 > 0$, $Q_2 > 0$, $R_1 > 0$, and $R_2 > 0$.

Calculating the derivative of Lyapunov-Krasovskii function and based on (13), it follows that

$$\begin{aligned} \dot{V}(t) = {} & 2x^T(t) P \dot{x}(t) + x^T(t) Q_1 x(t) + x^T(t - \eta_m) \cdot (Q_2 - Q_1) x(t - \eta_m) - x^T(t - \eta_M) Q_2 x(t - \eta_M) \\ & + \dot{x}^T(t) \left[\eta_m^2 R_1 + (\eta_M - \eta_m) R_2\right] \dot{x}(t) - \eta_m \int_{t-\eta_m}^{t} \dot{x}^T(\theta) R_1 \dot{x}(\theta) d_\theta \\ & + \int_{t-\eta_M}^{t-\eta_m} \dot{x}^T(s) R_2 \dot{x}(s) d_s + 2\left[x^T(t) M_1 + x^T(t - \eta(t)) M_2 + x^T(t - \eta_m) M_3 + x^T(t - \eta_M) M_4 + \dot{x}^T(s) M_5\right]. \end{aligned} \tag{21}$$

Based on Jensen's inequality, also used in [19], we have

$$-\eta_m \int_{t-\eta_m}^{t} \dot{x}^T(s) R_1 \dot{x}(s)\, d_s \le \left[x^T(t), x^T(t-\eta_m)\right] \cdot \begin{bmatrix} -R_1 & R_1 \\ R_1 & -R_1 \end{bmatrix} \begin{bmatrix} x(t) \\ x(t-\eta_m) \end{bmatrix},$$

$$-\int_{t-\eta_M}^{t-\eta_m} \dot{x}^T(s) R_2 \dot{x}(s)\, d_s \le \frac{1}{\eta_M - \eta_m}\left[x^T(t-\eta_m), x^T(t-\eta_M)\right] \cdot \begin{bmatrix} -R_2 & R_2 \\ R_2 & -R_2 \end{bmatrix} \begin{bmatrix} x(t-\eta_m) \\ x(t-\eta_M) \end{bmatrix}. \tag{22}$$

Applying Lemma 4 and Schur complement to inequality (16), we have

$$\begin{bmatrix} Q_1 + M_1 A - R_1 + S & A^T M_3{}^T + R_1 & A^T M_4{}^T & M_1\Theta + A^T M_2{}^T & A^T M_5{}^T - M_1 + P \\ * & Q_2 - Q_1 - R_1 - \dfrac{1}{\eta_M - \eta_m} R_2 & \dfrac{1}{\eta_M - \eta_m} R_2 & M_3\Theta & -M_3 \\ * & * & -Q_2 - \dfrac{1}{\eta_M - \eta_m} R_2 & M_4\Theta & -M_4 \\ * & * & * & M_2\Theta & \Theta^T M_4{}^T - M_2 \\ * & * & * & * & \Omega - M_5 \end{bmatrix} < 0, \tag{23}$$

where $\Theta = BKD_0(I + F(t))$.

From (21)–(23), based on Schur complement, it follows that

$$\dot{V}(t) \le -x^T(t) Sx(t) \le 0. \tag{24}$$

Therefore, system (13) is asymptotically stable and there exists $\lim_{t\to\infty} x(t) = 0$. And through the integral operation, we have

$$\int_0^\infty \dot{V}(t)\, d_t = V(\infty) - V(0) = -V(0) \le -\int_0^\infty x^T(t) Sx(t)\, d_t = -J. \tag{25}$$

So the inequality $J \le V(0)$ holds; submitting $t = 0$ to Lyapunov-Krasovskii function (19), the inequality (18) can be obtained. This completes the proof. □

Remark 6. The condition of stability is expressed with matrix inequality (16). It is worthy to point out that inequality (16) is not linear with respect to the gain matrix of the controller, so it is needed to be reformulated into LMIs via a change of variables.

Theorem 7. *Given a set of constants ξ_i ($i = 1, 2, 3, 4$), η_m, and η_M, if there exist a set of symmetric positive definite matrices $\widetilde{Q}_1$, $\widetilde{Q}_2$, $\widetilde{R}_1$, $\widetilde{R}_2$, and $\widetilde{S}$ and a constant $\varepsilon > 0$, as well as matrices $\widetilde{P} > 0$ and $X > 0$ and matrix Z satisfying LMI as*

$$\begin{bmatrix} \widetilde{\Pi} & \widetilde{\Xi} & \varepsilon^{-1}\widetilde{I} \\ * & \varepsilon^{-1} I & 0 \\ * & * & \varepsilon^{-1} I \end{bmatrix} < 0, \tag{26}$$

where

$$\widetilde{\Pi} = \begin{bmatrix} \widetilde{Q}_1 + \xi_1 AX^T + \xi_1 XA^T - \widetilde{R}_1 + \widetilde{S} & \xi_3 XA^T + \widetilde{R}_1 & \xi_4 XA^T & \xi_1 BZ + \xi_2 XA^T & XA^T - \xi_1 X^T + \widetilde{P} \\ * & \widetilde{Q}_2 - \widetilde{Q}_1 - \widetilde{R}_1 - \dfrac{1}{\eta_M - \eta_m}\widetilde{R}_2 & \dfrac{1}{\eta_M - \eta_m}\widetilde{R}_2 & \xi_3 BZ & -\xi_3 X^T \\ * & * & -\widetilde{Q}_2 - \dfrac{1}{\eta_M - \eta_m}\widetilde{R}_2 & \xi_4 BZ & -\xi_4 X^T \\ * & * & * & \xi_2 BZ + \xi_2 (BZ)^T & (BZ)^T - \xi_2 X^T \\ * & * & * & * & \widetilde{\Omega} - X - X^T \end{bmatrix}, \tag{27}$$

$$\widetilde{\Omega} = \eta_m{}^2 \widetilde{R}_1 + (\eta_M - \eta_m)\widetilde{R}_2,$$

$$\widetilde{\Xi} = \begin{bmatrix} \xi_1 BZ \\ \xi_3 BZ \\ \xi_4 BZ \\ \xi_2 BZ \\ BZ \end{bmatrix},$$

then system (13) is asymptotically stable with the guaranteed cost controller $K = ZX^{-T}D_0^{-1}$, where matrix D_0 is determined by the average value of data drift, and the cost function J satisfies

$$\begin{aligned} J \le\ & x^T(0)\, X^{-1}\widetilde{P}X^{-T}x(0) \\ & + \int_{-\eta_m}^{0} x^T(\alpha)\, X^{-1}\widetilde{Q}_1 X^{-T} x(\alpha)\, d\alpha \\ & + \int_{-\eta_M}^{-\eta_m} x^T(\alpha)\, X^{-1}\widetilde{Q}_2 X^{-T} x(\alpha)\, d\alpha \\ & + \eta_m \int_{-\eta_m}^{0}\int_{\theta}^{t} \dot{x}^T(\delta)\, X^{-1}\widetilde{R}_1 X^{-T}\dot{x}(\delta)\, d_\theta d_\delta \\ & + \int_{-\eta_M}^{-\eta_m}\int_{\tau}^{t} \dot{x}^T(\omega)\, X^{-1}\widetilde{R}_2 X^{-T}\dot{x}(\omega)\, d_\tau d_\omega. \end{aligned} \tag{28}$$

Proof. This Theorem is obtained through a suitable transformation on the basis of inequality (16) in Theorem 5. Firstly, we define $M_i = \xi_i M_5$ $(i = 1, 3, 4, 5)$ in (16). Obviously, (16) implies that $M_5 > 0$, so M_5 is nonsingular. Then, pre- and post-multiplying both sides of inequality (16) with $\text{diag}(\widetilde{X}, I, \varepsilon^{-1}I)$ and its transpose, and introducing new variables $XSX^T = \widetilde{S}$; $XPX^T = \widetilde{P}$; $XR_jX^T = \widetilde{R}_j$, $XQ_jX^T = \widetilde{Q}_j$ $(j = 1, 2)$; $KD_0X^T = Z$, it follows inquality (26), where $\widetilde{X} = \text{diag}(X, X, X, X, X)$ and $X = {M_5}^{-1}$. From the definition of D_0 in (9), we know D_0 is invertible, so K can be obtained by calculating $K = ZX^{-T}D_0^{-1}$. And it is easy to see that (26) and (28), respectively, imply (16) and (18). Therefore, from Theorem 5, we can complete the proof. □

4. Simulations

Example 8. Consider the linear system as follows:

$$\dot{x}(t) = \begin{bmatrix} 3 & 1 & 0 \\ -1 & -2 & 0 \\ 0 & 0 & -0.801 \end{bmatrix} x(t) + \begin{bmatrix} 1.02 \\ -0.1 \\ -0.2 \end{bmatrix} u(t). \tag{29}$$

For this simulation, we synthetically consider a time-varying delay by lumping the network-induced delay and packet dropouts together with upper bound $\eta_M = 0.61$ s and lower bound $\eta_m = 0.12$ s; namely, $0.12 \le \eta(t) \le 0.61$. Here time-varying sensor-to-controller data drift with the average value $D_0 = \text{diag}(0.6, 0.71, 0.6)$ is also considered, which is shown in Figure 2.

We choose the parameters $\xi_1 = 0.0002$, $\xi_2 = 11110.0001$, $\xi_3 = 1.005$, and $\xi_4 = -0.001$. By taking advantage of LMI toolbox and submitting these parameters above into inequality (26), we can obtain the guaranteed cost control gain

$$K = ZX^{-T}D_0^{-1} = \begin{bmatrix} -7.5329 & -0.3563 & 0.4067 \end{bmatrix} \quad \text{with } \varepsilon = 3.7545 \times 10^{16}. \tag{30}$$

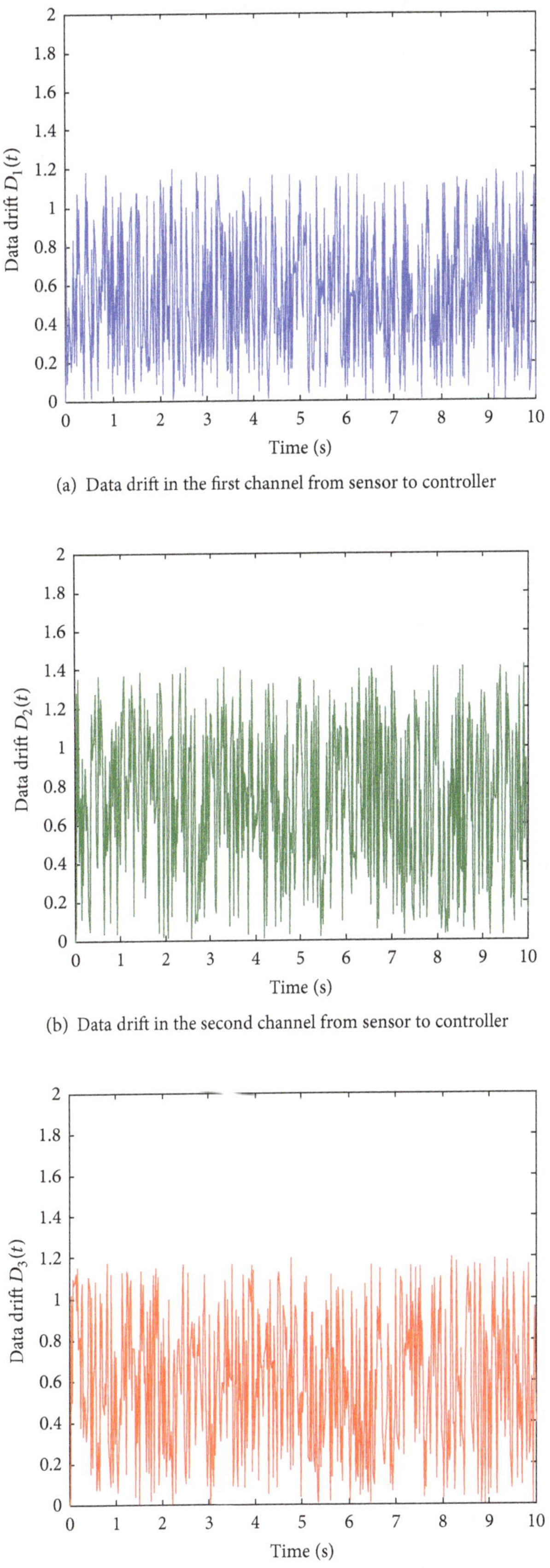

(a) Data drift in the first channel from sensor to controller

(b) Data drift in the second channel from sensor to controller

(c) Data drift in the third channel from sensor to controller

FIGURE 2: The time-varying data drift.

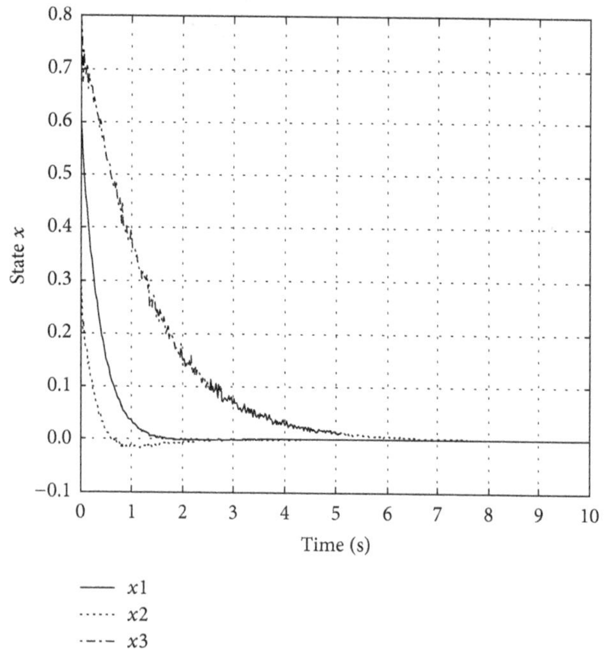

FIGURE 3: The state response curves of NCS with data drift.

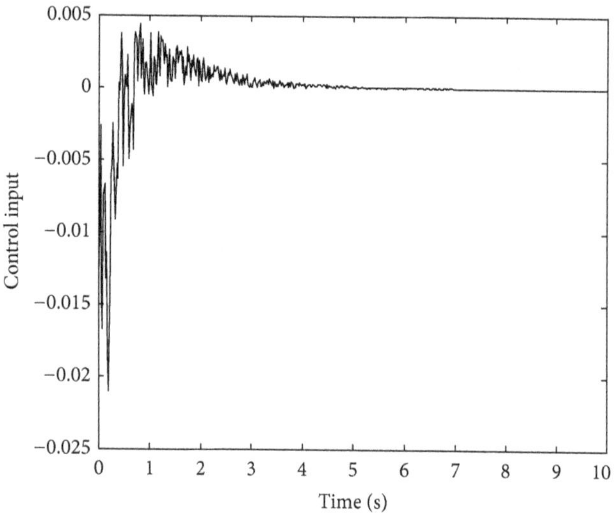

FIGURE 4: The control input of NCS with data drift.

The initial state of system is assumed $x(0) = [0.6\ \ 0.3\ \ 0.8]^T$, through the state response of NCS with the effect of data drift shown in Figure 3 and corresponding control input shown in Figure 4, we know all states of system get steady at 7.3 s.

In addition, we apply the method proposed by Luck [10] into the same problem. And the design of controller fails with $K = [0.4285\ \ -1.0243\ \ 0\ \ 0.4243]$, the response of system state is shown as Figure 5. Thus, it sufficiently demonstrates the effectiveness and feasibility of the guaranteed cost method proposed in this paper.

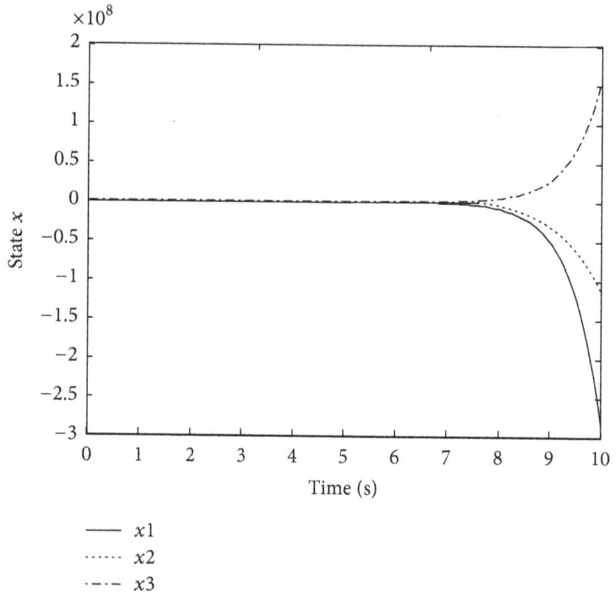

FIGURE 5: The state response curves of NCS with data drift.

To better illustrate the effectiveness of the method proposed in this paper, the following example is discussed.

Example 9. Consider the linear model of NCS as follows

$$\dot{x}(t) = \begin{bmatrix} -0.5 & 1.36 & 0 & 0.52 \\ -0.61 & 1.1 & 0 & 0.21 \\ 0.16 & 0.29 & -1.42 & 0 \\ 0 & 1 & -0.12 & 0.26 \end{bmatrix} x(t) + \begin{bmatrix} 1.37 \\ -1.1 \\ 1.55 \\ -0.73 \end{bmatrix} u(t). \tag{31}$$

For this simulation, we set the time-varying delay with upper bound $\eta_M = 0.15$ s and lower bound $\eta_m = 0.09$ s. And time-varying sensor-to-controller data drift with the average value $D_0 = \text{diag}(0.52, 0.61, 0.36, 0.65)$ is also considered.

By taking advantage of LMI toolbox and submitting these parameters above into inequality (26), we can obtain the guaranteed cost control gain $K = ZX^{-T}D_0^{-1} = [-1.8322\ \ -1.5264\ \ -0.3754\ \ -2.5736]$. The initial state of system is assumed $x(0) = [1.2\ \ 0.47\ \ 0.12\ \ -1.13]^T$; through the state response of NCS with the effect of data drift shown in Figure 6, we know all states of system get steady at 60 s. The guaranteed cost controller designed in this paper is able to make the NCS asymptotically stable even if affected by data drift. It sufficiently proves the effectiveness and feasibility of the method proposed in this paper.

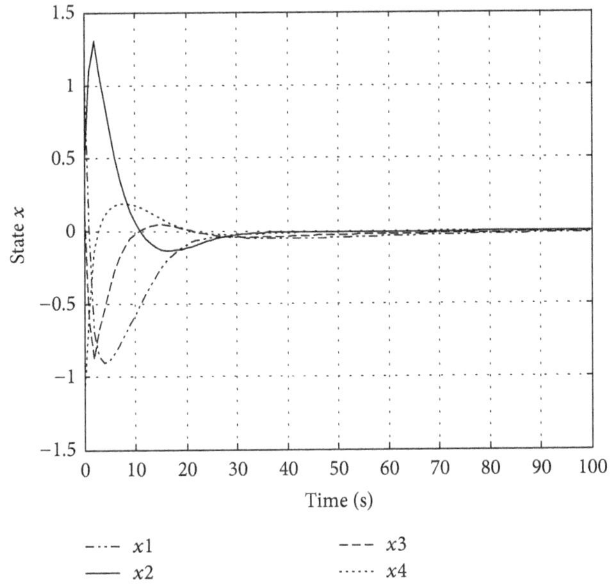

FIGURE 6: The state response curves of NCS with data drift.

5. Conclusions

The guaranteed cost control problem of a class of continuous-time networked control systems with data drift is investigated. With sensor-to-controller data drift considered, networked control system is modeled as a closed-loop system with time-varying parameters by lumping the network-induced delay and data-dropouts to a synthetically time-varying delay together. Moreover, by selecting an appropriate Lyapunov function, the guaranteed cost controller in terms of linear matrix inequality (LMI) is designed to cope with the effect of data drift and enhance the NCS's performance. And a simulation is given to prove the effectiveness and feasibility of the method. The convexity substitutions used in Theorem 7 ($M_i = \xi_i M_5$ $(i = 1, 3, 4, 5)$) may, in general, lead to significantly conservative results. Our next research task will be choosing more reasonable values of parameters ξ_i $(i = 1, 2, 3, 4)$ to reduce the conservatism further.

Conflict of Interests

The authors declare that there is no conflict of interests regarding the publication of this paper.

Acknowledgments

This work was partly supported by National Nature Science Foundation of China (61164014, 51375323, and 61563022), Major Program of Natural Science Foundation of Jiangxi Province, China (20152ACB20009), and Qing Lan Project of Jiangsu Province, China.

References

[1] J. Liu and D. Yue, "Event-triggering in networked systems with probabilistic sensor and actuator faults," *Information Sciences*, vol. 240, pp. 145–160, 2013.

[2] M. Jungers, E. B. Castelan, V. M. Moraes, and U. F. Moreno, "A dynamic output feedback controller for NCS based on delay estimates," *Automatica*, vol. 49, no. 3, pp. 788–792, 2013.

[3] Y. Xia, W. Xie, B. Liu, and X. Wang, "Data-driven predictive control for networked control systems," *Information Sciences*, vol. 235, pp. 45–54, 2013.

[4] Y. Zhang and J. Jiang, "Bibliographical review on reconfigurable fault-tolerant control systems," *Annual Reviews in Control*, vol. 32, no. 2, pp. 229–252, 2008.

[5] V. B. Kolmanovskii and J.-P. Richard, "Stability of some linear systems with delays," *IEEE Transactions on Automatic Control*, vol. 44, no. 5, pp. 984–989, 1999.

[6] C. Peng and T. C. Yang, "Event-triggered communication and H_∞ control co-design for networked control systems," *Automatica*, vol. 49, no. 5, pp. 1326–1332, 2013.

[7] H. Song, L. Yu, and W.-A. Zhang, "Networked H_∞ filtering for linear discrete-time systems," *Information Sciences*, vol. 181, no. 3, pp. 686–696, 2011.

[8] A. Ray and Y. Halevi, "Integrated communication and control systems: part II—design considerations," *Journal of Dynamic Systems, Measurement and Control*, vol. 110, no. 4, pp. 374–381, 1988.

[9] H. Chan and Ü. Özgüner, "Closed-loop control of systems over a communications network with queues," *International Journal of Control*, vol. 62, no. 3, pp. 493–510, 1995.

[10] R. Luck and A. Ray, "An observer-based compensator for distributed delays," *Automatica*, vol. 26, no. 5, pp. 903–908, 1990.

[11] J. Nilsson, *Real-time control systems with delays [Ph.D. thesis]*, Lund Institute of Technology, 1998.

[12] J. Xiong and J. Lam, "Stabilization of networked control systems with a logic ZOH," *IEEE Transactions on Automatic Control*, vol. 54, no. 2, pp. 358–363, 2009.

[13] K. Liu and E. Fridman, "Networked-based stabilization via discontinuous Lyapunov functionals," *International Journal of Robust and Nonlinear Control*, vol. 22, no. 4, pp. 420–436, 2012.

[14] X. Meng, J. Lam, and H. Gao, "Network-based H_∞ control for stochastic systems," *International Journal of Robust and Nonlinear Control*, vol. 19, no. 3, pp. 295–312, 2009.

[15] J. Li, Q. Zhang, H. Yu, and M. Cai, "Real-time guaranteed cost control of MIMO networked control systems with packet disordering," *Journal of Process Control*, vol. 21, no. 6, pp. 967–975, 2011.

[16] T. H. Lee, J. H. Park, D. H. Ji, O. M. Kwon, and S. M. Lee, "Guaranteed cost synchronization of a complex dynamical network via dynamic feedback control," *Applied Mathematics and Computation*, vol. 218, no. 11, pp. 6469–6481, 2012.

[17] X. Fang and J. Wang, "Stochastic observer-based guaranteed cost control for networked control systems with packet dropouts," *IET Control Theory & Applications*, vol. 2, no. 11, pp. 980–989, 2008.

[18] R. Wang, G.-P. Liu, W. Wang, D. Rees, and Y. B. Zhao, "Guaranteed cost control for networked control systems based on an improved predictive control method," *IEEE Transactions on Control Systems Technology*, vol. 18, no. 5, pp. 1226–1232, 2010.

[19] J.-S. Xie, B.-Q. Fan, J. Yang et al., "Guaranteed cost controller design of networked control systems with state delay," *Acta Automatica Sinica*, vol. 33, no. 2, pp. 170–174, 2007.

[20] L. Xie, H. Fang, and Y. Zheng, "Guaranteed cost control for networked control systems," *Journal of Control Theory and Applications*, vol. 2, no. 2, pp. 143–148, 2004 (Chinese).

[21] X. Li and X.-B. Wu, "Guaranteed cost fault-tolerant controller design of networked control systems under variable-period sampling," *Information Technology Journal*, vol. 8, no. 4, pp. 537–543, 2009.

[22] Y.-L. Wang and Q.-L. Han, "Modelling and controller design for discrete-time networked control systems with limited channels and data drift," *Information Sciences*, vol. 269, pp. 332–348, 2014.

[23] G. Garcia, J. Bernussou, and D. Arzelier, "Robust stabilization of discrete-time linear systems with norm-bounded time-varying uncertainty," *Systems and Control Letters*, vol. 22, no. 5, pp. 327–339, 1994.

16

Robust Fault Diagnosis Algorithm for a Class of Nonlinear Systems

Hai-gang Xu, Yue-feng Liao, and Xiao Han

Henan Mechanical and Electrical Engineering College, Xinxiang 453002, China

Correspondence should be addressed to Xiao Han; 1400651671@qq.com

Academic Editor: James Lam

A kind of robust fault diagnosis algorithm to Lipschitz nonlinear system is proposed. The novel disturbances constraint condition of the nonlinear system is derived by group algebra method, and the novel constraint condition can meet the system stability performance. Besides, the defined robust performance index of fault diagnosis observer guarantees the robust. Finally, the effectiveness of the algorithm proposed is proved in the simulations.

1. Introduction

The complicated systems as aircraft, missile systems, and control system [1–6] easily show faults; how to diagnose the fault is a difficult problem to handle. When the systems show fault and you cannot isolate it, the systems will be collapsed. Therefore, the fault diagnosis and fault tolerant technology are meaningful to enhance the system performances. And fault diagnosis technology is foundation of the fault tolerance; in another way, fault tolerance is realized by the fault estimations information. The original control laws will be regulated by the fault estimation information, so the system reliability will be improved, particularly the missile control system. In this paper, we aim at the missile attitude control system. Robust fault diagnosis methods are useful ways to solve the systems with the disturbances and they are also proved effective in systems applications.

The robust fault diagnosis observer is designed based on unknown input observer theory in [7–9], and state estimation errors decouple from disturbances. Most of papers make assumptions that the disturbances constraint condition is known; this assumption limits the algorithm applications.

Dealing with the deficiency on assumption that the disturbance is norm bounded, a novel constraint condition for disturbance is designed. Besides, the defined robust performance index of fault diagnosis observer guarantees the robust. Furthermore, the threshold is designed in fault decision section.

2. Problem Statement

Consider the system uncertainty and unknown input disturbances, system state-space model:

$$\begin{aligned}\dot{\mathbf{x}}(t) = {} & (\mathbf{A}+\Delta\mathbf{A})\,\mathbf{x}(t) + (\mathbf{B}+\Delta\mathbf{B})\,\mathbf{u}(t) \\ & + \mathbf{h}(\mathbf{x}(t),\mathbf{u}(t)) + \mathbf{g}_0(\mathbf{x}(t),\mathbf{u}(t),\mathbf{d}(t),t) \\ & + \mathbf{f}(t),\end{aligned} \quad (1)$$

$$\mathbf{y}(t) = \mathbf{C}\mathbf{x}(t). \quad (2)$$

$\mathbf{E}_1$, $\mathbf{E}_2$, $\mathbf{F}_1$, and $\mathbf{F}_2$ are known matrix. $\Delta\mathbf{A}$ and $\Delta\mathbf{B}$ are model errors.

Therefore, we can get another form as

$$\begin{aligned}\dot{\mathbf{x}}(t) = {} & \mathbf{A}\mathbf{x}(t) + \mathbf{B}\mathbf{u}(t) + \mathbf{h}(\mathbf{x}(t),\mathbf{u}(t)) \\ & + \mathbf{g}(\mathbf{x}(t),\mathbf{u}(t),\mathbf{d}(t),t) + \mathbf{f}(t),\end{aligned} \quad (3)$$

where $\mathbf{x}(0) = \mathbf{x}_0$ is initial value, $\mathbf{x}(t) \in \mathbf{R}^n$ is state, $\mathbf{u}(t) \in \mathbf{R}^p$ and $\mathbf{y}(t) \in \mathbf{R}^q$ are system input and output, $\mathbf{g} \in \mathbf{R}^n$ is unknown input, nonlinear functions $\mathbf{g}$ and $\mathbf{h}$ are continuous

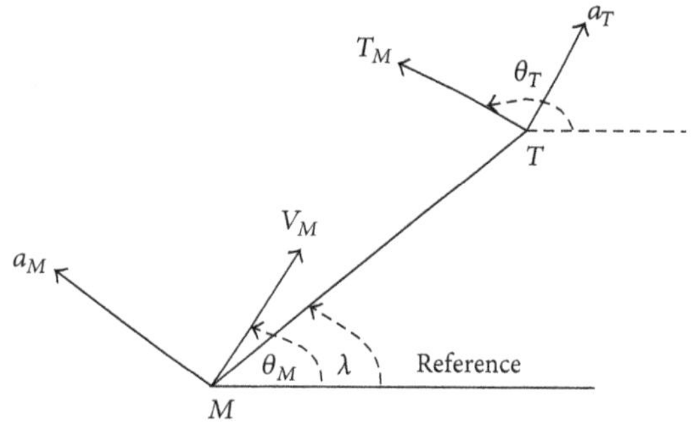

FIGURE 1: Missile and target geometry.

and differential, and $\mathbf{f}(t) \in \mathbf{R}^n$ is system fault. $\mathbf{A}$, $\mathbf{B}$, and $\mathbf{C}$ are known matrix.

The system state-space model can be obtained by Gronwall lemma [10–12]. Considering the complexity and particularity of missile control system, in the next two sections we perform the fault diagnosis algorithm from two aspects: stability and robust of the observer proposed.

Most of related papers assume that the constraint condition of external disturbances is norm bounded; however, a kind of nonlinear systems is unstable under the hypothetical constraint condition. Dealing with the limitations of disturbances constraint condition mentioned, a novel constraint condition of external disturbances is derived in Section 3; the systems hold stable under the condition proposed. Furthermore, the defined robust performance index of fault diagnosis observer guarantees the robust in Section 4.

The missile and target geometry are shown in Figures 1 and 2. The real system dynamics are described by the following differential equations:

$$\begin{aligned}
m\frac{dv}{dt} &= p\cos\alpha - X - mg\sin\theta,\\
mv\frac{d\theta}{dt} &= p\sin\alpha + Y - mg\cos\theta,\\
J_z\frac{d\omega_z}{dt} &= M_z,\\
\frac{dL}{dt} &= v\cos\theta,\\
\frac{dH}{dt} &= v\sin\theta,\\
\frac{d\vartheta}{dt} &= \omega_z,\\
\alpha &= \vartheta - \theta.
\end{aligned} \tag{4}$$

The system variables are Mach, longitudinal velocities and lateral velocities, control input, and so forth. The variables are demonstrated as follows.

3. Stability Analysis

Theorem 1. *There exists $M \geq 1$, $\omega < 0$, $t \geq 0$ to make nonlinear system (2) and (3) with external disturbances hold*

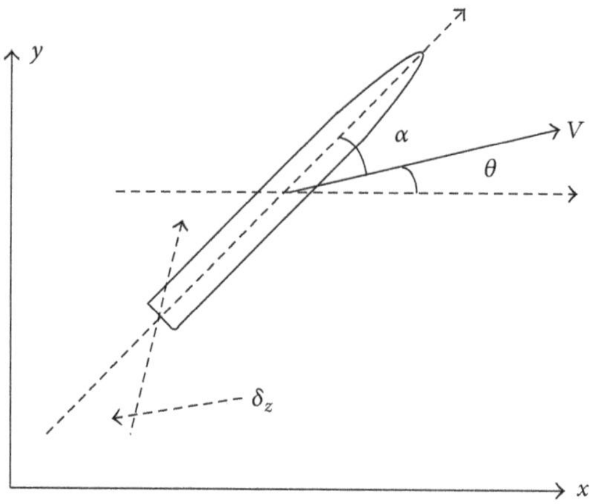

FIGURE 2: Air-frame axes.

stable, when the constraint condition of external disturbances is

$$\|\mathbf{g}(\mathbf{x}(t),\mathbf{u}(t),\mathbf{d}(t),t)\| < (\beta_1(t) - \beta_0(t)) - \frac{\omega\|\mathbf{x}(\tau)\|}{M} \tag{5}$$

and is defined as $\beta_0(t) = \|\mathbf{Bu}(t)\|$, $\beta_1(t) = \|\mathbf{h}(\mathbf{x}(t),\mathbf{u}(t))\|$.

Proof. Matrix $\mathbf{A}$ is Hurwitz matrix and $\mathbf{A}$ can generate asymptotic convergence linear semigroup ζ_t. Therefore, there exists $M \geq 1$, $\omega < 0$, $t \geq 0$ such that the inequality holds:

$$\|\zeta_t\| \leq M\exp(\omega t). \tag{6}$$

Fault-free mode, state-space description of system (2) and (3) with external disturbances is

$$\begin{aligned}
\mathbf{x}(t) = \mathbf{x}(0) + \int_0^t &[\mathbf{Ax}(\tau) + \mathbf{Bu}(\tau) + \mathbf{h}(\mathbf{x}(\tau),\mathbf{u}(\tau))\\
&+ \mathbf{g}(\mathbf{x}(\tau),\mathbf{u}(\tau),\mathbf{d}(\tau),\tau)]\,d\tau.
\end{aligned} \tag{7}$$

Therefore, $\exists$ stable linear semigroup ζ_t makes (8) hold:

$$\begin{aligned}
\mathbf{x}(t) = \zeta_t\mathbf{x}(0) + \zeta_{t-\tau}\int_0^t &[\mathbf{Bu}(\tau) + \mathbf{h}(\mathbf{x}(\tau),\mathbf{u}(\tau))\\
&+ \mathbf{g}(\mathbf{x}(\tau),\mathbf{u}(\tau),\mathbf{d}(\tau),\tau)]\,d\tau.
\end{aligned} \tag{8}$$

Apply the 2-norm to formula (8):

$$\begin{aligned}
\|\mathbf{x}(t)\| = \Big\|\zeta_t\mathbf{x}(0) + \zeta_{t-\tau}\int_0^t &[\mathbf{Bu}(\tau) + \mathbf{h}(\mathbf{x}(\tau),\mathbf{u}(\tau))\\
&+ \mathbf{g}(\mathbf{x}(\tau),\mathbf{u}(\tau),\mathbf{d}(\tau),\tau)]\,d\tau\Big\|.
\end{aligned} \tag{9}$$

The simplification form of formula (9) is

$$\begin{aligned}
\Xi = \mathbf{Bu}(\tau) &+ \mathbf{h}(\mathbf{x}(\tau),\mathbf{u}(\tau))\\
&+ \mathbf{g}(\mathbf{x}(\tau),\mathbf{u}(\tau),\mathbf{d}(\tau),\tau).
\end{aligned} \tag{10}$$

Initial value of state is $\mathbf{x}_0 = \mathbf{x}(0)$, we set $a = \|\mathbf{x}(0)\|$. From constraint of linear semigroup ζ_t in formula (6) and norm basic principle, the inequality constraint condition of formula (9) should be satisfied as follows:

$$\|\mathbf{x}(t)\| \le Ma\exp(\omega t) + \int_0^t \|M\exp[\omega(t-\tau)]\,\Xi\|\,d\tau. \quad (11)$$

Multiplied by $\exp(-\omega t)$ for two sides of formula (11):

$$\|\mathbf{x}(t)\|\exp(-\omega t) \le Ma + \int_0^t \|M\exp(-\omega\tau)\,\Xi\|\,d\tau,$$
$$\|\mathbf{x}(t)\|\exp(-\omega t) \le Ma + \int_0^t \frac{\|M\mathbf{x}(\tau)\exp(-\omega\tau)\,\Xi\|}{\|\mathbf{x}(\tau)\|}d\tau. \quad (12)$$

By Gronwall lemma,

$$\|\mathbf{x}(t)\| \le Ma\exp\left[\int_0^t\left(\omega+\frac{\|M\Xi\|}{\|\mathbf{x}(\tau)\|}\right)d\tau\right]. \quad (13)$$

For simplification, we define as follows:

$$\vartheta(t) = \int_0^t\left(\omega+\frac{\|M\Xi\|}{\|\mathbf{x}(\tau)\|}\right)d\tau. \quad (14)$$

The system is stable for $\forall t \to +\infty$, $\lim_{t\to+\infty}\vartheta(t) < |\varepsilon| < +\infty$ when there exists finite constant $|\varepsilon| < \infty$. There exists nonlinear semigroup π_t such that state-space of system is described by the following from formula (9):

$$\mathbf{x}(t) = \pi_t\mathbf{x}(0). \quad (15)$$

From formulas (13) and (14):

$$\|\mathbf{x}(t)\| = \|\pi_t\mathbf{x}(0)\| \le Ma\exp(\vartheta(t)). \quad (16)$$

Therefore, the nonlinear semigroup π_t is stable when $\vartheta(t) < 0$; in other words, system (2) and (3) with external disturbances is stable. As a result, the system holds stable when the following condition is satisfied:

$$\|\Xi\| < -\frac{\omega\|\mathbf{x}(\tau)\|}{M}. \quad (17)$$

Substitute the formula above into (10):

$$\|\mathbf{B}\mathbf{u}(t)+\mathbf{h}(\mathbf{x}(t),\mathbf{u}(t))+\mathbf{g}(\mathbf{x}(t),\mathbf{u}(t),\mathbf{d}(t),t)\| < -\frac{\omega\|\mathbf{x}(\tau)\|}{M}. \quad (18)$$

And then

$$\|\mathbf{B}\mathbf{u}(t)+\mathbf{h}(\mathbf{x}(t),\mathbf{u}(t))+\mathbf{g}(\mathbf{x}(t),\mathbf{u}(t),\mathbf{d}(t),t)\| > \|\mathbf{B}\mathbf{u}(t)\| - \|\mathbf{h}(\mathbf{x}(t),\mathbf{u}(t))\| + \|\mathbf{g}(\mathbf{x}(t),\mathbf{u}(t),\mathbf{d}(t),t)\|, \quad (19)$$

where, $\mathbf{u}(t)$ and $\mathbf{h}(\mathbf{x}(t),\mathbf{u}(t))$ are known. With definition $\beta_0(t) = \|\mathbf{B}\mathbf{u}(t)\|$, $\beta_1(t) = \|\mathbf{h}(\mathbf{x}(t),\mathbf{u}(t))\|$.

Substitute into the formula above:

$$\beta_0(t) - \beta_1(t) + \|\mathbf{g}(\mathbf{x}(t),\mathbf{u}(t),\mathbf{d}(t),t)\| < -\frac{\omega\|\mathbf{x}(\tau)\|}{M},$$
$$\|\mathbf{g}(\mathbf{x}(t),\mathbf{u}(t),\mathbf{d}(t),t)\| < (\beta_1(t)-\beta_0(t)) - \frac{\omega\|\mathbf{x}(\tau)\|}{M}. \quad (20)$$

System (2) and (3) with external disturbances is stable when the constraint condition of external disturbances satisfies the form of (20). Furthermore, if there exists $\beta_3 \in R^+$ make $(\beta_1(t)-\beta_0(t)) \le \beta_3\|\mathbf{x}(\tau)\|$, and therefore

$$\|\mathbf{g}(\mathbf{x}(t),\mathbf{u}(t),\mathbf{d}(t),t)\| < \left(\beta_3 - \frac{\omega}{M}\right)\|\mathbf{x}(\tau)\|. \quad (21)$$

The constraint condition of external disturbances satisfies the inequation

$$\frac{\|\mathbf{g}(\mathbf{x}(t),\mathbf{u}(t),\mathbf{d}(t),t)\|}{\|\mathbf{x}(\tau)\|} < \beta_3 - \frac{\omega}{M}. \quad (22)$$

The robustness performance index to external disturbances is defined as $\mathfrak{R}(\mathbf{A}) = \omega/M$. □

4. Robust Fault Diagnosis Algorithm

Theorem 2. *The robust fault diagnosis observer (24) of system (2) and (3) with fault $\beta_2 = \sup_{t\in[0,T]}\|\mathbf{f}(t)\|$ and external disturbances is asymptotic convergence. Therefore, the robust performance index of observer satisfies the inequality as follows:*

$$-\mathfrak{R}(\mathbf{A}-\mathbf{G}\mathbf{C}) > \lambda_1 + \frac{(\|\mathbf{g}(\mathbf{x}(\tau),\mathbf{u}(\tau),\mathbf{d}(\tau),\tau)\| + \beta_2)}{\|\mathbf{e}(\tau)\|}. \quad (23)$$

Proof. Construct robust fault diagnosis observer as follows for missile pitching motion control system (2) and (3) with fault and external disturbances:

$$\dot{\hat{\mathbf{x}}}(t) = \mathbf{A}\hat{\mathbf{x}}(t) + \mathbf{B}\mathbf{u}(t) + \mathbf{h}(\hat{\mathbf{x}}(t),\mathbf{u}(t)) + \mathbf{G}[\mathbf{y}(t)-\hat{\mathbf{y}}(t)],$$
$$\hat{\mathbf{y}}(t) = \mathbf{C}\hat{\mathbf{x}}(t). \quad (24)$$

System states estimation errors and observer residuals are

$$\mathbf{e}(t) = \mathbf{x}(t) - \hat{\mathbf{x}}(t),$$
$$\mathbf{r}(t) = \mathbf{y}(t) - \hat{\mathbf{y}}(t). \quad (25)$$

And therefore,

$$\begin{aligned}\dot{\mathbf{e}}(t) &= \dot{\mathbf{x}}(t) - \dot{\hat{\mathbf{x}}}(t)\\ &= \mathbf{A}\mathbf{x}(t) - \mathbf{A}\hat{\mathbf{x}}(t) + \mathbf{h}(\mathbf{x}(t), \mathbf{u}(t))\\ &\quad - \mathbf{h}(\hat{\mathbf{x}}(t), \mathbf{u}(t)) - \mathbf{G}[\mathbf{C}\mathbf{x}(t) - \mathbf{C}\hat{\mathbf{x}}(t)]\\ &\quad + \mathbf{g}(\mathbf{x}(t), \mathbf{u}(t), \mathbf{d}(t), t) + \mathbf{f}(t),\\ \dot{\mathbf{e}}(t) &= (\mathbf{A} - \mathbf{G}\mathbf{C})\,\mathbf{e}(t) + \mathbf{h}(\mathbf{x}(t), \mathbf{u}(t))\\ &\quad - \mathbf{h}(\hat{\mathbf{x}}(t), \mathbf{u}(t)) + \mathbf{g}(\mathbf{x}(t), \mathbf{u}(t), \mathbf{d}(t), t)\\ &\quad + \mathbf{f}(t),\\ \mathbf{r}(t) &= \mathbf{y}(t) - \hat{\mathbf{y}}(t) = \mathbf{C}\mathbf{e}(t).\end{aligned} \tag{26}$$

The states estimation errors of system (26) are $\mathbf{e}_0 = \mathbf{e}(0)$, $b = \|\mathbf{e}(0)\|$.

With definition: $\Psi = \mathbf{h}(\mathbf{x}(\tau), \mathbf{u}(\tau)) - \mathbf{h}(\hat{\mathbf{x}}(\tau), \mathbf{u}(\tau)) + \mathbf{f}(t) + \mathbf{g}(\mathbf{x}(\tau), \mathbf{u}(\tau), \mathbf{d}(\tau), \tau)$.

As a result, the states estimation errors are

$$\begin{aligned}\mathbf{e}(t) &= \exp[(\mathbf{A} - \mathbf{G}\mathbf{C})\,t]\,\mathbf{e}(0)\\ &\quad + \int_0^t \exp[(\mathbf{A} - \mathbf{G}\mathbf{C})(t-\tau)]\,\Psi\, d\tau.\end{aligned} \tag{27}$$

Apply the 2-norm to both sides of (27):

$$\begin{aligned}\|\mathbf{e}(t)\| &= \Big\|\exp[(\mathbf{A} - \mathbf{G}\mathbf{C})\,t]\,\mathbf{e}(0)\\ &\quad + \int_0^t \exp[(\mathbf{A} - \mathbf{G}\mathbf{C})(t-\tau)]\,\Psi\, d\tau\Big\|.\end{aligned} \tag{28}$$

The system matrix $\mathbf{A} - \mathbf{G}\mathbf{C}$ is Hurwitz matrix when system (26) is stable; therefore, it can generate a stable linear semigroup ζ_t.

Consequently, there exists $M \ge 1$, $\omega < 0$, $t \ge 0$ such that $\|\zeta_t\| \le M\exp(\omega t)$. Therefore, formula (28) fulfills the inequality as follows:

$$\|\mathbf{e}(t)\| \le Mb\exp(\omega t) + \int_0^t \|M\exp[\omega(t-\tau)]\,\Psi\|\, d\tau. \tag{29}$$

Multiplied by $\exp(-\omega t)$ for formula (29):

$$\begin{aligned}\|\mathbf{e}(t)\|\exp(-\omega t) &\le Mb + \int_0^t \|M\exp(-\omega\tau)\,\Psi\|\, d\tau,\\ \|\mathbf{e}(t)\|\exp(-\omega t) &\le Mb\\ &\quad + \int_0^t \frac{\|M\mathbf{e}(\tau)\exp(-\omega\tau)\,\Psi\|}{\|\mathbf{e}(\tau)\|} d\tau.\end{aligned} \tag{30}$$

Generally, the fault injected into the missile pitching motion control system is $\beta_2 = \sup_{t\in[0,T]}\|\mathbf{f}(t)\|$:

$$\begin{aligned}\|\Psi\| &\le \|\mathbf{h}(\mathbf{x}(\tau), \mathbf{u}(\tau)) - \mathbf{h}(\hat{\mathbf{x}}(\tau), \mathbf{u}(\tau))\| + \|\mathbf{f}(t)\|\\ &\quad + \|\mathbf{g}(\mathbf{x}(\tau), \mathbf{u}(\tau), \mathbf{d}(\tau), \tau)\|\\ &\le \lambda_1\|\mathbf{e}(\tau)\| + \beta_2\\ &\quad + \|\mathbf{g}(\mathbf{x}(\tau), \mathbf{u}(\tau), \mathbf{d}(\tau), \tau)\|,\\ \|\mathbf{e}(t)\| &\le Mb\exp\left\{\int_0^t \left[\omega + \frac{M\|\Psi\|}{\|\mathbf{e}(\tau)\|}\right] d\tau\right\}.\end{aligned} \tag{31}$$

Consequently, the fault diagnosis observer (24) is asymptotic convergence when the following formula holds:

$$\begin{aligned}&\omega + \frac{M\|\Psi\|}{\|\mathbf{e}(\tau)\|} < 0,\\ &-\frac{\omega}{M} > \lambda_1 + \frac{(\|\mathbf{g}(\mathbf{x}(\tau), \mathbf{u}(\tau), \mathbf{d}(\tau), \tau)\| + \beta_2)}{\|\mathbf{e}(\tau)\|}.\end{aligned} \tag{32}$$

The performance index $\mathfrak{R}(\mathbf{A} - \mathbf{G}\mathbf{C})$ of observer (24) satisfies the following constraint condition from stability theory:

$$\begin{aligned}-\mathfrak{R}(\mathbf{A} - \mathbf{G}\mathbf{C}) &> \lambda_1\\ &\quad + \frac{(\|\mathbf{g}(\mathbf{x}(\tau), \mathbf{u}(\tau), \mathbf{d}(\tau), \tau)\| + \beta_2)}{\|\mathbf{e}(\tau)\|}.\end{aligned} \tag{33}$$

Consequently, the fault diagnosis observer is asymptotic convergence when formula (33) holds. And then, fault diagnosis observer can be realized by robust performance index proposed. Consider

$$\begin{aligned}&\lambda_i\left[(\mathbf{A} - \mathbf{G}\mathbf{C}) + (\mathbf{A} - \mathbf{G}\mathbf{C})^T\right]\\ &\quad < -2\left\{\lambda_1 + \frac{(\|\mathbf{g}(\mathbf{x}(\tau), \mathbf{u}(\tau), \mathbf{d}(\tau), \tau)\| + \beta_2)}{\|\mathbf{e}(\tau)\|}\right\},\end{aligned} \tag{34}$$

where $\lambda_{max}(\cdot)$ and $\lambda_i(\cdot)$ represent the maximum and arbitrary eigenvalue for matrix $(\cdot)$; the gain matrix $\mathbf{G}$ of the observer can be solved by pole assignment when the robust performance index $\mathfrak{R}(\mathbf{A} - \mathbf{G}\mathbf{C})$ is given. □

5. Adaptive Threshold Design

Usually, compare the residuals with threshold to diagnose fault.

The states estimation errors are

$$\begin{aligned}\mathbf{e}(t) &= \mathbf{e}(0) + \int_0^t [\mathbf{h}(\mathbf{x}(\tau), \mathbf{u}(\tau)) - \mathbf{h}(\hat{\mathbf{x}}(\tau), \mathbf{u}(\tau))\\ &\quad + \mathbf{g}(\mathbf{x}(\tau), \mathbf{u}(\tau), \mathbf{d}(\tau), \tau)]\, d\tau.\end{aligned} \tag{35}$$

Therefore,

$$\begin{aligned}\mathbf{r}(t) &= \mathbf{C}\mathbf{e}(0) + \mathbf{C}\int_0^t [\mathbf{h}(\mathbf{x}(\tau), \mathbf{u}(\tau))\\ &\quad - \mathbf{h}(\hat{\mathbf{x}}(\tau), \mathbf{u}(\tau)) + \mathbf{g}(\mathbf{x}(\tau), \mathbf{u}(\tau), \mathbf{d}(\tau), \tau)]\, d\tau.\end{aligned} \tag{36}$$

Apply the 2-norm for formula (36):

$$\|\mathbf{r}(t)\| = \left\|\mathbf{Ce}(0) + \mathbf{C}\int_0^t [\mathbf{h}(\mathbf{x}(\tau),\mathbf{u}(\tau)) - \mathbf{h}(\hat{\mathbf{x}}(\tau),\mathbf{u}(\tau)) + \mathbf{g}(\mathbf{x}(\tau),\mathbf{u}(\tau),\mathbf{d}(\tau),\tau)]\,d\tau\right\| \quad (37)$$

with definition $b_c = \|\mathbf{C}\|$ and

$$\delta(\mathbf{h}) = \mathbf{h}(\mathbf{x}(\tau),\mathbf{u}(\tau)) - \mathbf{h}(\hat{\mathbf{x}}(\tau),\mathbf{u}(\tau)). \quad (38)$$

Therefore,

$$\|\mathbf{r}(t)\| \le bb_c + b_c\left\|\int_0^t [\delta(\mathbf{h}) + \mathbf{g}(\mathbf{x}(\tau),\mathbf{u}(\tau),\mathbf{d}(\tau),\tau)]\,d\tau\right\|. \quad (39)$$

We can get from (38)

$$\|\delta(\mathbf{h})\| \le \lambda_1\|\mathbf{x}(\tau) - \hat{\mathbf{x}}(\tau)\| = \lambda_1\|\mathbf{e}(\tau)\| = \frac{\lambda_1}{b_c}\|\mathbf{r}(\tau)\|. \quad (40)$$

As a result,

$$\|\mathbf{r}(t)\| \le bb_c + \int_0^t [\lambda_1\|\mathbf{r}(\tau)\| + b_c\|\mathbf{g}(\mathbf{x}(\tau),\mathbf{u}(\tau),\mathbf{d}(\tau),\tau)\|]\,d\tau. \quad (41)$$

We can get from (41)

$$\|\mathbf{r}(t)\| \le bb_c + \int_0^t \lambda_1 b_c\|\mathbf{e}(\tau)\|\,d\tau + \int_0^t b_c\left[\beta_1(\tau) - \beta_0(\tau) - \frac{\omega\|\mathbf{x}(\tau)\|}{M}\right]d\tau, \quad (42)$$

where maximum tolerant values of estimation errors $\mathbf{e}_{\max}$ and the system stable value $\mathbf{x}_{\text{sta}}$ are known:

$$\mathbf{e}_{\max} = \|\mathbf{e}_{\max}(t)\|, \qquad \lim_{t\to T}\|\mathbf{x}(t)\| \longrightarrow \mathbf{x}_{\text{sta}}. \quad (43)$$

It can be obtained from (42) that

$$\|\mathbf{r}(t)\| \le bb_c + \lambda_1 b_c\mathbf{e}_{\max}t + b_c\int_0^t\left[\beta_1(\tau) - \beta_0(\tau) - \frac{\omega x_{\max}}{M}\right]d\tau. \quad (44)$$

As a result, the adaptive threshold of the fault diagnosis observer designed is

$$\mathbf{J}_{\text{th}}(\mathbf{r}(t)) = bb_c + \left(\lambda_1 b_c\mathbf{e}_{\max} - \frac{\omega x_{\max}}{M}\right)t + b_c\int_0^t[\beta_1(\tau) - \beta_0(\tau)]\,d\tau. \quad (45)$$

6. Simulation

In order to verify the effectiveness of the algorithm proposed, the following simulations are performed.

6.1. Simulation Parameters. The differential equations of missile pitching motion control system are represented as follows [6].

The pitching moment: $M_z \approx \varphi_1(ma,\alpha) + \varphi_2(ma,\alpha)\delta_z + \varphi_3(\alpha,\omega_z,v)$. The missile aerodynamic parameters are as follows: $X(ma,\alpha)$, $Y(ma,\alpha)$, $\varphi_1(ma,\alpha)$, $\varphi_2(ma,\alpha)$, and $\varphi_3(\alpha,\omega_z,v)$. The force situation of missile with different flight states is different.

The missile empty weight is 230 kg, z-axis rotational inertia is $J_z = 247.26\,\text{kg}\cdot\text{m}^2$, and generator impulse thrust of missile attitude control system is $p = 2200\,\text{N}$. The initial location of missile in inertial coordinates system is $x_0 = 8530\,\text{m}$, $y = 11600\,\text{m}$; missile initial velocity is $v = 300\,\text{m/s}$, and trajectory pitching angle is $\theta = 0.536\,\text{rad}$.

The system disturbances are $d_i(t) = 0.01\sin(t)$, $i = 1,2,\ldots,7$. The constraint condition of the external disturbances can be derived by Section 3.

6.2. Performance Analysis for the Fault Diagnosis Algorithm. The supreme of missile external disturbances in the considered period under the Simulink condition from Theorem 1 is $\|\mathbf{g}\|_{\max} = 9.69507\times 10^5$. Therefore, not all of the disturbances can satisfy norm bounded constraint conditions and the prior constraint condition on external disturbances restricts the generality of fault diagnosis applications. The maximum value of estimation errors is defined as

$$\mathbf{e}_{\max} = [40\ \ 0.01\ \ 2\ \ 20\ \ 20\ \ 0.02\ \ 0.02]. \quad (46)$$

Therefore, $\|\mathbf{e}\|_{\max} = 1.202\times 10^3$. From Theorem 2,

$$\lambda_i\left[(\mathbf{A}-\mathbf{GC}) + (\mathbf{A}-\mathbf{GC})^T\right] < -806.5786. \quad (47)$$

The fault diagnosis observer designed is asymptotic convergence robust fault diagnosis observer when the poles are placed at the left plane of −806.5786. It should be noticed that the gain matrix $\mathbf{G}$ is not unique because of the poles selected; in this paper, the poles are settled on

$$\mathbf{p} = [-850+5\text{i}, -850-5\text{i}, -880+6\text{i}, -880-6\text{i}, -900+10\text{i}, -900-10\text{i}, -1000]. \quad (48)$$

The gain matrix $\mathbf{G}$ of the observer can be get through the pole assignment.

As a result, the missile trajectory character is depicted in Figure 3 and substitute the gain matrix G into the observer (26) and then the residual effect is depicted in Figure 4 by Simulink. Without loss of generality, just take the 3rd channel residual of fault diagnosis as an example to research and assume the missile attitude control system.

Residual is asymptotic convergence and therefore it has the robustness to disturbances from Figure 4. The good convergence of residual illustrates that the algorithm proposed is effective. Furthermore, the smooth residual curve also

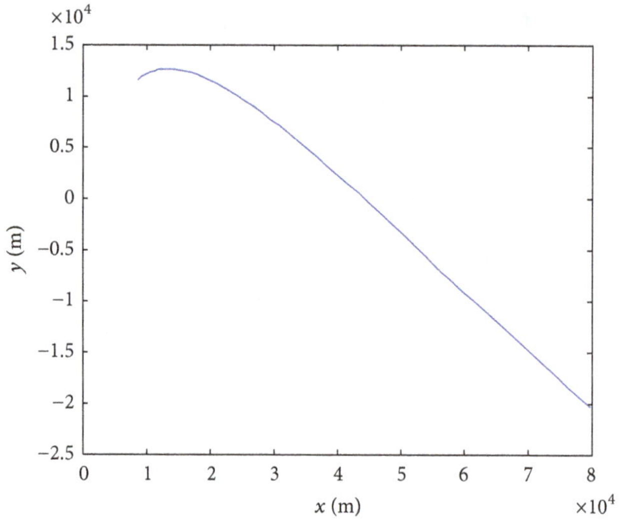

FIGURE 3: The missile trajectory character.

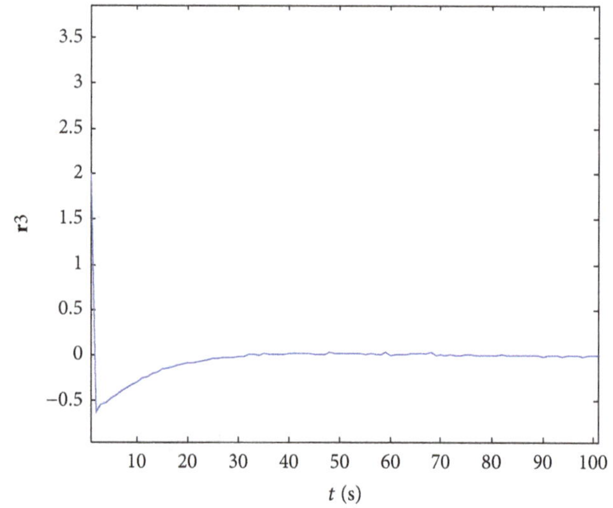

FIGURE 4: The 3rd channel residual of the observer.

illustrates that the disturbances constraint condition which can satisfy the system stability is reasonable and the defined robust performance index is practicable.

7. Conclusion

In this paper, novel disturbances constraint condition is derived to improve the limitation that external disturbance is norm bounded. And then, the novel constraint condition can meet the system stability. Besides, the defined robust performance index of fault diagnosis observer guarantees the robust. In decision-making unit, adaptive threshold is designed. Finally, simulation results show the effectiveness of the algorithm proposed.

Conflict of Interests

The authors declare that there is no conflict of interests regarding the publication of this paper.

References

[1] H. Buschek, "Design and flight test of a robust autopilot for the IRIS-T air-to-air missile," *Control Engineering Practice*, vol. 11, no. 5, pp. 551–558, 2003.

[2] Q. Hu, B. Xiao, and M. I. Friswell, "Fault tolerant control with H_∞ performance for attitude tracking of flexible spacecraft," *IET Control Theory and Applications*, vol. 6, no. 10, pp. 1388–1399, 2012.

[3] S. Chaib, D. Boutat, A. Banali, and F. Kratz, "Invertibility of switched nonlinear systems. Application to missile faults reconstruction," in *Proceedings of the 46th IEEE Conference on Decision and Control (CDC '07)*, pp. 3239–3244, IEEE, New Orleans, La, USA, December 2007.

[4] M. A. Wael and Q. Quan, "Robust hybrid control for ballistic missile longitudinal autopilot," *Chinese Journal of Aeronautics*, vol. 24, no. 6, pp. 777–788, 2011.

[5] A. Tsourdos and B. A. White, "Adaptive flight control design for nonlinear missile," *Control Engineering Practice*, vol. 13, no. 3, pp. 373–382, 2005.

[6] G. M. Siouris, *Missile Guidance and Control Systems*, Springer, New York, NY, USA, 2004.

[7] J. Zarei and J. Poshtan, "Sensor fault detection and diagnosis of a process using unknown input observer," *Mathematical and Computational Applications*, vol. 16, no. 1, pp. 31–42, 2011.

[8] C. Hajiyev and F. Caliskan, "Sensor and control surface/actuator failure detection and isolation applied to F-16 flight dynamic," *Aircraft Engineering and Aerospace Technology*, vol. 77, no. 2, pp. 152–160, 2005.

[9] H. R. Karimi, M. Zapateiro, and N. Luo, "A linear matrix inequality approach to robust fault detection filter design of linear systems with mixed time-varying delays and nonlinear perturbations," *Journal of the Franklin Institute*, vol. 347, no. 6, pp. 957–973, 2010.

[10] A. Pazy, *Semi-Groups of Linear Operators and Applications to Partial Differential Equations*, Springer, New York, NY, USA, 1983.

[11] P. A. Ioannou and J. Sun, *Robust Adaptive Control*, Prentice Hall, New York, NY, USA, 1996.

[12] H.-M. Qian, Z.-D. Fu, J.-B. Li, and L.-L. Yu, "Robust fault diagnosis algorithm for a class of Lipschitz system with unknown exogenous disturbances," *Measurement*, vol. 46, no. 8, pp. 2324–2334, 2013.

17

Robust Adaptive Output Feedback Control Scheme for Chaos Synchronization with Input Nonlinearity

Xiaomeng Li,[1] Zhanshan Zhao,[1] Jing Zhang,[2,3] and Meixia Zhu[1]

[1]*School of Computer Science & Software Engineering, Tianjin Polytechnic University, Tianjin 300387, China*
[2]*School of Textiles, Tianjin Polytechnic University, Tianjin 300387, China*
[3]*Tianjin Vocational Institute, Tianjin 300410, China*

Correspondence should be addressed to Zhanshan Zhao; zhzhsh127@163.com

Academic Editor: Hung-Yuan Chung

This paper proposes a robust adaptive output feedback control strategy which can automatically regulate control gain for chaos synchronization. Chaotic systems with input nonlinearities, delayed nonlinear coupling, and external disturbance can achieve synchronization by applying this strategy. Utilizing Lyapunov method and LMI technique, the conditions ensuring chaos synchronization are obtained. Finally, simulations are given to show the effectiveness of our control strategy.

1. Introduction

Chaos synchronization has attracted a lot of interest due to its wide engineering application in various areas like secure communication [1–4], neural networks [5, 6], electronic engineering [7], and so on [8, 9]. Consider the fact that chaotic system is a class of nonlinear dynamical system which sensitively depends on initial conditions. It is necessary to solve the problem of chaos synchronization.

In the practical physical systems, physical limitations will lead to state nonlinearity and input nonlinearity, such as sector [10–12], saturation [13–15], and dead zones [16–19]. Considering state nonlinearity, [20] proposed an adaptive control strategy for multirate networked nonlinear systems. In [21], fuzzy control method has been applied to a class of nonlinear system. Filter design and H_∞ performance for nonlinear networked systems have been researched in [22]. A novel sliding mode observer approach has been proposed for a class of stochastic systems in [23]. Moreover, it should not be ignored that effect of nonlinear control inputs can cause serious degradation of synchronization performance even nonsynchronous. Thus, nonlinear control inputs should be considered in synchronization controller design of chaotic systems. Unfortunately, in [24, 25], input nonlinearity has not been considered.

As a source of nonsynchronism, time-varying delay has to be faced in many engineering synchronization systems, such as chemical processes [26, 27] and pneumatic systems [28]. Therefore, designing a controller for time-varying delay systems is necessary [29].

Recently, considering nonlinearly coupled chaotic systems, [30] proposed a state feedback controller to achieve synchronization. However, the input nonlinearity was not considered. To the authors' knowledge, synchronization for coupled chaotic systems with input nonlinearity and time delays has been rarely mentioned. Furthermore, in real application, only the output state is available. Therefore, it is necessary to design a synchronization controller in output feedback form.

Motivated by the previous discussions, we propose an adaptive output feedback controller to make chaotic systems synchronize. Input nonlinearities, nonlinear coupling, and time-varying delay have been taken into account. By utilizing Lyapunov method and LMI technique, the conditions ensuring synchronization are obtained.

In the rest of this paper, Section 2 provides systems and problem description. Then a robust adaptive output feedback control strategy is proposed for chaos synchronization in Section 3. In Section 4, simulations are given to demonstrate effectiveness of this control strategy. Conclusion are collected in Section 5.

2. Problem Formulation

Consider chaotic systems as follows:

$$\begin{aligned}\dot{x}_m(t) &= Ax_m(t) + A_c x_m(t - d_1(t)) + Bf(t, x_m(t)) \\ &\quad + B_c g(t, x_m(t - d_1(t))) \\ &\quad - B_d k(t, x_s(t - d_2(t))), \\ y_m(t) &= Cx_m(t)\end{aligned} \tag{1}$$

$$\begin{aligned}\dot{x}_s(t) &= Ax_s(t) + A_c x_s(t - d_1(t)) + Bf(t, x_s(t)) \\ &\quad + B_c g(t, x_s(t - d_1(t))) \\ &\quad - B_d k(t, x_m(t - d_2(t))) + D\omega(t) \\ &\quad + E\Lambda(u(t)), \\ y_s(t) &= Cx_s(t),\end{aligned} \tag{2}$$

where $x_m \in \mathbb{R}^n$, $x_s \in \mathbb{R}^n$, $y_m \in \mathbb{R}^m$, and $y_s \in \mathbb{R}^m$ are the state vector and output vector for the drive and response system, respectively. $f(t,x), g(t,x), k(t,x) : \mathbb{R}^n \times \mathbb{R}^n \to \mathbb{R}^m$ represent nonlinear vectors. $u(t) = [u_1(t) \cdots u_m(t)]^T \in \mathbb{R}^m$ is the control input vector; $\omega(t)$ denotes the external disturbance; time-varying delay $d_1(t)$ and $d_2(t)$ satisfy

$$d_{11} \le \dot{d}_1(t) \le d_{12}, \tag{3}$$

$$d_{21} \le \dot{d}_2(t) \le d_{22}. \tag{4}$$

$\Lambda(u(t)) = [\lambda_1(u_1(t)) \cdots \lambda_m(u_m(t))]^T$ is representing the nonlinear control input vector which satisfies the following inequality:

$$\nu_i(t)\lambda_i(\nu_i(t)) \ge \chi_i(\nu_i(t))^2. \tag{5}$$

χ_i is an unknown positive constant satisfying $\chi^* = \min \chi_i$. Constant matrices $A, A_c, B, B_c, B_d, C, D, E$ have appropriate dimensions.

Synchronization error can be defined as $e(t) = x_s(t) - x_m(t)$. Using (1) and (2), synchronization error can be obtained:

$$\begin{aligned}\dot{e}(t) &= Ae(t) + A_c e(t - d_1(t)) + B\Psi(t) + B_c\Upsilon(t, d_1) \\ &\quad + B_d\Xi(t, d_2) + D\omega(t) + E\Lambda(u(t)), \\ y_e(t) &= Ce(t),\end{aligned} \tag{6}$$

where

$$\begin{aligned}\Psi(t) &= f(t, x_s(t)) - f(t, x_m(t)), \\ \Upsilon(t, d_1) &= g(t, x_s(t - d_1(t))) - g(t, x_m(t - d_1(t))), \\ \Xi(t, d_2) &= k(t, x_s(t - d_2(t))) - k(t, x_m(t - d_2(t))).\end{aligned} \tag{7}$$

The objective is to make drive and response systems synchronize. Obviously, if $e(t) \to 0$, then $x_s(t) - x_m(t) \to 0$ and it means that system (1) and (2) is synchronized.

To obtain the synchronization conditions, the following lemma and assumptions will be used during the proof.

Lemma 1. *If matrix* $H = \begin{bmatrix} H_{11} & H_{12} \\ H_{21} & H_{22}\end{bmatrix}$, *where* H_{11} *and* H_{22} *are square matrices, then the following inequalities are equivalent:*

(1) $H < 0$;

(2) $H_{11} < 0,\ H_{22} - H_{12}^T H_{11}^{-1} H_{12} < 0$;

(3) $H_{22} < 0,\ H_{11} - H_{12} H_{22}^{-1} H_{12}^T < 0$.

Assumption 2. The nonlinear function $\Psi(t)$, $\Upsilon(t, d_1)$, and $\Xi(t, d_2)$ satisfy the global Lipschitz condition:

$$\begin{aligned}\|\Psi(t)\| &\le L_1 \|x_s(t) - x_m(t)\|, \\ \|\Upsilon(t, d_1)\| &\le L_2 \|x_s(t - d_1(t)) - x_m(t - d_1(t))\|, \\ \|\Xi(t, d_2)\| &\le L_3 \|x_s(t - d_2(t)) - x_m(t - d_2(t))\|.\end{aligned} \tag{8}$$

Assumption 3. Matrix $P > 0$ and satisfies the following equation:

$$E^T P = C. \tag{9}$$

3. Robust Adaptive Controller Design Based on LMI

In order to make drive and response systems synchronize, the following adaptive controller is considered:

$$u(t) = -\frac{\nu}{2} Ce(t), \tag{10}$$

where ν is an adaptive control gain and is adjusted by the following adaptation law:

$$\dot{\nu} = \chi^* \rho \|Ce(t)\|^2, \quad \nu(0) > 0, \tag{11}$$

where ρ is a positive parameter.

By applying of the adaptive controller, synchronization errors will converge to zero asymptotically. In Theorems 4 and 5 the main results will be presented.

Theorem 4. *Consider the drive and response system (1) and (2) under $\omega(t)=0$. By application of the adaptive control law (10) and (11), if existing symmetric and positive definite matrices P, W_1, W_2, and a scalar $\alpha>0$, satisfying the following LMI:*

$$\begin{bmatrix} \Delta & PA_c & 0 & PB & PB_c & PB_d & L_1^T & 0 & 0 \\ * & -(1-d_{12})W_1 & 0 & 0 & 0 & 0 & 0 & L_2^T & 0 \\ * & * & -(1-d_{22})W_2 & 0 & 0 & 0 & 0 & 0 & L_3^T \\ * & * & * & -I & 0 & 0 & 0 & 0 & 0 \\ * & * & * & * & -I & 0 & 0 & 0 & 0 \\ * & * & * & * & * & -I & 0 & 0 & 0 \\ * & * & * & * & * & * & -I & 0 & 0 \\ * & * & * & * & * & * & * & -I & 0 \\ * & * & * & * & * & * & * & * & -I \end{bmatrix} < 0, \tag{12}$$

where

$$\Delta = PA + A^TP - \alpha PEE^TP + W_1 + W_2, \tag{13}$$

the system (1) and (2) is synchronized.

Proof. We consider the following Lyapunov-Krasovskii functional:

$$V = e^T(t)Pe(t) + \int_{t-d_1(t)}^{t} e^T(v)W_1e(v)\,dv + \int_{t-d_2(t)}^{t} e^T(v)W_2e(v)\,dv + \frac{1}{2}\rho^{-1}\tilde{\nu}^2, \tag{14}$$

where $\tilde{\nu}=\nu^*-\nu$. ν^* and ρ are positive constants.

The derivative of V can be calculated as follows:

$$\begin{aligned}\dot V &= 2e^T(t)P\dot e(t) + e^T(t)W_1e(t) - \left(1-\dot d_1(t)\right)e^T(t-d_1(t))W_1e(t-d_1(t)) + e^T(t)W_2e(t) - \left(1-\dot d_2(t)\right)e^T(t-d_2(t))W_2e(t-d_2(t)) - \rho^{-1}\dot\nu\tilde\nu \\ &= 2e^T(t)P\left[Ae(t) + A_ce(t-d_1(t)) + B\Psi(t) + B_c\Upsilon(t,d_1) + B_d\Xi(t,d_2) + E\Lambda(u(t))\right] + e^T(t)\cdot W_1e(t) - \left(1-\dot d_1(t)\right)e^T(t-d_1(t))W_1e(t-d_1(t)) + e^T(t)W_2e(t) - \left(1-\dot d_2(t)\right)e^T(t-d_2(t))W_2e(t-d_2(t)) - \rho^{-1}\dot\nu\tilde\nu.\end{aligned} \tag{15}$$

Incorporating (3) and (4), we can get

$$\begin{aligned}\dot V \le{}& e^T(t)\left(PA + A^TP - \alpha PEE^TP + W_1 + W_2\right)e(t) + \alpha\left\|E^TPe(t)\right\|^2 + 2e^T(t)PA_ce(t-d_1(t)) \\ &+ 2e^T(t)PB\Psi(t) + 2e^T(t)PB_c\Upsilon(t,d_1) + 2e^T(t)PB_d\Xi(t,d_2) + 2e^T(t)PE\Lambda(u(t)) \\ &- (1-d_{12})e^T(t-d_1(t))W_1e(t-d_1(t)) - (1-d_{22})e^T(t-d_2(t))W_2e(t-d_2(t)) - \rho^{-1}\dot\nu\tilde\nu.\end{aligned} \tag{16}$$

Using (8) leads to

$$\begin{aligned}\dot V \le{}& e^T(t)\left(PA + A^TP - \alpha PEE^TP + W_1 + W_2\right)e(t) + \alpha\left\|E^TPe(t)\right\|^2 + 2e^T(t)PA_ce(t-d_1(t)) \\ &+ 2e^T(t)PB\Psi(t) + 2e^T(t)PB_c\Upsilon(t,d_1) + 2e^T(t)PB_d\Xi(t,d_2) + 2e^T(t)PE\Lambda(u(t)) \\ &- (1-d_{12})e^T(t-d_1(t))W_1e(t-d_1(t)) - (1-d_{22})e^T(t-d_2(t))W_2e(t-d_2(t)) \\ &- \rho^{-1}\dot\nu\tilde\nu + e^T(t)L_1^TL_1e(t) + e^T(t-d_1(t))L_2^TL_2e(t-d_1(t)) \\ &+ e^T(t-d_2(t))L_3^TL_3e(t-d_2(t)) - \Psi^T(t)\Psi(t) - \Upsilon^T(t,d_1)\Upsilon(t,d_1) - \Xi^T(t,d_2)\Xi(t,d_2).\end{aligned} \tag{17}$$

Assume $Ce(t) = Z(t)_{m\times 1}$; $Z_n(t)$ is the nth element of $Z(t)$. By using (11), it is easy to prove that $\nu > 0$. Based on (5), we need to consider the following two cases:

(1) When $Z_n(t) > 0$, we have $u_n(t) < 0$. Multiplying $Z_n(t) > 0$ and dividing $u_n(t) < 0$ by both sides of (5), we can obtain that $Z_n(t)\lambda_n(\nu_n(t)) \le \chi_n Z_n(t)\nu_n(t)$.

(2) When $Z_n(t) < 0$, we have $u_n(t) > 0$. Multiplying $Z_n(t) < 0$ and dividing $u_n(t) > 0$ by both sides of (5), we can obtain that $Z_n(t)\lambda_n(\nu_n(t)) \le \chi_n Z_n(t)\nu_n(t)$.

Form the previous discussion, we can prove that the following relation always holds:

$$Z_n(t)\lambda_n(\nu_n(t)) \le \chi_n Z_n(t)\nu_n(t). \tag{18}$$

Using Assumption 3 and $\chi^* = \min \chi_n$ we have

$$\begin{aligned} 2e^T(t)PE\Lambda(\nu(t)) &= 2\sum_{n=1}^{m} Z_n(t)\lambda_n(\nu_n(t)) \\ &\le 2\sum_{n=1}^{m} \chi_n Z_n(t)\nu_n(t) \\ &\le -\chi^*\nu\|Ce(t)\|^2. \end{aligned} \tag{19}$$

Let $\alpha = \chi^*\nu^*$; incorporating the previous result (19), we can obtain

$$\begin{aligned} \dot V \le{}& e^T(t) \\ &\cdot\left(PA + A^TP - \alpha PEE^TP + W_1 + W_2 + L_1^TL_1\right)e(t) \\ &+ \chi^*\nu^*\|Ce(t)\|^2 + 2e^T(t)PA_ce(t-d_1(t)) \\ &+ 2e^T(t)PB\Psi(t) + 2e^T(t)PB_c\Upsilon(t,d_1) + 2e^T(t) \\ &\cdot PB_d\Xi(t,d_2) - \chi^*\nu\|Ce(t)\|^2 - (1-d_{12}) \\ &\cdot e^T(t-d_1(t))W_1e(t-d_1(t)) - (1-d_{22}) \\ &\cdot e^T(t-d_2(t))W_2e(t-d_2(t)) - \chi^*(\nu^*-\nu) \\ &\cdot\|Ce(t)\|^2 + e^T(t-d_1(t))L_2^TL_2e(t-d_1(t)) \\ &+ e^T(t-d_2(t))L_3^TL_3e(t-d_2(t)) - \Psi^T(t)\Psi(t) \\ &- \Upsilon^T(t,d_1)\Upsilon(t,d_1) - \Xi^T(t,d_2)\Xi(t,d_2) = e^T(t) \\ &\cdot\left(PA + A^TP - \alpha PEE^TP + W_1 + W_2 + L_1^TL_1\right)e(t) \\ &+ e^T(t-d_1(t))\left(L_2^TL_2 - (1-d_{12})W_1\right) \\ &\cdot e(t-d_1(t)) + e^T(t-d_2(t)) \\ &\cdot\left(L_3^TL_3 - (1-d_{22})W_2\right)e(t-d_2(t)) + 2e^T(t) \\ &\cdot PA_ce(t-d_1(t)) + 2e^T(t)PB\Psi(t) + 2e^T(t) \\ &\cdot PB_c\Upsilon(t,d_1) + 2e^T(t)PB_d\Xi(t,d_2) - \Psi^T(t)\Psi(t) \\ &- \Upsilon^T(t,d_1)\Upsilon(t,d_1) - \Xi^T(t,d_2)\Xi(t,d_2) \end{aligned} \tag{20}$$

which further can be written as $\dot V \le \xi^T\Gamma_1\xi$, where

$$\xi^T = \left[e^T(t)\ \ e^T(t-d_1(t))\ \ e^T(t-d_2(t))\ \ \Psi^T(t)\ \ \Upsilon^T(t,d_1)\ \ \Xi^T(t,d_2)\right]^T, \tag{21}$$

$$\Gamma_1 = \begin{bmatrix} \Delta_1 & PA_c & 0 & PB & PB_c & PB_d \\ * & \Delta_2 & 0 & 0 & 0 & 0 \\ * & * & \Delta_3 & 0 & 0 & 0 \\ * & * & * & -I & 0 & 0 \\ * & * & * & * & -I & 0 \\ * & * & * & * & * & -I \end{bmatrix} < 0, \tag{22}$$

where

$$\begin{aligned} \Delta_1 &= PA + A^TP - \alpha PEE^TP + L_1^TL_1 + W_1 + W_2, \\ \Delta_2 &= L_2^TL_2 - (1-d_{12})W_1, \\ \Delta_3 &= L_3^TL_3 - (1-d_{22})W_2. \end{aligned} \tag{23}$$

Using Lemma 1, (22) can be transformed to (12), which completed the proof of Theorem 4. □

In fact, disturbances always come from surrounding environment and communication channel. So, it is significant to solve the robustness problem in practical application. Theorem 5 will tackle this issue.

Theorem 5. *Consider the drive and response system (1) and (2) under $\omega(t) \ne 0$. By application of the adaptive control law (10) and (11), for given γ, if existing symmetric and positive definite matrices P, W_1, W_2, and a scalar $\alpha > 0$, satisfy the following LMI:*

$$\begin{bmatrix} \Pi & PA_c & 0 & PB & PB_c & PB_d & PD & L_1^T & 0 & 0 \\ * & -(1-d_{12})W_1 & 0 & 0 & 0 & 0 & 0 & 0 & L_2^T & 0 \\ * & * & -(1-d_{22})W_2 & 0 & 0 & 0 & 0 & 0 & 0 & L_3^T \\ * & * & * & -I & 0 & 0 & 0 & 0 & 0 & 0 \\ * & * & * & * & -I & 0 & 0 & 0 & 0 & 0 \\ * & * & * & * & * & -I & 0 & 0 & 0 & 0 \\ * & * & * & * & * & * & -\gamma^2 I & 0 & 0 & 0 \\ * & * & * & * & * & * & * & -I & 0 & 0 \\ * & * & * & * & * & * & * & * & -I & 0 \\ * & * & * & * & * & * & * & * & * & -I \end{bmatrix} < 0, \tag{24}$$

where

$$\Pi = PA + A^T P - \alpha PEE^T P + W_1 + W_2 + C^T C, \tag{25}$$

the attention rate γ for H_∞ synchronization in the disturbance situation can be achieved.

Proof. With zero initial condition, let us introduce

$$J = \int_o^\infty \left[y_e^T(t)\, y_e(t) - \gamma^2 \omega^T(t)\, \omega(t) \right] dt \le 0. \tag{26}$$

For $\omega(t) \ne 0$, the following function can be obtained:

$$J \le \int_o^\infty \left[\dot{V}_d + y_e^T(t)\, y_e(t) - \gamma^2 \omega^T(t)\, \omega(t) \right] dt \le 0. \tag{27}$$

From the proof of Theorem 4, $\dot{V}_d$ can be obtained:

$$\begin{aligned} \dot{V}_d = {} & e^T(t) \\ & \cdot \left(PA + A^T P - \alpha PEE^T P + L_1^T L_1 + W_1 + W_2 \right) \\ & \cdot e(t) + e^T(t - d_1(t)) \left(L_2^T L_2 - (1 - d_{12}) W_1 \right) \\ & \cdot e(t - d_1(t)) + e^T(t - d_2(t)) \\ & \cdot \left(L_3^T L_3 - (1 - d_{22}) W_2 \right) e(t - d_2(t)) + 2e^T(t) \\ & \cdot PA_c e(t - d_1(t)) + 2e^T(t)\, PB\Psi(t) + 2e^T(t) \\ & \cdot PB_c \Upsilon(t, d_1) + 2e^T(t)\, PB_d \Xi(t, d_2) + 2e^T(t) \\ & \cdot PD\omega(t) - \Psi^T(t)\Psi(t) - \Upsilon^T(t, d_1)\Upsilon(t, d_1) \\ & - \Xi^T(t, d_2)\, \Xi(t, d_2). \end{aligned} \tag{28}$$

Let

$$\xi_d^T = \left[e^T(t) \;\; e^T(t - d_1(t)) \;\; e^T(t - d_2(t)) \;\; \omega^T(t) \;\; \Psi^T(t) \;\; \Upsilon^T(t, d_1) \;\; \Xi^T(t, d_2) \right]^T. \tag{29}$$

Equation (28) can be written as

$$J \le \xi_d^T \Gamma_2 \xi_d^T, \tag{30}$$

where

$$\Gamma_2 = \begin{bmatrix} \Pi_1 & PA_c & 0 & PB & PB_c & PB_d & PD \\ * & \Pi_2 & 0 & 0 & 0 & 0 & 0 \\ * & * & \Pi_3 & 0 & 0 & 0 & 0 \\ * & * & * & -I & 0 & 0 & 0 \\ * & * & * & * & -I & 0 & 0 \\ * & * & * & * & * & -I & 0 \\ * & * & * & * & * & 0 & -\gamma^2 I \end{bmatrix}, \tag{31}$$

$$\Pi_1 = PA + A^T P - \alpha PEE^T P + L_1^T L_1 + W_1 + W_2 + C^T C, \tag{32}$$

$$\Pi_2 = L_2^T L_2 - (1 - d_{12}) W_1, \quad \Pi_3 = L_3^T L_3 - (1 - d_{22}) W_2. \tag{33}$$

Using Lemma 1, (31) can be transformed to (24), which completed the proof of Theorem 5. □

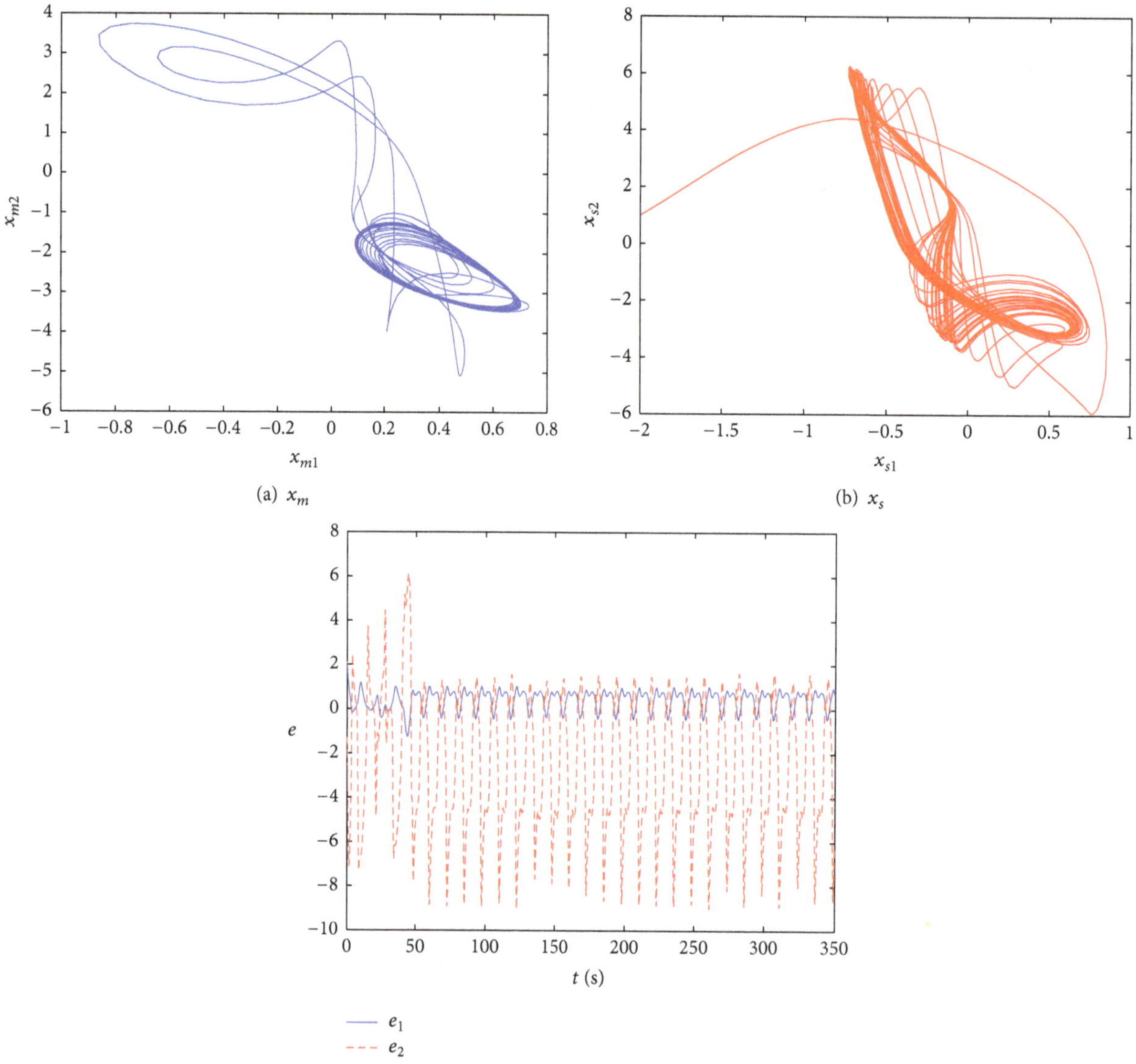

(a) x_m (b) x_s (c) $e(t)$

FIGURE 1: Behavior of the drive and response systems without any control: (a) phase trajectory of drive system, (b) phase trajectory of response system, and (c) synchronization errors.

4. Simulation Results

Consider the drive and response system (1) and (2) with parameters:

$$A = \begin{bmatrix} -1 & 0 \\ 0 & -1 \end{bmatrix},$$

$$A_c = \begin{bmatrix} 0 & 0 \\ 0 & 0 \end{bmatrix},$$

$$B = \begin{bmatrix} 2 & -0.1 \\ -5 & 4.5 \end{bmatrix},$$

$$B_c = \begin{bmatrix} -1.5 & -0.1 \\ -0.2 & -4 \end{bmatrix},$$

$$B_d = \begin{bmatrix} 0.001 & 0 \\ 0 & 0.001 \end{bmatrix},$$

$$C = \begin{bmatrix} 1.5 & 0 \\ 0 & 0.5 \end{bmatrix},$$

$$E = \begin{bmatrix} 2 & 0 \\ 0 & 2 \end{bmatrix},$$

$$D = \begin{bmatrix} 1 & 0 \\ 0 & 1 \end{bmatrix}.$$

$$f(t, x(t)) = \tanh(x(t)),$$

$$g(t, x(t - d_1(t))) = \tanh(x(t - d_1(t))),$$

$$h(t, x(t - d_2(t))) = x(t - d_2(t))\sin(t),$$

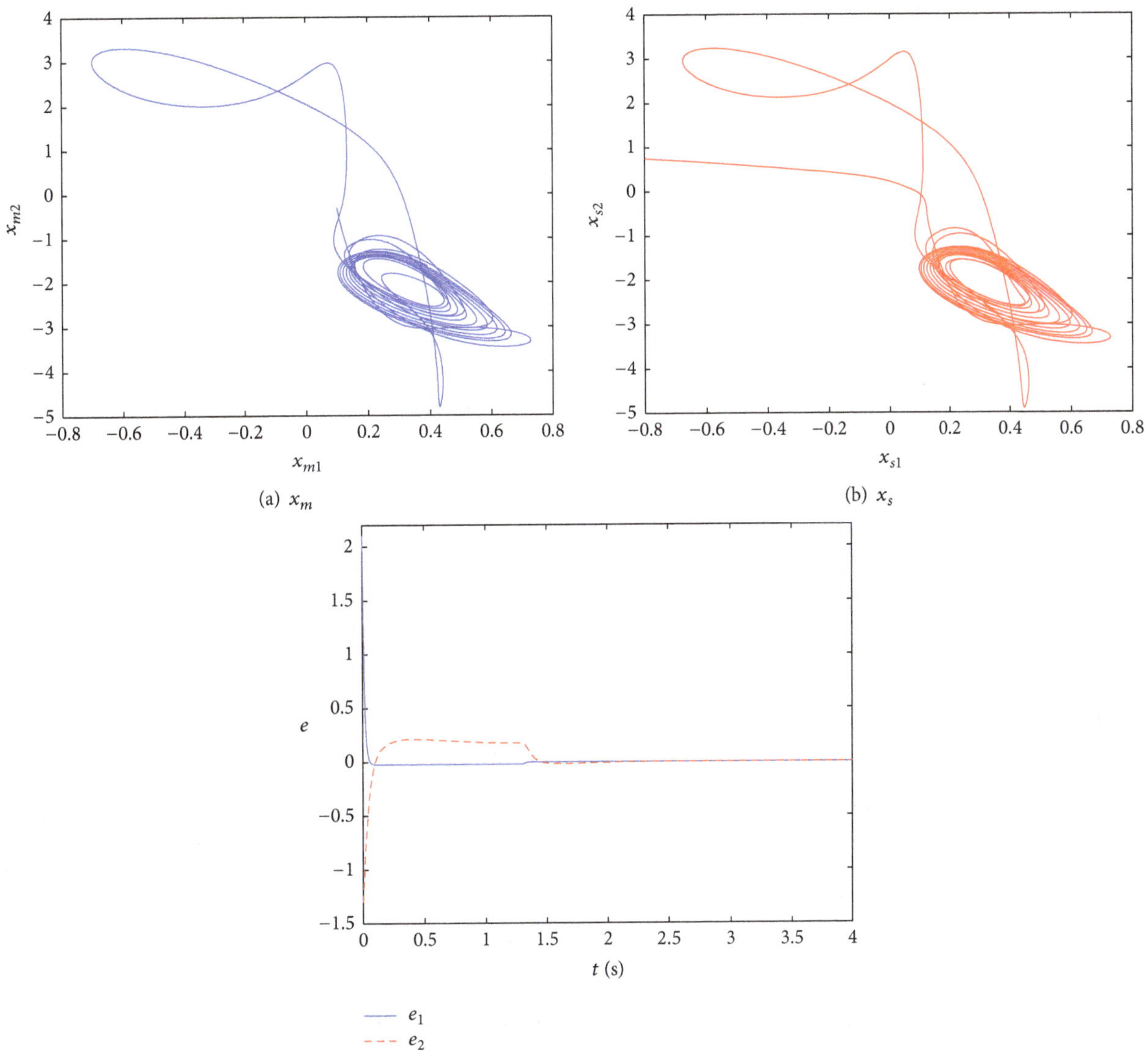

(c) $e(t)$

FIGURE 2: Behavior of the drive and response systems with controller: (a) phase trajectory of drive system, (b) phase trajectory of response system, and (c) synchronization errors.

$$\begin{gathered}
d_1(t) = 1 + 0.3\sin(t), \\
d_2(t) = 1 - 0.02\sin(10t), \\
\begin{bmatrix} \lambda(u_1(t)) \\ \lambda(u_2(t)) \end{bmatrix} = \begin{bmatrix} 1 + 0.4\sin(u_1(t))\,u_1(t) \\ 1.2 + 0.2\cos(u_2(t))\,u_2(t) \end{bmatrix}, \\
\omega(t) = [0.3\sin(100t) \;\; 0.5\sin(110t)]^T .
\end{gathered} \tag{34}$$

Initial conditions are chosen as

$$\begin{gathered}
(x_{m1}(0), x_{m2}(0)) = (0.1, -0.3), \\
(x_{s1}(0), x_{s2}(0)) = (-2, 1), \\
\chi^* \rho = 5, \\
\gamma = 0.6.
\end{gathered} \tag{35}$$

When $\omega(t) = 0$, using the LMI given in Theorem 4 (12), we can obtain

$$\begin{gathered}
W_1 = \begin{bmatrix} 29.1148 & 0.2902 \\ 0.2902 & 3.7117 \end{bmatrix}, \\
W_2 = \begin{bmatrix} 27.6971 & 0.2399 \\ 0.2399 & 2.9992 \end{bmatrix}, \\
P = \begin{bmatrix} 0.7500 & 0 \\ 0 & 0.2500 \end{bmatrix}, \\
\alpha = 46.6242.
\end{gathered} \tag{36}$$

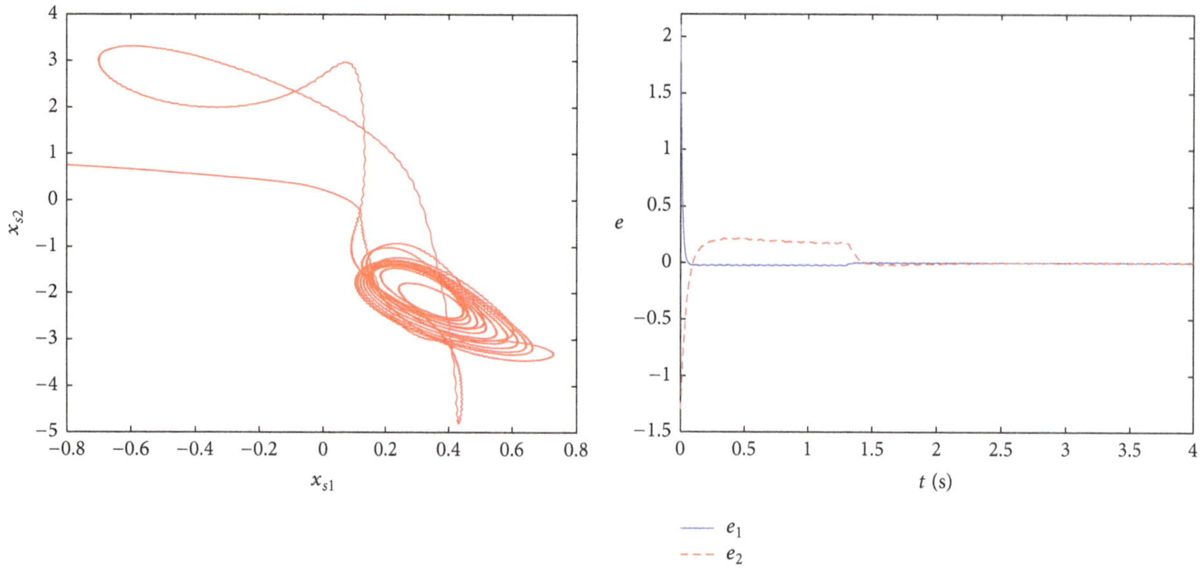

(a) x_s

(b) $e(t)$

FIGURE 3: Behavior of the drive and response systems with controller and disturbance: (a) phase trajectory of response system and (b) synchronization errors.

When $\omega(t) \neq 0$, using the LMI given in Theorem 5 (24), we can obtain

$$W_1 = \begin{bmatrix} 48.1311 & 0.2457 \\ 0.2457 & 6.9022 \end{bmatrix},$$
$$W_2 = \begin{bmatrix} 32.2655 & 0.2254 \\ 0.2254 & 6.8700 \end{bmatrix},$$
$$P = \begin{bmatrix} 0.7500 & 0 \\ 0 & 0.2500 \end{bmatrix}, \qquad (37)$$
$$\alpha = 77.6427.$$

The phase trajectory of system (1) and (2) without any control is shown in Figures 1(a) and 1(b). The error between them is shown in Figure 1(c). It is obvious in Figure 1 that the systems are nonsynchronous. After application of the proposed controller, the phase trajectory of system (1) and (2) with nonlinear control inputs is shown in Figures 2(a) and 2(b). The error between them is shown in Figure 2(c). It is obvious in Figure 2 that the systems are synchronous. Figure 3(a) illustrates the phase trajectory of response system with nonlinear control inputs and disturbances after applying the proposed controller. Figure 3(b) illustrates that the error signal tends to zero in a short time, in spite of the disturbances.

5. Conclusion

We investigate the synchronization problem of chaotic systems with nonlinear control inputs. A robust adaptive controller has been established. By applying this controller, control gain can be regulated automatically and the synchronization of chaotic systems can be achieved. Then, considering external disturbances, we propose a new H_∞ synchronization method for chaotic systems. From the above simulation results, we can find that the error signal tends to zero in a short time. Therefore our control strategy is effective in synchronizing chaotic systems.

Conflict of Interests

The authors declare that there is no conflict of interests regarding the publication of this paper.

Acknowledgment

The project is supported by the National Natural Science Foundation of China (Grant nos. 61503280, 61403278, and 61272006).

References

[1] W. Al-Hussaibi, "Effect of filtering on the synchronization and performance of chaos-based secure communication over Rayleigh fading channel," *Communications in Nonlinear Science and Numerical Simulation*, vol. 26, no. 1–3, pp. 87–97, 2015.

[2] M. Chadli and I. Zelinka, "Chaos synchronization of unknown inputs Takagi-Sugeno fuzzy: application to secure communications," *Computers & Mathematics with Applications*, vol. 68, no. 12, pp. 2142–2147, 2014.

[3] J. Q. Yang, Y. T. Chen, and F. Zhu, "Singular reduced-order observer-based synchronization for uncertain chaotic systems subject to channel disturbance and chaos-based secure communication," *Applied Mathematics and Computation*, vol. 229, pp. 227–238, 2014.

[4] Z. Y. He, K. Li, L. X. Yang, and Y. H. Shi, "A robust digital secure communication scheme based on sporadic coupling chaos

synchronization," *IEEE Transactions on Circuits and Systems I: Fundamental Theory and Applications*, vol. 47, no. 3, pp. 397–403, 2000.

[5] M. Kalpana, P. Balasubramaniam, and K. Ratnavelu, "Direct delay decomposition approach to synchronization of chaotic fuzzy cellular neural networks with discrete, unbounded distributed delays and Markovian jumping parameters," *Applied Mathematics and Computation*, vol. 254, pp. 291–304, 2015.

[6] D. Meng, "Neural networks adaptive synchronization for four-dimension energy resource system with unknown dead zones," *Neurocomputing*, vol. 151, no. 3, pp. 1495–1499, 2015.

[7] J. J. Ohtsubo, "Chaos synchronization and chaotic signal masking in semiconductor lasers with optical feedback," *IEEE Journal of Quantum Electronics*, vol. 38, no. 9, pp. 1141–1154, 2002.

[8] J. Awrejcewicz, A. V. Krysko, V. Dobriyan, I. V. Papkova, and V. A. Krysko, "Chaotic and synchronized dynamics of non-linear Euler-Bernoulli beams," *Computers & Structures*, vol. 155, pp. 85–96, 2015.

[9] A. Göksua, U. E. Kocamazb, and Y. Uyarogluc, "Synchronization and control of chaos in supply chain management," *Computers & Industrial Engineering*, vol. 86, pp. 107–115, 2015.

[10] X. J. Lu, H.-X. Li, and M. H. Huang, "Stability and robust design using a sector nonlinearity approach for nonlinear manufacturing systems," *Mechanism and Machine Theory*, vol. 82, pp. 115–127, 2014.

[11] T. F. Liu, Z.-P. Jiang, and D. J. Hill, "A sector bound approach to feedback control of nonlinear systems with state quantization," *Automatica*, vol. 48, no. 1, pp. 145–152, 2012.

[12] Y. D. Pan, K. D. Kumar, G. J. Liu, and K. Furuta, "Design of variable structure control system with nonlinear time-varying sliding sector," *IEEE Transactions on Automatic Control*, vol. 54, no. 8, pp. 1981–1986, 2009.

[13] N. A. Saeed, W. A. El-Ganini, and M. Eissa, "Nonlinear time delay saturation-based controller for suppression of nonlinear beam vibrations," *Applied Mathematical Modelling*, vol. 37, no. 20-21, pp. 8846–8864, 2013.

[14] M. Eissa, A. Kandil, W. A. El-Ganaini, and M. Kamel, "Vibration suppression of a nonlinear magnetic levitation system via time delayed nonlinear saturation controller," *International Journal of Non-Linear Mechanics*, vol. 72, pp. 23–41, 2015.

[15] Y.-L. Huang and C.-K. Sun, "Nonlinear saturation behaviors of high-speed p i n photodetectors," *Journal of Lightwave Technology*, vol. 18, no. 2, pp. 203–212, 2000.

[16] T. P. Zhang and S. S. Ge, "Adaptive dynamic surface control of nonlinear systems with unknown dead zone in pure feedback form," *Automatica*, vol. 44, no. 7, pp. 1895–1903, 2008.

[17] Z. Q. Zhang, S. Y. Xu, and B. Y. Zhang, "Exact tracking control of nonlinear systems with time delays and dead-zone input," *Automatica*, vol. 52, pp. 272–276, 2015.

[18] C. X. Hu, B. Yao, and Q. F. Wang, "Performance-oriented adaptive robust control of a class of nonlinear systems preceded by unknown dead zone with comparative experimental results," *IEEE/ASME Transactions on Mechatronics*, vol. 18, no. 1, pp. 178–189, 2013.

[19] S. C. Tong and Y. M. Li, "Adaptive fuzzy output feedback tracking backstepping control of strict-feedback nonlinear systems with unknown dead zones," *IEEE Transactions on Fuzzy Systems*, vol. 20, no. 1, pp. 168–180, 2012.

[20] T. Wang, H. Gao, and J. Qiu, "A combined adaptive neural network and nonlinear model predictive control for multirate networked industrial process control," *IEEE Transactions on Neural Networks and Learning Systems*, 2015.

[21] H. Y. Li, C. W. Wu, P. Shi, and Y. B. Gao, "Control of nonlinear networked systems with packet dropouts: interval type-2 fuzzy model-based approach," *IEEE Transactions on Cybernetics*, vol. 45, no. 11, pp. 2378–2389, 2014.

[22] H. Li, C. Wu, L. Wu, H.-K. Lam, and Y. Gao, "Filtering of interval type-2 fuzzy systems with intermittent measurements," *IEEE Transactions on Cybernetics*, 2015.

[23] H. Y. Li, H. J. Gao, P. Shi, and X. D. Zhao, "Fault-tolerant control of Markovian jump stochastic systems via the augmented sliding mode observer approach," *Automatica*, vol. 50, no. 7, pp. 1825–1834, 2014.

[24] J. H. Park, D. H. Ji, S. C. Won, and S. M. Lee, "H_∞ synchronization of time-delayed chaotic systems," *Applied Mathematics and Computation*, vol. 204, no. 1, pp. 170–177, 2008.

[25] J. H. Kim and H. B. Park, "H^∞ state feedback control for generalized continuous/discrete time-delay system," *Automatica*, vol. 35, no. 8, pp. 1443–1451, 1999.

[26] S. C. Xu and J. Bao, "Distributed control of plant-wide chemical processes with uncertain time-delays," *Chemical Engineering Science*, vol. 84, pp. 512–532, 2012.

[27] C. H. Zhang, J. N. He, Y. L. Li, X. Y. Li, and P. Li, "Ignition delay times and chemical kinetics of diethoxymethane/O_2/Ar mixtures," *Fuel*, vol. 154, pp. 346–351, 2015.

[28] M.-W. Hong, C.-L. Lin, and B.-M. Shiu, "Stabilizing network control for pneumatic systems with time-delays," *Mechatronics*, vol. 19, no. 3, pp. 399–409, 2009.

[29] D. Karimipour, S. Pourdehi, and P. Karimaghaee, "Adaptive unstable periodic orbit stabilization of uncertain time-delayed chaotic systems subjected to input nonlinearity," *Systems & Control Letters*, vol. 61, no. 12, pp. 1168–1174, 2012.

[30] M. H. Zaheer, M. Rehan, G. Mustafa, and M. Ashraf, "Delay-range-dependent chaos synchronization approach under varying time-lags and delayed nonlinear coupling," *ISA Transactions*, vol. 53, no. 6, pp. 1716–1730, 2014.

18

AUV-Based Plume Tracking: A Simulation Study

Awantha Jayasiri,[1] Raymond G. Gosine,[1] George K. I. Mann,[1] and Peter McGuire[2]

[1] *Faculty of Engineering and Applied Science, Memorial University of Newfoundland, St. John's, NL, Canada A1B3X5*
[2] *C-CORE, Captain Robert A. Bartlett Building, Morrissey Road, St. John's, NL, Canada A1B3X5*

Correspondence should be addressed to Awantha Jayasiri; awanthas@yahoo.com

Academic Editor: Kalyana C. Veluvolu

This paper presents a simulation study of an autonomous underwater vehicle (AUV) navigation system operating in a GPS-denied environment. The AUV navigation method makes use of underwater transponder positioning and requires only one transponder. A multirate unscented Kalman filter is used to determine the AUV orientation and position by fusing high-rate sensor data and low-rate information. The paper also proposes a gradient-based, efficient, and adaptive novel algorithm for plume boundary tracking missions. The algorithm follows a centralized approach and it includes path optimization features based on gradient information. The proposed algorithm is implemented in simulation on the AUV-based navigation system and successful boundary tracking results are obtained.

1. Introduction

Most plume tracking and detection systems reported in the literature are based on surface dynamic oceanographic features. These are detected by satellites and their images are preprocessed for selecting regions of interest to generate optimal tracking sequences [1, 2]. As automated data collection is becoming more prevalent, optimum path planning and trajectory designs for autonomous underwater vehicles (AUVs) are becoming more important, since those planning approaches are required to navigate the AUV for collecting information. In order to track evolving features of interest in the ocean using predictive ocean models, several waypoint selection algorithms are developed and experimentally tested in [3]. Design and control of trajectories for AUVs to obtain optimal data collection are presented in [4]. The AUVs rely on GPS data for accurate position fixes. However, due to the dielectric contrast and high dielectric loss factor of seawater compared to air, most of the strength of the GPS signal is reflected back or attenuated and the AUV has to surface occasionally for position update. Therefore, in situations like under-ice oil spills and deep-sea exploration, the detection and tracking tasks need to be performed without GPS support and rely on a very low frequency acoustic communication channel.

A review of the state-of-the-art AUV navigation techniques is presented in [5, 6], along with a brief comparison of their mission-based suitability. The methods discussed include inertial, acoustic, and geophysical AUV navigation. To this extent, obtaining GPS-based surface fixes as well as utilizing a Long Base Line (LBL), Short Base Line (SBL), or Ultra Short Base Line (USBL) system has been the standard practice for AUV navigation [7]. These systems have similar deployment and transponder positioning challenges. Inverted USBL configuration, where the USBL array is located on the vehicle, interrogates transponders placed in known positions [7]. The study reported in [8] discusses underwater transponder positioning (UTP), which requires only one transducer due to the tight coupling with the vehicle's Inertial Navigation System (INS), and serves as an alternative approach.

In this paper, we use the inverted USBL configuration, which enables AUV positioning to be single-referenced and hence greatly reduces the complex operational logistics. A GPS-capable transponder, which is suspended from surface ice or a platform, is considered. The transponder transmits its position with each ping to the AUV and provides a means to position update in a GPS-denied undersea environment. It is also assumed that several receivers are placed along the body of the AUV on a noncoplanar configuration, keeping

FIGURE 1: REMUS AUV (image credit: Kongsberg Maritime).

a sufficient space between each of the receivers. The transponder and AUV's clocks are synchronized and the transponder broadcasts a unique signal with a known delay [9, 10]. Upon arrival of the transmitted signal to the AUV, a single trip travel time for each of the receivers is recorded. The recorded travel times are then used for the measurement update in state estimation. In between transponder broadcasts, short-term dead-reckoning is utilized based on high-rate INS data.

In operation conditions such as rough bathymetry or limited sensor range, where DVL bottom-track data is unavailable, the AUV's dynamic model-based position information is incorporated for position estimation [11–13]. A Linear Quadratic Regulator (LQR) is implemented based on the estimated states and further improved by adding a Proportional-Integral (PI) Controller for rudder control. The proposed approach is implemented in the nonlinear dynamic model of the REMUS AUV (Figure 1) presented in [14].

The paper has the following contributions:

(1) a multirate unscented Kalman filter is employed for sensor fusion in AUV localization;

(2) a novel efficient adaptive plume boundary tracking algorithm is developed using gradient information;

(3) numerous simulation results are presented to verify the approach.

The rest of the paper is organized as follows. In Section 2, an introduction on sensor package is given and the AUV navigation algorithm is presented. In Section 3, simulated results of AUV navigation are discussed. In Section 4, a plume boundary tracking algorithm is presented with its simulation results. The paper concludes by reporting the future research directions. A preliminary version of this paper appears in proceedings of the 27th Canadian Conference on Electrical and Computer Engineering (CCECE) [15]. In this version we focused on providing a detailed description of the work and more simulation results.

2. Sensor and System Modeling

2.1. Preliminaries and Notation. According to [16], $n(=6)$ Degrees Of Freedom (DOF) model of AUV dynamics and kinematics can be derived as

$$F = B(v)\,u_{\text{in}} - (C(v)\,v + D(v)\,v + g(q)) = M\dot{v} \quad (1)$$

$$\dot{\eta} = J(q)\,v, \quad (2)$$

where $M_{n\times n}$ matrix represents the inertia of the vehicle and hydrodynamic added-mass; $C_{n\times n}$ matrix includes rigid body Coriolis and centrifugal components as well as added-mass derivatives corresponding to the velocity coupling;

TABLE 1: Sensor characteristics.

Sensor type	Measurand	Frequency	Noise, σ
Accelerometer	Specific force	100 Hz	0.1 m/s^2
Gyroscope	Angular velocity	100 Hz	0.005 rad/s
Attitude sensors 1 and 2	Roll & pitch	10 Hz	0.05 rad
Attitude sensor 3	Yaw	10 Hz	0.2 rad
Pressure sensor	Pressure	5 Hz	200 Pa
DVL	Linear velocity	5 Hz	0.05 m/s
Acoustic receivers	Time delay	500 kHz	$1/\sqrt{3}\,\mu s$

$D_{n\times n}$ matrix includes energy dissipative terms due to relative motion between vehicle and surrounding fluid; $g_{n\times 1}$ combines gravitational and buoyancy forces; and $B_{n\times m}$ is the thruster control matrix, where m is the number of thrusters. Furthermore, $v = [u^T \ \omega^T]^T \in \mathbb{R}^6$, where $u = [u_1 \ u_2 \ u_3]^T$ and $\omega = [\omega_1 \ \omega_2 \ \omega_3]^T$ are body-fixed linear and angular velocities. The total force matrix is represented as F. Also, $\eta = [p^T \ q^T]^T$, where $p = [x \ y \ z]^T$ is the position vector relative to the inertial reference frame origin and $q = [\phi \ \theta \ \psi]^T$ is the vector of Euler angles, which are roll, pitch, and yaw, respectively. $u_{\text{in}} \in \mathbb{R}^m$ is the control input of the thrusters. The Jacobian $J = \begin{bmatrix} J_1(q) & 0_{3\times3} \\ 0_{3\times3} & J_2(q) \end{bmatrix}$, where J_1 is the coordinate transform matrix from u to $\dot{p}$ and J_2 relates ω to Euler rate vector $\dot{q}$. These are computed as follows:

$$J_1 = \begin{bmatrix} c\psi c\theta & -s\psi c\phi + c\psi s\theta s\phi & s\psi s\phi + c\psi c\phi s\theta \\ s\psi c\theta & c\psi c\phi + s\phi s\theta s\psi & -c\psi s\phi + s\theta s\psi c\phi \\ -s\theta & c\theta s\phi & c\theta c\phi \end{bmatrix}$$
$$J_2 = \begin{bmatrix} 1 & s\phi t\theta & c\phi t\theta \\ 0 & c\phi & -s\phi \\ 0 & \dfrac{s\phi}{c\theta} & \dfrac{c\phi}{c\theta} \end{bmatrix}, \quad (3)$$

where $s \Rightarrow \sin$, $c \Rightarrow \cos$, and $t \Rightarrow \tan$, with $0 \le \phi < 2\pi$, $-\pi/2 < \theta < \pi/2$, and $0 \le \psi < 2\pi$. We assume that $\theta \ne \pm\pi/2$ as otherwise J_2 approaches a singularity. If such an operation is required, quaternion-based attitude representation can be adopted as in [17]. The coefficient matrices we used in this work are based on the REMUS AUV dynamic model [14]. Our AUV has a typical set of sensors as mentioned in Table 1. The noise in each acoustic receiver is the quantization error introduced by the analog to digital converter of the system [18]. The transponder emits signals at 1 Hz and the control loop runs at 100 Hz. The AUV navigation system including USBL array is depicted in Figure 2.

2.2. Sensor Modeling

2.2.1. IMU Measurements. The Inertial Measurement Unit (IMU) sensor consists of accelerometers and gyroscopes (gyros), which measure specific forces and angular rates in

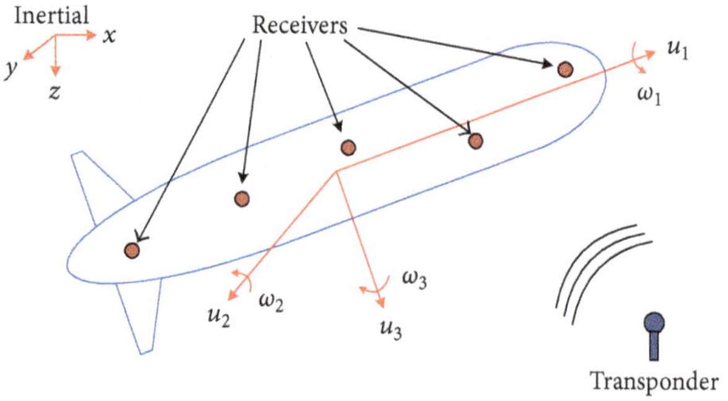

FIGURE 2: AUV navigation system adopted from [19].

the body-fixed coordinate system. The accelerometer reading Z_a can be modeled as [20]

$$Z_a = \dot{u} + [\omega\times]\, u + J_1^T \vec{g} + b_a + n_a, \tag{4}$$

where $[\omega\times]$ is the skew symmetric matrix cross product form of the vector ω, which is given as

$$[\omega\times] = \begin{bmatrix} 0 & -\omega_3 & \omega_2 \\ \omega_3 & 0 & -\omega_1 \\ -\omega_2 & \omega_1 & 0 \end{bmatrix}, \tag{5}$$

where $\vec{g} = [0\ \ 0\ \ 9.81\ \ \mathrm{m/s^2}]^T$ is the gravity vector and $b_a = 0.05\,\mathrm{m/s^2}$ is the bias of the reading with $\dot{b}_a = 0$. The measurement noise is distributed as $n_a \sim \mathcal{N}(0, \sigma_a^2 I_3)$, with σ_a^2 as the variance for each direction. The gyro reading Z_g is modeled as [20]

$$Z_g = \omega + b_g + n_g, \tag{6}$$

where $b_g = 0.01\,\mathrm{rad/s}$ is the bias of the reading with $\dot{b}_g = 0$. The noise of the reading is distributed as $n_g \sim \mathcal{N}(0, \sigma_g^2 I_3)$, with σ_g^2 as the variance on each axis.

2.2.2. Attitude Measurements. Attitude sensors composed of magnetometers and compasses measure roll, pitch, and yaw angles. These sensors are modeled as $Z_{\phi\theta\psi}$ [20]:

$$Z_{\phi\theta\psi} = q + n_{\phi\theta\psi}, \tag{7}$$

where the sensor noise is distributed as $n_{\phi\theta\psi} \sim \mathcal{N}(0, Q_{\phi\theta\psi})$ with $Q_{\phi\theta\psi} = \mathrm{diag}(\sigma_\phi^2, \sigma_\theta^2, \sigma_\psi^2)$ having $\sigma_\phi^2, \sigma_\theta^2, \sigma_\psi^2$ as the variances of ϕ, θ, ψ measurements.

2.2.3. Pressure Sensor Reading. A model for the pressure sensor is given as [20]

$$Z_{\mathrm{pr}} = \rho_{\mathrm{sw}} \vec{g}^T \left(p + J_1 l_{\mathrm{pr}}\right) + n_{\mathrm{pr}}, \tag{8}$$

where Z_{pr} is the sensor reading, ρ_{sw} is the seawater density, and l_{pr} is the sensor location in body frame. The sensor noise is distributed as $n_{\mathrm{pr}} \sim \mathcal{N}(0, \sigma_{\mathrm{pr}}^2)$ with σ_{pr}^2 as the variance.

2.2.4. DVL Sensor Reading. The DVL update (in processed form) is modeled as

$$Z_{\mathrm{dvl}} = u + n_{\mathrm{dvl}}, \tag{9}$$

where Z_{dvl} is the DVL reading and the sensor noise is distributed as $n_{\mathrm{dvl}} \sim \mathcal{N}(0, Q_{\mathrm{dvl}})$ with $Q_{\mathrm{dvl}} = \sigma_{\mathrm{dvl}}^2 I_3$ having σ_{dvl}^2 as the variance in each direction.

2.2.5. Velocity Estimation Using Dynamic Model. When DVL bottom-track data is unavailable, the body-fixed linear velocities are calculated using AUV's nonlinear dynamic model and used in the measurement update process:

$$Z_{\mathrm{mod}} = u + n_{\mathrm{mod}}, \tag{10}$$

where Z_{mod} is an estimate for u calculated using (1) and $n_{\mathrm{mod}} \sim \mathcal{N}(0, Q_{\mathrm{mod}})$. The noise matrix Q_{mod} for this estimate can be calculated recursively as

$$Q_{\mathrm{mod}} = G_t Q_{\mathrm{mod}} G_t^T + V_t M_t V_t^T, \tag{11}$$

where $G_t = I_3 + M^{-1}[\partial F/\partial u]dt$ and $V_t = [\partial u/\partial u_{\mathrm{in}}]$. The matrix $M_t = 0.01 I_3$ represents the noise in control space, which includes the noises in rudder and elevator angle control and the noise in thruster force.

Therefore, the linear velocity measurement $Z_u = Z_{\mathrm{dvl}}$ or Z_{mod} and its noise covariance $Q_u = Q_{\mathrm{dvl}}$ or Q_{mod} accordingly.

2.2.6. Time of Arrival Measured Using USBL. The time of arrival (TOA) of the acoustic wave from the transponder to the ith receiver r_i $(i = 1, \ldots, 5)$ is modeled as follows:

$$Z_{t_i} = \frac{\left\lfloor f \left\| \left(p + J_1 l_{r_i}\right) - \mathrm{Tr}_{xyz} \right\| / V_s \right\rceil}{f} + n_{r_i}, \tag{12}$$

where Z_{t_i} is the counted reading, f is the sampling frequency, l_{r_i} is the position of r_i on the AUV body, Tr_{xyz} is the transponder position, V_s is the speed of sound in the seawater, and n_{r_i} is the quantization noise. Note that $n_{r_i} \sim \mathcal{U}(-1/(2f), 1/(2f))$. In this work we assume $V_s = 1500\,\mathrm{m/s}$. However, in practical implementations V_s can be computed using Conductivity Temperature Depth (CTD) sensor readings [21]. Furthermore, we assume that TOA is disturbed by a noise of $n_{t_i} \sim \mathcal{N}(0, Q_t)$ with $Q_t = \sigma_{\mathrm{TOA}}^2 I_5$ and $\sigma_{\mathrm{TOA}} = 1/\sqrt{3}\,\mu\mathrm{s}$, having σ_{TOA}^2 as the noise variance.

A tightly coupled approach was adopted to infer the range and angle of arrival information. This is performed by feeding an array of TOA measurements to the estimation program. Let Z_t be the vector that includes all TOA measurements as

$$Z_t = \left[Z_{t_1}, \ldots, Z_{t_5}\right]^T. \tag{13}$$

2.3. System Modeling. Incorporating the kinematics and measurements, the system equations can be written as follows [20]:

$$\begin{aligned}\dot{q} &= J_2\left(Z_g - b_g\right),\\ \dot{b}_g &= 0,\\ \dot{p} &= J_1 u,\\ \dot{u} &= Z_a - [\omega\times]\,u - J_1^T\vec{g} - b_a,\\ \dot{b}_a &= 0.\end{aligned} \tag{14}$$

The state vector consists of 15 states:

$$y = \begin{bmatrix} q^T & b_g^T & p^T & u^T & b_a^T \end{bmatrix}^T. \tag{15}$$

The process noise covariance matrix is given as

$$R = \begin{bmatrix} J_2\left(\sigma_g^2 I_3\right) J_2^T dt^2 & 0 & 0 \\ 0 & 0 & 0 \\ 0 & J_1\left(\sigma_u^2\right) J_1^T dt^2 & 0 \\ 0 & \left(\sigma_a^2 I_3\right) dt^2 & 0 \\ 0 & 0 & 0 \end{bmatrix}. \tag{16}$$

The measurement model is written as $Z = h + Q$, where Z is the measurement vector, h is the measurement function, and Q is the measurement noise covariance. They are given as follows:

$$\begin{aligned} Z &= \begin{bmatrix} Z_{\text{pr}} & Z_t & Z_{\phi\theta\psi} & Z_u \end{bmatrix}^T,\\ h &= \begin{bmatrix} \left.\begin{matrix} \rho_{\text{sw}}\vec{g}^T\left(p + J_1 l_{\text{pr}}\right) \\ \dfrac{\lfloor f\left\|\left(p + J_1 l_{r_i}\right) - \text{Tr}_{xyz}\right\|/V_s \rfloor}{f}, \; i \in [1,5] \end{matrix}\right\}(=h_1) \\ q(=h_2) \\ u(=h_3) \end{bmatrix},\\ Q &= \begin{bmatrix} \sigma_{\text{pr}}^2 & 0 & 0 & 0 \\ 0 & Q_t & 0 & 0 \\ 0 & 0 & Q_{\phi\theta\psi} & 0 \\ 0 & 0 & 0 & Q_u \end{bmatrix}. \end{aligned} \tag{17}$$

3. State Estimation, Control, and Navigation

Based on (14), the state vector is propagated through time and a multirate unscented Kalman filter (UKF) is implemented for state estimation. The filter update process runs asynchronously as a response to the measurement readings. Once the transmitted signal is received by all of the receivers, the system performs the position update. If the difference in time of arrival of the signal between the first and last receiver is very small, then the AUV movement at that time is neglected.

3.1. Controller Implementation. The navigation system is decoupled for ease of control implementation, assuming negligible coupling effects between vertical and horizontal planes. AUV movement in these two planes is governed by the control of its elevator angle (δ_e) and rudder angle (δ_r), respectively. Two controllers are developed based on estimated states and the linearised versions of state equations to control δ_e and δ_r. Mechanical constraints on these angles are such that $-\pi/3 \le \delta_e, \delta_r \le \pi/3$.

δ_e is controlled by C_{δ_e}:

$$\begin{aligned} C_{\delta_e} &= \text{LQR}_e \\ &= ek_1 u_3 + ek_2\omega_2 + ek_3\theta + ek_4 z + ek_5\int z_{\text{err}}, \end{aligned} \tag{18}$$

where LQR_e is the Linear Quadratic Regulator (LQR) controller for elevator angle, $ek_1, \ldots, ek_5$ are LQR gains, and $z_{\text{err}} = z - \text{WP}_z$ (i.e., the difference between z and current waypoint z coordinate).

δ_r is controlled by C_{δ_r}, which combines LQR and Proportional-Integral (PI) Controllers:

$$\begin{aligned} C_{\delta_r} &= \text{LQR}_r + \text{PI}_r,\\ \text{LQR}_r &= rk_1 u_2 + rk_2\omega_3 + rk_3\psi_{\text{err}},\\ \text{PI}_r &= k_p d_\perp + k_i\int d_\perp, \end{aligned} \tag{19}$$

where rk_1, rk_2, rk_3 are LQR gains and k_p, k_i are proportional and integral gains. Also, $\psi_{\text{err}} = \psi - \psi_d$, where

$$\psi_d = \tan^{-1}\left(\frac{\text{WP}_y - y}{\text{WP}_x - x}\right), \tag{20}$$

and $d_\perp$ is the perpendicular distance from the AUV position to the assigned path x, y plane. After several iterations to reduce settling time, overshoot, and steady state errors, the optimized controller gains were found to be

$$\begin{aligned} ek_1 &= 0.5110,\\ ek_2 &= -4.1867,\\ ek_3 &= -6.6161,\\ ek_4 &= 3.1698,\\ ek_5 &= 0.0035.\\ rk_1 &= 0.3137,\\ rk_2 &= -0.9071,\\ rk_3 &= -1.1180,\\ k_p &= 0.01,\\ k_i &= 0.00001. \end{aligned} \tag{21}$$

AUV reaching the waypoint is determined by

$$\left\|\text{WP}_{xyz} - p\right\| < 1\,\text{m}. \tag{22}$$

3.2. An Algorithm for Localization and Autonomous Navigation. Incorporating the state estimation and control strategies, an algorithm is developed for AUV localization and navigation and is shown in Algorithm 1.

One advantage of this approach is that gyroscope readings are not treated as measurements and hence ω is not included in the state vector. Consequently, the dimension of the state vector is reduced. In fact, ω is used for the state-based controller implementation. However, since the gyros have very low noise ($\sigma = 0.005$ rad/s), we assume that eliminating the bias error from gyro readings will provide ω with sufficient accuracy.

(1) Initialize with: y_0 and its covariance P_0
(2) **for** $i \in (1, \ldots, \mathit{number\ of}$ WP) **do**
(3)   **for** $k \in (1, \ldots, \infty)$ **do**
(4)   Read IMU measurements
(5)   Calculate sigma points
(6)   Compute y_k^-, P_k^- using y_{k-1}^+, P_{k-1}^+ and (14)
(7)   Redraw sigma points
(8)   **if** *Attitude measurements are available* **then**
(9)   Read $Z_{\phi\theta\psi}$
(10)   **if** *Pressure measurement is available* **then**
(11)   Read Z_{pr}
(12)   **if** DVL *bottom-track is available* **then**
(13) $Z_u = Z_{\mathrm{DVL}}, Q_u = Q_{\mathrm{DVL}}$
(14)   **else**
(15) $Z_u = Z_{\mathrm{mod}}, Q_u = Q_{\mathrm{mod}}$
(16)   **if** Tr *signal is received by all receivers* **then**
(17)   Construct Z_t
(18)   Compute y_k^+, P_k^+
(19)   Implement C_{δ_s} and C_{δ_r} using (18) and (19)
(20)   Check the condition in (22)
(21)   **if** WP *is achieved* **then**
(22)   Exit and start from the next WP

ALGORITHM 1: AUV localization and navigation algorithm.

3.3. Simulation Setup. Navigation simulations are performed assuming a 2400 m × 1000 m × 10 m three-dimensional space. The ocean current velocity is assumed as V_C (m/s) = $[0.2\ \ 0.4\ \ 0.3]^T$ in x, y, and z directions, respectively. The mission is to perform a lawnmower-type navigation in the environment under five different cases:

(1) navigation without acoustic-based position fixes,

(2) navigation with position aiding and DVL dropout,

(3) dynamic model-aided navigation when DVL data is unavailable,

(4) navigation with transponder dropout and no velocity aiding,

(5) navigation with incremental transponder dropouts.

The start and end positions of the mission are given as 1200 m, 0 m, and −10 m and −1200 m, 1000 m, and −10 m in inertial x, y, z directions, respectively. While navigating between these two positions, a 10 m step change occurs in vertical (z) direction.

3.4. Bias Estimation. The predefined bias values of the accelerometers and gyroscopes are estimated. This is depicted in Figure 3. Figures 3(a)–3(c) show the estimates of accelerometer bias in directions u_1–u_3, respectively. Figures 3(d)–3(f) show gyro bias estimates, respectively, in ω_1–ω_3. All the bias estimates converged quickly within less than 30 s and remained constants thereafter.

3.5. Navigation without Acoustic-Based Position Fixes. Simulation is performed without transponder aiding for x, y position fixes throughout the run. However, velocity aiding is provided by DVL measurements. The results are shown in Figure 4. Figures 4(b)–4(d) show the error in AUV position estimates of inertial x, y, z directions, respectively. The errors in x, y positions are still low due to the accurate localization based on low noise DVL measurements. The error in z is still very low as the estimation of z is updated based on pressure sensor readings.

3.6. Navigation with DVL Dropout. Navigation simulation is performed with a DVL data dropout over a 5000 s period, but transponder aiding is provided throughout the run. During the DVL dropout period, (9) and (10) are not used and the lines (12)–(15) in Algorithm 1 are not implemented in the controller. The results are shown in Figure 5. In Figure 5(a), the blue path shows the navigation in normal conditions where DVL data is available. The red path from A to B shows the time where DVL data is unavailable.

3.7. Model-Aided Navigation under DVL Dropout. In this case, the AUV's dynamic model-based information is incorporated in the DVL dropout period. Figures 6(a) and 6(b) depict the 2D and 3D plots of the trajectory, respectively. Note that from A to B the algorithm uses model-based data and successful navigation is performed. Figures 7(a)–7(c) show the errors associated with this model-aided navigation which occurs between 0.5×10^4 s and 1×10^4 s. Figures 8(a)–8(c) show the errors in Euler angles under model-aided navigation from A to B shown in Figure 6. The errors in ϕ and θ are very low and under the range of ±0.02 rad. However, error in ψ is under ±0.05 rad. The computation of Euler angles is not affected by the DVL dropout.

3.8. Navigation with Transponder Dropout and No Velocity Aiding. Navigation simulation is performed with the transponder dropout and no velocity aiding for 5000 s period. The 2D plot of the trajectory is shown in Figure 9(a) and errors in x, y, and z are shown in Figures 9(b), 9(c), and 9(d), respectively. Note that, with no velocity aiding, the AUV quickly diverges from the assigned path resulting in high errors in x and y. Once the velocity aiding is established again (in this case the DVL reading) the AUV converges and errors in x, y are decreased.

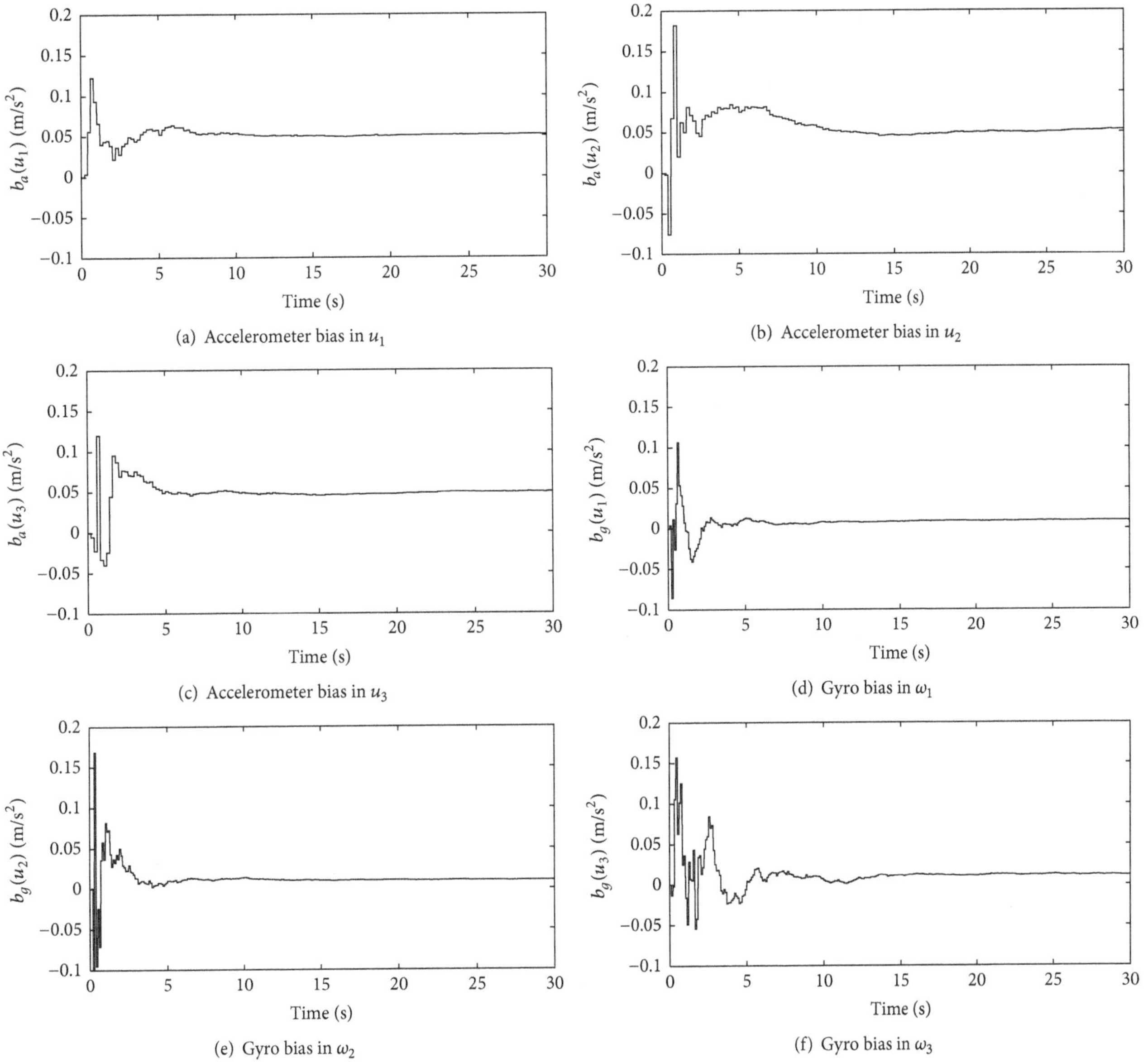

(a) Accelerometer bias in u_1

(b) Accelerometer bias in u_2

(c) Accelerometer bias in u_3

(d) Gyro bias in ω_1

(e) Gyro bias in ω_2

(f) Gyro bias in ω_3

FIGURE 3: Estimated accelerometer and gyro biases.

3.9. Navigation with Incremental Transponder Dropouts. In this case, we consider a scenario where transponder dropout occurs at every 50 s and consequent loss of x, y position fixes. The simulation run is limited to 10000 s. Figures 10(a) and 10(b) show the errors in range and azimuth from the AUV to the transponder position, respectively. Figures 10(c) and 10(d) show the errors in inertial x and y directions due to the transponder dropouts. It can be observed that, even with incremental loss of x, y position, the estimated horizontal position is maintained with low error bounds.

3.10. Discussion on the AUV Navigation Results. We have performed numerous simulations using the AUV localization and navigation algorithm. The AUV localization is performed by employing an unscented Kalman filter, where the true nonlinear system is used to capture the correct mean and covariance to the 3rd order, providing better performance than standard extended Kalman filter based approaches [22]. The localization scheme yields satisfactory performance even with the transponder dropouts.

In Figures 5(b) and 5(c) increased errors in x, y positions can be observed when DVL dropout occurs. At this period the AUV failed to perform a smooth navigation. Also, in Figure 5(d) there is a small reduction of the error in inertial z at the DVL dropout period. At the DVL drop the observability matrix is not full rank. Hence, to preserve the observability the dimensions are reduced. As a result, the cross-covariance (which is introduced by DVL measurements) of filter covariance matrix is eliminated and a better representation of the uncertainty can be obtained in inertial z direction, which may cause the error reduction. However, in practice the error in z is mainly caused by the depth controller, which is mostly

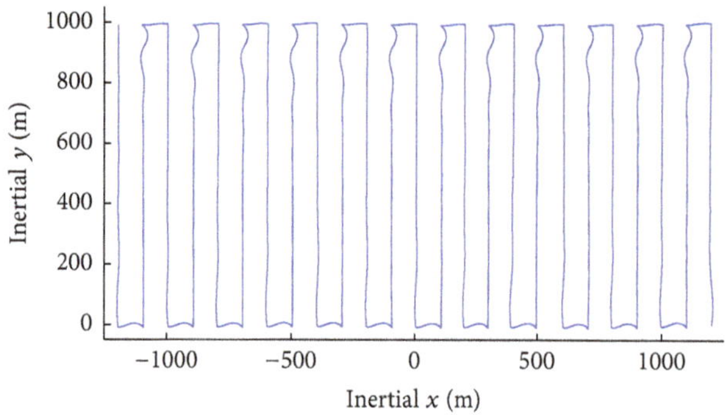

(a) 2D plot of navigation with DVL support

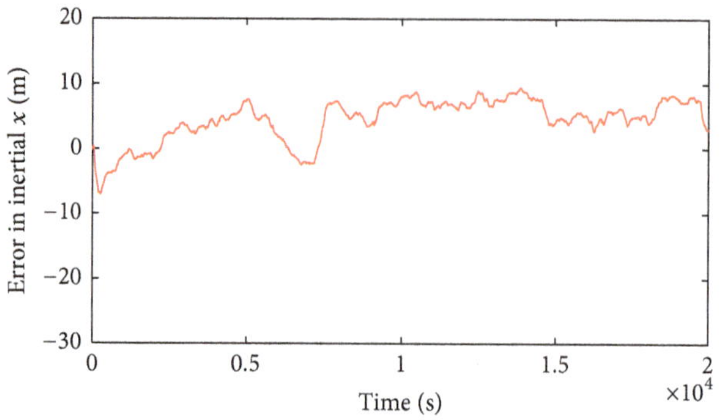

(b) Error in x with DVL support

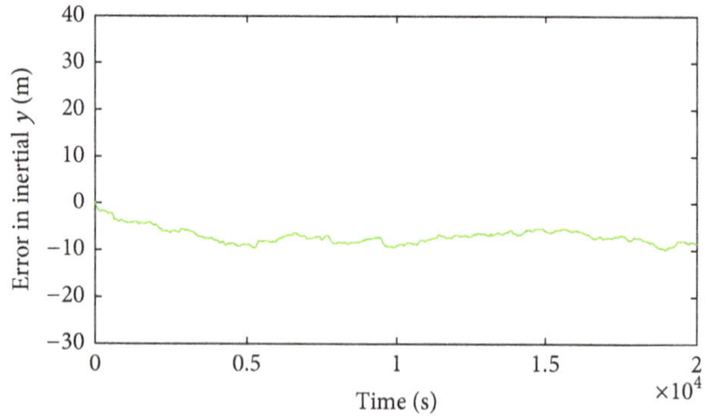

(c) Error in y with DVL support

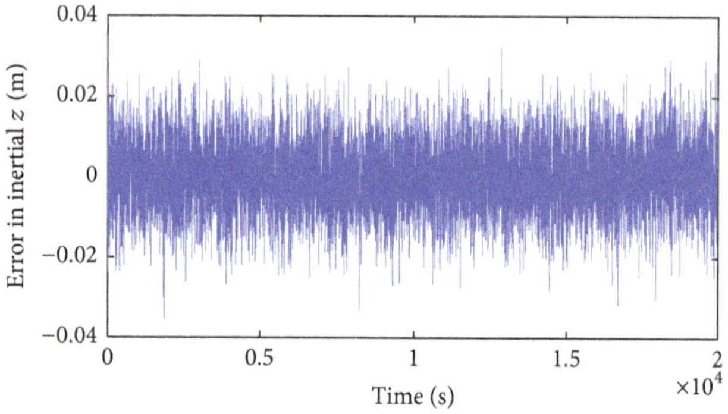

(d) Error in z with DVL support

FIGURE 4: Navigation with DVL support and without x, y position fixes.

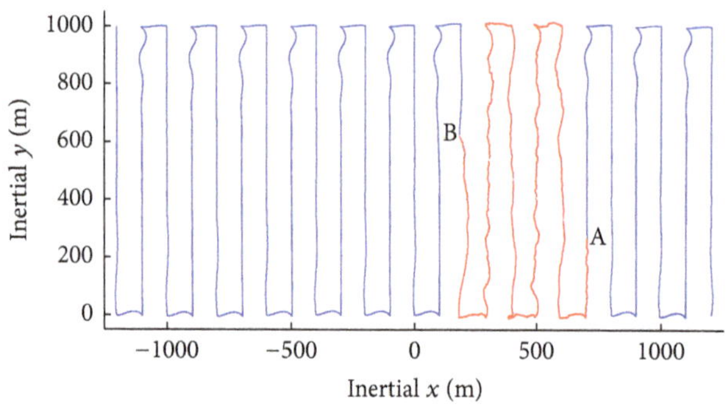

(a) 2D plot of navigation with DVL dropout from A to B

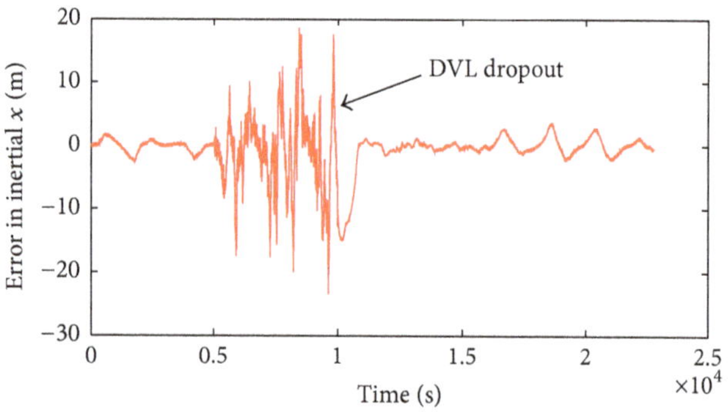

(b) Error in x with DVL dropout for 5000 s

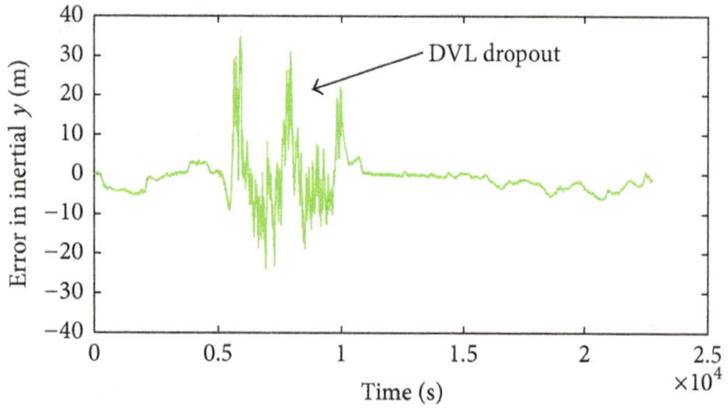

(c) Error in y with DVL dropout for 5000 s

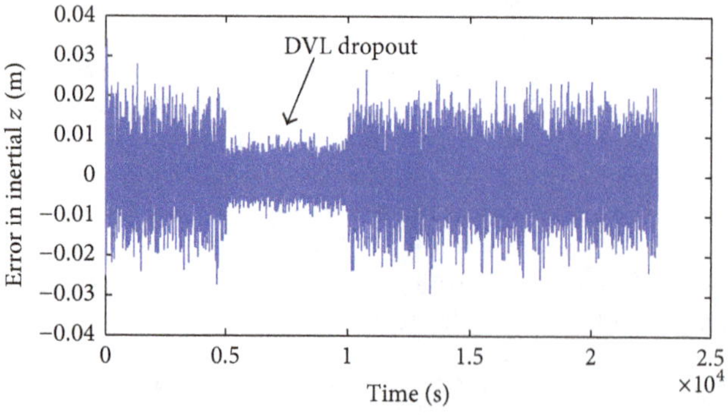

(d) Error in z with DVL dropout for 5000 s

FIGURE 5: Navigation with DVL dropout for 5000 s.

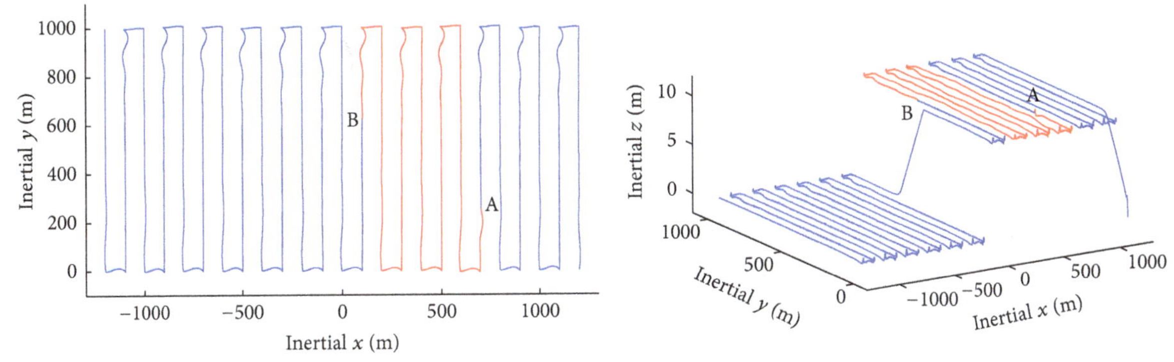

(a) 2D plot (b) 3D plot

Figure 6: 2D and 3D plots under model-aided navigation from A to B.

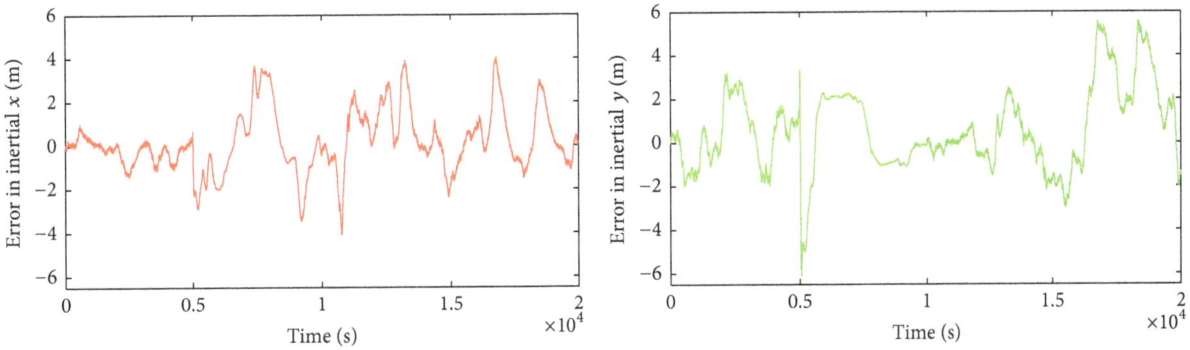

(a) Error in x with model-aided navigation within 0.5–1.0 $\times 10^4$ s

(b) Error in y with model-aided navigation within 0.5–1.0 $\times 10^4$ s

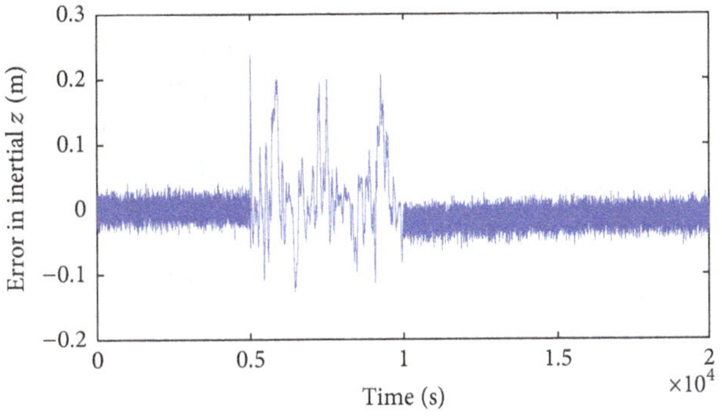

(c) Error in z with model-aided navigation within 0.5–1.0 $\times 10^4$ s

Figure 7: Errors of x, y, z in model-aided navigation within 0.5–1.0 $\times 10^4$ s (A to B) in Figure 6.

decoupled from the horizontal motion and makes decisions based readings from only the pressure sensor.

In the model-aided navigation under DVL dropout, an increment of the error in inertial z direction can be observed at the DVL dropout period on Figure 7(c). This can be due to the fact that the new cross-covariance, which is introduced by using the model-aided information in the DVL dropout period, degrades the filter performance, hence leading to higher error in z.

The rudder angle control is achieved by combining LQR with PI controller. As a result, the AUV was able to withstand the strong currents (V_C (m/s) = $[0.2\ \ 0.4\ \ 0.3]^T$) and successful performances are shown in all cases.

4. Adaptive Plume Tracking

According to Algorithm 2, AUV navigation is performed and information is collected about the plume boundary based on *in situ* fluorometer readings. Incorporating that information, an estimate on dispersion of the plume can be achieved. Table 2 depicts the methods and limitations of the state-of-the-art plume tracking strategies used in the

input: Fluorometer readings Fr_i $(i = 1, \ldots, n)$, threshold (T_f)
output: Adaptive boundary tracking algorithm
(1) **while** $\mathrm{Fr}_i < T_f$ **do**
(2) Navigate the AUV through pre-defined path.
(3) **if** AUV *reaches to last* WP **then**
(4) Could not locate the plume...! exit
(5) **while** *boundary track is not complete* **do**
(6) **if** $\mathrm{Fr}_i > T_f$ **then**
(7) Plume boundary is detected: log (x_t, y_t)
(8) Move AUV until $i = k_1$ $(\Delta a \propto k_1)$
(9) Calculate the gradient g_t using (x_t, y_t) and (x_{t-1}, y_{t-1})
(10) Set $\psi = g_t$
(11) Calculate section length $\Delta b \propto 1/\|g_t - g_{t-1}\|$ and move AUV
(12) Set $\psi = \pi/2$ and move
(13) **if** $\mathrm{Fr}_i < T_f$ **then**
(14) Repeat from line (8) to (12)
(15) **if** $i > k_2$ **then**
(16) Plume cannot be located...!
(17) Calculate the center point (x_c, y_c) using $x_1, y_1, \ldots, x_t, y_t$
(18) Set ψ to (x_c, y_c) and move
(19) Repeat from line (6)

ALGORITHM 2: Plume boundary tracking algorithm.

literature. In this work, we developed an efficient adaptive plume tracking algorithm to track the plume boundary and investigate its dispersion, with path optimization. An initial bounding box is assigned based on some prior knowledge (such as remote sensing data) and it is assumed that the oil plume is located inside the bounding box. The tracking algorithm navigates the AUV through the predefined search path and fluorometer readings are incorporated for detecting the plume boundary. Once a fluorometer reading Fr_i exceeds a predefined threshold T_f, a plume can be detected with a higher concentration. At this point, the AUV reaches the first boundary point and the algorithm switches to the boundary detecting mode.

The proposed approach is depicted in Figure 11 and the steps followed are described in Algorithm 2. Let k_1 be a predefined number of consecutive fluorometer readings to ensure the AUV is inside the plume boundary. Also, let k_2 be another predefined number of consecutive fluorometer readings, which are lower than T_f, to ensure that the plume cannot be located. The transect Δa is proportional to k_1. The same can be applied when the AUV is outside the boundary. Furthermore, Δb is used for rapid plume mapping and it helps to improve the efficiency of the algorithm. To calculate Δb, the system needs the two latest consecutive gradient values (as stated in line (11) of Algorithm 2). These consecutive gradient values are obtained from the four consecutive logged (x, y) positions. The calculation of Δb is mentioned in line (9) of Algorithm 2. When the plume boundary is relatively smooth, the two consecutive gradients are nearly equal and the section length Δb can be long. This is better for smooth sections of the plume boundary. Also when the plume boundary seems irregular, those consecutive gradients will be different and Δb will be short comparatively. As a result, the system can map the complex, irregular-shaped section of the boundary.

Lines (15)–(18) in Algorithm 2 ensure that the plume is tracked all the time. Here, k_2 controls the distance the AUV moves before it turns to the center point direction. The condition in line (5) is evaluated based on a predefined distance between start and end points (points A and B in Figure 11) and terminates the plume detection mode.

One limitation of the approach is that the accuracy of plume boundary coverage entirely depends on the sensitivity of the fluorometer and the predefined k_1, k_2 values. The performance on "noisy" plume boundaries may be improved by adaptively changing the k_1, k_2 values and adding an array of fluorometers.

4.1. Plume with an Ellipse-Shaped Boundary. A horizontal plume dispersion with a simple ellipse shape is considered. The major and minor axis lengths of the ellipse are 100 m and 150 m. The angle shift from the major axis is $\pi/3$. Figure 12(a) shows the true and detected plume boundaries under DVL-aided and dropout (dynamic model-based) navigation. Figures 12(b) and 12(c) show the 3D view of the navigation and the error in plume detection in x, y, z directions, respectively. Point A is the place where the AUV looses the DVL data. Although instant high errors can be observed in inertial x, y, z, those errors slowly converge due to the help of model-aided navigation.

TABLE 2: Different plume tracking strategies in literature.

Article	Number of agents	Properties	Limitations
[23]	Multiple	Uses CUSUM algorithm, presents the boundary estimation problem as a HMM, and recast as an optimization problem	Considered only ellipse-shaped, no path optimization, simulation only, intervehicle communication issues not addressed, and AUV dynamics not considered
[24]	Multiple	Decentralized, gradient-free algorithm, convergent and stable	AUV dynamics not considered, high computation cost, intervehicle communication issues not addressed, and only very simple plume shapes considered
[3]	Single, multiple	Cooperative, generating polygons to follow based on ocean model predictions, simulation, and practical implementation	Trajectory based on (roughly) approximated polygons, temporal constraints not considered, unable to react to the fast moving features, and ignoring the dynamics of the glider
[25]	Single	Behavior-based approach for plume mapping, subsumption architecture, showing experimental results	Limitations of behavior-based approach, no description on adaptive mapping, and implementing simple preplanned lawn-mower strategy
[26]	Single	Uses colored dissolved organic matter (CDOM) sensor for planed missions, adaptive planning using *in situ* current, and temperature measurements, and gets the 3D track of the AUV, practical implementation	No information on path optimization and no adaptive tracking
[27]	Single	Based on peak-capture algorithm, it generates a sawtooth trajectory and uses depth information and practical implementation	No information on path optimization
[28]	Multiple	Adaptive behavior-based system, acoustic communication within AUVs, representing the plume using Fourier orders when reconstructing	No information on path optimization and communication overhead
[29]	Single	Uses a plume indicator function and real-time implementation and uses adaptive transects; transect length depends on number of consecutive samples; distance between transects is a percentage of previous samples	Less path optimization, not using an AUV, no information about the convergence, and coverage of the used algorithm
[30]	Single	Uses remote sensing data to detect hotspots, uses surface current to project plumes spatiotemporally, and runs in a lawnmower type pattern, practical implementation	No path optimization, only using predefined pattern, and no adaptive tracking
Proposed	Single	Path optimization based on gradient information, adaptive plume tracking, and centralized approach	Only simulation results, no comparison data available, relying on remote sensing data for locating the plume region initially, low performance to noisy plume boundaries, and using only one fluorometer

4.2. Plume Boundary Modeled Using Fourier Orders. A rough estimate of the horizontal dispersion of a real plume can be obtained using Fourier orders of the form [28]

$$R_k = \sum_{i=0}^{k} A_i \cos(i\theta + \phi_i) + R_u, \quad (23)$$

where R_k is the radial distance to the plume boundary, $k(=10)$ is the highest Fourier order of the series, $R_u(= 50\,\text{m})$ is the undisturbed plume radius, $\theta(= [0, 2\pi))$ is the angles about the center of the plume evaluated for each degree, $A_i(= (+25R_u/2k, -25R_u/2k))$ is the radial amplitude perturbation, and $\phi_i(= [-\pi, +\pi])$ is the phase shift of ith order. Figures 13(a)

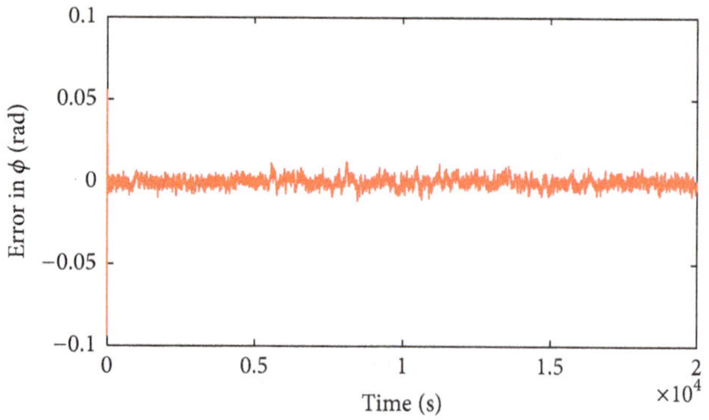

(a) Error in ϕ with model-aided navigation within 0.5–1.0 $\times 10^4$ s

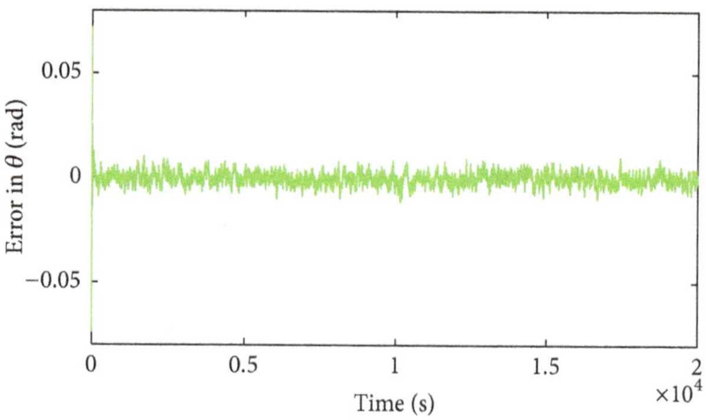

(b) Error in θ with model-aided navigation within 0.5–1.0 $\times 10^4$ s

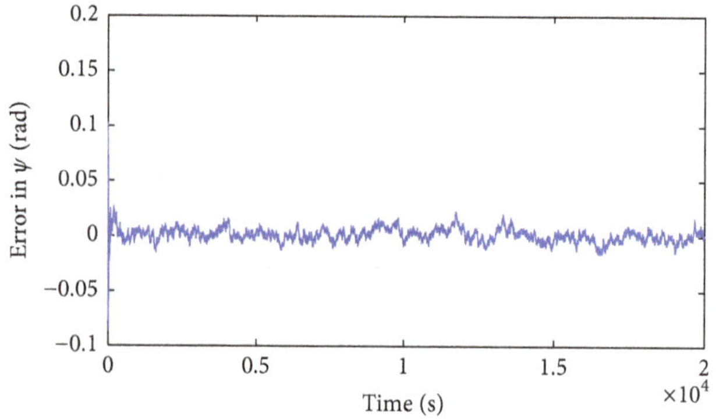

(c) Error in ψ with model-aided navigation within 0.5–1.0 $\times 10^4$ s

FIGURE 8: Errors of Euler angles in model-aided navigation within 0.5–1.0 $\times 10^4$ s (A to B) in Figure 6.

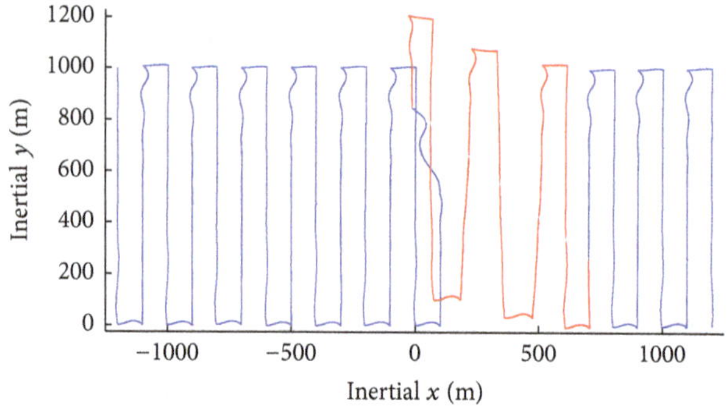

(a) x, y plot of navigation with transponder dropout for 5000 s

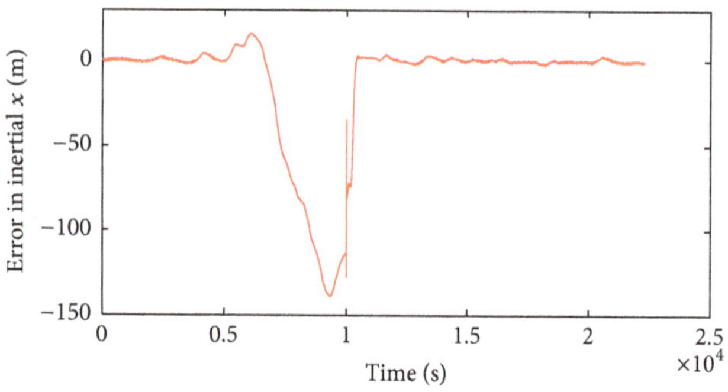

(b) Error in x for navigation with transponder dropout for 5000 s

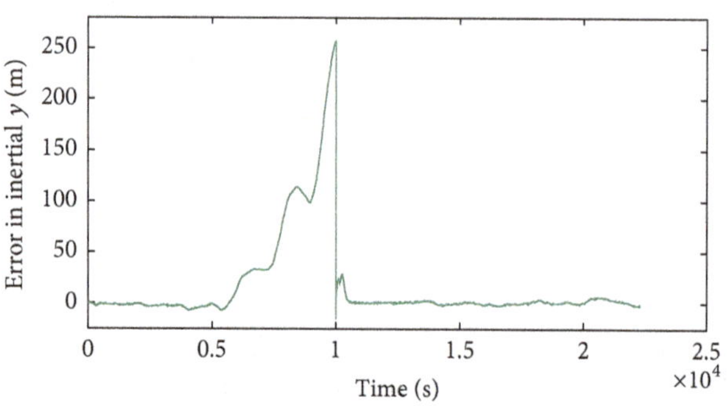

(c) Error in y for navigation with transponder dropout for 5000 s

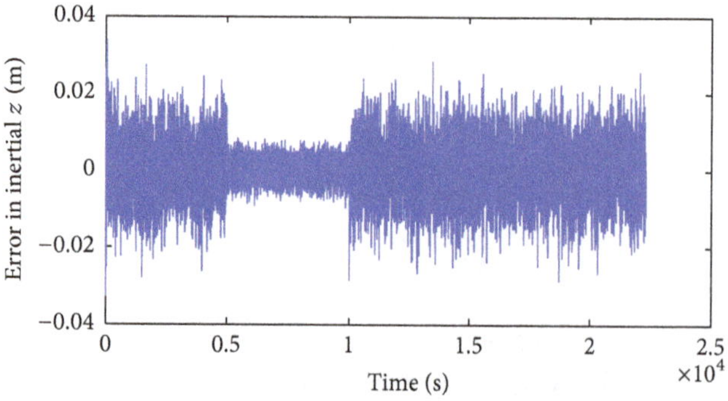

(d) Error in z for navigation with transponder dropout for 5000 s

FIGURE 9: Navigation with a transponder dropout and no velocity aiding for 5000 s.

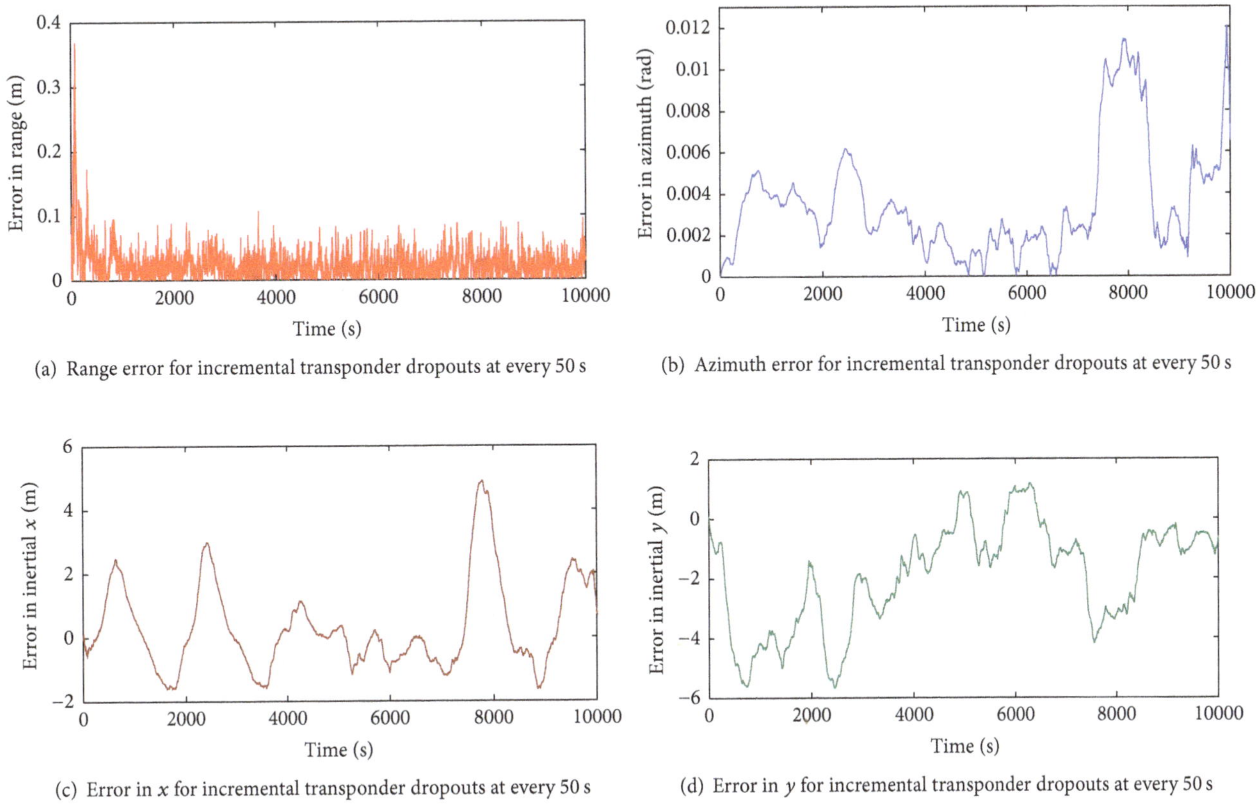

(a) Range error for incremental transponder dropouts at every 50 s

(b) Azimuth error for incremental transponder dropouts at every 50 s

(c) Error in x for incremental transponder dropouts at every 50 s

(d) Error in y for incremental transponder dropouts at every 50 s

FIGURE 10: Navigation with incremental transponder dropouts at every 50 s.

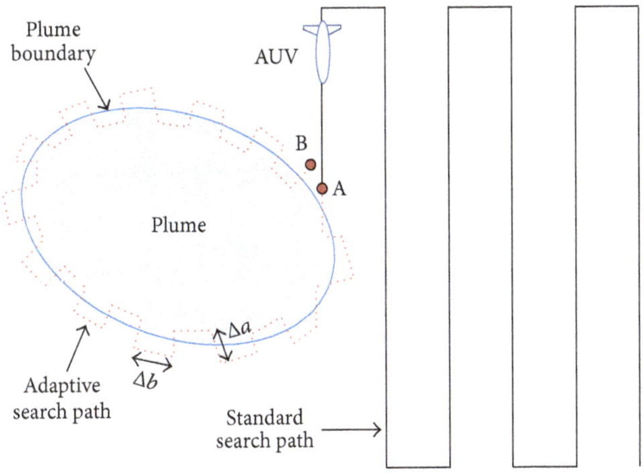

FIGURE 11: Plume boundary tracking.

and 13(b) show the dispersion of the plume and the detected plume boundary in 2D and 3D view. Figure 13(c) shows the error in plume detection in x, y, z directions. Note that the section length Δb is reduced due to the irregularity of the plume dispersion.

4.3. Discussion on the Developed Plume Tracking Algorithm. The proposed approach was able to successfully track a more realistic plume boundary demonstrating its path optimization and adaptive features. This can be observed in Figure 13(a). From sections A-B where the irregularity of the plume boundary is lower, the AUV crosses the plume more loosely. Also at the places where the irregularity seems higher (sections C-D or E-F) AUV crosses the plume tightly. Also, the AUV navigation plan is not predefined and adaptive according to the shape of the plume boundary.

Certainly there are lots of facts that could influence the limitations of the algorithm. The measurement uncertainty of

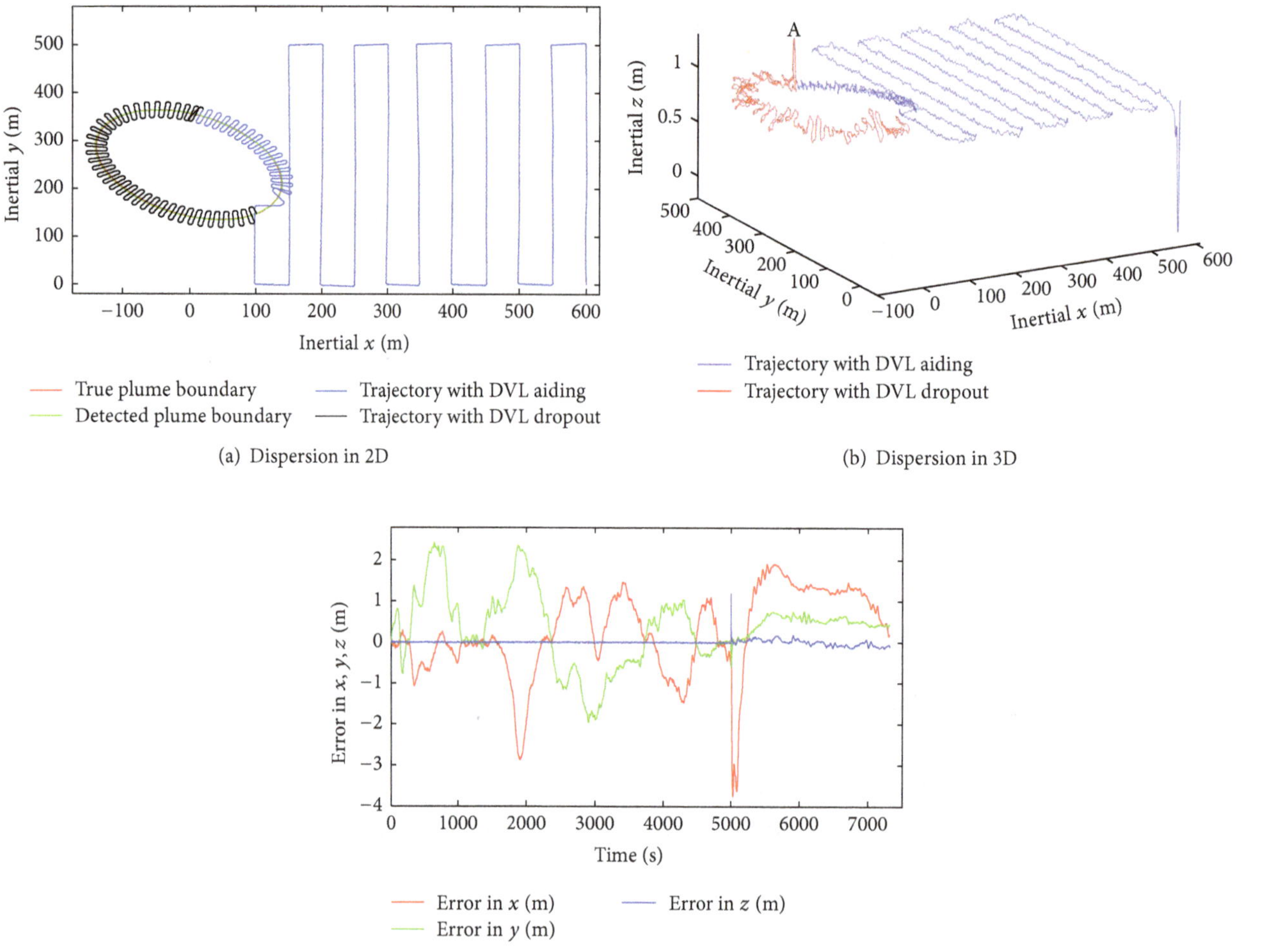

(a) Dispersion in 2D

(b) Dispersion in 3D

(c) Error in x, y, z. At 5000 s DVL drop occurs and model-aided navigation starts

FIGURE 12: Plume with an ellipse-shaped boundary.

the oil sensor is not considered and the operation is assumed as an ON/OFF type. Consequently, those uncertainties can affect the accuracy of the detection/tracking of the plume boundary. The accuracy of the plume tracking system can be improved by adding an array of fluorometers to better estimate the oil concentration. Another challenge is the noisy plume boundaries where the gradient information based on fluorometer readings cannot be established accurately enough. Consequently, the AUV could fail to generate a reliable boundary tracking. Furthermore, the accuracy of the boundary tracking specially with irregular gradients is limited by the physical AUV dynamics and actuator constraints. Moreover, due to the medium density changes in a plume area the behavior of the acoustic could be complicated. This can introduce screening effects that severely impair the acoustic system and the AUV may fail to estimate x, y positions accurately. However, at this work we did not model the medium density changes and its effect in acoustics. We assumed that the acoustic channel behaves the same both in and out of the plume. We believe it can be an interesting future work.

5. Conclusion

In this work we have developed a multirate UKF algorithm for AUV localization in a GPS-denied undersea environment. Furthermore, an adaptive plume detection and tracking system is developed. The proposed tracking algorithm uses gradient information. The algorithms are implemented in simulations and successful results are obtained.

The adaptive plume tracking system developed in this work is not limited to tracking oil plumes and it can be also used to track other types of plumes such as biological and chemical. As a future work we expect extending the proposed system to track 3D dynamic plumes and implementing it in real time to investigate most of the issues present in physical scenarios, which are not examined in detail with the current simulation work. Moreover, extending this system for multi-AUV navigation missions such as implementing a cooperative AUV network for under-ice oil plume detection and tracking would be an interesting research study. This will decrease the overall time for feature tracking and improve the usage and sharing of information,

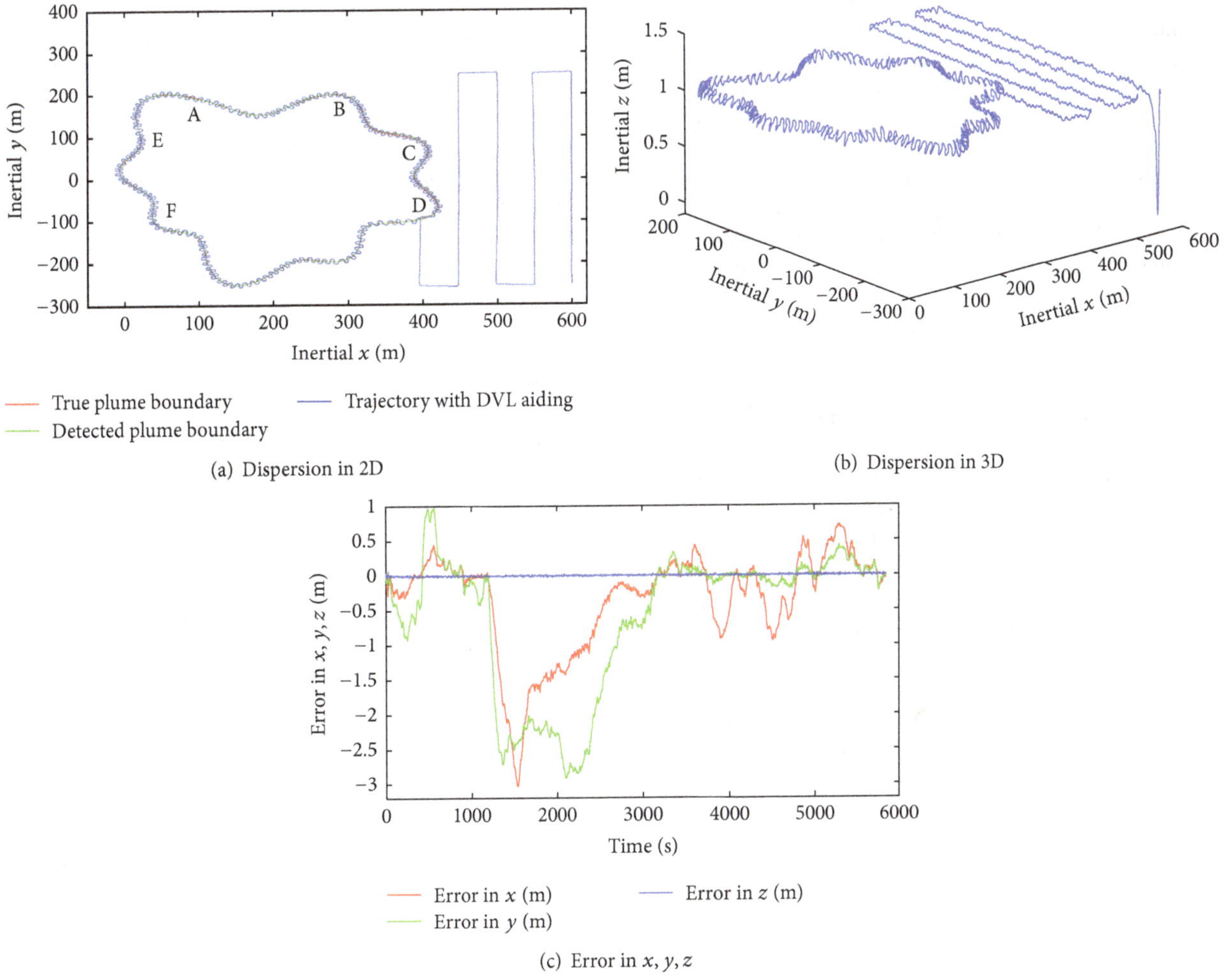

FIGURE 13: Plume boundary modeled using Fourier orders.

which lead to a better representation of the plume boundary.

Conflict of Interests

The authors declare that there is no conflict of interests regarding the publication of this paper.

References

[1] J. Das, F. Py, T. Maughan et al., "Simultaneous tracking and sampling of dynamic oceanographic features with autonomous underwater vehicles and lagrangian drifters," in *Experimental Robotics: The 12th International Symposium on Experimental Robotics*, vol. 79 of *Springer Tracts in Advanced Robotics*, pp. 541–555, Springer, Berlin, Germany, 2014.

[2] J. Das, K. Rajan, S. Frolov et al., "Towards marine bloom trajectory prediction for AUV mission planning," in *Proceedings of the IEEE International Conference on Robotics and Automation (ICRA '10)*, pp. 4784–4790, IEEE, Anchorage, Alaska, USA, May 2010.

[3] R. N. Smith, Y. Chao, P. P. Li, D. A. Caron, B. H. Jones, and G. S. Sukhatme, "Planning and implementing trajectories for autonomous underwater vehicles to track evolving ocean processes based on predictions from a regional ocean model," *International Journal of Robotics Research*, vol. 29, no. 12, pp. 1475–1497, 2010.

[4] N. E. Leonard, D. A. Paley, F. Lekien, R. Sepulchre, D. M. Fratantoni, and R. E. Davis, "Collective motion, sensor networks, and ocean sampling," *Proceedings of the IEEE*, vol. 95, no. 1, pp. 48–74, 2007.

[5] L. Paull, S. Saeedi, M. Seto, and H. Li, "AUV navigation and localization: a review," *IEEE Journal of Oceanic Engineering*, vol. 39, no. 1, pp. 131–149, 2014.

[6] L. Stutters, H. Liu, C. Tiltman, and D. J. Brown, "Navigation technologies for autonomous underwater vehicles," *IEEE Transactions on Systems, Man and Cybernetics Part C: Applications and Reviews*, vol. 38, no. 4, pp. 581–589, 2008.

[7] M. Morgado, P. Oliveira, C. Silvestre, and J. F. Vasconcelos, "Embedded vehicle dynamics aiding for USBL/INS underwater navigation system," *IEEE Transactions on Control Systems Technology*, vol. 22, no. 1, pp. 322–330, 2014.

[8] O. Hegrenas, K. Gade, O. K. Hagen, and P. E. Hagen, "Underwater transponder positioning and navigation of autonomous underwater vehicles," in *Proceedings of the MTS/IEEE Biloxi—Marine Technology for Our Future: Global and Local Challenges (OCEANS '09)*, pp. 1–7, IEEE, Biloxi, Miss, USA, October 2009.

[9] R. M. Eustice, L. L. Whitcomb, H. Singh, and M. Grund, "Experimental results in synchronous-clock one-way-travel-time acoustic navigation for autonomous underwater vehicles," in *Proceedings of the IEEE International Conference on Robotics and Automation (ICRA '07)*, pp. 4257–4264, Roma, Italy, April 2007.

[10] M. F. Fallon, M. Kaess, H. Johannsson, and J. J. Leonard, "Efficient AUV navigation fusing acoustic ranging and side-scan sonar," in *Proceedings of the IEEE International Conference on Robotics and Automation (ICRA '11)*, pp. 2398–2405, IEEE, Shanghai, China, May 2011.

[11] A. V. Inzartsev, *Underwater Vehicles*, InTech, Rijeka, Croatia, 2009.

[12] Ø. Hegrenæs, O. Hallingstad, and K. Gade, "Towards model-aided navigation of underwater vehicles," *Modeling, Identification and Control*, vol. 28, no. 4, pp. 113–123, 2007.

[13] Ø. Hegrenaes and O. Hallingstad, "Model-aided INS with sea current estimation for robust underwater navigation," *IEEE Journal of Oceanic Engineering*, vol. 36, no. 2, pp. 316–337, 2011.

[14] T. Prestero, *Verification of a six-degree of freedom simulation mode for the REMUS autonomous underwater vehicle [Master's Thesis]*, Massachusetts Institute of Technology, Cambridge, Mass, USA, 2001.

[15] A. Jayasiri, R. G. Gosine, G. K. I. Mann, and P. McGuire, "Simulation of aided AUV navigation and adaptive plume tracking," in *Proceedings of the IEEE 27th Canadian Conference on Electrical and Computer Engineering (CCECE '14)*, pp. 1–6, IEEE, Toronto, Canada, May 2014.

[16] T. I. Fossen, *Guidance and Control of Ocean Vehicles*, John Wiley & Sons, 1994.

[17] J. Diebel, "Representing attitude: euler angles, unit quaternions, and rotation vectors," Tech. Rep., Stanford University, Stanford, Calif, USA, 2006.

[18] M. K. L. Zaworski, D. Chaberski, and M. Zielinski, "Quantization error in time-to-digital converters," *Metrology and Measurement Systems*, vol. 19, no. 1, pp. 115–122, 2012.

[19] M. Morgado, P. Oliveira, C. Silvestre, and J. Vasconcelos, "Improving aiding techniques for usbl tightly-coupled inertial navigation system," in *Proceedings of the 17th IFAC World Coggress*, pp. 15 973–15 978, Seoul, South Korea, July 2008.

[20] P. A. Miller, J. A. Farrell, Y. Zhao, and V. Djapic, "Autonomous underwater vehicle navigation," *IEEE Journal of Oceanic Engineering*, vol. 35, no. 3, pp. 663–678, 2010.

[21] G. Grenon, P. E. An, S. M. Smith, and A. J. Healey, "Enhancement of the inertial navigation system for the morpheus autonomous underwater vehicles," *IEEE Journal of Oceanic Engineering*, vol. 26, no. 4, pp. 548–560, 2001.

[22] E. Wan and R. Van der Merwe, "The unscented kalman filter for nonlinear estimation," in *Proceedings of the Adaptive Systems for Signal Processing, Communications, and Control Symposium (AS-SPCC '00)*, pp. 153–158, IEEE, Alberta, Canada, 2000.

[23] Z. Jin and A. L. Bertozzi, "Environmental boundary tracking and estimation using multiple autonomous vehicles," in *Proceedings of the 46th IEEE Conference on Decision and Control (CDC '07)*, pp. 4918–4923, IEEE, New Orleans, La, USA, December 2007.

[24] M. Kemp, A. Bertozzi, and D. Marthaler, "Multi-UUV perimeter surveillance," in *Proceedings of the Autonomous Underwater Vehicles (IEEE/OES '14)*, pp. 102–107, Sebasco, Me, USA, June 2004.

[25] Y. Tian, W. Li, A. Zhang, and J. Yu, "Behavior-based control of an autonomous underwater vehicle for adaptive plume mapping," in *Proceedings of the 2nd International Conference on Intelligent Control and Information Processing (ICICIP '11)*, vol. 2, pp. 719–724, IEEE, Harbin, China, July 2011.

[26] P. Rogowski, E. Terrill, M. Otero, L. Hazard, and W. Middleton, "Mapping ocean outfall plumes and their mixing using autonomous underwater vehicles," *Journal of Geophysical Research-Oceans*, vol. 117, no. 7, Article ID C07016, pp. 1–12, 2012.

[27] Y. Zhang, R. S. McEwen, J. P. Ryan et al., "A peak-capture algorithm used on an autonomous underwater vehicle in the 2010 Gulf of Mexico oil spill response scientific survey," *Journal of Field Robotics*, vol. 28, no. 4, pp. 484–496, 2011.

[28] S. Petillo, H. Schmidt, and A. Balasuriya, "Constructing a distributed AUV network for underwater plume-tracking operations," *International Journal of Distributed Sensor Networks*, vol. 2012, Article ID 191235, 12 pages, 2012.

[29] C. J. Cannell, A. S. Gadre, and D. J. Stilwell, "Boundary tracking and rapid mapping of a thermal plume using an autonomous vehicle," in *Proceedings of the OCEANS*, pp. 1–6, IEEE, Boston, Mass, USA, September 2006.

[30] J. Das, K. Rajan, S. Frolov et al., "Towards marine bloom trajectory prediction for AUV mission planning," in *Proceedings of the IEEE International Conference on Robotics and Automation (ICRA '10)*, pp. 4784–4790, Anchorage, Alaska, USA, May 2010.

19

New Smith Internal Model Control of Two-Motor Drive System Based on Neural Network Generalized Inverse

Guohai Liu, Jun Yuan, Wenxiang Zhao, and Yaojie Mi

School of Electrical and Information Engineering, Jiangsu University, Zhenjiang 212013, China

Correspondence should be addressed to Wenxiang Zhao; zwx@ujs.edu.cn

Academic Editor: Qiao Zhang

Multimotor drive system is widely applied in industrial control system. Considering the characteristics of multi-input multioutput, nonlinear, strong-coupling, and time-varying delay in two-motor drive systems, this paper proposes a new Smith internal model (SIM) control method, which is based on neural network generalized inverse (NNGI). This control strategy adopts the NNGI system to settle the decoupling issue and utilizes the SIM control structure to solve the delay problem. The NNGI method can decouple the original system into several composite pseudolinear subsystems and also complete the pole-zero allocation of subsystems. Furthermore, based on the precise model of pseudolinear system, the proposed SIM control structure is used to compensate the network delay and enhance the interference resisting the ability of the whole system. Both simulation and experimental results are given, verifying that the proposed control strategy can effectively solve the decoupling problem and exhibits the strong robustness to load impact disturbance at various operations.

1. Introduction

Multimotor drive system is a multi-input multioutput (MIMO), nonlinear, strongly coupled, and time-varying delay system, which has been widely applied in modern industry. High precision coordinated control performance can improve quality and productivity of products, such as electric vehicles and rail transits [1–4]. In order to meet the requirements of industrial applications, the delay problem of the system must be solved, and then the coupling variables of this system can be decoupled.

In a multimotor drive system, decoupling speed and tension is a key issue. Traditional decoupling control methods strongly depend on mathematical model of the system, such as PID control, sliding mode control [5], feed-forward control [6], and adaptive fuzzy control [7]. Moreover, the structure characteristics or parameters are easily influenced by load variation and interference factor. So, it is hard to obtain accurate system mathematical model. A new control strategy based on support vector machine (SVM) theory has been proposed to decouple current and rotational speed of a permanent-magnet (PM) motor system [8]. Recently, neural network generalized inverse (NNGI) is used to decouple strong-coupling controlled variables [9–11]. Torque and flux components of five-phase PM motors were decoupled by multiple-reference-frame transformation, in which the artificial neural network (ANN) controller was trained online to adapt system uncertainties [12]. Based on a single artificial neuron requiring no offline training, an intelligent speed controller for the PM motor was proposed to adapt to various drives without extensive knowledge of motor behavior [13]. Since the analysis model of the traveling-wave ultrasonic motor is difficult to obtain, a generalized regression neural-network-based model is developed to solve the problem and the transfer function identification is no longer required [14]. An ANN-based estimator is implemented to eliminate mechanical sensors and then realize the sensorless control of a PM motor [15]. By using a pseudolinear composite system, NNGI can transform a MIMO nonlinear system into several single-input single-output linear systems, in which an accurate mathematic model is not required.

In addition, the demand of real-time transmission is high because a large transmission delay will result in uncontrollable system in practice. So, an effective control strategy is necessary to implement the delay compensation. Traditional Smith predictor control is an effective method to conquer the influence of system delay [16, 17]. However, this control method is sensitive to the errors of estimate model and external disturbance. An inaccurate estimate model or a strong interference will affect control quality and even lead to system instability. According to the strong robustness and anti-interference abilities of internal model control [18], the control structure is widely used in many applications. A strategy based on internal model control was proposed for a matrix converter-based PMSM drive system to reduce the adverse impact on drive performance caused by nonlinear output characteristics of matrix converter in the case of input voltage disturbance [19]. The speed regulation problem for a PM motor drive was settled by a fuzzy adaptive law based internal model control scheme, and the effectiveness of the proposed methods has been verified [20]. An internal model control with a conditional integrator was proposed for the robust output regulation of a DC/DC buck converter [21]. In the work, Smith predictor control is codesigned with internal model control structure, and then the Smith internal model (SIM) control structure is formed. This SIM control incorporates the advantages of both Smith estimate control and internal model control, which can not only reduce the effect of network delay, but also improve the robustness of system. However, since it is hard to obtain the estimate model in multimotor system, the application of SIM control is very limited. On the basis of NNGI method, the accurate model of generalized pseudolinear subsystem will be obtained and the estimate model can be acquired without mismatch. Therefore, the delay compensation can be implemented successfully.

The purpose of this paper is to propose a new control method, namely, NNGI-based Smith internal model (NNGI-SIM) control, to improve the operating performance of a two-motor drive system. This NNGI-SIM control can avoid the mismatch between prediction model and original model. The theory of NNGI decoupling method is introduced first; then, the two-motor drive system is decoupled by NNGI method; the accurate model of generalized pseudolinear subsystem will be obtained and the mismatch condition will be conquered. After the theory of SIM control is deduced, the NNGI system is structured and codesigned with SIM control strategy to form the new NNGI-SIM control and then applied in two-motor drive system.

The structure of this paper is as follows: in Section 1, the multimotor drive system will be introduced. In Section 2, the mathematical model of two-motor drive system will be deduced. In Section 3, the new NNGI-SIM control will be formed and applied in two-motor drive system. In Sections 4 and 5, decoupling effect and robustness under load disturbance of the proposed two-motor drive will be verified by both simulations and experiments. Finally, conclusions will be drawn in Section 6.

2. Mathematical Model

A two-motor synchronous system is shown in Figure 1; it consists of two three-phase asynchronous induction motors, in which motor 1 is the master motor and motor 2 is the slave one. The belt-pulley is installed on the motor shaft, and both motors are combined by transmission belt on the pulley one. Floating rollers strain the belt to increase the friction between the belt and the driving rollers. Two motors operate synchronously by adjusting the action between speed and tension.

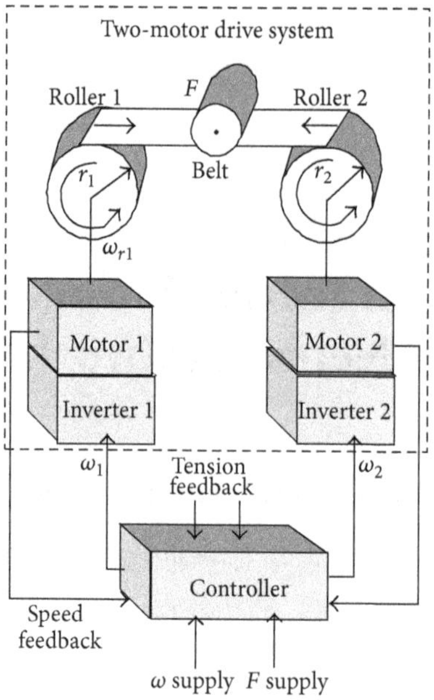

FIGURE 1: Physical model diagram of two-motor drive system.

When both motors operate in the vector control mode, the rotor flux is considered to be invariant. Then, the mathematic model of the system can be written as

$$\begin{aligned}\dot{\omega}_{r1} &= \frac{n_{p1}}{J_1}\left[(\omega_1-\omega_{r1})\frac{n_{p1}T_{r1}}{L_{r1}}\psi_{r1}^2-(T_{L1}+r_1F)\right],\\ \dot{\omega}_{r2} &= \frac{n_{p2}}{J_2}\left[(\omega_2-\omega_{r2})\frac{n_{p2}T_{r2}}{L_{r2}}\psi_{r2}^2-(T_{L2}-r_2F)\right],\\ \dot{F} &= \frac{K}{T}\left(\frac{1}{n_{p1}}r_1k_1\omega_{r1}-\frac{1}{n_{p2}}r_2k_2\omega_{r2}\right)-\frac{F}{T},\end{aligned}\quad (1)$$

where ω_i and ω_{ri} are the synchronous angular speed and the electric angular speed, F is the tension of the belt, n_{pi} is the pole-pairs number of the number i motor, J_i, ψ_{ri}, and L_{ri} are rotor inertia, rotor flux, and rotor self-inductance, T_{ri} is the electromagnetic time constant, T_{Li} is the load torque, $K = E/v$ is the transfer coefficient, $T = L_0/Av$ is the time constant of tension variation, E is Young's modulus of elasticity, v is the expected line speed, L_0 is the distance between racks, A is the section area, r_i is the radius of belt-pulley, and k_i is the speed ratio ($i = 1, 2$), respectively.

3. Controller Design

3.1. NNGI System. Generalized inverse system can realize the inverse mapping relationship from the output to the input of original system. By setting the original order, derivatives and their linear combination of expected output of original system can be used as the input of the generalized inverse

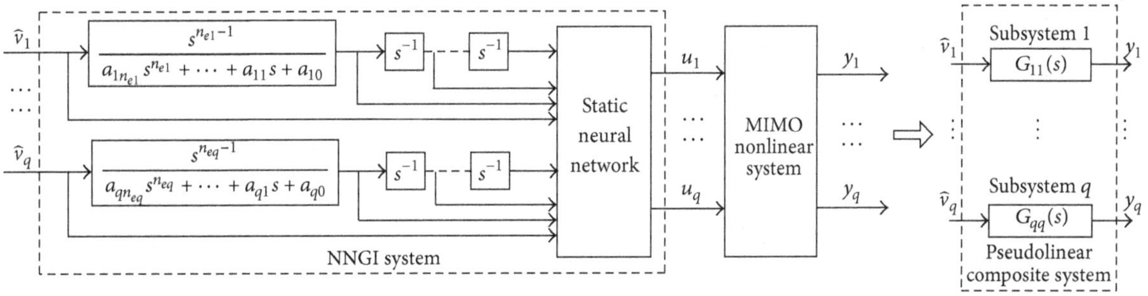

FIGURE 2: NNGI pseudolinear composite subsystem equivalent chart.

system. Then, the output can be used as the input of original system to drive the original system, generating the expected output [22].

As a MIMO nonlinear system which has q-dimensional input vector $u = (u_1, u_2, \ldots, u_q)^T \in R^q$ and q-dimensional output vector $y = (y_1, y_2, \ldots, y_q)^T \in R^q$, the differential equation can be expressed as

$$F\left(y^{(\varepsilon)T} Y, u^{(\sigma)T}, U\right) = 0, \tag{2}$$

where

$$\begin{aligned}
\varepsilon &= \left(\varepsilon_1, \ldots, \varepsilon_q\right)^T, \\
\sigma &= \left(\sigma_1, \ldots, \sigma_q\right)^T, \\
y^{(\varepsilon)} &= \left(y_1^{(\varepsilon_1)}, \ldots, y_q^{(\varepsilon_q)}\right), \\
u^{(\sigma)} &= \left(u_1^{(\sigma_1)}, \ldots, u_q^{(\sigma_q)}\right), \\
Y &= \left(y_1^{(\varepsilon_1-1)}, \ldots, y_1, \ldots, y_q^{(\varepsilon_q-1)}, \ldots, y_q\right)^T, \\
U &= \left(u_1^{(\sigma_1-1)}, \ldots, u_1, \ldots, u_q^{(\sigma_q-1)}, \ldots, u_q\right)^T.
\end{aligned} \tag{3}$$

ε_j and σ_j are the highest derivative orders of input u_j and output y_j.

If the relative order of the system $\alpha = (\alpha_1, \alpha_2, \ldots, \alpha_q)^T$ is existent and the sum of relative order $\sum_{j=1}^{q} \alpha_j = n$ (n is the order of the system), the state equation of the system can be transformed into differential equation. Hence, when the relative order is equal to the original order, the generalized inverse of the system can be directly obtained as

$$u = \overline{\phi}\left(\overline{Y}, \overline{v}\right), \tag{4}$$

where

$$\begin{aligned}
\overline{Y} &= \left(y_1, y_1^{(1)}, \ldots, y_1^{(\alpha_1-1)}, \ldots, y_q, y_q^{(1)}, \ldots, y_q^{(\alpha_q-1)}\right)^T, \\
\overline{v} &= \left(\overline{v_1}, \overline{v_2}, \ldots, \overline{v_q}\right)^T, \\
\overline{v_j} &= a_{j0} y_j + a_{j1} y_j^{(1)} + \cdots + a_{j(\alpha_j-1)} y_j^{(\alpha_j-1)} + a_{j\alpha_j} y_j^{(\alpha_j)} \\
&\qquad (j = 1 \sim q).
\end{aligned} \tag{5}$$

By connecting the generalized inverse system in series, the coupled original system can be decoupled into several generalized pseudolinear composite subsystems and the transfer function of the subsystem can be expressed as

$$\begin{aligned}
\mathbf{G}(s) &= \operatorname{diag}\left(G_{11}(s), \ldots, G_{qq}(s)\right) \\
&= \operatorname{diag}\left(\frac{1}{a_{1\alpha_1} s^{\alpha_1} + a_{1(\alpha_1-1)} s^{(\alpha_1-1)} + \cdots + a_{11} s + a_{10}}, \right. \\
&\qquad \left. \ldots, \frac{1}{a_{q\alpha_q} s^{\alpha_q} + a_{q(\alpha_q-1)} s^{(\alpha_q-1)} + \cdots + a_{q1} s + a_{q0}}\right).
\end{aligned} \tag{6}$$

In practice, the structure of MIMO nonlinear system is complicated and it is hard to obtain its accurate mathematical model. Even if the existence of generalized inverse system has been proved, the expression of generalized inverse of the system cannot be derived for various occasions. To solve this problem, an ANN is used to approximate the original system. This method is independent of the accurate mathematical model, thus enhancing the popularization of inverse system algorithm. The NNGI pseudolinear composite subsystem equivalent chart is shown in Figure 2.

3.2. NNGI Construction for Two-Motor Drive System. According to the mathematic model of two-motor drive system in (1), the existence of generalized inverse system will be proved. Variables are chosen as follows.

State variable is

$$x = \left[x_1, x_2, x_3\right]^T = \left[\omega_{r1}, \omega_{r2}, F\right]^T; \tag{7}$$

control variable is

$$u = \left[u_1, u_2\right]^T = \left[\omega_1, \omega_2\right]^T; \tag{8}$$

output variable is

$$y = \left[y_1, y_2\right]^T = \left[\omega_{r1}, F\right]^T. \tag{9}$$

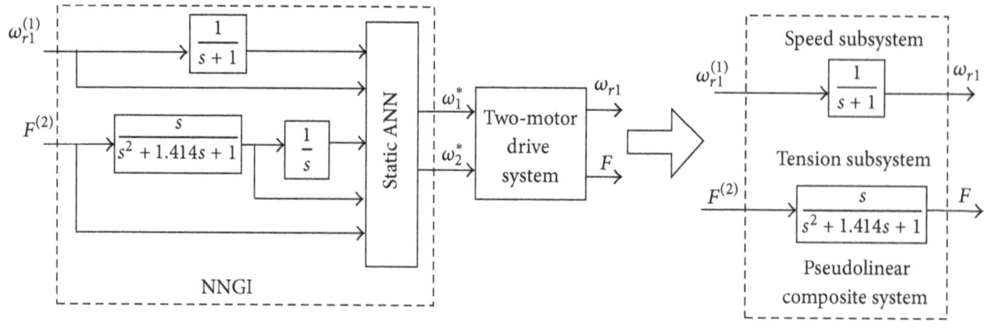

FIGURE 3: Pseudolinear composite system.

Then, the mathematic model of two-motor drive system can be expressed as

$$\dot{x} = f(x,u) = \begin{Bmatrix} \frac{n_{p1}}{J_1}\left[(u_1 - x_1)\frac{n_{p1}T_{r1}}{L_{r1}}\psi_{r1}^2 - (T_{L1} + r_1x_3)\right] \\ \frac{n_{p2}}{J_2}\left[(u_2 - x_2)\frac{n_{p2}T_{r2}}{L_{r2}}\psi_{r2}^2 - (T_{L2} - r_2x_3)\right] \\ \frac{K}{T}\left(\frac{1}{n_{p1}}r_1k_1x_1 - \frac{1}{n_{p2}}r_2k_2x_2\right) - \frac{x_3}{T} \end{Bmatrix}. \quad (10)$$

According to the generalized inverse theory, the corresponding Jacobi matrix can be described as

$$J(x,u) = \begin{bmatrix} \frac{\partial y_1^{(1)}}{\partial u_1} & \frac{\partial y_1^{(1)}}{\partial u_2} \\ \frac{\partial y_2^{(2)}}{\partial u_1} & \frac{\partial y_2^{(2)}}{\partial u_2} \end{bmatrix} = \begin{bmatrix} \frac{n_{p1}^2T_{r1}\psi_{r1}^2}{J_1L_{r1}} & 0 \\ \frac{Kr_1k_1n_{p1}T_{r1}\psi_{r1}^2}{TJ_1L_{r1}} & -\frac{Kr_2k_2n_{p2}T_{r2}\psi_{r2}^2}{TJ_2L_{r2}} \end{bmatrix}, \quad (11)$$

$$\text{Det}(J(x,u)) = \frac{Kr_2k_2n_{p1}^2n_{p2}T_{r1}T_{r2}\psi_{r1}^2\psi_{r2}^2}{TJ_1J_2L_{r1}L_{r2}}.$$

When $\psi_{r1} \neq 0$ and $\psi_{r2} \neq 0$, the Jacobi matrix is nonsingular. The relative order of the system is $\alpha = (\alpha_1, \alpha_2) = (1,2)$, $\alpha_1 + \alpha_2 = 3$, and it is equal to the order of the system. In addition, the original order of the system is $n_e = (n_{e1}, n_{e2}) = (1,2)$, $n_{e1} + n_{e2} = 3$, and it is the same as relative order. So, the inverse of the system is existent and can be expressed as

$$u = \overline{\phi}(\{y_1, \dot{y}_1, y_2, \dot{y}_2, \ddot{y}_2\}, \overline{v}), \quad (12)$$

where

$$\overline{v} = (\overline{v_1}, \overline{v_2})^T,$$
$$\overline{v_1} = a_{10}y_1 + a_{11}y_1^{(1)}, \quad (13)$$
$$\overline{v_2} = a_{20}y_2 + a_{21}y_2^{(1)} + a_{22}y_2^{(2)}.$$

In (12), to keep the stability of the open-loop system, $a_{10} = 1$, $a_{11} = 1$; $a_{20} = 1$, $a_{21} = 1.414$, and $a_{22} = 1$. Then, the transfer function of the subsystem is

$$G(s) = \text{diag}(G_{11}, G_{22}) = \text{diag}\left(\frac{1}{s+1}, \frac{1}{s^2 + 1.414s + 1}\right). \quad (14)$$

By connecting the generalized inverse system in series, the original system can be decoupled and simplified into a first-order subsystem and a second-order subsystem. As a result, the composite pseudolinear system is derived and shown in Figure 3.

Simultaneously, by adjusting the parameters a_{10}, a_{11}, a_{20}, a_{21}, and a_{22}, the zeros and poles of pseudolinear subsystems can be developed exactly. Two generalized pseudolinear subsystems have neither right half-plane zero, and then the original system will obtain open-loop stabilization after it is in series with the NNGI system. The NNGI open-loop control diagram is shown in Figure 4.

3.3. Profibus-DP Network Delay. Profibus communication uses the main token polling mechanism; each site on the bus connects with another in the bus way. Master station is composed of control center such as computer and PLC, while slave station consists of actuators such as inverter and sensors. Each site on the Profibus has its own logical address; a loop called logical loop is formed between the master stations. A token generated during the operation of the frame will poll along the logical link between the master stations. One master station is allowed to poll along other slave stations only when a token reaches this master station, and then the data transmission will be finished. While the process from sending task information of site to receiving the task information by

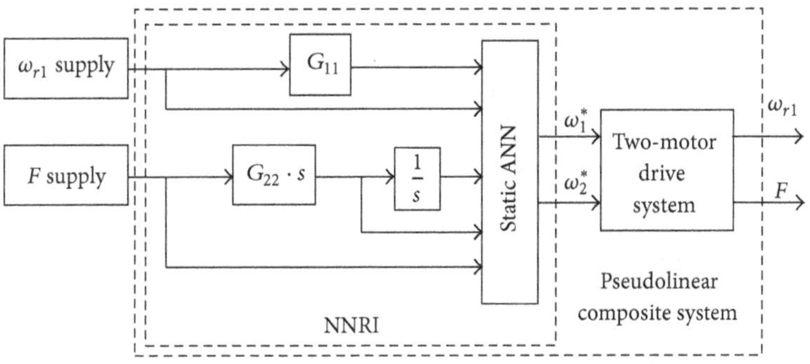

FIGURE 4: NNGI open-loop control diagram.

target is too time intensive, the Profibus-DP network delay is generated.

Profibus-DP network delay is indeterminate in practice. However the maximum delay can be calculated and it can be a reference for the delay of ideal internal model e^{-ms} [23, 24]. The approximate expression can be written as

$$R_h = B + \text{int}\left[\frac{n_h}{(n+1)}\right] \cdot (T_{TR} + T_{chm} + \tau) + \xi_h, \quad (15)$$

where B stands for the longest initial blocking, n_h is the number of high priority request messages added to the outgoing queue at critical instant, T_{TR} is the target token rotation time, T_{chm} is the maximum time of executing a high priority message transmission, τ is the token latency, $n+1$ is the number of high priority messages which are transmitted during the maximum token holding time, and ξ_h is the time between the finish of the last maximum token holding time and the end of target message transmission. There are three possible cases and the expression can be written as

$$\xi_h = \begin{cases} -\tau, & n_h = \text{int}\left[\frac{n_h}{n+1}\right] \cdot (n+1), \\ T_{chm}, & n_h = \text{int}\left[\frac{n_h}{n+1}\right] \cdot (n+1) + 1, \\ \left(n_h - \text{int}\left[\frac{n_h}{n+1}\right] \cdot (n+1)\right) \cdot T_{chm} + \tau, & n_h > \text{int}\left[\frac{n_h}{n+1}\right] \cdot (n+1) + 1. \end{cases} \quad (16)$$

3.4. SIM Control. According to the previous discussions, two-motor drive system is a time-varying delay system. In order to settle the time delay issue, a SIM control is introduced into the controller design. The SIM control incorporates the advantages of Smith estimate control and internal model control. This control strategy is a delay compensation method, which is to separate the delay part of controlled object and connect in series with the system transfer function. It is an effective way to inhibit the effect of time delay. Moreover, SIM control can also restrain the influence of unmeasured disturbances and improve system robustness.

The structure of the SIM control is shown in Figure 5, in which $G_P(s)$ and $G_m(s)$ are the actual object model and the prediction internal model, respectively. e^{-Ps} and e^{-ms} are the related delay of controlled object and prediction internal model, respectively. $G_c(s)$ is the controller. $D(s)$ is the external disturbance. $R(s)$ and $Y(s)$ are the input and the output of the system, respectively.

3.4.1. Delay Compensation Analysis. According to Figure 5, the output can be written as

$$\begin{aligned} Y(s) &= \frac{G_c(s) G_P(s) e^{-Ps}}{1 + G_c(s) G_m(s) + G_c(s)\left[G_P(s) e^{-Ps} - G_m(s) e^{-ms}\right]} \cdot R(s) \\ &+ \frac{1 + G_c(s) G_m(s) - G_c(s) G_m(s) e^{-ms}}{1 + G_c(s) G_m(s) + G_c(s)\left[G_P(s) e^{-Ps} - G_m(s) e^{-ms}\right]} \cdot D(s). \end{aligned} \quad (17)$$

The closed-loop transfer function of control system is

$$\begin{aligned} G(s) &= \frac{Y(s)}{R(s)} \\ &= \frac{G_c(s) G_P(s) e^{-Ps}}{1 + G_c(s) G_m(s) + G_c(s)\left[G_P(s) e^{-Ps} - G_m(s) e^{-ms}\right]}. \end{aligned} \quad (18)$$

When the prediction internal model is accurate, $G_m(s)e^{-ms}$ is equal to $G_P(s)e^{-Ps}$, and the transfer function can be written as

$$G(s) = \frac{G_c(s) G_P(s)}{1 + G_c(s) G_P(s)} e^{-Ps}. \quad (19)$$

Hence, there is no delay part in the denominator of closed-loop transfer function and the delay is totally compensated.

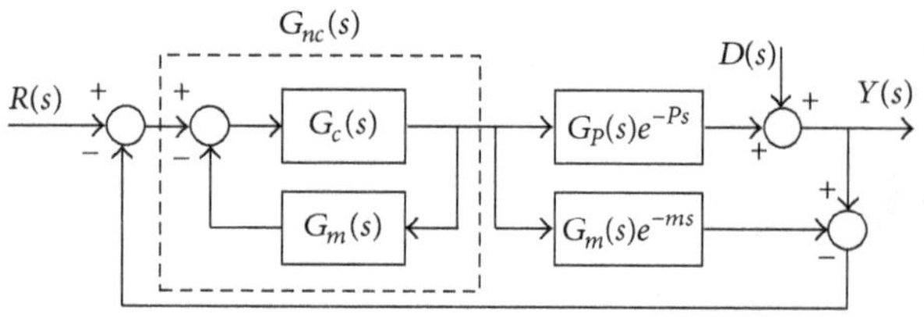

FIGURE 5: SIM control diagram.

3.4.2. *Fixed Value Tracking Ability.* According to Figure 5, when the external disturbance $D(s)$ is zero, then the error of closed-loop system can be written as

$$E(s) = R(s) - Y(s) = \frac{1 - G_{nc}(s) G_m(s) e^{-ms}}{1 + G_{nc}(s)\left(G_P(s) e^{-Ps} - G_m(s) e^{-ms}\right)} R(s). \quad (20)$$

In consideration of modeling error of actual system, the controlled object can be described with multiplicative uncertainty:

$$G_P(s) e^{-Ps} = G_m(s) e^{-ms}\left(1 + l_m(s)\right), \quad (21)$$

where $l_m(s)$ represents the uncertainty and it is bounded.

The structure in dashed box is represented by $G_{nc}(s)$. A low pass filter is added to improve the capacity of resisting disturbance. The filter is represented as $F(s)$, and then $G_{nc}(s)$ can be written as

$$G_{nc}(s) = G_m^{-1}(s) F(s). \quad (22)$$

Therefore, the transfer function of low pass filter is

$$F(s) = \frac{1}{(\lambda s + 1)^n}, \quad (23)$$

where λ describes filter parameter and n stands for the order of filter.

Then, the error of closed-loop system can be simplified as

$$E(s) = \frac{(\lambda s + 1)^n - e^{-ms}}{(\lambda s + 1)^n + l_m(s) e^{-ms}} R(s). \quad (24)$$

The limit of $l_m(s)$ can be expressed as

$$\lim_{s\to 0} l_m(s) = C, \quad (25)$$

where C is a constant.

When the input signal is step signal,

$$R(s) = \frac{R_0}{s}, \quad (26)$$

then, the limit of $E(s)$ is

$$e(\infty) = \lim_{s\to 0} sE(s) = \lim_{s\to 0} s\left(\frac{(\lambda s + 1)^n - e^{-ms}}{(\lambda s + 1)^n + l_m(s) e^{-ms}} \cdot \frac{R_0}{s}\right) = 0. \quad (27)$$

3.4.3. *Anti-Interference Ability.* According to Figure 5, when the input $R(s)$ is zero, then the error of closed-loop system can be written as

$$E(s) = R(s) - Y(s) = \frac{1 - G_{nc}(s) G_m(s) e^{-ms}}{1 + G_{nc}(s)\left(G_P(s) e^{-Ps} - G_m(s) e^{-ms}\right)} (-D(s)). \quad (28)$$

Similar to the analysis of fixed value tracking ability, the error of closed-loop system can be simplified as

$$E(s) = \frac{(\lambda s + 1)^n - e^{-ms}}{(\lambda s + 1)^n + l_m(s) e^{-ms}} (-D(s)). \quad (29)$$

When the external disturbance is step signal,

$$D(s) = \frac{D_0}{s}, \quad (30)$$

then, the limit of $E(s)$ is

$$e(\infty) = \lim_{s\to 0} sE(s) = \lim_{s\to 0} s\left(-\frac{(\lambda s + 1)^n - e^{-ms}}{(\lambda s + 1)^n + l_m(s) e^{-ms}} \cdot \frac{D_0}{s}\right) = 0. \quad (31)$$

In conclusion, the SIM control structure can satisfactorily solve delay problem. What is more, this control structure also has good input signal tracking ability and anti-interference ability.

3.5. *NNGI-SIM Control.* As mentioned above, SIM control can remove the delay of forward network from the closed-loop and set it as a gain block before the output. As a result, the delay on the return network path can be totally eliminated. Meanwhile, SIM control can also enhance robustness and anti-interference ability of the system with the internal model control structure. However, the accuracy of prediction model $G_m(s)$ is crucial. When there is a mismatch between prediction model and actual model, the control quality will be significantly deteriorated. This problem can be solved by NNGI system. After connecting with NNGI system, the original system is decoupled into two subsystems, namely, a first-order speed one and a second-order tension one. The accurate mathematical model of two subsystems can be obtained after pole-zero allocation of the pseudolinear subsystems, and then the internal model can be predicted. Then, the mismatch condition will be avoided. The diagram of NNGI-SIM control strategy of the two-motor drive system is shown in Figure 6.

In Figure 5, the function of the construction in dashed box can be expressed as

$$G_{nc}(s) = \frac{G_c(s)}{1 + G_c(s) G_m(s)}. \quad (32)$$

According to (22) and (32), $G_c(s)$ can be calculated as

$$G_c(s) = \frac{G_m^{-1}(s) F(s)}{1 - F(s)}. \quad (33)$$

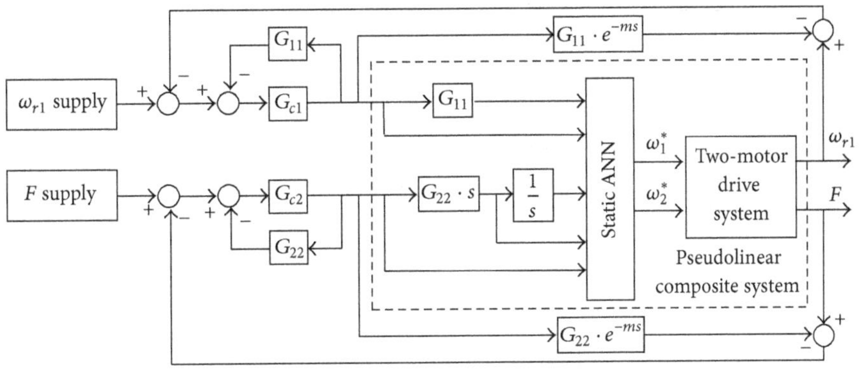

FIGURE 6: NNGI-SIM control diagram.

In a two-motor NNGI system, the original system is separated into two subsystems. For the speed subsystem, the transfer function is $G_{11} = 1/(s+1)$ and the order of filter n is one. The expression of $G_{c1}(s)$ can be transformed into the structure of classical PI controller, in which K_{P1} is $1/\lambda_1$ and T_{I1} is 1. According to (23) and (33), $G_{c1}(s)$ can be written as

$$G_{c1}(s) = \frac{G_{11}^{-1}(s) F_1(s)}{1 - F_1(s)} = \frac{1}{\lambda_1}\left(1 + \frac{1}{s}\right). \tag{34}$$

For the tension subsystem, according to (14), the transfer function is $G_{22} = 1/(s^2 + 1.414s + 1)$ and the order of filter n is one; then $G_{c2}(s)$ can be written as

$$G_{c2}(s) = \frac{G_{22}^{-1}(s) F_2(s)}{1 - F_2(s)} = \frac{s^2 + 1.414s + 1}{\lambda_2 s}. \tag{35}$$

The expression of $G_{c2}(s)$ can be transformed into the structure of classical PID controller:

$$G_{c2}(s) = \frac{1.414}{\lambda_2}\left(1 + \frac{1}{1.414s} + \frac{1}{1.414}s\right), \tag{36}$$

where K_{P2} is $1.414/\lambda_2$, T_{I2} is 1.414, and T_{D2} is 1/1.414.

4. Simulation

4.1. Procedure of Simulation. In order to verify the decoupling ability of the proposed NNGI system, feed-forward neural network is adopted to approximate inversion system. The design procedure of NNGI pseudolinear composite system is briefly given as follows.

4.1.1. Sample Collection. Since the parameters of the original system model are unknown or variable, the selected signals should stimulate the dynamic and static characters of original system sufficiently. So, the square wave signal is chosen to verify dynamic performance of the proposed motor drive in each frequency band. Differentiators in the NNGI system can be obtained by using five-point numerical derivative algorithm, in which high computing accuracy is offered. The speed value ranges from 100 r/min to 600 r/min and tension value ranges from 100 N to 600 N.

4.1.2. Neural Network Training. According to (12), $\{y_1, y_1 + \dot{y}_1, y_2, \dot{y}_2, \ddot{y}_2 + 1.414\dot{y}_2 + y_2\}$ are chosen as the input and u is chosen as the output to train the NNGI system. Before training neural network, the sampled data should be normalized and break off both ends to improve convergence precision. Furthermore, the neural network adopts 5-15-2 3-layer structure; the activation functions of hidden layer and output layer choose tansig and pureline functions, respectively. The weight and the threshold of the neural network should be adjusted by using the Levenberg-Marquardt algorithm until convergence precision attains the expectant effect. The output of neural network should be antinormalized back to the original range. Then, the training of NNGI system is over.

4.1.3. Integration of Pseudolinear Composite System. After putting the NNGI system in front of the original system, the pseudolinear composite system is constructed. A strong coupling and nonlinear system is separated into two pseudolinear composite subsystems. By adopting the SIM control structure, two independent control loops are formed to achieve the aim of decoupling control.

A MIMO nonlinear system is decoupled by NNGI method, and then the accurate model of generalized pseudolinear subsystem can be obtained, in which the controllers' parameters depend on the accurate mathematical model. The robustness of the controllers is mainly determined by the filters, and the first-order filter is considered as the best controller. After a lot of simulation experiments, the filter parameters λ_1 and λ_2 are chosen to be 0.2 and 1.2, respectively, and then the filters can meet the system requirements. Hence, the controller parameters of two pseudolinear subsystems can be optimized. According to the analysis of previous discussions, the expressions of G_{c1} and G_{c2} can be written as the classical PID controller. According to (34) and (36), K_{P1} is 5 and T_{I1} is 1 in the speed subsystem, K_{P2} is 1.2, T_{I2} is 0.8, and T_{D2} is 0.8 in the tension subsystem. The simulation model is developed according to Figure 6. Based on the NNGI-SIM control strategy, the coupling issue can be settled and the delay of Profibus-DP network can also be compensated.

4.2. Comparative Discussions. In order to evaluate the proposed NNGI-SIM control system, ω_{r1} is set to track square

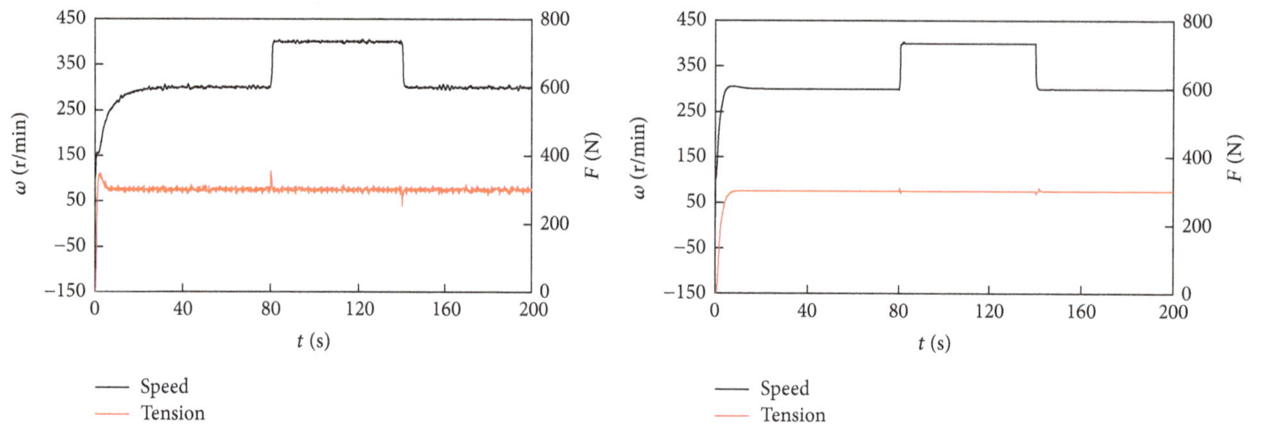

(a) PID control strategy

(b) NNGI-SIM control strategy

FIGURE 7: Simulated responses when ω_{r1} suddenly changes.

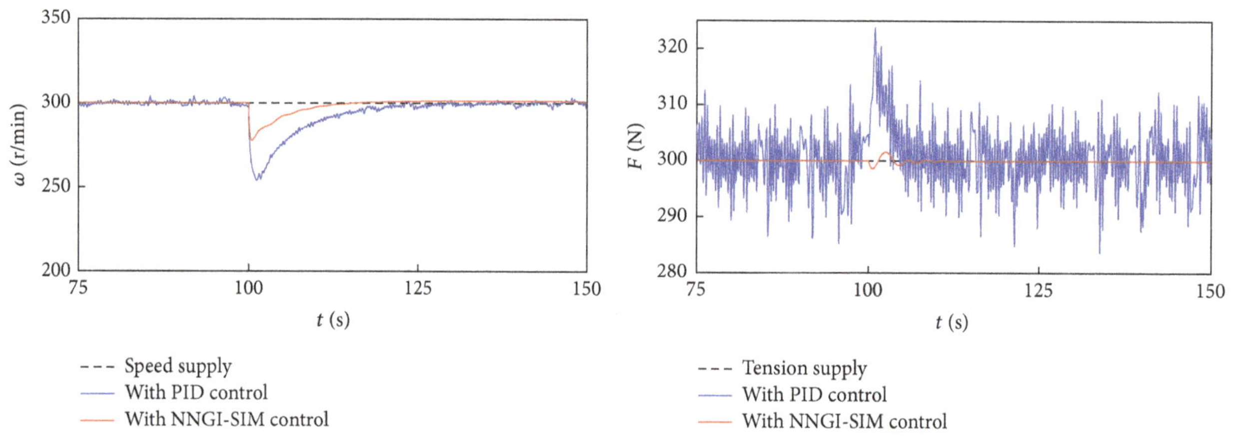

(a) Speed responses

(b) Tension responses

FIGURE 8: Simulated responses for sudden increased load impact.

wave response and F is set to constant 300 N. ω_{r1} increases from 200 r/min to 250 r/min at 80 s and decreases from 250 r/min to 200 r/min at 140 s. As shown in Figure 7, due to the effect of system delay, the responses of traditional PID control have high frequency shiver during the whole simulation. Large fluctuations of the tension responses occur at the constant of speed sudden changing. Therefore, it cannot effectively decouple speed and tension. Moreover, the response speed is slow and transient time is too long. By contrast, with the NNGI-SIM control structure, the shiver problem has been solved. Furthermore, the tension fluctuation caused by the sudden change of speed is restrained to a great extent. The disturbance caused by speed impact is significantly reduced. Hence, this two-motor system is effectively coupled; the speed of motors and tension will not be influenced by each other.

In order to evaluate the restraining disturbance ability of the proposed control strategy, ω_{r1} and F are set to be constants of 300 r/min and 300 N, respectively. Figure 8 is the sudden load increase from 2 Nm to 5 Nm at 100 s, while Figure 9 is the sudden load decrease from 5 Nm to 2 Nm at 100 s. At the PID control operation, speed and tension have large fluctuations and both of them are significantly affected by load jump. In contrast, at the proposed control operation, the fluctuation of speed is smoothed and tension is immune to this load change, indicating that the proposed system has a strong antidisturbance ability. Furthermore, the fluctuation issue caused by transmission delay has been settled completely. Since SIM control structure can separate the delay part of controlled object and connect in series with the system transfer function, the operation of system is immune to delay. So, the tension responses by NNGI-SIM control are relatively smooth and the delay of SIM control structure is obviously compensated.

5. Experimental Verification

The experimental platform of multimotor drive system is designed as shown in Figure 10, by which the number of the test motors can be up to three. The whole system includes industrial computer, Siemens S7-300 PLC, Siemens micro-master vector inverters, 2.2 kW three-phase asynchronous

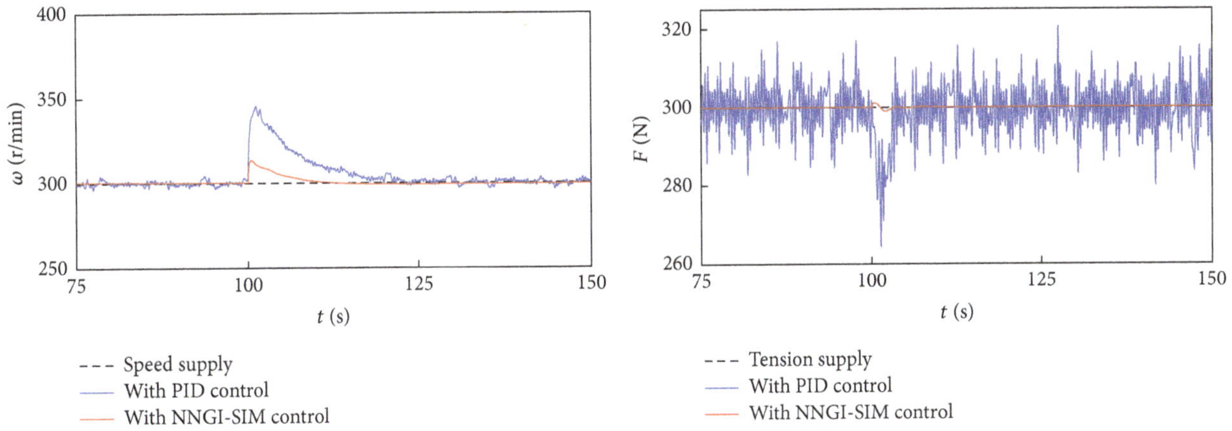

(a) Speed responses

(b) Tension responses

FIGURE 9: Simulated responses for sudden decreased load impact.

FIGURE 10: Experimental platform.

motors, photoelectric encoder, magnetic powder brakes, and tension sensor.

In the experimental platform, S7-300 PLC is the main controller which sends the frequency signal to the inverters. In addition, the inverters drive motors in vector control mode and the motors connect conveyor belt with roller.

The PLC controls two inverters through Profibus-DP communication which is the main way of message transmission delay. According to (15) and (16), this study sets the data transfer rate of Profibus-DP network as 1.5 Mbps, and then the data bit period is equal to 0.667 us. Besides, n_h is chosen to be 20 and T_{TR} is fixed to 8 ms, and then B is equal to 2.18 ms, T_{chm} is equal to 1.27 ms, τ is approximately equal to 0.04 ms, $n$is equal to 6, and ξ_h is equal to 7.66 ms in the third case.

Hence, according to (15), the maximum delay of Profibus-DP network is

$$R_h = B + \mathrm{int}\left[\frac{n_h}{n+1}\right] \cdot \left(T_{\mathrm{TR}} + T_{\mathrm{chm}} + \tau\right) + \xi_h = 28.46\,\mathrm{ms}. \qquad (37)$$

The hardware experiment condition is similar to simulation one, of which the procedure can be divided into three parts, namely, sample collection, training offline, and control implementation. However, differently, the experimental motor will be affected by external disturbances and mechanical vibration, hence resulting in the fluctuations of response speed and tension. To eliminate these harmful effects, antipulse interference average filter is applied to filter the real-time sampled signals. This filter method collects N sets of date continually, removes the maximum and the minimum, and takes the average of the remaining date as the output of the filter. The feedback signal after filtering has no significant fluctuation and is smooth. It is conducive to improve curve-fitting condition and convergence precision in neural network training.

This work implements the feed-forward neural networks with instruction list of Siemens PLC. After transferring the trained weights and thresholds to the corresponding register units, the NNGI system is accomplished in S7-300 PLC platform. Then, by adopting the SIM control structure to deal with the problem of Profibus-DP network delay, the whole closed-loop system is formed.

In order to verify the decoupling effectiveness, the tension supply of 300 N is chosen and the speed supply is set to suddenly change from 300 r/min to 400 r/min. The responses of the experiment system with PID control and with NNGI-SIM control are compared in Figure 11. As shown in Figure 11(a), the tension response fluctuates dramatically when the speed suddenly changes. Thus, the strong coupling problem cannot be solved by the traditional PID control. Moreover, the transmission delay leads to the fluctuation of responses, which agrees with the simulated one. Comparatively, in Figure 11(b), with the sudden change of speed from 300 r/min to 400 r/min at 80 s, the jitter has been reduced, so the sudden change of speed has less influence on tension response. It is obvious that the tension and the speed are successfully decoupled by using the proposed control strategy. Hence, the NNGI-SIM control is superior to PID control in the two-motor drive system with network delay.

The measured responses of speed and tension under sudden load impact are shown in Figures 12 and 13. In Figure 12, the motor starts at no load and then increases to 3 Nm suddenly. By contrast, in Figure 13, the load decreases from 3 Nm to 0 Nm.

As shown in Figures 12 and 13, the proposed control strategy reduces the impact on speed and tension caused by the sudden application of load. The fluctuation of tension is mitigated effectively under NNGI-SIM control. However,

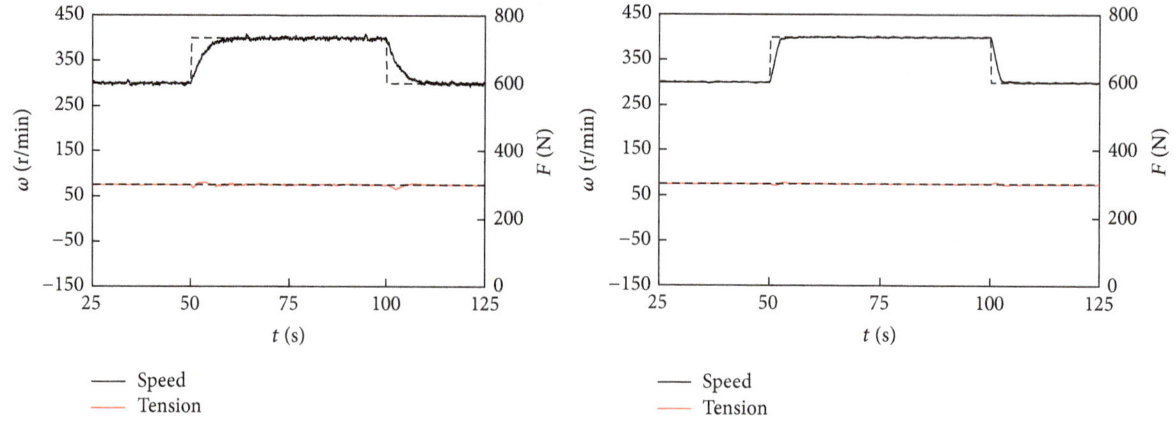

(a) PID control strategy

(b) NNGI-SIM control strategy

FIGURE 11: Experimental results when ω_{r1} suddenly changes.

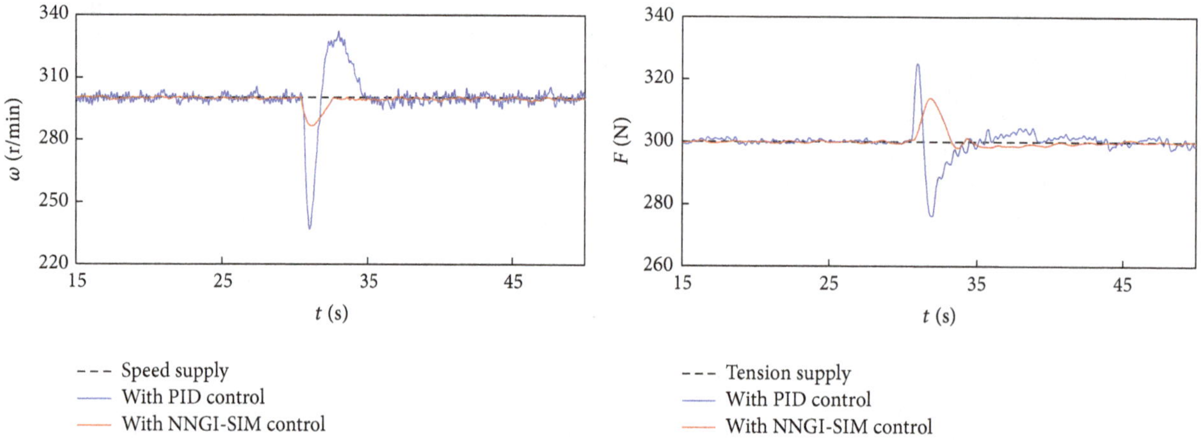

(a) Speed responses

(b) Tension responses

FIGURE 12: Experimental results for sudden increased load impact.

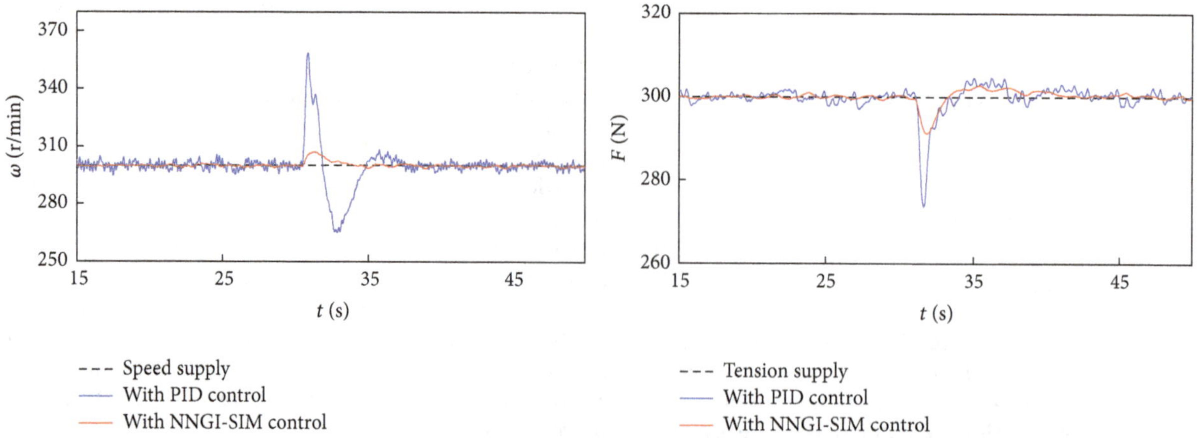

(a) Speed responses

(b) Tension responses

FIGURE 13: Experimental results for sudden decreased load impact.

the improvement of speed response is relatively less because the magnetic powder brake has direct effects on motor shaft. Overall, the NNGI-SIM control strategy achieves the good dynamic and static performances, and it has strong robustness to the load disturbances.

6. Conclusion

In this paper, a new control strategy has been proposed to solve the Profibus-DP network delay problem and decouple the speed and the tension of the two-motor drive system by incorporating NNGI and SIM control. The NNGI strategy has been used to transform the MIMO two-motor drive system to several single-input single-output systems, and then the speed and tension are decoupled validly. By adopting the SIM control structure to compose a closed-loop system, the delay issue has been compensated. The designed system can provide good starting and tracking capabilities. The utilities of the proposed control strategy have been verified by both the simulation and the experiments at various operations. The proposed control strategy has a great application prospect in the field of multimotor control system.

Conflict of Interests

The authors declare that there is no conflict of interests regarding the publication of this paper.

Acknowledgments

This work was supported by the National Natural Science Foundation of China (51277194 and 61273154), by the Key Project of Natural Science Foundation of Jiangsu Higher Education Institutions (15KJA470002), by the Research Fund for 333 Project of Jiangsu Province (BRA2015302), and by the Priority Academic Program Development of Jiangsu Higher Education Institutions.

References

[1] R. Cao, C. Mi, and M. Cheng, "Quantitative comparison of flux-switching permanent-magnet motors with interior permanent magnet motor for EV, HEV, and PHEV applications," *IEEE Transactions on Magnetics*, vol. 48, no. 8, pp. 2374–2384, 2012.

[2] G. Liu, J. Yang, W. Zhao, J. Ji, Q. Chen, and W. Gong, "Design and analysis of a new fault-tolerant permanent-magnet vernier machine for electric vehicles," *IEEE Transactions on Magnetics*, vol. 48, no. 11, pp. 4176–4179, 2012.

[3] Y. Du, K. T. Chau, M. Cheng, Y. Wang, and J. Li, "A linear doubly-salient HTS machine for wave energy conversion," *IEEE Transactions on Applied Superconductivity*, vol. 21, no. 3, pp. 1109–1113, 2011.

[4] W. Zhao, M. Cheng, J. Ji, R. Cao, Y. Du, and F. Li, "Design and analysis of a new fault-tolerant linear permanent-magnet motor for maglev transportation applications," *IEEE Transactions on Applied Superconductivity*, vol. 22, no. 3, 2012.

[5] W.-B. Lin, C.-A. Chen, and H.-K. Chiang, "Design and implementation of a sliding mode controller using a Gaussian radial basis function neural network estimator for a synchronous reluctance motor speed drive," *Electric Power Components and Systems*, vol. 39, no. 6, pp. 548–562, 2011.

[6] R.-L. Lin, P.-Y. Yeh, and C.-H. Liu, "Positive feed-forward control scheme for distributed power conversion system with multiple voltage sources," *IEEE Transactions on Power Electronics*, vol. 27, no. 7, pp. 3186–3194, 2012.

[7] R. H. Du, Y. F. Wu, W. Chen, and Q. Chen, "Adaptive fuzzy speed control for permanent magnet synchronous motor servo systems," *Electric Power Components and Systems*, vol. 42, no. 8, pp. 798–807, 2014.

[8] G. Liu, L. Chen, W. Zhao, Y. Jiang, and L. Qu, "Internal model control of permanent magnet synchronous motor using support vector machine generalized inverse," *IEEE Transactions on Industrial Informatics*, vol. 9, no. 2, pp. 890–898, 2013.

[9] G. Liu, K. Yu, and W. Zhao, "Neural network based internal model decoupling control of three-motor drive system," *Electric Power Components and Systems*, vol. 40, no. 14, pp. 1621–1638, 2012.

[10] D. Yu, F. Liu, P.-Y. Lai, and A. Wu, "Nonlinear dynamic compensation of sensors using inverse-model-based neural network," *IEEE Transactions on Instrumentation and Measurement*, vol. 57, no. 10, pp. 2364–2376, 2008.

[11] Y. Zhang, D. Guo, and Z. Li, "Common nature of learning between back-propagation and hopfield-type neural networks for generalized matrix inversion with simplified models," *IEEE Transactions on Neural Networks and Learning Systems*, vol. 24, no. 4, pp. 579–592, 2013.

[12] L. Guo and L. Parsa, "Model reference adaptive control of five-phase IPM motors based on neural network," *IEEE Transactions on Industrial Electronics*, vol. 59, no. 3, pp. 1500–1508, 2012.

[13] C. B. Butt and M. A. Rahman, "Untrained artificial neuron-based speed control of interior permanent-magnet motor drives over extended operating speed range," *IEEE Transactions on Industry Applications*, vol. 49, no. 3, pp. 1146–1153, 2013.

[14] T.-C. Chen and C.-H. Yu, "Generalized regression neural-network-based modeling approach for traveling-wave ultrasonic motors," *Electric Power Components and Systems*, vol. 37, no. 6, pp. 645–657, 2009.

[15] S. A. R. Kashif and M. A. Saqib, "Sensorless control of a permanent magnet synchronous motor using artificial neural network based estimator—an application of the four-switch three-phase inverter," *Electric Power Components and Systems*, vol. 42, no. 1, pp. 1–12, 2014.

[16] S. Alcántara, A. Ibeas, J. A. Herrera, R. Vilanova, and C. Pedret, "Multi-model smith predictor based control of multivariable systems with uncertain bounded external delays," *IEEE Latin America Transactions*, vol. 7, no. 1, pp. 42–53, 2009.

[17] C.-L. Lai, P.-L. Hsu, and W. X. Zhao, "Design the remote control system with the time-delay estimator and the adaptive smith predictor," *IEEE Transactions on Industrial Informatics*, vol. 6, no. 1, pp. 73–80, 2010.

[18] D. E. Rivera, M. Morari, and S. Skogestad, "Internal model control: PID controller design," *Industrial & Engineering Chemistry Process Design and Development*, vol. 25, no. 1, pp. 252–265, 1986.

[19] C. Xia, Y. Yan, P. Song, and T. Shi, "Voltage disturbance rejection for matrix converter-based PMSM drive system using internal model control," *IEEE Transactions on Industrial Electronics*, vol. 59, no. 1, pp. 361–372, 2012.

[20] S. Li and H. Gu, "Fuzzy adaptive internal model control schemes for PMSM speed-regulation system," *IEEE Transactions on Industrial Informatics*, vol. 8, no. 4, pp. 767–779, 2012.

[21] X. Wei, K. M. Tsang, and W. L. Chan, "DC/DC buck converter using internal model control," *Electric Power Components and Systems*, vol. 37, no. 3, pp. 320–330, 2009.

[22] X. Z. Dai, D. He, T. Zhang, and K. Zhang, "ANN generalised inversion for the linearisation and decoupling control of nonlinear systems," *IEE Proceedings: Control Theory and Applications*, vol. 150, no. 3, pp. 267–277, 2003.

[23] S. Vitturi, "Stochastic model of the profibus DP cycle time," *IEE Proceedings—Science, Measurement and Technology*, vol. 151, no. 5, pp. 335–342, 2004.

[24] K. C. Lee, S. Lee, and M. H. Lee, "Remote fuzzy logic control of networked control system via Profibus-DP," *IEEE Transactions on Industrial Electronics*, vol. 50, no. 4, pp. 784–792, 2003.

20

Robust Fault Estimation for a Class of T-S Fuzzy Singular Systems with Time-Varying Delay via Improved Delay Partitioning Approach

Chao Sun,[1,2] FuLi Wang,[1,3] and XiQin He[2]

[1]*College of Information Science and Engineering, Northeastern University, Shenyang, Liaoning 110819, China*
[2]*College of Sciences, University of Science and Technology Liaoning, Anshan, Liaoning 114051, China*
[3]*State Key Laboratory of Synthetical Automation of Process Industries, Northeastern University, Shenyang, Liaoning 110819, China*

Correspondence should be addressed to Chao Sun; chao_sun@163.com

Academic Editor: Kalyana C. Veluvolu

The problem of delay-dependent robust fault estimation for a class of Takagi-Sugeno (T-S) fuzzy singular systems is investigated. By decomposing the delay interval into two unequal subintervals and with a new and tighter integral inequality transformation, an improved delay-dependent stability criterion is given in terms of linear matrix inequalities (LMIs) to guarantee that the fuzzy singular system with time-varying delay is regular, impulse-free, and stable firstly. Then, based on this criterion, by considering the system fault as an auxiliary disturbance vector and constructing an appropriate fuzzy augmented system, a fault estimation observer is designed to ensure that the error dynamic system is regular, impulse-free, and robustly stable with a prescribed H_∞ performance satisfied for all actuator and sensor faults simultaneously, and the obtained fault estimates can practically better depict the size and shape of the faults. Finally, numerical examples are given to show the effectiveness of the proposed approach.

1. Introduction

The demand for increased productivity leads to more challenging operating conditions for many modern engineering systems. The issue of fault detection and isolation (FDI) algorithms in dynamic systems and their applications to a wide range of industrial processes have been an active research area over the past two decades, as can be seen, in survey papers [1] for linear systems, [2] for multimodels representation and [3, 4] for nonlinear systems. By using FDI procedures, the reliability can be achieved by fault-tolerant control, which relies on early detection and isolation of faults. So FDI have become a popular topic and received considerable attention. However, it is generally difficult, in practice engineering, to obtain the exact information of the size of system fault from an FDI strategy only because of the existence of model uncertainties, time delays, and disturbances [5]. As pointed out in [6], accurate and timely fault estimation is an important antecedent for satisfactory control reconfiguration. Therefore, the problem of fault estimation has stirred renewed research interest, and a variety of fault estimation approaches have been developed in the literatures; see, for example, [7–11] and the references therein.

On the other hand, singular systems have been extensively studied in the past years due to the fact that singular systems better describe physical systems than state-space ones, especially the T-S fuzzy singular systems, because they can combine the flexibility of fuzzy logic theory and fruitful linear singular system theory into a unified framework to approximate complex nonlinear singular systems. In fact, singular systems can be found in electrical circuits, economic systems, moving robots, and many other systems. Recently, many results about fault estimation have been reported on analysis and design of singular systems. For instance, by using online learning methodology, [12] proposed a fault estimation method for continuous-time nonlinear singular systems. Reference [7] used generalized unknown input observer to deal with the robust fault detection problem for linear singular systems. Reference [13] deals with actuator fault estimation for a class of discrete-time linear parameter-varying singular systems. Reference [14] designed a robust fault detection filter for a class of nonlinear singular systems described by linear parameter-varying form with global Lipschitz term. However, it is known that time delays are frequently encountered in various engineering and communication systems, and a time delay in a dynamical system is often a primary source of instability and performance degradation. Therefore, it is important to develop fault estimation methods for time delay singular systems. But the fault diagnosis for singular systems with time-varying delay has not been well investigated yet [7, 12–14]. More recently, [15] proposed a fault detection, isolation, and estimation scheme via unknown input proportional integral observers for linear descriptor systems. Reference [8] investigated fault detection for discrete-time switched singular systems with time-varying delays. It should be noticed that [8] deals with discrete-time switched singular systems while this paper focuses on the fuzzy continuous-time case. In [16], discrete-time T-S systems with sensor faults are first formulated as a descriptor representation, and then a fault detection filter is designed based on the obtained descriptor system. However, this paper studies fault detection for regular systems by using the technique of descriptor systems. In [17], the author proposed a k-step fault estimation method for T-S fuzzy time delay system that only deals with regular systems. Moreover, our paper considers how to estimate the actuator and sensor faults simultaneously while attenuating the influence of the disturbance noise, which is not considered in [17]. It is known that singular system representation is a generalization of the regular system. Therefore, the proposed method is more general than that in [16, 17].

The aim of this paper is to develop a robust fault estimation method for a class of T-S fuzzy singular systems with time delays. The basic idea is to construct an augmented system by taking the fault as auxiliary disturbance vector and then design a fault estimation observer based on this augmented system. The main contribution of the proposed method lies in the following aspects. First, without ignoring any useful terms in the Lyapunov-Krasovskii functional (LKF), by decomposing the delay interval into two unequal subintervals, a new LKF is constructed; then the free weighting matrices approach is introduced to develop a new delay-dependent stability criteria, which ensure that the considered system is regular, impulse-free, and stable. Compared with some existing results, the approach to be proposed in this paper can be expected to give better results. Second, a new robust fault estimation observer with a novel structure is proposed for T-S fuzzy singular systems with time delays and actuator and sensor faults simultaneously, which is the main contribution of this paper. The proposed observer can be designed by solving a set of linear matrix inequalities and to attenuate the effect of unknown disturbance, fault variation on fault estimation. The effectiveness of the method is illustrated by some numerical examples.

The rest of this paper is organized as follows. The system description and preliminaries are presented in Section 2. Section 3 presents the main results on new stability criteria of fuzzy singular systems with time-varying delays and robust fault estimation observer design scheme. In Section 4, simulation results of numerical examples are presented to demonstrate the effectiveness and merits of the proposed methods. Finally, Section 5 concludes the paper.

Notations. Throughout the paper, $\mathbb{R}^n$ denotes the n-dimensional real Euclidean space; I denotes the identity matrix; the superscripts T and -1 stand for the matrix transpose and inverse, respectively; notation $X > 0 (X \geq 0)$ means that matrix X is real symmetric positive definite (positive semidefinite); $\|\cdot\|$ is the spectral norm. If not explicitly stated, all matrices are assumed to have compatible dimensions for algebraic operations. The symbol "$*$" stands for matrix block induced by symmetry; $\text{sym}(X)$ stands for $X + X^T$.

2. System Description and Preliminaries

Consider a nonlinear singular system which can be represented by the following extended T-S fuzzy time delay model with external disturbance, actuator and sensor faults, simultaneously.

Plant Rule i. If $\xi_1(t)$ is M_{i1} and ... and $\xi_p(t)$ is M_{ip}, then

$$\begin{aligned} E\dot{x}(t) &= A_i x(t) + A_{\tau i} x(t-\tau(t)) + B_i u(t) + B_{fi} f(t) \\ &\quad + B_{di} d(t), \\ y(t) &= C_i x(t) + C_{\tau i} x(t-\tau(t)) + D_i u(t) + D_{fi} f(t) \\ &\quad + D_{di} d(t), \\ x(t) &= \phi_i(t), \quad \forall t \in [-\tau_2, 0], \ i = 1, 2, \ldots, r, \end{aligned} \tag{1}$$

where $x(t) \in \mathbb{R}^n$ is the state vector, $u(t) \in \mathbb{R}^q$ denotes the input vector, and $y(t) \in \mathbb{R}^l$ stands for the system output vector. $d(t) \in \mathbb{R}^m$ is the exogenous disturbance input that belongs to $L_2[0, \infty)$; $f(t) \in \mathbb{R}^q$ represents the possible fault. The matrix $E \in \mathbb{R}^{n \times n}$ is a constant matrix, which may be singular; that is, $\text{rank}(E) = g \leq n$. A_i, $A_{\tau i}$, B_i, B_{fi}, B_{di}, C_i, $C_{\tau i}$, D_i, D_{fi}, and D_{di} are constant real matrices of appropriate dimensions. It is assumed that the pairs (E, A_i, C_i) are of full column rank, where $i = 1, 2, \ldots, r$. $\xi_1(t), \ldots, \xi_p(t)$ are the premise variables, M_{ij} $(i = 1, 2, \ldots, r,\ j = 1, 2, \ldots, p)$ are fuzzy sets, and $\phi_i(t)$ is a vector-valued initial continuous function defined on the interval $[-\tau_2, 0]$. In this paper, it is also assumed that the premise variables do not depend on the input variables $u(t)$; $\tau(t)$ is the time-varying delay and satisfies

$$\begin{aligned} \tau_1 &\leq \tau(t) \leq \tau_2, \\ 0 &\leq \dot{\tau}(t) \leq \tau_D. \end{aligned} \tag{2}$$

Then, by fuzzy blending, the overall fuzzy singular system model is given by

$$\begin{aligned} E\dot{x}(t) &= \sum_{i=1}^{r} \mu_i(\xi(t)) \{A_i x(t) + A_{\tau i} x(t-\tau(t)) \\ &\quad + B_i u(t) + B_{fi} f(t) + B_{di} d(t)\}, \end{aligned}$$

$$y(t) = \sum_{i=1}^{r} \mu_i(\xi(t)) \{C_i x(t) + C_{\tau i} x(t-\tau(t)) + D_i u(t) + D_{fi} f(t) + D_{di} d(t)\},$$
$$x(t) = \sum_{i=1}^{r} \mu_i(\xi(t)) \phi_i(t), \quad \forall t \in [-\tau_2, 0], \tag{3}$$

where the fuzzy basis functions are given by

$$\mu_i(\xi(t)) = \frac{\beta_i(\xi(t))}{\sum_{j=1}^{r} \beta_j(\xi(t))}, \quad \beta_i(\xi(t)) = \prod_{i=1}^{p} M_{ij}(\xi(t)) \tag{4}$$

where $M_{ij}(\xi_j(t))$ represents the grade of membership of $\xi_j(t)$ in M_{ij}. Here, it is easy to find that for all $t > 0$, we have

$$\beta_i(\xi(t)) \geq 0, \quad (i = 1, 2, \ldots, r), \quad \sum_{j=1}^{r} \beta_j(\xi(t)) > 0$$
$$\mu_i(\xi(t)) \geq 0, \quad (i = 1, 2, \ldots, r), \quad \sum_{j=1}^{r} \mu_j(\xi(t)) = 1 \tag{5}$$

For convenience of notations, in the sequel, we denote

$$A(t) = \sum_{i=1}^{r} \mu_i(\xi(t)) A_i,$$
$$A_\tau(t) = \sum_{i=1}^{r} \mu_i(\xi(t)) A_{\tau i},$$
$$B(t) = \sum_{i=1}^{r} \mu_i(\xi(t)) B_i,$$
$$B_f(t) = \sum_{i=1}^{r} \mu_i(\xi(t)) B_{fi},$$
$$B_d(t) = \sum_{i=1}^{r} \mu_i(\xi(t)) B_{di},$$
$$C(t) = \sum_{i=1}^{r} \mu_i(\xi(t)) C_i, \tag{6}$$
$$C_\tau(t) = \sum_{i=1}^{r} \mu_i(\xi(t)) C_{\tau i},$$
$$D(t) = \sum_{i=1}^{r} \mu_i(\xi(t)) D_i,$$
$$D_f(t) = \sum_{i=1}^{r} \mu_i(\xi(t)) D_{fi},$$
$$D_d(t) = \sum_{i=1}^{r} \mu_i(\xi(t)) D_{di},$$
$$\phi(t) = \sum_{i=1}^{r} \mu_i(\xi(t)) \phi_i(t).$$

Then, we can rewrite system (3) as

$$E\dot{x}(t) = A(t)x(t) + A_\tau(t)x(t-\tau(t)) + B(t)u(t) + B_f(t)f(t) + B_d(t)d(t),$$
$$y(t) = C(t)x(t) + C_\tau(t)x(t-\tau(t)) + D(t)u(t) + D_f(t)f(t) + D_d(t)d(t), \tag{7}$$
$$x(t) = \phi(t), \quad t \in [-\tau_2, 0].$$

Before proceeding further, we will introduce some definitions and assumptions to be needed in the development of main results throughout this paper. Consider an unforced singular time delay system described by

$$E\dot{x}(t) = Ax(t) + A_\tau x(t-\tau(t))$$
$$x(t) = \phi(t), \quad t \in [-\tau, 0]. \tag{8}$$

Definition 1 (see [18]). (1) The pair (E, A) is said to be regular if $\det(zE - A)$ is not identically zero.

(2) The pair (E, A) is said to be impulse-free if $\deg(\det(zE - A)) = \operatorname{rank}(E)$.

(3) The pair (E, A) is said to be stable, if all roots of $\det(zE - A) = 0$ lie inside the unit disk with center at the origin.

(4) The delayed singular system (8) is said to be admissible if the pair (E, A) is regular, impulse-free, and stable.

Definition 2 (see [19]). (1) The singular system (8) is said to be regular and impulse-free if the pair (E, A) is regular and impulse-free.

(2) The singular system (8) is said to be asymptotically stable, if for any $\varepsilon > 0$, there exists a scalar $\delta(\varepsilon) > 0$ such that, for any compatible initial conditions, $\phi(t)$ with $\sup_{-\tau(t)\leq t\leq 0}\|\phi(t)\| < \delta(\varepsilon)$; the solution $x(t)$ of (8) satisfies $\|x(t)\| < \varepsilon$ for $t \geq 0$ and $\lim_{t\to\infty} x(t) = 0$.

The singular time delay system (8) may have an impulsive solution. However, the regularity and nonimpulse of (E, A) guarantee the existence and uniqueness of impulse-free solution to (8) on $[0, \infty)$.

Lemma 3 (see [20]). *If a functional $V : C_n[-\tau, 0] \to \mathbb{R}$ is continuous and $x(t, \phi)$ is a solution to (8), one defines $\dot{V}(\phi) = \lim_{h\to 0^+} \sup(1/h)(V(x(t+h, \phi) - V(\phi)))$. Denote the system parameters of (8) as*

$$(E, A, A_\tau) = \left(\begin{bmatrix} I_g & 0 \\ 0 & 0 \end{bmatrix}, \begin{bmatrix} A_{11} & A_{12} \\ A_{21} & A_{22} \end{bmatrix}, \begin{bmatrix} A_{\tau 11} & A_{\tau 12} \\ A_{\tau 21} & A_{\tau 22} \end{bmatrix} \right). \tag{9}$$

Assume that the singular system (8) is regular and impulse-free, A_{22} is invertible, and $\rho(A_{22}^{-1}A_{\tau 22}) < 1$. Then, system (8) is stable if there exists positive numbers α, μ, ν, and a continuous function; $V : C_n[-\tau, 0] \to \mathbb{R}$, such that

$$\mu \|\phi_1(0)\|^2 \leq V(\phi) \leq \nu \|\phi\|^2,$$
$$\dot{V}(x_t) \leq -\alpha \|x_t\|^2, \tag{10}$$

where $x_t = x(t+\theta)$ with $\theta \in [-\tau, 0]$ and $\phi = [\phi_1^T \ \phi_2^T]$ with $\phi_1 \in \mathbb{R}^q$.

In order to address the main results, the following assumptions are made.

Assumption 4. Matrices E and C_i satisfy the following rank condition:

$$\operatorname{rank}\begin{bmatrix} E \\ C_i \end{bmatrix} = n. \quad (11)$$

Assumption 5. The triple matrix (E, A_i, C_i) is R-detectable [21] for $\forall i = 1, 2, \ldots, r$; that is,

$$\operatorname{rank}\begin{bmatrix} sE - A_i \\ C_i \end{bmatrix} = n, \quad s \in C_+. \quad (12)$$

Remark 6. Both Assumptions 4 and 5 are necessary conditions for the existence of the designed observer in the latter section. Similar assumptions can be also found in [21] and the references therein. Meanwhile, for T-S fuzzy system description (3), we can see that a general time-varying delay fuzzy singular system is considered in this paper, including possible actuator, sensor faults, and exogenous disturbance input simultaneously.

3. Main Results

3.1. Delay-Dependent Stability. In this subsection, we suggest developing a delay-dependent stability condition for the nominal unforced fuzzy singular system of (7), which can be written as

$$\begin{aligned} E\dot{x}(t) &= A(t)x(t) + A_\tau(t)x(t-\tau(t)) + B_d(t)d(t), \\ y(t) &= C(t)x(t) + C_\tau(t)x(t-\tau(t)) + D_d(t)d(t), \quad (13) \\ x(t) &= \phi(t), \quad t \in [-\tau_2, 0]. \end{aligned}$$

Theorem 7. *For the given τ_1, τ_2, τ_D, the free fuzzy singular system (13) with $d(t) = 0$ is admissible for any time-varying delay $\tau(t)$ satisfying (2), if there exists a nonsingular matrix P, symmetric positive-definite matrices $Q_1 > 0$, $Q_2 > 0$, $S_1 > 0$, $S_2 > 0$, and $R > 0$, and any appropriately dimensioned matrices $V = [V_1\ V_2\ V_3\ V_4\ V_5]$ and $W = [W_1\ W_2\ W_3\ W_4\ W_5]$, such that the following set of inequalities hold:*

$$E^TP = P^TE \geq 0, \quad (14)$$

$$\begin{bmatrix} \Psi_i & (\tau_2-\tau_1)V^T \\ * & -(\tau_2-\tau_1)R \end{bmatrix} < 0, \quad (15)$$

$$\begin{bmatrix} \Psi_i & (\tau_2-\tau_1)W^T \\ * & -(\tau_2-\tau_1)R \end{bmatrix} < 0, \quad (16)$$

where

$$\Psi_i = \Psi_{1i} + (\tau_2-\tau_1)\Gamma_1^TR\Gamma_1,$$

$$\Psi_{1i} = \begin{bmatrix} \Psi_{11i} & -V_1^TE & V_1^TE - W_1^TE & W_1^TE & P^TA_{\tau i} \\ * & \Psi_{22} & \Psi_{23} & W_2^TE - E^TV_4 & -E^TV_5 \\ * & * & \Psi_{33} & \Psi_{34} & E^TV_5 - E^TW_5 \\ * & * & * & \Psi_{44} & E^TW_5 \\ * & * & * & * & -\varepsilon(1-\tau_D)Q_1 \end{bmatrix} \quad (17)$$

with

$$\begin{aligned} \Psi_{11i} &= P^TA_i + A_i^TP + \varepsilon Q_1 + Q_2, \\ \Psi_{22} &= S_1 - Q_2 - V_2^TE - E^TV_2, \\ \Psi_{23} &= V_2^TE - W_2^TE - E^TV_3, \\ \Psi_{33} &= S_2 - S_1 + \operatorname{sym}(V_3^TE - W_3^TE), \quad (18) \\ \Psi_{34} &= W_3^TE - E^TV_4 - E^TW_4, \\ \Psi_{44} &= -S_2 + W_4^TE + E^TW_4. \end{aligned}$$

Proof. The proof of this theorem is divided into two parts. The first one is concerned with the regularity and the impulse-free characterizations, and the second one treats the stability property of system (13).

Since $\operatorname{rank}(E) = g \leq n$, there must exist two invertible matrices $G \in \mathbb{R}^{n\times n}$ and $H \in \mathbb{R}^{n\times n}$ such that

$$\widetilde{E} = GEH = \begin{bmatrix} I_g & 0 \\ 0 & 0 \end{bmatrix}. \quad (19)$$

Similar to (19), we define

$$\widetilde{A}_i = GA_iH = G\begin{bmatrix} A_{i11} & A_{i12} \\ A_{i21} & A_{i22} \end{bmatrix}H = \begin{bmatrix} \widetilde{A}_{i11} & \widetilde{A}_{i12} \\ \widetilde{A}_{i21} & \widetilde{A}_{i22} \end{bmatrix}, \quad i = 1, 2, \ldots, r, \quad (20)$$

$$\widetilde{P} = G^{-T}PH = G^{-T}\begin{bmatrix} P_{11} & P_{12} \\ P_{21} & P_{22} \end{bmatrix}H = \begin{bmatrix} \widetilde{P}_{11} & \widetilde{P}_{12} \\ \widetilde{P}_{21} & \widetilde{P}_{22} \end{bmatrix}.$$

Since $\Psi_i < 0$ and $Q_1 > 0$ and $Q_2 > 0$ and $R > 0$, we can formulate the following inequality easily:

$$F = A_i^TP + P^TA_i < 0. \quad (21)$$

Then, pre- and postmultiplying $F < 0$ by H^T and H, respectively, (21) yields

$$\widetilde{F} = \widetilde{A}_i^T\widetilde{P} + \widetilde{P}^T\widetilde{A}_i = \begin{bmatrix} \widetilde{F}_{11} & \widetilde{F}_{12} \\ * & \widetilde{A}_{i22}^T\widetilde{P}_{22} + \widetilde{P}_{22}^T\widetilde{A}_{i22} \end{bmatrix} < 0. \quad (22)$$

Since $\widetilde{F}_{11}$ and $\widetilde{F}_{12}$ are irrelevant to the results of the following discussion, the real expression of these two variables is omitted here. From (22), it is easy to see that

$$\widetilde{A}_{i22}^T \widetilde{P}_{22} + \widetilde{P}_{22}^T \widetilde{A}_{i22} < 0. \tag{23}$$

Since $\mu_i(\xi(t)) \geq 0$ and $\sum_{i=1}^r \mu_i(\xi(t)) = 1$, we have

$$\sum_{i=1}^r \mu_i(\xi(t)) \left(\widetilde{A}_{i22}^T \widetilde{P}_{22} + \widetilde{P}_{22}^T \widetilde{A}_{i22} \right) < 0. \tag{24}$$

This implies that $\sum_{i=1}^r \mu_i(\xi(t))\widetilde{A}_{i22}$ is nonsingular. Therefore, the unforced fuzzy singular system (13) is regular and impulse-free.

Next, we will show the stability of system (13). If conditions (15)-(16) hold, we have

$$\sum_{i=1}^r \mu_i \begin{bmatrix} P^T A_i + A_i^T P + \varepsilon Q_1 + Q_2 & P^T A_{\tau i} \\ * & -\varepsilon (1-\tau_D) Q_1 \end{bmatrix} < 0. \tag{25}$$

Premultiplying and postmultiplying the preceding inequality by

$$\begin{bmatrix} 0 & I & 0 & 0 \\ 0 & 0 & 0 & I \end{bmatrix} \tag{26}$$

and its transposes, respectively, we obtain

$$\begin{bmatrix} \operatorname{sym}\left(P_{22}^T \left(\sum_{i=1}^r \mu_i A_{i22} \right) \right) + \varepsilon Q_{122} + Q_{222} & P_{22}^T \left(\sum_{i=1}^r \mu_i A_{\tau i22} \right) \\ * & -\varepsilon (1-\tau_D) Q_{122} \end{bmatrix} < 0. \tag{27}$$

which implies that

$$\begin{bmatrix} \operatorname{sym}\left(P_{22}^T \left(\sum_{i=1}^r \mu_i A_{i22} \right) \right) + \varepsilon Q_{122} & P_{22}^T \left(\sum_{i=1}^r \mu_i A_{\tau i22} \right) \\ * & -\varepsilon (1-\tau_D) Q_{122} \end{bmatrix} < 0. \tag{28}$$

It follows from $(1,1)$-block of (28) that A_{i22} is invertible. Then, premultiplying and postmultiplying the preceding inequality by

$$\left[-\left(\sum_{i=1}^r \mu_i A_{\tau i22} \right)^T \left(\sum_{i=1}^r \mu_i A_{i22} \right)^{-T} \quad I \right] < 0 \tag{29}$$

and its transpose, respectively, yield

$$\begin{aligned} &\left(\left(\sum_{i=1}^r \mu_i A_{i22} \right)^{-1} \left(\sum_{i=1}^r \mu_i A_{\tau i22} \right) \right)^T \\ &\cdot Q_{122} \left(\left(\sum_{i=1}^r \mu_i A_{i22}^{-1} \right) \left(\sum_{i=1}^r \mu_i A_{\tau i22} \right) \right) - (1-\tau_D) \\ &\cdot Q_{122} < 0 \end{aligned} \tag{30}$$

which shows that $\rho((\sum_{i=1}^r \mu_i A_{i22})^{-1}(\sum_{i=1}^r \mu_i A_{\tau i22})) < 1$ holds for all allowable μ_i.

Now, let us choose the following Lyapunov-Krasovskii function as

$$V(t) = V_1(t) + V_2(t) + V_3(t), \tag{31}$$

where

$$\begin{aligned} V_1(t) &= x^T(t) E^T P x(t), \\ V_2(t) &= \varepsilon \int_{t-\tau(t)}^t x^T(s) Q_1 x(s)\, ds \\ &\quad + \int_{t-\tau_1}^t x^T(s) Q_2 x(s)\, ds \\ &\quad + \int_{t-\tau_p}^{t-\tau_1} x^T(s) S_1 x(s)\, ds \\ &\quad + \int_{t-\tau_2}^{t-\tau_p} x^T(s) S_2 x(s)\, ds, \\ V_3(t) &= \int_{-\tau_2}^{-\tau_1} \int_{t+\theta}^t \dot{x}^T(s) E^T R E \dot{x}(s)\, ds\, d\theta, \end{aligned} \tag{32}$$

where the unknown matrices P, $Q_1 > 0$, $Q_2 > 0$, $S_1 > 0$, $S_2 > 0$, and $R > 0$ are to be determined, and $\tau_p = \tau_1 + \rho\delta$, $\delta = \tau_2 - \tau_1$, and $0 < \rho < 1$. Then, the time derivatives of $V(t)$ along the trajectories of the dynamic systems (13) satisfy

$$\begin{aligned} \dot{V}_1(t) &= x^T(t) \left(P^T A(t) + A^T(t) P \right) x(t) \\ &\quad + 2x^T(t) P^T A_\tau(t) x(t-\tau(t)), \\ \dot{V}_2(t) &= \varepsilon x^T(t) Q_1 x(t) \\ &\quad - \varepsilon (1-\dot{\tau}(t)) x^T(t-\tau(t)) Q_1 x(t-\tau(t)) \\ &\quad + x^T(t) Q_2 x(t) - x^T(t-\tau_1) Q_2 x(t-\tau_1) \\ &\quad + x^T(t-\tau_1) S_1 x(t-\tau_1) \\ &\quad - x^T(t-\tau_p) S_1 x(t-\tau_p) \\ &\quad + x^T(t-\tau_p) S_2 x(t-\tau_p) \\ &\quad - x^T(t-\tau_2) S_2 x(t-\tau_2), \\ \dot{V}_3(t) &= (\tau_2 - \tau_1)(E\dot{x}(t))^T R (E\dot{x}(t)) \\ &\quad - \int_{t-\tau_2}^{t-\tau_1} \dot{x}^T(s) E^T R E \dot{x}(s)\, ds \end{aligned}$$

$$= (\tau_2 - \tau_1)(E\dot{x}(t))^T R(E\dot{x}(t)) - \int_{t-\tau_p}^{t-\tau_1} \dot{x}^T(s) E^T RE\dot{x}(s)\,ds - \int_{t-\tau_2}^{t-\tau_p} \dot{x}^T(s) E^T RE\dot{x}(s)\,ds. \quad (33)$$

Denoting $\Gamma_1 = [A(t)\ 0\ 0\ 0\ A_\tau(t)]$, $\zeta_1^T(t) = [x^T(t)\ x^T(t-\tau_1)\ x^T(t-\tau_p)\ x^T(t-\tau_2)\ x^T(t-\tau(t))]$ and $\beta_1^T(t,s) = [\zeta_1^T(t)\ (E\dot{x}(s))^T]$, and from Newton-Leibniz formula, we can easily obtain that

$$-2\zeta_1^T(t) V^T\left[Ex(t-\tau_1) - Ex(t-\tau_p) - \int_{t-\tau_p}^{t-\tau_1} E\dot{x}(s)\,ds\right] = 0,$$
$$-2\zeta_1^T(t) W^T\left[Ex(t-\tau_p) - Ex(t-\tau_2) - \int_{t-\tau_2}^{t-\tau_p} E\dot{x}(s)\,ds\right] = 0. \quad (34)$$

Therefore, a straightforward computation gives

$$\begin{aligned}
\dot{V}(t) &= x^T(t)\left[P^T A(t) + A^T(t) P + \varepsilon Q_1 + Q_2\right] x(t) + 2x^T(t) P^T A_\tau(t) x(t-\tau(t)) - \varepsilon(1-\dot{\tau}(t)) x^T(t \\
&- \tau(t)) Q_1 x(t-\tau(t)) + x^T(t-\tau_1)(S_1 - Q_2) x(t - \tau_1) + x^T(t-\tau_p)(S_2 - S_1) x(t-\tau_p) + x^T(t \\
&- \tau_2)(-S_2) x(t-\tau_2) + (\tau_2 - \tau_1)(E\dot{x}(t))^T R(E\dot{x}(t)) - \int_{t-\tau_p}^{t-\tau_1} \dot{x}^T(s) E^T RE\dot{x}(s)\,ds - \int_{t-\tau_2}^{t-\tau_p} \dot{x}^T(s) \\
&\cdot E^T RE\dot{x}(s)\,ds + \int_{t-\tau_p}^{t-\tau_1} 2\zeta_1^T(t) V^T E\dot{x}(s)\,ds + \int_{t-\tau_2}^{t-\tau_p} 2\zeta_1^T(t) W^T E\dot{x}(s)\,ds - 2\zeta_1^T(t) \\
&\cdot V^T E[0\ I\ -I\ 0\ 0]\zeta_1(t) - 2\zeta_1^T(t) \cdot W^T E[0\ 0\ I\ -I\ 0]\zeta_1(t) \le \frac{1}{\tau_2 - \tau_1} \\
&\cdot \int_{t-\tau_p}^{t-\tau_1} \left[\zeta_1^T(t)\Psi(t)\zeta_1(t) + 2(\tau_2 - \tau_1)\zeta_1^T(t) V^T E\dot{x}(s) - (\tau_2 - \tau_1)\dot{x}^T(s) E^T RE\dot{x}(s)\right] ds + \frac{1}{\tau_2 - \tau_1} \\
&\cdot \int_{t-\tau_2}^{t-\tau_p} \left[\zeta_1^T(t)\Psi(t)\zeta_1(t) + 2(\tau_2 - \tau_1)\zeta_1^T(t) W^T E\dot{x}(s) - (\tau_2 - \tau_1)\dot{x}^T(s) E^T RE\dot{x}(s)\right] ds = \frac{1}{(\tau_2 - \tau_1)} \\
&\cdot \int_{t-\tau_p}^{t-\tau_1} \beta_1^T(t,s) \begin{bmatrix} \Psi(t) & (\tau_2 - \tau_1) V^T \\ * & -(\tau_2 - \tau_1) R \end{bmatrix} \beta_1(t,s)\,ds + \frac{1}{(\tau_2 - \tau_1)} \int_{t-\tau_2}^{t-\tau_p} \beta_1^T(t,s) \\
&\cdot \begin{bmatrix} \Psi(t) & (\tau_2 - \tau_1) W^T \\ * & -(\tau_2 - \tau_1) R \end{bmatrix} \beta_1(t,s)\,ds.
\end{aligned} \quad (35)$$

Then, if conditions (15)-(16) hold, there exists $\alpha > 0$ such that $\dot{V}(x_t) < \alpha\|x_t\|$. By Lemma 3, we conclude that the unforced fuzzy singular system (13) is stable. This completes the proof. □

Remark 8. In some existing literature, for example, [22, 23], some delay-dependent criteria are given in terms of LMIs to guarantee that the fuzzy singular system is admissible by using LKF approach and integral inequality, such as Lemma 2 in [22] and Lemma 2.3 in [23]. However, in the proof of our result of Theorem 7, we use one identical equality to estimate the upper bound of the derivative of $V(t)$ without any model transformation. Moreover, it is interesting to mention that this study presents criteria based on the free weighting matrix method, in which the bounding techniques on some cross product terms are not involved [11, 23]. The major feature of this method is to reduce the conservatism engendered by the system transformations and the bounding techniques.

Remark 9. For time delay systems, the Lyapunov functional candidate always involves the integral term $\int_{-\tau}^{0}\int_{t+\theta}^{t} \dot{x}^T(s)R\dot{x}(s)ds\,d\theta$, and the derivative of it was estimated as $\tau\dot{x}^T(t)R\dot{x}(t) - \int_{t-\tau}^{t} \dot{x}^T(s)R\dot{x}(s)ds$. However, the term $\int_{t-\tau}^{t} \dot{x}^T(s)R\dot{x}(s)ds$ was ignored [25–27], or some useful negative integral term was lost; see, for example, [28, 29]. Instead, in this paper all those terms $\int_{t-\tau_p}^{t-\tau_1} \dot{x}^T(s)R\dot{x}(s)ds$ and $\int_{t-\tau_2}^{t-\tau_p} \dot{x}^T(s)R\dot{x}(s)ds$, which contain a great amount of useful information about systems, are preserved. Therefore, it is obvious to see that this method will lead to less conservatives than the existing ones in [25–29]. Furthermore, the introduction of parameter ε ($\varepsilon \ge 0$) indicates that Lemma 3 can be suitable for time-varying delay $\tau(t)$ being unknown or not differentiable; that is, in the case of time-varying delay $\tau(t)$ not differentiable, one can set $\varepsilon = 0$.

In order to estimate system faults, the following fault estimation observer is constructed:

$$\begin{aligned} E\dot{\hat{x}}(t) &= A(t)\hat{x}(t) + A_\tau(t)\hat{x}(t-\tau(t)) + B(t)u(t) \\ &\quad + B_f(t)\hat{f}(t) - L(t)(\hat{y}(t) - y(t)), \\ \hat{y}(t) &= C(t)\hat{x}(t) + C_\tau(t)\hat{x}(t-\tau(t)) + D(t)u(t) \\ &\quad + D_f(t)\hat{f}(t), \\ \dot{\hat{f}}(t) &= -F(t)(\hat{y}(t) - y(t)), \end{aligned} \tag{36}$$

where $\hat{x}(t) \in \mathbb{R}^n$ is the observer state, $\hat{y}(t) \in \mathbb{R}^l$ is the observer output, and $\hat{f}(t) \in \mathbb{R}^q$ is an estimate of fault $f(t)$. The objective is to design the appropriate dimension gain matrices $L(t) \in \mathbb{R}^{n\times l}$ and $F(t) \in \mathbb{R}^{q\times l}$ and estimate the fault despite the presence of the disturbance and state delay, where $L(t) = \sum_{i=1}^{r}\mu_i(\xi(t))L_i$ and $F(t) = \sum_{i=1}^{r}\mu_i(\xi(t))F_i$. Let us define $e_x(t) = \hat{x}(t) - x(t)$, $e_y(t) = \hat{y}(t) - y(t)$, $e_f(t) = \hat{f}(t) - f(t)$, and $e^T(t) = [e_x^T(t), e_f^T(t)]$, $\omega^T(t) = [d^T(t), \dot{f}^T(t)]$; then the error dynamic systems are deduced from (7) and (36) as follows:

$$\begin{aligned} \overline{E}\dot{e}(t) &= \left[\overline{A}(t) - \overline{L}(t)\overline{C}(t)\right]e(t) \\ &\quad + \left[\overline{A}_\tau(t) - \overline{L}(t)\overline{C}_\tau(t)\right]e(t-\tau(t)) \\ &\quad + \left[\overline{L}(t)\overline{D}_d(t) - \overline{B}_d(t)\right]\omega(t), \\ e_y(t) &= \overline{C}(t)e(t) + \overline{C}_\tau(t)e(t-\tau(t)) - \overline{D}_d(t)\omega(t), \end{aligned} \tag{37}$$

where

$$\begin{aligned} \overline{E} &= \begin{bmatrix} E & 0 \\ 0 & I_q \end{bmatrix}, \\ \overline{A}(t) &= \begin{bmatrix} A(t) & B_f(t) \\ 0 & 0 \end{bmatrix}, \\ \overline{A}_\tau(t) &= \begin{bmatrix} A_\tau(t) & 0 \\ 0 & 0 \end{bmatrix}, \\ \overline{B}_d(t) &= \begin{bmatrix} B_d(t) & 0 \\ 0 & I_q \end{bmatrix}, \\ \overline{L}(t) &= \begin{bmatrix} L(t) \\ F(t) \end{bmatrix}, \\ \overline{C}(t) &= [C(t) \;\; D_f(t)], \\ \overline{C}_\tau(t) &= [C_\tau(t) \;\; 0], \\ \overline{D}_d(t) &= [D_d(t) \;\; 0]. \end{aligned} \tag{38}$$

Therefore, H_∞ robust fault estimation observer design problem to be addressed in this paper can be formulated as follows: (i) The error dynamic system (37) with $\omega(t) = 0$ is admissible for any time delay satisfying (2); (ii) for a given scalar γ, the following H_∞ performance is satisfied:

$$\int_0^L \left\|e_f(t)\right\|^2 dt \le \gamma^2 \int_0^L \|\omega(t)\|^2 dt \tag{39}$$

for all $L > 0$ and $\omega(t) \in L_2[0,\infty)$ under zero initial conditions.

For simplicity, we introduce the following vectors:

$$\begin{aligned} \zeta_2^T(t) &= \left[e^T(t) \;\; e^T(t-\tau_1) \;\; e^T(t-\tau_p) \;\; e^T(t-\tau_2) \;\; e^T(t-\tau(t)) \;\; \omega^T(t)\right], \\ \Gamma_2 &= \left[A(t) - \overline{L}(t)\overline{C}(t) \;\; 0 \;\; 0 \;\; 0 \;\; \overline{A}_\tau(t) - \overline{L}(t)\overline{C}_\tau(t) \;\; \overline{L}(t)\overline{D}_d(t) - \overline{B}_d(t)\right]. \end{aligned} \tag{40}$$

Then, the state of error dynamics (37) can be rewritten as $\overline{E}\dot{e}(t) = \Gamma_2\zeta_2(t)$.

Remark 10. From error dynamics (37), we can see that the new matrices $\overline{A}(t)$, $\overline{A}_\tau(t)$, $\overline{C}(t)$, $\overline{C}_\tau(t)$, $\overline{B}_d(t)$, and $\overline{D}_d(t)$ are known matrices, while the matrices $\overline{L}(t)$ contain two matrices $L(t)$ and $F(t)$ that have to be designed. Therefore, the proposed robust fault estimation observer design is converted to the problem of seeking the gain matrix $\overline{L}(t)$.

Next, a fuzzy augmented fault estimation observer design method under H_∞ performance is proposed to achieve robust fault estimation by following lemma under time-varying state delay.

Lemma 11. *For the given positive scalars τ_1, τ_2, τ_D, and γ, the error dynamic system (37) is admissible with $\omega(t) = 0$ while satisfying a prescribed H_∞ performance (39), if there exist appropriately dimensional matrices P, $Q_1 > 0$, $Q_2 > 0$, $S_1 > 0$, $S_2 > 0$, $R > 0$, and $\overline{L}(t)$ and free weighting matrices $V = [V_1 \; V_2 \; V_3 \; V_4 \; V_5 \; V_6]$ and $W = [W_1 \; W_2 \; W_3 \; W_4 \; W_5 \; W_6]$, such that the following inequalities hold:*

$$\overline{E}^T P = P^T\overline{E} \ge 0, \tag{41}$$

$$\Delta_1(t) = \begin{bmatrix} \Phi(t) & (\tau_2 - \tau_1)V^T \\ * & -(\tau_2 - \tau_1)R \end{bmatrix} < 0, \tag{42}$$

$$\Delta_2(t) = \begin{bmatrix} \Phi(t) & (\tau_2 - \tau_1)W^T \\ * & -(\tau_2 - \tau_1)R \end{bmatrix} < 0, \tag{43}$$

where

$$\Phi(t) = \Phi_1(t) + (\tau_2 - \tau_1)\Gamma_2^T R \Gamma_2,$$

$$\Phi_1(t) = \begin{bmatrix} \Phi_{11}(t) & -V_1^T E & V_1^T E - W_1^T E & W_1^T E & \Phi_{15}(t) & \Phi_{16}(t) \\ * & \Phi_{22} & \Phi_{23} & W_2^T E - E^T V_4 & -E^T V_5 & -E^T V_6 \\ * & * & \Phi_{33} & \Phi_{34} & E^T V_5 - E^T W_5 & E^T V_6 - E^T W_6 \\ * & * & * & \Phi_{44} & E^T W_5 & E^T W_6 \\ * & * & * & * & -\varepsilon(1-\tau_D)Q_1 & 0 \\ * & * & * & * & * & -\gamma^2 I \end{bmatrix} \quad (44)$$

with

$$\Phi_{11}(t) = \text{sym}\left(P^T\left(\overline{A}(t) - \overline{L}(t)\overline{C}(t)\right)\right) + \varepsilon Q_1 + Q_2 + \overline{I}_q \overline{I}_q^T,$$

$$\Phi_{15}(t) = P^T\left(\overline{A}_\tau(t) - \overline{L}(t)\overline{C}_\tau(t)\right),$$

$$\Phi_{16}(t) = P^T\left(\overline{L}(t)\overline{D}_d(t) - \overline{B}_d(t)\right),$$

$$\overline{I}_q^T = [0 \;\; I_q],$$

$$\Phi_{22} = S_1 - Q_2 - V_2^T E - E^T V_2,$$

$$\Psi_{23} = V_2^T E - W_2^T E - E^T V_3,$$

$$\Psi_{33} = S_2 - S_1 + \text{sym}\left(V_3^T E - W_3^T E\right),$$

$$\Psi_{34} = W_3^T E - E^T V_4 - E^T W_4,$$

$$\Psi_{44} = -S_2 + W_4^T E + E^T W_4. \quad (45)$$

Proof. First, we show that the error dynamic system (37) with $\omega(t) = 0$ is regular and impulse-free. Since $\overline{E}$ is singular and $\text{rank}(\overline{E}) = g + q$, there always exist two nonsingular matrices $\widehat{G} \in \mathbb{R}^{(n+q)\times(n+q)}$ and $\widehat{U} \in \mathbb{R}^{(n+q)\times(n+q)}$ such that

$$\widehat{G}\overline{E}\widehat{H} = \begin{bmatrix} I_{g+q} & 0 \\ 0 & 0 \end{bmatrix}. \quad (46)$$

Accordingly, denote

$$\widehat{G}\left(\overline{A}(t) - \overline{L}(t)\overline{C}(t)\right)\widehat{H} = \widehat{G}\widehat{A}(t)\widehat{H} = \begin{bmatrix} \widehat{A}_{11}(t) & \widehat{A}_{12}(t) \\ \widehat{A}_{21}(t) & \widehat{A}_{22}(t) \end{bmatrix}, \quad (47)$$

$$\widehat{G}^{-T} P \widehat{H} = \begin{bmatrix} \widehat{P}_{11} & \widehat{P}_{12} \\ \widehat{P}_{21} & \widehat{P}_{22} \end{bmatrix}.$$

From (41) and using the expressions in (46)-(47), it is easy to obtain that $\widehat{P}_{12} = 0$. Since (42)-(43) hold, we have $\Phi(t) = \Phi_1(t) + (\tau_2 - \tau_1)\Gamma_2^T R \Gamma_2 < 0$. Moreover, noting $R > 0$ and by Schur complement, we can get $\Phi_{11}(t) = \text{sym}(P^T(\overline{A}(t) - \overline{L}(t)\overline{C}(t))) + \varepsilon Q_1 + Q_2 + \overline{I}_q \overline{I}_q^T < 0$. Then, premultiplying and postmultiplying $\Phi_{11}(t) < 0$ by $\widehat{H}^T$ and $\widehat{H}$, respectively, we have $\widehat{A}_{22}^T(t)\widehat{P}_{22} + \widehat{P}_{22}^T\widehat{A}_{22}(t) < 0$, which implies that $\widehat{A}_{22}(t)$ is nonsingular, and thus the pair $(\overline{E}, \widehat{A}(t))$ is regular and impulse-free. Hence, the error dynamic system (37) is regular and impulse-free for any time delay $\tau(t)$ satisfying (2) when inequalities (41)–(43) hold. Next, we will prove that the error dynamic system (37) is stable with H_∞ performance. To this end, the Lyapunov-Krasovskii functional candidate is constructed as follows:

$$\begin{aligned} V(t) &= e^T(t)\overline{E}^T P e(t) + \varepsilon \int_{t-\tau(t)}^{t} e^T(s) Q_1 e(s)\,ds \\ &+ \int_{t-\tau_1}^{t} e^T(s) Q_2 e(s)\,ds \\ &+ \int_{t-\tau_p}^{t-\tau_1} e^T(s) S_1 e(s)\,ds \\ &+ \int_{t-\tau_2}^{t-\tau_p} e^T(s) S_2 e(s)\,ds \\ &+ \int_{-\tau_2}^{-\tau_1}\int_{t+\theta}^{t} \dot{e}^T(s)\overline{E}^T R \overline{E}\dot{e}(s)\,ds\,d\theta, \end{aligned} \quad (48)$$

where the unknown matrices P, $Q_1 > 0$, $Q_2 > 0$, $S_1 > 0$, $S_2 > 0$, and $R > 0$ are to be determined. Then, the time derivatives of $V(t)$ along the trajectories of the error dynamic system (37) satisfy

$$\begin{aligned} \dot{V}(t) &= e^T(t)\left[P^T\left(\overline{A}(t) - \overline{L}(t)\overline{C}(t)\right)\right. \\ &\left. + \left(\overline{A}(t) - \overline{L}(t)\overline{C}(t)\right)^T P + \varepsilon Q_1 + Q_2\right] e(t) \\ &+ 2e^T(t) P^T\left(\overline{A}_\tau(t) - \overline{L}(t)\overline{C}_\tau(t)\right) e(t - \tau(t)) \\ &+ 2e^T(t) P^T\left(\overline{L}(t)\overline{D}_d(t) - \overline{B}_d(t)\right)\omega(t) - \varepsilon(1 \end{aligned}$$

$$- \dot{\tau}(t)) e^T(t-\tau(t)) Q_1 e(t-\tau(t)) + e^T(t-\tau_1)(S_1 - Q_2) e(t-\tau_1) + e^T(t-\tau_\rho)(S_2 - S_1) e(t-\tau_\rho) + e^T(t-\tau_2)(-S_2) e(t-\tau_2) + (\tau_2-\tau_1)(\overline{E}\dot{e}(t))^T \cdot R(\overline{E}\dot{e}(t)) - \int_{t-\tau_2}^{t-\tau_1} \dot{e}^T(s)\overline{E}^T R\overline{E}\dot{e}(s)\,ds. \quad (49)$$

Denoting $\beta_2^T(t,s) = [\zeta_2^T(t)\ \ (\overline{E}\dot{e}(s))^T]$, and from Newton-Leibniz formula, a straightforward computation gives

$$\begin{aligned}
&\dot{V}(t) + e_f^T(t) e_f(t) - \gamma^2 \omega^T(t)\omega(t) = \dot{V}(t) + e^T(t) \cdot \overline{I}_q \overline{I}_q^T e(t) - \gamma^2\omega^T(t)\omega(t) = e^T(t)\big[P^T(\overline{A}(t) - \overline{L}(t)\overline{C}(t)) + (\overline{A}(t) - \overline{L}(t)\overline{C}(t))^T P + \varepsilon Q_1 + Q_2 \\
&+ \overline{I}_q\overline{I}_q^T\big] e(t) + 2e^T(t)P^T(\overline{A}_\tau(t) - \overline{L}(t)\overline{C}_\tau(t)) e(t - \tau(t)) + 2e^T(t)P^T(\overline{L}(t)\overline{D}_d(t) - \overline{B}_d(t))\omega(t) \\
&- \varepsilon(1-\dot{\tau}(t)) e^T(t-\tau(t))Q_1 e(t-\tau(t)) + e^T(t - \tau_1)(S_1 - Q_2) e(t-\tau_1) + e^T(t-\tau_\rho)(S_2 - S_1) e(t - \tau_\rho) + e^T(t-\tau_2)(-S_2)e(t-\tau_2) - \gamma^2\omega^T(t)\omega(t) \\
&+ (\tau_2-\tau_1)(\overline{E}\dot{e}(t))^T R(\overline{E}\dot{e}(t)) - \int_{t-\tau_\rho}^{t-\tau_1} \dot{e}^T(s) \cdot \overline{E}^T R\overline{E}\dot{e}(s)\,ds - \int_{t-\tau_2}^{t-\tau_\rho} \dot{e}^T(s)\overline{E}^T R\overline{E}\dot{e}(s)\,ds \\
&- 2\zeta_2^T(t) V^T \overline{E}[0\ \ I\ \ -I\ \ 0\ \ 0\ \ 0]\zeta_2(t) + \int_{t-\tau_\rho}^{t-\tau_1} 2\zeta_2^T(t)V^T\overline{E}\dot{e}(s)\,ds - 2\zeta_2^T(t) \cdot W^T\overline{E}[0\ \ 0\ \ I\ \ -I\ \ 0\ \ 0]\zeta_2(t) + \int_{t-\tau_2}^{t-\tau_\rho} 2\zeta_2^T(t) \cdot W^T\overline{E}\dot{e}(s)\,ds \le \frac{1}{\tau_2-\tau_1} \\
&\cdot \int_{t-\tau_\rho}^{t-\tau_1}\big[\zeta_2^T(t)\Phi(t)\zeta_2(t) + 2(\tau_2-\tau_1)\zeta_2^T(t)V^T\overline{E}\dot{e}(s) - (\tau_2-\tau_1)\dot{e}^T(s)\overline{E}^T R\overline{E}\dot{e}(s)\big]\,ds + \frac{1}{\tau_2-\tau_1} \\
&\cdot \int_{t-\tau_2}^{t-\tau_\rho}\big[\zeta_2^T(t)\Phi(t)\zeta_2(t) + 2(\tau_2-\tau_1)\zeta_2^T(t)W^T\overline{E}\dot{e}(s) - (\tau_2-\tau_1)\dot{e}^T(s)\overline{E}^T R\overline{E}\dot{e}(s)\big]\,ds = \frac{1}{\tau_2-\tau_1} \\
&\cdot \int_{t-\tau_\rho}^{t-\tau_1}\beta_2^T(t,s)\begin{bmatrix}\Phi(t) & (\tau_2-\tau_1)V^T \\ * & -(\tau_2-\tau_1)R\end{bmatrix}\beta_2(t,s)\,ds + \frac{1}{\tau_2-\tau_1}\int_{t-\tau_2}^{t-\tau_\rho}\beta_2^T(t,s) \cdot \begin{bmatrix}\Phi(t) & (\tau_2-\tau_1)W^T \\ * & -(\tau_2-\tau_1)R\end{bmatrix}\beta_2(t,s)\,ds.
\end{aligned} \quad (50)$$

If (42)-(43) hold, one has $\dot{V}(t) + e_f^T(t)e_f(t) - \gamma^2\omega^T(t)\omega(t) < 0$. By noticing $V(L) \ge 0$ and $V(0) = 0$ under zero initial conditions, we can conclude that (39) holds for all $L > 0$ and any nonzero $\omega(t) \in L_2[0,\infty)$.

On the other hand, under conditions (42)-(43), by choosing the same Lyapunov function as (48) and following the similar line in the earlier deduction under conditions (42)-(43), then substitute A_i and $A_{\tau i}$ by $(\overline{A}(t) - \overline{L}(t)\overline{C}(t))$ and $(\overline{A}_\tau(t) - \overline{L}(t)\overline{C}_\tau(t))$; we can easily obtain that the time derivative of $V(t)$ along the solution of error dynamics (37) with $\omega(t) = 0$ satisfies $\dot{V}(e_t) < \alpha\|e_t\|$ by using Theorem 7, which indicates the stability of system (37). This complete the proof. □

In the following, we will focus on the design of observer based on Lemma 11 and provide a new sufficient condition for the existence of robust fault estimation observer for fuzzy singular time delay system (3).

Theorem 12. *For the given positive scalars τ_1, τ_2, τ_D, δ, and γ, the error dynamic system (37) is admissible with $\omega(t) = 0$ while satisfying a prescribed H_∞ performance (39), if there exist appropriately dimensional matrices P, $Q_1 > 0$, $Q_2 > 0$, $S_1 > 0$, $S_2 > 0$, $R > 0$, and Y_i and free weighting matrices $V = [V_1\ V_2\ V_3\ V_4\ V_5\ V_6]$ and $W = [W_1\ W_2\ W_3\ W_4\ W_5\ W_6]$, such that the following inequalities hold:*

$$\overline{E}^T P = P^T\overline{E} \ge 0, \quad (51)$$

$$\Xi_{ii} < 0 \quad i = 1, 2, \ldots, r, \quad (52)$$

$$\Xi_{ij} + \Xi_{ji} \le 0 \quad 1 \le i < j \le r, \quad (53)$$

$$\Pi_{ii} < 0 \quad i = 1, 2, \ldots, r, \quad (54)$$

$$\Pi_{ij} + \Pi_{ji} \le 0 \quad 1 \le i < j \le r, \quad (55)$$

where

$$\Xi_{ij} = \begin{bmatrix} \Phi_{ij} & (\tau_2-\tau_1)V^T & \sqrt{(\tau_2-\tau_1)}\overline{\Gamma}_{2ij}^T \\ * & -(\tau_2-\tau_1)R & 0 \\ * & * & -2\delta\overline{E}^T P + \delta^2 R \end{bmatrix},$$

$$\Pi_{ij} = \begin{bmatrix} \Phi_{ij} & (\tau_2-\tau_1)W^T & \sqrt{(\tau_2-\tau_1)}\overline{\Gamma}_{2ij}^T \\ * & -(\tau_2-\tau_1)R & 0 \\ * & * & -2\delta\overline{E}^T P + \delta^2 R \end{bmatrix}, \quad (56)$$

where

$$\Phi_{ij} = \begin{bmatrix} \Phi_{11ij} & -V_1^T\overline{E} & V_1^T\overline{E} - W_1^T\overline{E} & W_1^T\overline{E} & \Phi_{15ij} & \Phi_{16ij} \\ * & \Phi_{22} & \Phi_{23} & W_2^T\overline{E} - \overline{E}^T V_4 & -\overline{E}^T V_5 & -\overline{E}^T V_6 \\ * & * & \Phi_{33} & \Phi_{34} & \overline{E}^T V_5 - \overline{E}^T W_5 & \overline{E}^T V_6 - \overline{E}^T W_6 \\ * & * & * & \Phi_{44} & \overline{E}^T W_5 & \overline{E}^T W_6 \\ * & * & * & * & -\varepsilon(1-\tau_D)Q_1 & 0 \\ * & * & * & * & * & -\gamma^2 I \end{bmatrix} \quad (57)$$

with

$$\Phi_{11ij} = \mathrm{sym}\left(P^T\overline{A}_i - Y_i\overline{C}_j\right) + \varepsilon Q_1 + Q_2 + \overline{I}_q\overline{I}_q^T,$$

$$\Phi_{15ij} = P^T\overline{A}_{\tau i} - Y_i\overline{C}_{\tau j},$$

$$\Phi_{16ij} = Y_i\overline{D}_{dj} - P^T\overline{B}_{di},$$

$$\Phi_{22} = S_1 - Q_2 - V_2^T E - E^T V_2,$$

$$\Psi_{23} = V_2^T E - W_2^T E - E^T V_3,$$

$$\Psi_{33} = S_2 - S_1 + \mathrm{sym}\left(V_3^T E - W_3^T E\right),$$

$$\Psi_{34} = W_3^T E - E^T V_4 - E^T W_4,$$

$$\Psi_{44} = -S_2 + W_4^T E + E^T W_4,$$

$$\overline{I}_q^T = [0 \ \ I_q],$$

$$\overline{\Gamma}_{2ij} = \overline{E}\left[P^T\overline{A}_i - Y_i\overline{C}_j \ \ 0 \ \ 0 \ \ 0 \ \ P^T\overline{A}_{\tau i} - Y_i\overline{C}_{\tau j} \ \ Y_i\overline{D}_{dj} - P^T\overline{B}_{di}\right]. \quad (58)$$

Then the observer gain matrices can be obtained as $\overline{L}_i = \left[\begin{smallmatrix} L_i \\ F_i \end{smallmatrix}\right] = P^{-T}Y_i$.

Proof. For any scalar δ, it follows from the fact $(\delta R - P)R^{-1}(\delta R - P) \geq 0$ that $-PR^{-1}P \leq -2\delta P + \delta^2 R$. By Schur complement theorem, we can conclude that (42)-(43) hold if the following inequalities hold:

$$\begin{bmatrix} \Phi_1(t) & (\tau_2-\tau_1)V^T & \sqrt{(\tau_2-\tau_1)}\Gamma_2^T P\overline{E}^T \\ * & -(\tau_2-\tau_1)R & 0 \\ * & * & -2\delta\overline{E}P^T + \delta^2 R \end{bmatrix} < 0,$$

$$\begin{bmatrix} \Phi_1(t) & (\tau_2-\tau_1)W^T & \sqrt{(\tau_2-\tau_1)}\Gamma_2^T P\overline{E}^T \\ * & -(\tau_2-\tau_1)R & 0 \\ * & * & -2\delta\overline{E}P^T + \delta^2 R \end{bmatrix} < 0, \quad (59)$$

where $\Phi_1(t)$ is defined in Lemma 11. Then, if (52)–(55) hold and with the changes of variables as $Y_i = P^T L_i$, we have

$$\Delta_1(t) = \sum_{i=1}^{r}\mu_i^2(\xi(t))\,\Xi_{ii} + \sum_{i=1}^{r}\sum_{i<j}^{r}\mu_i(\xi(t))\,\mu_j(\xi(t))\left(\Xi_{ij} + \Xi_{ji}\right) < 0,$$

$$\Delta_2(t) = \sum_{i=1}^{r}\mu_i^2(\xi(t))\,\Pi_{ii} + \sum_{i=1}^{r}\sum_{i<j}^{r}\mu_i(\xi(t))\,\mu_j(\xi(t))\left(\Pi_{ij} + \Pi_{ji}\right) < 0 \quad (60)$$

which imply that the error dynamics (37) are stable with $\omega(t) = 0$ while satisfying the prescribed H_∞ performance (39) by Lemma 11. The proof is completed. □

Remark 13. Theorem 12 provides a criterion for designing H_∞ robust fault estimation observer of fuzzy singular time delay systems, which guarantees the stability of the resulting dynamic error system with H_∞ performance $\gamma > 0$. As the delay term $\tau(t)$ is not simply enlarged, the proposed conditions are less conservative. Moreover, the proposed method is not only able to better depict the size and shape of the actuator fault but also able to estimate the sensor faults simultaneously.

Remark 14. Note that conditions (52)–(55) are LMIs. This indicates that the conditions (52)-(55) can be included as an optimization variable problems, which can be exploited to reduce the attenuation level bound. Then, the minimum attenuation level of H_∞ performance can be obtained by the mincx function of Matlab toolbox. From the practical point of view, it is interesting to find an estimation law, which minimizes the disturbance rejection level γ for the error dynamic system. This can be done by solving a convex optimization problem **P**: min ϑ subject to (51)–(55) with $\vartheta = \gamma^2$.

Remark 15. In dealing with time-varying faults, there may be a time delay between the fault estimation and the system fault. This phenomenon results from the influence of fault variation. Theoretically, the attenuation level $\gamma_{\min}$ can be minimized so that the fault estimation is insensitive to the fault variation. However, the cost is that the fault estimation becomes less robust to disturbance noise. Therefore, the attenuation level of $\gamma_{\min}$, the fault variation, and disturbance are a trade-off.

4. Numerical Examples

In this section, three examples are given to show the effectiveness of our results. All the numerical results are calculated via the Yalmip toolbox of Matlab.

Example 1. Consider a continuous fuzzy singular system composed of two rules and the following system matrices [22, 23]:

$$E = \begin{bmatrix} 1 & 0 & 0 & 0 \\ 0 & 1 & 0 & 0 \\ 0 & 0 & 1 & 0 \\ 0 & 0 & 0 & 0 \end{bmatrix},$$
$$A_1 = \begin{bmatrix} -3 & 0 & 0 & 0.2 \\ 0 & -4 & 0.1 & 0 \\ 0 & 0 & -0.1 & 0 \\ 0.1 & 0.1 & -0.2 & -0.2 \end{bmatrix},$$
$$A_2 = \begin{bmatrix} -2 & 0 & 0 & -0.2 \\ 0 & -2.5 & -0.1 & 0 \\ 0 & -0.2 & -0.3 & 0 \\ 0.1 & 0.1 & -0.2 & -0.2 \end{bmatrix},$$
$$A_{\tau 1} = \begin{bmatrix} -0.5 & 0 & 0 & 0 \\ 0 & -1 & 0 & 0 \\ 0 & 0.1 & -0.2 & 0 \\ 0 & 0 & 0 & 0 \end{bmatrix},$$
$$A_{\tau 2} = \begin{bmatrix} -0.5 & 0 & 0 & 0 \\ 0 & -1 & 0 & 0 \\ 0 & 0.1 & -0.5 & 0 \\ 0 & 0 & 0 & 0 \end{bmatrix}. \quad (61)$$

Assume that the delay $\tau(t)$ satisfies (2) and set $\rho = 0.5$. The obtained results are listed in Table 1, where $N(m)$ stands for the total number of decision variables. Table 1 tabulates a comparison of the maximum allowable upper delay bound τ_2 for a prescribed τ_D. It can be seen from the table that our results are marked better than those obtained by the method in [22, 23]. Moreover, it is worth mentioning that the method proposed in this paper uses fewer number of LMIs scalar variables and fewer number of LMIs for stability computation; thus our method is more computationally efficient for improving the upper bound of delay; the stability criterion we derived is less conservative than those reported in the aforementioned papers.

Example 2. Consider the following singular time delay system:

$$\begin{bmatrix} 1 & 0 \\ 0 & 0 \end{bmatrix} \dot{x}(t) = \begin{bmatrix} 0.6341 & 0.5413 \\ -0.6121 & -1.1210 \end{bmatrix} x(t) + \begin{bmatrix} -0.4500 & 0 \\ 0 & -0.1210 \end{bmatrix} x(t-h). \quad (62)$$

To compare with the existing results, we assume that $\tau_D = 0$ and $\rho = 0.5$. Table 2 lists the comparison results on the maximum allowed time delay τ_2 via the methods in [10, 20, 22–24] and Theorem 7 in this paper. From the comparison result we can see that the stability criterion we derived by using free weighting matrix approach in this work is less conservative than those reported in [10, 20, 22–24].

Example 3. Consider the following fuzzy singular system with time-varying delay:

$$\begin{aligned} E\dot{x}(t) &= \sum_{i=1}^{2} \mu_i(\xi(t)) \{A_i x(t) + A_{\tau i} x(t-\tau(t)) \\ &\quad + B_i u(t) + B_{fi} f(t) + B_{di} d(t)\}, \\ y(t) &= \sum_{i=1}^{2} \mu_i(\xi(t)) \{C_i x(t) + C_{\tau i} x(t-\tau(t)) \\ &\quad + D_{fi} f(t) + D_{di} d(t)\}, \end{aligned} \quad (63)$$

where

$$E = \begin{bmatrix} 1 & 0 & 0 \\ 0 & 1 & 0 \\ 0 & 0 & 0 \end{bmatrix},$$
$$A_1 = \begin{bmatrix} -0.60 & 1 & -1 \\ -0.65 & -0.5 & 0.2 \\ 0.74 & -4 & -1 \end{bmatrix},$$
$$A_2 = \begin{bmatrix} -0.60 & 1 & -0.8 \\ -0.65 & -0.50 & 0.4 \\ 0.82 & -6.37 & -1 \end{bmatrix},$$

TABLE 1: Allowable upper bound of τ_2 for various τ_D in Example 1.

τ_D	0.1	0.35	0.6	0.85	0.9	0.95	$N(m)$
Theorem 1 in [22]	3.3623	2.9810	2.6010	1.8330	1.3080	—	24
Theorem 3.1 in [23]	3.3685	3.1560	3.1510	3.0760	2.6750	2.0780	19
Theorem 7	4.1230	3.9891	3.7104	3.5247	3.4450	3.3702	15

TABLE 2: Comparison of maximum allowed delays τ_2 in Example 2.

[10]	[20]	[24]	[22]	[23]	Theorem 7
—	2.1328	2.1372	2.1372	2.3393	3.1324

TABLE 3: Minimum index γ for various δ in Example 3 with $\tau_D = 0.2$.

δ	0.5	1	2	5	10
$\tau = 0.5$	1.6322	1.4559	1.3732	1.3465	1.5532

$$A_{\tau 1} = \begin{bmatrix} 0.07 & 0.02 & -0.01 \\ -0.09 & 0 & 0.02 \\ 0.07 & 0.05 & 0 \end{bmatrix},$$
$$A_{\tau 2} = \begin{bmatrix} 0.07 & 0.12 & -0.01 \\ -0.08 & 0 & 0.03 \\ 0.12 & 0.04 & 0 \end{bmatrix},$$
$$B_i = B_{fi} = \begin{bmatrix} 0.7 & 0 & 0 \end{bmatrix},$$
$$D_{di} = 0.2, \qquad (i = 1, 2),$$
$$B_{d1} = B_{d2} = \begin{bmatrix} 0.1 & 0.1 & 0.1 \end{bmatrix},$$
$$C_1 = C_2 = \begin{bmatrix} -0.2 & 0.5 & -0.15 \end{bmatrix},$$
$$C_{t1} = C_{t2} = \begin{bmatrix} -0.02 & 0.05 & -0.15 \end{bmatrix}. \tag{64}$$

Here, we consider the case where $\delta = 2$ and the time-varying delay is given as $\tau(t) = 0.3 + 0.2\sin(t)$, and a straightforward calculation gives $\tau_1 = 0.1$, $\tau_2 = 0.5$, and $\tau_D = 0.2$. By solving the conditions in (51)–(55), we obtain that the achieved $\gamma_{\min} = 1.3140$, and the feasible solution is (due to space consideration, we do not list all the matrices here)

$$P = \begin{bmatrix} 10.2168 & 11.6131 & 0.0000 & -4.2840 \\ 11.6131 & 32.7665 & 0.0000 & -2.8547 \\ 0.0000 & 0.0000 & 1.4842 & 0.0000 \\ -4.2840 & -2.8547 & 0.0000 & 2.1990 \end{bmatrix},$$

$$Y_1 = \begin{bmatrix} 4.0957 \\ 50.1306 \\ -7.9107 \\ -3.6154 \end{bmatrix}, \qquad Y_2 = \begin{bmatrix} 4.2943 \\ 48.5194 \\ -10.5807 \\ -3.7476 \end{bmatrix}. \tag{65}$$

The associate fault estimation observer gains in (36) are

$$L_1 = \begin{bmatrix} -20.3217 \\ 5.7954 \\ -5.3299 \end{bmatrix},$$
$$F_1 = -33.7106,$$
$$L_2 = \begin{bmatrix} -19.8103 \\ 5.6276 \\ -7.1289 \end{bmatrix},$$
$$F_2 = -32.9922. \tag{66}$$

According to Theorem 12, we can consider different δ to find the minimum index γ for the given $\tau_D = 0.2$; see [25, 28] for more details. The corresponding results are summarized in Table 3.

In order to illustrate the performance of robust fault estimation observer in dealing with fuzzy singular systems with time delay, first, an abrupt fault is simulated. It is assumed that the abrupt fault $f(t)$ is created as

$$f(t) = \begin{cases} 0 & 0 \le t < 10 \\ -10 & 10 \le t \le 30. \end{cases} \tag{67}$$

For simulation purpose, we choose the membership functions for Rules 1 and 2 to be $\mu_1(\xi(t)) = 1/(1+\exp(x_1(t)+0.5))$ and $\mu_2(\xi(t)) = 1 - \mu_1(\xi(t))$ with the initial state condition as $x(0) = \begin{bmatrix} -1.571 & -1.356 & 1.279 \end{bmatrix}^T$ while the initial estimate is $\hat{x}(0) = \begin{bmatrix} 0 & 0 & 0 \end{bmatrix}^T$. Meanwhile, it is assumed that $d(t)$ is band-limited white noise with power 0.1 and sampling time 0.01s. Then, the fault estimation result of the robust fault estimation observer is depicted in Figure 1. Therein, the fault is depicted by red dash-and-dot line, and the fault estimation

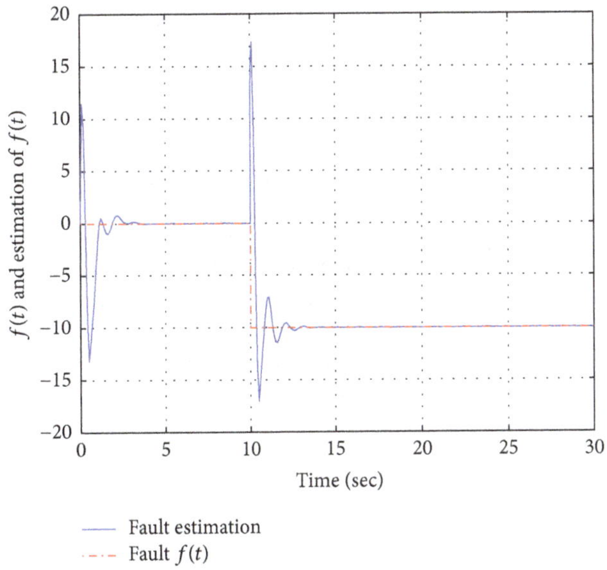

FIGURE 1: Fault estimation result of the robust fault estimation observer in abrupt fault case $f(t)$.

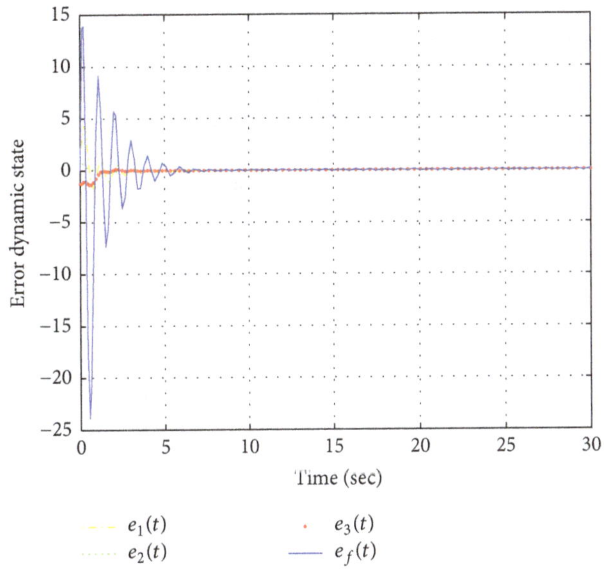

FIGURE 2: Response curves of error dynamics $e_1(t)$, $e_2(t)$, $e_3(t)$, and $e_f(t)$ in abrupt fault case $f(t)$.

is represented by the blue solid one. As shown in Figure 1, the robust fault estimation observer is insensitive to the model disturbance. Moreover, although there is estimation error, the fault estimate can quickly track the fault. This illustrates the fast convergence rate of the fault estimation observer in the face of initial estimation error. The simulation results shown in Figure 1 obviously illustrate that the proposed fault estimation has a good performance to estimate fault, and the error dynamic system is also stable in Figure 2. To illustrate the performance of robust fault estimation observer

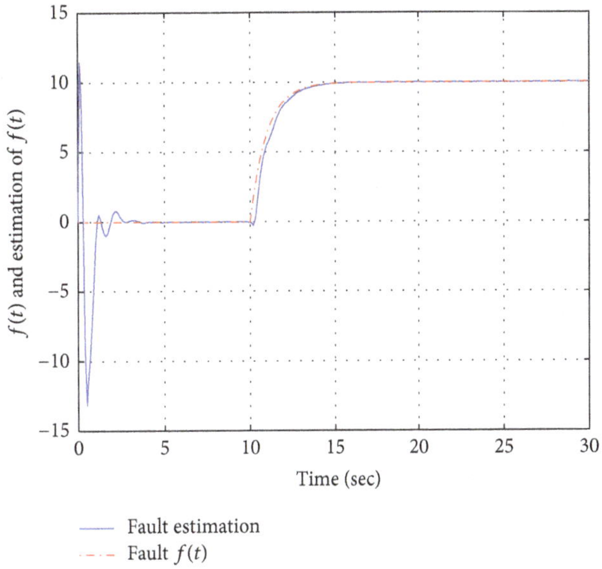

FIGURE 3: Fault estimation result of the robust fault estimation observer in fault $f_1(t)$.

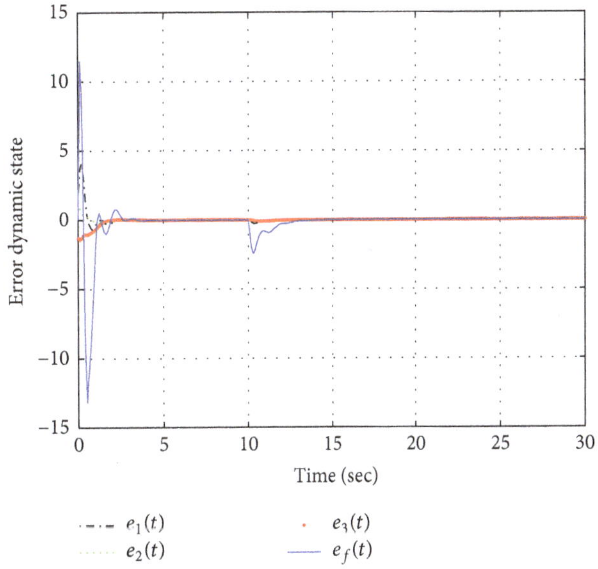

FIGURE 4: Response curves of error dynamics $e_1(t)$, $e_2(t)$, $e_3(t)$, and $e_f(t)$ in fault $f_1(t)$.

design method, the following time-varying fault is further considered:

$$f_1(t) = \begin{cases} 0 & 0 \le t < 10 \\ 10\left(1 - e^{-(t-5)}\right) & 10 \le t \le 20, \end{cases}$$
$$f_2(t) = \begin{cases} 0 & 0 \le t < 10 \\ 0.5\sin(0.8(t-5)) & 10 \le t \le 20. \end{cases} \quad (68)$$

In this situation, the fault estimation result is depicted in Figures 3 and 5. It can be seen from Figures 3 and 5 that the fault is estimated with satisfactory accuracy and

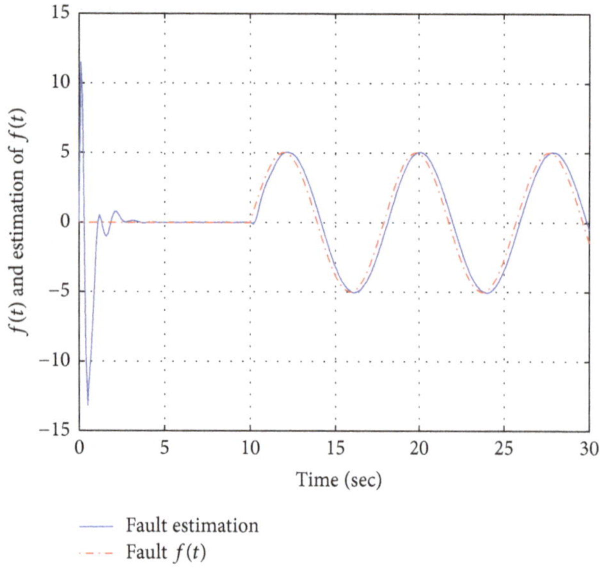

FIGURE 5: Fault estimation result of the robust fault estimation observer in fault $f_2(t)$.

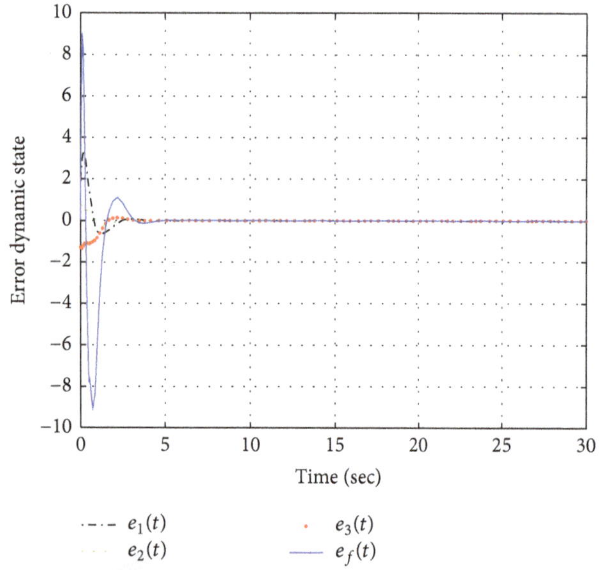

FIGURE 6: Response curves of error dynamics $e_1(t)$, $e_2(t)$, $e_3(t)$, and $e_f(t)$ in fault $f_2(t)$.

rapidity. The state of the error dynamic system is also stable in Figures 4 and 6. In [13], actuator fault estimation observer is designed for discrete-time linear parameter-varying descriptor systems, but the systems with time-varying delay case are not considered. In [9], pole assignment is used to ensure the fault estimation convergence speed while our method utilizes H_∞ technique to attenuate the effect of fault variation. From Figures 3 and 5, it can be seen that these methods have similar fault convergence speed. Nevertheless, [9] only deals with regular systems not with time delay. Moreover, our method considers the fuzzy singular system with actuator and sensor faults simultaneously. Therefore, the proposed method is more general than that in [9, 13].

5. Conclusion

In this paper, the robust fault estimation problem for T-S fuzzy singular systems with time-varying delays is considered. By considering the fault as an auxiliary disturbance vector, based on the Lyapunov theorem and improved delay partitioning method with free weighting matrix approach, we give some less conservative criteria, which guarantee that the considered system is regular, impulse-free, and stable, while limiting the influence of disturbance despite the presence of actuator and sensor faults simultaneously. This paper proposes a novel fault estimation observer and presents an LMI-based design method for the fuzzy singular system. Finally, some numerical examples are used to demonstrate the effectiveness and performance of the proposed method.

Conflict of Interests

The authors declare that there is no conflict of interests regarding the publication of this paper.

Acknowledgments

This work is supported in part by the National Natural Science Foundation of China (Grant nos. 61533007, 61374146 and 61174215), Project 863 of China (Grant no. 2011AA060204), and IAPI Fundamental Research Funds (Grant no. 2013ZCX02-04).

References

[1] D. Theilliol, H. Noura, and J.-C. Ponsart, "Fault diagnosis and accommodation of a three-tank system based on analytical redundancy," *ISA Transactions*, vol. 41, no. 3, pp. 365–382, 2002.

[2] M. Rodrigues, D. Theilliol, M. Adam-Medina, and D. Sauter, "A fault detection and isolation scheme for industrial systems based on multiple operating models," *Control Engineering Practice*, vol. 16, no. 2, pp. 225–239, 2008.

[3] J. Bokor and Z. Szabó, "Fault detection and isolation in nonlinear systems," *Annual Reviews in Control*, vol. 33, no. 2, pp. 113–123, 2009.

[4] P. Mhaskar, C. McFall, A. Gani, P. D. Christofides, and J. F. Davis, "Isolation and handling of actuator faults in nonlinear systems," *Automatica*, vol. 44, no. 1, pp. 53–62, 2008.

[5] M. Liu, X. Cao, and P. Shi, "Fault estimation and tolerant control for fuzzy stochastic systems," *IEEE Transactions on Fuzzy Systems*, vol. 21, no. 2, pp. 221–229, 2013.

[6] Y. Zhang and J. Jiang, "Bibliographical review on reconfigurable fault-tolerant control systems," *Annual Reviews in Control*, vol. 32, no. 2, pp. 229–252, 2008.

[7] G. R. Duan, D. Howe, and R. J. Patton, "Robust fault detection in descriptor linear systems via generalized unknown input observers," *International Journal of Systems Science*, vol. 33, no. 5, pp. 369–377, 2002.

[8] J. Lin, S. Fei, Z. Gao, and J. Ding, "Fault detection for discrete-time switched singular time-delay systems: an average dwell

time approach," *International Journal of Adaptive Control and Signal Processing*, vol. 27, no. 7, pp. 582–609, 2013.

[9] K. Zhang, B. Jiang, and P. Shi, "Fault estimation observer design for discrete-time takagi-sugeno fuzzy systems based on piecewise lyapunov functions," *IEEE Transactions on Fuzzy Systems*, vol. 20, no. 1, pp. 192–200, 2012.

[10] E. K. Boukas, "Singular linear systems with delay: $\mathscr{H}_\infty$ stabilization," *Optimal Control Applications and Methods*, vol. 28, no. 4, pp. 259–274, 2007.

[11] M. Wu, Y. He, J.-H. She, and G.-P. Liu, "Delay-dependent criteria for robust stability of time-varying delay systems," *Automatica*, vol. 40, no. 8, pp. 1435–1439, 2004.

[12] A. T. Vemuri, M. M. Polycarpou, and A. R. Ciric, "Fault diagnosis of differential-algebraic systems," *IEEE Transactions on Systems, Man, and Cybernetics Part A: Systems and Humans.*, vol. 31, no. 2, pp. 143–152, 2001.

[13] Z. H. Wang, M. Rodrigues, D. Theilliol, and Y. Shen, "Actuator fault estimation observer design for discrete-time linear parameter-varying descriptor systems," *International Journal of Adaptive Control and Signal Processing*, vol. 29, no. 2, pp. 242–258, 2015.

[14] B. Boulkroune, S. Halabi, and A. Zemouche, "H_-/H_∞ fault detection filter for a class of nonlinear descriptor systems," *International Journal of Control*, vol. 86, no. 2, pp. 253–262, 2013.

[15] H. Hamdi, M. Rodrigues, C. Mechmeche, D. Theilliol, and N. B. Braiek, "Fault detection and isolation in linear parameter-varying descriptor systems via proportional integral observer," *International Journal of Adaptive Control and Signal Processing*, vol. 26, no. 3, pp. 224–240, 2012.

[16] M. Chadli, A. Abdo, and S. X. Ding, "H_-/H_∞ fault detection filter design for discrete-time Takagi-Sugeno fuzzy system," *Automatica*, vol. 49, no. 7, pp. 1996–2005, 2013.

[17] S.-J. Huang and G.-H. Yang, "Fault tolerant controller design for T-S fuzzy systems with time-varying delay and actuator faults: a k-step fault-estimation approach," *IEEE Transactions on Fuzzy Systems*, vol. 22, no. 6, pp. 1526–1540, 2014.

[18] S. Xu and J. Lam, *Robust Control and Filtering of Singular Systems*, Springer, Berlin, Germany, 2006.

[19] S. Xu, P. Van Dooren, R. Stefan, and J. Lam, "Robust stability and stabilization for singular systems with state delay and parameter uncertainty," *IEEE Transactions on Automatic Control*, vol. 47, no. 7, pp. 1122–1128, 2002.

[20] E. Fridman, "Stability of linear descriptor systems with delay: a Lyapunov-based approach," *Journal of Mathematical Analysis and Applications*, vol. 273, no. 1, pp. 24–44, 2002.

[21] D. Koenig and S. Mammar, "Design of proportional-integral observer for unknown input descriptor systems," *IEEE Transactions on Automatic Control*, vol. 47, no. 12, pp. 2057–2062, 2002.

[22] H. Zhang, Y. Shen, and G. Feng, "Delay-dependent stability and H_∞ control for a class of fuzzy descriptor systems with time-delay," *Fuzzy Sets and Systems*, vol. 160, no. 12, pp. 1689–1707, 2009.

[23] K. Mourad, S. Mansour, and T. Ahmed, "Delay-dependent stability and robust $L_2 - L_\infty$ control for a class of fuzzy descriptor systems with time-varying delay," *International Journal of Robust and Nonlinear Control*, vol. 23, no. 3, pp. 284–304, 2013.

[24] F. Yang and Q. L. Zhang, "Delay-dependent H_∞ control for linear descriptor systems with delay in state," *Journal of Control Theory and Applications*, vol. 3, no. 1, pp. 76–84, 2005.

[25] S. J. Huang, X. Q. He, and N. N. Zhang, "New results on H_∞ filter design for nonlinear systems with time delay via T-S fuzzy models," *IEEE Transactions on Fuzzy Systems*, vol. 19, no. 1, pp. 193–199, 2011.

[26] C. Lin, Q.-G. Wang, T. H. Lee, and B. Chen, "H_∞ filter design for nonlinear systems with time-delay through T-S fuzzy model approach," *IEEE Transactions on Fuzzy Systems*, vol. 16, no. 3, pp. 739–746, 2008.

[27] C.-H. Lien and K.-W. Yu, "Robust control for Takagi-Sugeno fuzzy systems with time-varying state and input delays," *Chaos, Solitons & Fractals*, vol. 35, no. 5, pp. 1003–1008, 2008.

[28] Y. K. Su, B. Chen, C. Lin, and H. G. Zhang, "A new fuzzy H_∞ filter design for nonlinear continuous-time dynamic systems with time-varying delays," *Fuzzy Sets and Systems*, vol. 160, no. 24, pp. 3539–3549, 2009.

[29] J. H. Zhang, Y. Q. Xia, and R. Tao, "New results on H_∞ filtering for fuzzy time-delay systems," *IEEE Transactions on Fuzzy Systems*, vol. 17, no. 1, pp. 128–137, 2009.

21

Indefinite LQ Optimal Control with Terminal State Constraint for Discrete-Time Uncertain Systems

Yuefen Chen[1,2] and Minghai Yang[2]

[1]*School of Science, Nanjing University of Science and Technology, Nanjing 210094, China*
[2]*College of Mathematics and Information Science, Xinyang Normal University, Xinyang 464000, China*

Correspondence should be addressed to Yuefen Chen; yfchen@xynu.edu.cn

Academic Editor: Petko Petkov

Uncertainty theory is a branch of mathematics for modeling human uncertainty based on the normality, duality, subadditivity, and product axioms. This paper studies a discrete-time LQ optimal control with terminal state constraint, whereas the weighting matrices in the cost function are indefinite and the system states are disturbed by uncertain noises. We first transform the uncertain LQ problem into an equivalent deterministic LQ problem. Then, the main result given in this paper is the necessary condition for the constrained indefinite LQ optimal control problem by means of the Lagrangian multiplier method. Moreover, in order to guarantee the well-posedness of the indefinite LQ problem and the existence of an optimal control, a sufficient condition is presented in the paper. Finally, a numerical example is presented at the end of the paper.

1. Introduction

The linear quadratic (LQ) optimal control problem has been pioneered by Kalman [1] for deterministic systems, which is extended to stochastic systems by Wonham [2], and has rapid development in both theory and application [3]. Usually, it is an assumption that the control weighting matrix in the cost is strictly definite. For stochastic LQ optimal control, it is first revealed in [4] that even if the state and control weighting matrices are indefinite the corresponding problem may be still well-posed, which evoked a series of subsequent researches in continuous time [5] and in discrete-time [6]. In fact, some constraints are of considerable importance in many physical systems; the system state and control input are always subject to various constraints, so the constrained stochastic LQ issue has a concrete application background. For that reason, some researchers discussed stochastic LQ optimal problems with indefinite control weights and constraints [7, 8].

As is well known, these stochastic optimal control problems have been well studied by probability theory which is based on a large number of sample sizes. Sometimes, no samples are available to estimate the probability distribution. For such situation, we have to invite some domain experts to evaluate the belief degree that each event will occur. In order to rationally deal with belief degrees, uncertainty theory was established by Liu [9] in 2007 and refined by Liu [10] in 2010. Nowadays, uncertainty theory has become a new branch of mathematics for modeling indeterminate phenomena, which has been well developed and applied in a wide variety of real problems: option pricing problem [11], facility location problem [12], inventory problem [13], assignment problem [14], and production control problem [15].

Based on the uncertainty theory, Zhu [16] proposed an uncertain optimal control model in 2010 and gave an equation of optimality as a counterpart of Hamilton-Jacobi-Bellman equation. After that, some uncertain optimal control problems have been solved. As such, Sheng and Zhu [17] investigated an optimistic value model of uncertain optimal control problem; Yan and Zhu [18] established an uncertain optimal control model for switched systems. Inspired by the preceding work, we will tackle an indefinite LQ optimal control with terminal state constraint for discrete-time uncertain systems, which is a constrained uncertain optimal control problem. The rest of the paper is organized as follows. Section 2 collects some preliminary results. In

Section 3, an indefinite LQ optimal control with terminal state constraint is discussed. We present a general expression for the optimal control set in Section 4. A numerical example is applied in Section 5 to demonstrate the effectiveness of the model. We conclude the paper in Section 6.

For convenience, throughout the paper, we adopt the following notations: $\mathbf{R}^n$ is the real n-dimensional Euclidean space; $\mathbf{R}^{m\times n}$ is the set of all $m\times n$ matrices; M^τ is the transpose of matrix M; and $\mathrm{tr}(M)$ is the trace of a square matrix M. Moreover, $M>0$ (resp., $M\ge 0$) means that $M=M^\tau$ and M is positive (resp., positive semidefinite) definite.

2. Some Preliminaries

In this section, we introduce some useful definitions about uncertainty theory and Moore-Penrose pseudoinverse of a matrix.

Let Γ be a nonempty set, and let $\mathscr{L}$ be a σ-algebra over Γ. Each element Λ in $\mathscr{L}$ is called an event. An *uncertain measure* was defined by Liu [9] via the following three axioms.

Axiom 1 (normality axiom). $\mathscr{M}\{\Gamma\}=1$ for the universal set Γ.

Axiom 2 (duality axiom). $\mathscr{M}\{\Lambda\}+\mathscr{M}\{\Lambda^c\}=1$ for any event Λ.

Axiom 3 (subadditivity axiom). For every countable sequence of events $\Lambda_1,\Lambda_2,\ldots$, we have

$$\mathscr{M}\left\{\bigcup_{i=1}^{\infty}\Lambda_i\right\}\le\sum_{i=1}^{\infty}\mathscr{M}\{\Lambda_i\}. \tag{1}$$

The triplet $(\Gamma,\mathscr{L},\mathscr{M})$ is called an *uncertainty space*. Furthermore, Liu [19] defined a product uncertain measure by the product axiom.

Axiom 4 (product axiom). Let $(\Gamma_k,\mathscr{L}_k,\mathscr{M}_k)$ be uncertainty spaces for $k=1,2,\ldots$. Then, the product uncertain measure $\mathscr{M}$ on the product σ-algebra satisfies

$$\mathscr{M}\left\{\prod_{k=1}^{\infty}\Lambda_k\right\}=\bigwedge_{k=1}^{\infty}\mathscr{M}_k\{\Lambda_k\}, \tag{2}$$

where Λ_k are arbitrarily chosen events from $\mathscr{L}_k$ for $k=1,2,\ldots$, respectively.

An *uncertain variable* is defined by Liu [9] as a function ξ from an uncertainty space $(\Gamma,\mathscr{L},\mathscr{M})$ to the set of real numbers such that $\{\xi\in B\}$ is an event for any Borel set B. In addition, an *uncertainty distribution* of ξ is defined as

$$\Phi(x)=\mathscr{M}\{\gamma\in\Gamma\mid\xi(\gamma)\le x\}, \tag{3}$$

for any real number x.

Independence is an important concept in uncertainty theory. The uncertain variables $\xi_1,\xi_2,\ldots,\xi_m$ are said to be *independent* (Liu [19]) if

$$\mathscr{M}\left\{\bigcap_{i=1}^{m}(\xi_i\in B_i)\right\}=\min_{1\le i\le m}\mathscr{M}\{\xi_i\in B_i\} \tag{4}$$

for any Borel sets $B_1,B_2,\ldots,B_n$ of real numbers.

An uncertain variable ξ is called *linear* (Liu [9]) if it has a linear uncertainty distribution

$$\Phi(x)=\begin{cases}0, & \text{if } x\le a,\\ \dfrac{(x-a)}{(b-a)}, & \text{if } a\le x\le b,\\ 1, & \text{if } x\ge b\end{cases} \tag{5}$$

denoted by $\mathscr{L}(a,b)$, where a and b are real numbers with $a<b$.

Let ξ be an uncertain variable. Then, the *expected value* (Liu [9]) of ξ is defined by

$$E[\xi]=\int_0^{+\infty}\mathscr{M}\{\xi\ge r\}\,\mathrm{d}r-\int_{-\infty}^{0}\mathscr{M}\{\xi\le r\}\,\mathrm{d}r \tag{6}$$

provided that at least one of the two integrals is finite.

Remark 1. For numbers a and b, $E[a\xi+b\eta]=aE[\xi]+bE[\eta]$ if ξ and η are independent uncertain variables. Generally speaking, the expected value operator is not necessarily linear if the independence is not assumed.

Remark 2. Let

$$\boldsymbol{\xi}=\begin{pmatrix}\xi_{11} & \xi_{12} & \cdots & \xi_{1q}\\ \xi_{21} & \xi_{22} & \cdots & \xi_{2q}\\ \cdots & \cdots & \cdots & \cdots\\ \xi_{p1} & \xi_{p2} & \cdots & \xi_{pq}\end{pmatrix}, \tag{7}$$

where ξ_{ij} are uncertain variables for $i=1,2,\ldots,p$, $j=1,2,\ldots,q$. The expected value of $\boldsymbol{\xi}$ is provided by

$$E[\boldsymbol{\xi}]=\begin{pmatrix}E[\xi_{11}] & E[\xi_{12}] & \cdots & E[\xi_{1q}]\\ E[\xi_{21}] & E[\xi_{22}] & \cdots & E[\xi_{2q}]\\ \cdots & \cdots & \cdots & \cdots\\ E[\xi_{p1}] & E[\xi_{p2}] & \cdots & E[\xi_{pq}]\end{pmatrix}. \tag{8}$$

Lemma 3 (Penrose [20]). *Let a matrix $M\in\mathbf{R}^{m\times n}$ be given. Then, there exists a unique matrix $M^+\in\mathbf{R}^{n\times m}$ such that*

$$\begin{aligned}MM^+M&=M,\\ M^+MM^+&=M^+,\\ (MM^+)^\tau&=MM^+,\\ (M^+M)^\tau&=M^+M.\end{aligned} \tag{9}$$

The matrix M^+ is called the Moore-Penrose pseudoinverse of M.

Lemma 4 (Penrose [20]). *Let matrices L, M, and N be given with appropriate sizes. Then, the matrix equation $LXM=N$ has a solution X if and only if $LL^+NMM^+=N$. Moreover, any solution to $LXM=N$ is represented by $X=L^+NM^+ +Y-L^+LYMM^+$, where Y is a matrix with an appropriate size.*

3. Indefinite LQ Optimal Control with Constraints

3.1. Problem Statement. Consider the following indefinite LQ optimal control with terminal state constraint for discrete-time uncertain systems:

$$\inf_{\substack{\mathbf{u}_k\\0\le k\le N-1}} J(\mathbf{x}_0,\mathbf{u})$$

$$= \sum_{k=0}^{N-1} E\left[\mathbf{x}_k^\tau Q_k \mathbf{x}_k + \mathbf{u}_k^\tau R_k \mathbf{u}_k\right] + E\left[\mathbf{x}_N^\tau Q_N \mathbf{x}_N\right] \tag{10}$$

$$\text{subject to}\quad \mathbf{x}_{k+1} = A_k\mathbf{x}_k + B_k\mathbf{u}_k + \lambda_k\left(A_k\mathbf{x}_k + B_k\mathbf{u}_k\right)\xi_k,$$

$$k = 0,1,\ldots,N-1,\ \mathbf{x}(0) = \mathbf{x}_0$$

$$E\left[\mathbf{x}_N^\tau \mathbf{x}_N\right] = c,$$

where $0 \le |\lambda_k| \le 1$, state $\mathbf{x}_k \in \mathbf{R}^n$, control input $\mathbf{u}_k \in \mathbf{R}^m$, $k = 0,1,\ldots,N-1$, and $\mathbf{x}_0 \in \mathbf{R}^n$ is a given crisp vector. Denote $\mathbf{u} = (\mathbf{u}_0,\mathbf{u}_1,\ldots,\mathbf{u}_{N-1})$. Moreover, $Q_0,Q_1,\ldots,Q_N$ and $R_0,R_1,\ldots,R_{N-1}$ are real symmetric matrices with appropriate dimensions. In addition, $c \ge 0$ is a constant; the coefficients $A_0,A_1,\ldots,A_{N-1}$ and $B_0,B_1,\ldots,B_{N-1}$ are crisp matrices having appropriate dimensions determined from context. Besides, the noises $\xi_0,\xi_1,\ldots,\xi_{N-1}$ are independent linear uncertain variables $\mathscr{L}(-1,1)$ with the distribution

$$\Phi(x) = \begin{cases} 0, & \text{if } x \le -1,\\ \dfrac{(x+1)}{2}, & \text{if } -1 \le x \le 1,\\ 1, & \text{if } x \ge 1.\end{cases} \tag{11}$$

In this paper, the weighting matrices in the objective functional are not required to be definite. Therefore, problem (10) is an indefinite LQ optimal control problem. Next, we give the following definitions.

Definition 5. The indefinite LQ problem (10) is called well-posed if

$$V(\mathbf{x}_0) = \inf_{\substack{\mathbf{u}_k\\0\le k\le N-1}} J(\mathbf{x}_0,\mathbf{u}) > -\infty,\quad \forall \mathbf{x}_0 \in \mathbf{R}^n. \tag{12}$$

Definition 6. A well-posed problem is called solvable, if, for $\mathbf{x}_0 \in \mathbf{R}^n$, there is a control sequence $(\mathbf{u}_0^*,\mathbf{u}_1^*,\ldots,\mathbf{u}_{N-1}^*)$ that achieves $V(\mathbf{x}_0)$. In this case, the control $(\mathbf{u}_0^*,\mathbf{u}_1^*,\ldots,\mathbf{u}_{N-1}^*)$ is called an optimal control sequence.

3.2. An Equivalent Problem. Next, we transform the uncertain LQ optimal control problem (10) into an equivalent deterministic LQ optimal control problem which is subject to a matrix difference equation constraint.

Let $X_k = E[\mathbf{x}_k\mathbf{x}_k^\tau]$. Since state $\mathbf{x}_k \in \mathbf{R}^n$, $\mathbf{x}_k\mathbf{x}_k^\tau$ is $n\times n$ matrix whose elements are uncertain variables, and X_k is a symmetric crisp matrix ($k = 0,1,\ldots,N$). Denote $\mathbf{K} = (K_0,K_1,\ldots,K_{N-1})$, where K_i are matrices for $i = 0,1,\ldots,N-1$.

Theorem 7. *If the indefinite LQ problem (10) is solvable by a feedback control*

$$\mathbf{u}_k = K_k\mathbf{x}_k, \tag{13}$$

where K_k are constant crisp matrices, then it is equivalent to the following deterministic optimal control problem:

$$\min_{\substack{K_k\\0\le k\le N-1}} J(X_0,\mathbf{K}) = \sum_{k=0}^{N-1}\operatorname{tr}\left[\left(Q_k + K_k^\tau R_k K_k\right)X_k\right] + \operatorname{tr}\left[Q_N X_N\right]$$

$$\textit{subject to}\quad X_{k+1} = \left(1+\frac{1}{3}\lambda_k^2\right)\left(A_kX_kA_k^\tau + A_kX_kK_k^\tau B_k^\tau + B_kK_kX_kA_k^\tau + B_kK_kX_kK_k^\tau B_k^\tau\right), \tag{14}$$

$$X_0 = \mathbf{x}_0\mathbf{x}_0^\tau,$$

$$\operatorname{tr}\left[X_N\right] = c,$$

for $k = 0,1,\ldots,N-1$.

Proof. Assume that the indefinite LQ problem (10) is solvable by a feedback control

$$\mathbf{u}_k = K_k\mathbf{x}_k, \tag{15}$$

for $k = 0,1,\ldots,N-1$. Let $X_k = E[\mathbf{x}_k\mathbf{x}_k^\tau]$ for $k = 0,1,\ldots,N$. Then, we have

$$\begin{aligned} X_{k+1} &= E\left[\mathbf{x}_{k+1}\mathbf{x}_{k+1}^\tau\right]\\ &= E\{\left[A_k + B_kK_k + \lambda_k\left(A_k + B_kK_k\right)\xi_k\right]\\ &\quad\cdot \mathbf{x}_k\mathbf{x}_k^\tau\left[A_k^\tau + K_k^\tau B_k^\tau + \lambda_k\left(A_k^\tau + K_k^\tau B_k^\tau\right)\xi_k\right]\}\\ &= A_kX_kA_k^\tau + A_kX_kK_k^\tau B_k^\tau + B_kK_kX_kA_k^\tau\\ &\quad + B_kK_kX_kK_k^\tau B_k^\tau + E\left[U_k\xi_k + V_k\xi_k^2\right], \end{aligned} \tag{16}$$

where

$$U_k = 2\lambda_k\left(A_kX_kA_k^\tau + A_kX_kK_k^\tau B_k^\tau + B_kK_kX_kA_k^\tau + B_kK_kX_kK_k^\tau B_k^\tau\right)$$

$$V_k = \lambda_k^2 (A_k X_k A_k^\tau + A_k X_k K_k^\tau B_k^\tau + B_k K_k X_k A_k^\tau + B_k K_k X_k K_k^\tau B_k^\tau). \quad (17)$$

Then, we obtain that $\lambda_k U_k = 2V_k$. Because ξ_k and ξ_k^2 are not independent, we know that

$$E[U_k\xi_k + V_k\xi_k^2] \neq U_k E[\xi_k] + V_k E[\xi_k^2]. \quad (18)$$

We will deal with (18) as follows.

(i) If $V_k = \mathbf{0}$, we obtain

$$E[U_k\xi_k + V_k\xi_k^2] = E[U_k\xi_k] = U_k E[\xi_k] = \mathbf{0}. \quad (19)$$

(ii) If $V_k \neq \mathbf{0}$, we know that $\lambda_k \neq 0$ and $|2/\lambda_k| \geq 2$. According to Example 2 in [21], we have

$$E[U_k\xi_k + V_k\xi_k^2] = E\left[\frac{2}{\lambda_k}V_k\xi_k + V_k\xi_k^2\right] = V_k E\left[\frac{2}{\lambda_k}\xi_k + \xi_k^2\right] = \frac{1}{3}V_k. \quad (20)$$

Therefore, we have

$$E[U_k\xi_k + V_k\xi_k^2] = \frac{1}{3}V_k. \quad (21)$$

Substituting (21) into (16) produces the following state matrix:

$$X_{k+1} = \left(1 + \frac{1}{3}\lambda_k^2\right)(A_k X_k A_k^\tau + A_k X_k K_k^\tau B_k^\tau + B_k K_k X_k A_k^\tau + B_k K_k X_k K_k^\tau B_k^\tau). \quad (22)$$

The associated cost function reduces to

$$\min_{\substack{K_k \\ 0\leq k\leq N-1}} J(X_0, \mathbf{K}) = \min_{\substack{K_k \\ 0\leq k\leq N-1}} \sum_{k=0}^{N-1} \mathrm{tr}[(Q_k + K_k^\tau R_k K_k) X_k] + \mathrm{tr}[Q_N X_N], \quad (23)$$

and the constraint $E[\mathbf{x}_N^\tau \mathbf{x}_N] = c$ becomes $\mathrm{tr}[X_N] = c$. □

Remark 8. Obviously, if problem (10) has a linear feedback optimal control solution $\mathbf{u}_k^* = K_k^* \mathbf{x}_k$ $(k = 0, 1, \ldots, N-1)$, then K_k^* $(k = 0, 1, \ldots, N-1)$ is the optimal solution of problem (14).

3.3. A Necessary Condition for State Feedback Control. In this subsection, a necessary condition for the optimal linear state feedback control with deterministic gains to the indefinite LQ problem (10) is obtained by applying the deterministic matrix maximum principle [22].

Theorem 9. *If the indefinite LQ problem (10) is solvable by a feedback control*

$$\mathbf{u}_k = K_k \mathbf{x}_k, \quad (24)$$

where K_k are constant crisp matrices, then there exist symmetric matrices H_k and a nonnegative $\gamma \in \mathbf{R}^1$ solving the following constrained difference equation:

$$\begin{aligned} H_k &= Q_k + \left(1 + \frac{1}{3}\lambda_k^2\right) A_k^\tau H_{k+1} A_k - M_k^\tau L_k^+ M_k, \\ L_k L_k^+ M_k - M_k &= 0, \\ L_k &= R_k + \left(1 + \frac{1}{3}\lambda_k^2\right) B_k^\tau H_{k+1} B_k \geq 0, \\ M_k &= \left(1 + \frac{1}{3}\lambda_k^2\right) B_k^\tau H_{k+1} A_k, \\ H_N &= Q_N + \gamma I, \end{aligned} \quad (25)$$

for $k = 0, 1, \ldots, N-1$. Moreover,

$$K_k = -L_k^+ M_k + Y_k - L_k^+ L_k Y_k \quad (26)$$

with $Y_k \in \mathbf{R}^{m\times n}$, $k = 0, 1, \ldots, N-1$, being any given crisp matrices.

Proof. Assume that the indefinite LQ problem (10) is solvable by

$$\mathbf{u}_k = K_k \mathbf{x}_k, \quad (27)$$

where the matrices K_k $(k = 0, 1, \ldots, N-1)$ are viewed as the control to be determined. It is obvious that K_k is also the optimal solution of problem (14) which is deterministic LQ optimal control problem. Hence, we can apply the matrix Lagrangian multiplier method to solve problem (14).

Let matrices H_{k+1} $(k = 0, 1, \ldots, N-1)$ be the Lagrange multipliers of $\mathbf{h}_{k+1}(X_k, K_k)$ $(k = 0, 1, \ldots, N-1)$, and let $\gamma \in \mathbf{R}^1$ be the Lagrange multiplier of $g(X_N) = 0$. Then, the Lagrange function is formed as

$$\mathscr{L} = J(X_0, \mathbf{K}) + \sum_{k=0}^{N-1} \mathrm{tr}[H_{k+1}\mathbf{h}_{k+1}(X_k, K_k)] + \gamma g(X_N), \quad (28)$$

where

$$J(X_0, \mathbf{K}) = \sum_{k=0}^{N-1} \mathrm{tr}[(Q_k + K_k^\tau R_k K_k) X_k] + \mathrm{tr}[Q_N X_N]$$

$$\mathbf{h}_{k+1}(X_k, K_k) = \left(1 + \frac{1}{3}\lambda_k^2\right)(A_k X_k A_k^\tau + A_k X_k K_k^\tau B_k^\tau + B_k K_k X_k A_k^\tau + B_k K_k X_k K_k^\tau B_k^\tau) - X_{k+1}, \quad (29)$$

$$g(X_N) = \mathrm{tr}[X_N] - c.$$

According to the first-order necessary conditions for optimality [22], we have

$$\frac{\partial \mathscr{L}}{\partial K_k} = 0 \quad (k = 0, 1, \ldots, N-1), \tag{30}$$

$$H_k = \frac{\partial \mathscr{L}}{\partial X_k} \quad (k = 0, 1, \ldots, N-1), \tag{31}$$

$$H_N = Q_N + \gamma I. \tag{32}$$

Based on the partial rule of gradient matrices [22], (30) can be transformed into

$$\begin{aligned}&\left[R_k + \left(1 + \frac{1}{3}\lambda_k^2\right) B_k^\tau H_{k+1} B_k\right] K_k \\ &\quad + \left(1 + \frac{1}{3}\lambda_k^2\right) B_k^\tau H_{k+1} A_k = 0.\end{aligned} \tag{33}$$

Let

$$\begin{aligned}L_k &= R_k + \left(1 + \frac{1}{3}\lambda_k^2\right) B_k^\tau H_{k+1} B_k, \\ M_k &= \left(1 + \frac{1}{3}\lambda_k^2\right) B_k^\tau H_{k+1} A_k.\end{aligned} \tag{34}$$

Then, (33) can be rewritten as $L_k K_k + M_k = 0$. Applying Lemma 4, we have $L_k L_k^+ M_k = M_k$, and

$$K_k = -L_k^+ M_k + Y_k - L_k^+ L_k Y_k, \quad Y_k \in \mathbf{R}^{m\times n}. \tag{35}$$

For (31), according to

$$H_k = \frac{\partial \mathscr{L}}{\partial X_k} \quad (k = 0, 1, \ldots, N-1), \tag{36}$$

we have

$$\begin{aligned}H_k &= Q_k + \left(1 + \frac{1}{3}\lambda_k^2\right) A_k^\tau H_{k+1} A_k \\ &\quad + K_k^\tau \left[R_k + \left(1 + \frac{1}{3}\lambda_k^2\right) B_k^\tau H_{k+1} B_k\right] K_k \\ &\quad + \left(1 + \frac{1}{3}\lambda_k^2\right) A_k^\tau H_{k+1} B_k K_k \\ &\quad + \left(1 + \frac{1}{3}\lambda_k^2\right) K_k^\tau B_k^\tau H_{k+1} A_k.\end{aligned} \tag{37}$$

Substituting (35) into (37), we obtain

$$H_k = Q_k + \left(1 + \frac{1}{3}\lambda_k^2\right) A_k^\tau H_{k+1} A_k - M_k^\tau L_k^+ M_k. \tag{38}$$

Consider the objective functional

$$\begin{aligned}J(\mathbf{x}_0, \mathbf{u}) &= \sum_{k=0}^{N-1} E\left[\mathbf{x}_k^\tau Q_k \mathbf{x}_k + \mathbf{u}_k^\tau R_k \mathbf{u}_k\right] + E\left[\mathbf{x}_N^\tau Q_N \mathbf{x}_N\right] \\ &= \sum_{k=0}^{N-1} E\left\{\left[\mathbf{x}_k^\tau Q_k \mathbf{x}_k + \mathbf{u}_k^\tau R_k \mathbf{u}_k\right] + E\left[\mathbf{x}_{k+1}^\tau H_{k+1} \mathbf{x}_{k+1}\right]\right. \\ &\quad \left. - E\left[\mathbf{x}_k^\tau H_k \mathbf{x}_k\right]\right\} + E\left[\mathbf{x}_N^\tau Q_N \mathbf{x}_N\right] - E\left[\mathbf{x}_N^\tau H_N \mathbf{x}_N\right] \\ &\quad + \mathbf{x}_0^\tau H_0 \mathbf{x}_0 = \sum_{k=0}^{N-1} \left\{\operatorname{tr}\left[\left(Q_k + K_k^\tau R_k K_k\right) X_k\right]\right. \\ &\quad \left. + \operatorname{tr}\left[H_{k+1} X_{k+1}\right] - \operatorname{tr}\left[H_k X_k\right]\right\} + \operatorname{tr}\left[\left(Q_N - H_N\right)\right. \\ &\quad \left. \cdot X_N\right] + \mathbf{x}_0^\tau H_0 \mathbf{x}_0.\end{aligned} \tag{39}$$

Since $X_{k+1} = (1 + (1/3)\lambda_k^2)(A_k X_k A_k^\tau + A_k X_k K_k^\tau B_k^\tau + B_k K_k X_k A_k^\tau + B_k K_k X_k K_k^\tau B_k^\tau)$, the objective functional can be rewritten as

$$\begin{aligned}J(X_0, \mathbf{K}) &= \sum_{k=0}^{N-1} \left\{\operatorname{tr}\left[\left(Q_k + K_k^\tau R_k K_k\right) + \left(1 + \frac{1}{3}\lambda_k^2\right)\right.\right. \\ &\quad \cdot \left(A_k^\tau H_{k+1} A_k + K_k^\tau B_k^\tau H_{k+1} A_k + A_k^\tau H_{k+1} B_k K_k\right. \\ &\quad \left.\left.\left. + K_k^\tau B_k^\tau H_{k+1} B_k K_k\right) - H_k\right] X_k\right\} + \operatorname{tr}\left[\left(Q_N - H_N\right)\right. \\ &\quad \left. \cdot X_N\right] + \mathbf{x}_0^\tau \mathrm{H}_0 \mathbf{x}_0 = \sum_{k=0}^{N-1} \operatorname{tr}\left\{\left[Q_k\right.\right. \\ &\quad \left. + \left(1 + \frac{1}{3}\lambda_k^2\right) A_k^\tau H_{k+1} A_k - H_k\right] + \left(1 + \frac{1}{3}\lambda_k^2\right) \\ &\quad \cdot K_k^\tau B_k^\tau H_{k+1} A_k + \left(1 + \frac{1}{3}\lambda_k^2\right) A_k^\tau H_{k+1} B_k K_k \\ &\quad \left. + K_k^\tau \left[R_k + \left(1 + \frac{1}{3}\lambda_k^2\right) B_k^\tau H_{k+1} B_k\right] K_k\right\} X_k \\ &\quad + \operatorname{tr}\left[\left(Q_N - H_N\right) X_N\right] + \mathbf{x}_0^\tau H_0 \mathbf{x}_0 \\ &= \sum_{k=0}^{N-1} \operatorname{tr}\left[M_k^\tau L_k^+ M_k + K_k^\tau M_k + M_k^\tau K_k + K_k^\tau L_k K_k\right] \\ &\quad \cdot X_k + \operatorname{tr}\left[\left(Q_N - H_N\right) X_N\right] + \mathbf{x}_0^\tau H_0 \mathbf{x}_0.\end{aligned} \tag{40}$$

By applying (32) and Lemma 3, a completion of square implies

$$\begin{aligned}&J(X_0, \mathbf{K}) \\ &\quad = \sum_{k=0}^{N-1} \operatorname{tr}\left[\left(K_k + L_k^+ M_k\right)^\tau L_k \left(K_k + L_k^+ M_k\right) X_k\right] - c\gamma \\ &\quad + \mathbf{x}_0^\tau H_0 \mathbf{x}_0.\end{aligned} \tag{41}$$

We assert that L_k $(k = 0, 1, \ldots, N-1)$ must satisfy

$$L_k = R_k + \left(1 + \frac{1}{3}\lambda_k^2\right) B_k^\tau H_{k+1} B_k \geq 0. \tag{42}$$

If it is not so, there is an L_p for $p \in \{0, 1, \ldots, N-1\}$ with a negative eigenvalue λ. Denote the unitary eigenvector with respect to λ as $\mathbf{v}_\lambda$ (i.e., $\mathbf{v}_\lambda^\tau \mathbf{v}_\lambda = 1$ and $L_p \mathbf{v}_\lambda = \lambda \mathbf{v}_\lambda$). Let $\delta \neq 0$ be an arbitrary scalar and construct a control sequence $\tilde{\mathbf{u}} = (\tilde{\mathbf{u}}_1, \tilde{\mathbf{u}}_2, \ldots, \tilde{\mathbf{u}}_{N-1})$ as follows:

$$\tilde{\mathbf{u}}_k = \begin{cases} -L_k^+ M_k \mathbf{x}_k, & k \neq p, \\ \delta |\lambda|^{-1/2} \mathbf{v}_\lambda - L_k^+ M_k \mathbf{x}_k, & k = p. \end{cases} \tag{43}$$

The associated cost functional becomes

$$\begin{aligned} J(\mathbf{x}_0, \tilde{\mathbf{u}}) &= \sum_{k=0}^{N-1} \operatorname{tr}\left[\left(\widetilde{K}_k + L_k^+ M_k\right)^\tau L_k \left(\widetilde{K}_k + L_k^+ M_k\right) X_k\right] - c\gamma + \mathbf{x}_0^\tau H_0 \mathbf{x}_0 \\ &= \sum_{k=0}^{N-1} E\left[\left(\tilde{\mathbf{u}}_k + L_k^+ M_k \mathbf{x}_k\right)^\tau L_k \left(\tilde{\mathbf{u}}_k + L_k^+ M_k \mathbf{x}_k\right)\right] - c\gamma + \mathbf{x}_0^\tau H_0 \mathbf{x}_0 \\ &= \left[\frac{\delta}{|\lambda|^{1/2}} \mathbf{v}_\lambda\right]^\tau L_p \left[\frac{\delta}{|\lambda|^{1/2}} \mathbf{v}_\lambda\right] - c\gamma + \mathbf{x}_0^\tau H_0 \mathbf{x}_0 \\ &= -\delta^2 - c\gamma + \mathbf{x}_0^\tau H_0 \mathbf{x}_0. \end{aligned} \tag{44}$$

Let $\delta \to \infty$. Then, $J(\mathbf{x}_0, \tilde{\mathbf{u}}) \to -\infty$, which contradicts the well-posedness of problem (10). □

3.4. Special Cases. We have obtained that $L_k \geq 0$ in the constrained difference equation (25) of Theorem 9. The following corollaries are special cases of the above result if we have $L_k > 0$ and $L_k = 0$.

Corollary 10. *The indefinite LQ problem (10) is uniquely solvable if and only if $L_k > 0$ for $k = 0, 1, \ldots, N-1$. Moreover, the unique optimal control is given by*

$$\mathbf{u}_k = -L_k^{-1} M_k \mathbf{x}_k, \quad k = 0, 1, \ldots, N-1. \tag{45}$$

Proof. By using Theorem 9, we immediately obtain the corollary. □

Corollary 11. *If $L_k = 0$ for $k = 0, 1, \ldots, N-1$, then any admissible control of the indefinite LQ problem (10) is optimal and the constrained difference equation (25) reduces to the following linear system:*

$$\begin{aligned} &H_k = Q_k + \left(1 + \frac{1}{3}\lambda_k^2\right) A_k^\tau H_{k+1} A_k, \\ &R_k + \left(1 + \frac{1}{3}\lambda_k^2\right) B_k^\tau H_{k+1} B_k = 0, \\ &B_k^\tau H_{k+1} A_k = 0, \\ &H_N = Q_N + \gamma I, \end{aligned} \tag{46}$$

for $k = 0, 1, \ldots, N-1$.

Proof. Letting $L_k = 0$ in (25), it is easy to obtain the linear system (46). Letting $L_k = 0$ in (41), (41) is simplified as

$$J(\mathbf{x}_0, \mathbf{u}) = -c\gamma + \mathbf{x}_0^\tau H_0 \mathbf{x}_0, \tag{47}$$

which implies that $V(\mathbf{x}_0) = -c\gamma + \mathbf{x}_0^\tau H_0 \mathbf{x}_0$ for any admissible control. Then, any admissible control of the indefinite LQ problem (10) is optimal. □

3.5. Well-Posedness of the Indefinite LQ Problem. In the following, it is shown that the solvability of the constrained difference equation (25) is sufficient for the well-posedness of the indefinite LQ problem and the existence of an optimal control. Moreover, any optimal control can be represented explicitly as a linear state feedback by the solution of (25).

Theorem 12. *The indefinite LQ problem (10) is well-posed if there exist symmetric matrices H_k and $\gamma \in \mathbf{R}^1$ satisfying the constrained difference equation (25). Moreover, the optimal control is given by*

$$\mathbf{u}_k = -\left[R_k + \left(1 + \frac{1}{3}\lambda_k^2\right) B_k^\tau H_{k+1} B_k\right]^+ \cdot \left[\left(1 + \frac{1}{3}\lambda_k^2\right) B_k^\tau H_{k+1} A_k\right] \mathbf{x}_k, \quad k = 0, 1, \ldots, N-1. \tag{48}$$

Furthermore, the optimal cost of the indefinite LQ problem (10) is

$$V(\mathbf{x}_0) = \mathbf{x}_0^\tau H_0 \mathbf{x}_0 - c\gamma. \tag{49}$$

Proof. Let H_k and $\gamma \in \mathbf{R}^1$ satisfy (25). Then,

$$\begin{aligned} J(\mathbf{x}_0, \mathbf{u}) &= \sum_{k=0}^{N-1} E\left[\mathbf{x}_k^\tau Q_k \mathbf{x}_k + \mathbf{u}_k^\tau R_k \mathbf{u}_k\right] + E\left[\mathbf{x}_N^\tau Q_N \mathbf{x}_N\right] \\ &= \sum_{k=0}^{N-1} \{E\left[\mathbf{x}_k^\tau Q_k \mathbf{x}_k + \mathbf{u}_k^\tau R_k \mathbf{u}_k\right] + E\left[\mathbf{x}_{k+1}^\tau H_{k+1} \mathbf{x}_{k+1}\right] \\ &\quad - E\left[\mathbf{x}_k^\tau H_k \mathbf{x}_k\right]\} + E\left[\mathbf{x}_N^\tau Q_N \mathbf{x}_N\right] - E\left[\mathbf{x}_N^\tau H_N \mathbf{x}_N\right] \\ &\quad + \mathbf{x}_0^\tau H_0 \mathbf{x}_0 = \sum_{k=0}^{N-1} \{\operatorname{tr}\left[\left(Q_k + K_k^\tau R_k K_k\right) X_k\right] \\ &\quad + \operatorname{tr}\left[H_{k+1} X_{k+1}\right] - \operatorname{tr}\left[H_k X_k\right]\} + \operatorname{tr}\left[\left(Q_N - H_N\right) \cdot X_N\right] + \mathbf{x}_0^\tau H_0 \mathbf{x}_0 \\ &= \sum_{k=0}^{N-1} \operatorname{tr}\left\{\left[Q_k + \left(1 + \frac{1}{3}\lambda_k^2\right) A_k^\tau H_{k+1} A_k - H_k\right]\right. \end{aligned}$$

$$+\left(1+\frac{1}{3}\lambda_k^2\right)K_k^\tau B_k^\tau H_{k+1}A_k$$

$$+\left(1+\frac{1}{3}\lambda_k^2\right)A_k^\tau H_{k+1}B_kK_k$$

$$+K_k^\tau\left[R_k+\left(1+\frac{1}{3}\lambda_k^2\right)B_k^\tau H_{k+1}B_k\right]K_k\Big\}X_k$$

$$+\operatorname{tr}\left[\left(Q_N-H_N\right)X_N\right]+\mathbf{x}_0^\tau H_0\mathbf{x}_0$$

$$=\sum_{k=0}^{N-1}\operatorname{tr}\left[M_k^\tau L_k^+M_k+K_k^\tau M_k+M_k^\tau K_k+K_k^\tau L_kK_k\right]$$

$$\cdot X_k+\operatorname{tr}\left[\left(Q_N-H_N\right)X_N\right]+\mathbf{x}_0^\tau H_0\mathbf{x}_0. \tag{50}$$

By applying Lemma 3, a completion of square implies

$$J\left(X_0,\mathbf{K}\right)=\sum_{k=0}^{N-1}\operatorname{tr}\left[\left(K_k+L_k^+M_k\right)^\tau L_k\left(K_k+L_k^+M_k\right)X_k\right]+\operatorname{tr}\left[\left(Q_N-H_N\right)X_N\right]+\mathbf{x}_0^\tau H_0\mathbf{x}_0. \tag{51}$$

Since $L_k\geq 0$, from (51), we can easily deduce that the cost function of problem (10) is bounded from below by

$$V\left(\mathbf{x}_0\right)=\operatorname{tr}\left[\left(Q_N-H_N\right)X_N\right]+\mathbf{x}_0^\tau H_0\mathbf{x}_0>-\infty,\quad \forall\mathbf{x}_0\in\mathbf{R}^n. \tag{52}$$

Hence, the indefinite LQ problem (10) is well-posed. It is clear that it is solvable by the feedback control

$$\mathbf{u}_k=-K_k\mathbf{x}_k=-L_k^+M_k\mathbf{x}_k,\quad k=0,1,\ldots,N-1. \tag{53}$$

Furthermore, by using $\operatorname{tr}[X_N]=c$ and $H_N=Q_N+\gamma I$ which we have obtained in Theorems 7 and 9, (52) indicates that the optimal value of problem (10) equals

$$V\left(\mathbf{x}_0\right)=\mathbf{x}_0^\tau H_0\mathbf{x}_0-c\gamma. \tag{54}$$

□

4. General Expression for the Optimal Control Set

In this part, we will present a general expression for the optimal control set based on the solution to (25).

Theorem 13. *Assume that H_k $(k=0,1,\ldots,N-1)$ and $\gamma\geq 0\in\mathbf{R}^1$ solves the constrained difference equation (25). A sufficient and necessary condition that $\mathbf{u}_k$ is in the set of all optimal feedback controls for indefinite LQ problem (10) is that*

$$\mathbf{u}_k=-\left(L_k^+M_k+Y_k-L_k^+L_kY_k\right)\mathbf{x}_k+Z_k-L_k^+M_kZ_k,\quad k=0,1,\ldots,N-1, \tag{55}$$

where $Y_k\in\mathbf{R}^{m\times n}$ and $Z_k\in\mathbf{R}^m$ are arbitrary variables with appropriate size.

Proof.

Sufficiency. According to the same calculation as in Theorem 9, we have

$$J\left(\mathbf{x}_0,\mathbf{u}\right)=\sum_{k=0}^{N-1}E\left[\mathbf{x}_k^\tau Q_k\mathbf{x}_k+\mathbf{u}_k^\tau R_k\mathbf{u}_k\right]+E\left[\mathbf{x}_N^\tau Q_N\mathbf{x}_N\right]$$

$$=\sum_{k=0}^{N-1}\operatorname{tr}\left\{\left[Q_k+\left(1+\frac{1}{3}\lambda_k^2\right)A_k^\tau H_{k+1}A_k-H_k\right]\right.$$

$$+\left(1+\frac{1}{3}\lambda_k^2\right)K_k^\tau B_k^\tau H_{k+1}A_k+\left(1+\frac{1}{3}\lambda_k^2\right)$$

$$\cdot A_k^\tau H_{k+1}B_kK_k+K_k^\tau\left[R_k+\left(1+\frac{1}{3}\lambda_k^2\right)B_k^\tau H_{k+1}B_k\right] \tag{56}$$

$$\cdot K_k\Big\}X_k-c\gamma+\mathbf{x}_0^\tau H_0\mathbf{x}_0=\sum_{k=0}^{N-1}E\left[\mathbf{x}_k^\tau\left(M_k^\tau L_k^+M_k\right.\right.$$

$$\left.\left.+K_k^\tau M_k+M_k^\tau K_k+K_k^\tau L_kK_k\right)\mathbf{x}_k\right]-c\gamma+\mathbf{x}_0^\tau H_0\mathbf{x}_0$$

$$=\sum_{k=0}^{N-1}E\left[\mathbf{x}_k^\tau M_k^\tau L_k^+M_k\mathbf{x}_k+2\mathbf{x}_k^\tau M_k^\tau\mathbf{u}_k+\mathbf{u}_k^\tau L_k\mathbf{u}_k\right]$$

$$-c\gamma+\mathbf{x}_0^\tau H_0\mathbf{x}_0.$$

By denoting $T_k^1=-(Y_k-L_k^+L_kY_k)$ and $T_k^2=-(Z_k-L_k^+L_kZ_k)$, we obtain

$$L_kT_k^1=0,\qquad L_kT_k^2=0. \tag{57}$$

According to (56) and (57), we obtain

$$J\left(\mathbf{x}_0,\mathbf{u}\right)=\sum_{k=0}^{N-1}E\left[\mathbf{u}_k+\left(L_k^+M_k+T_k^1\right)\mathbf{x}_k+T_k^2\right]^\tau\cdot L_k\left[\mathbf{u}_k+\left(L_k^+M_k+T_k^1\right)\mathbf{x}_k+T_k^2\right]-c\gamma+\mathbf{x}_0^\tau H_0\mathbf{x}_0. \tag{58}$$

As $L_k\geq 0$, we know that the control $\mathbf{u}_k=-[(L_k^+M_k+T_k^1)\mathbf{x}_k+T_k^2]$ minimizes $J(\mathbf{x}_0,\mathbf{u})$ with the optimal value $-c\gamma+\mathbf{x}_0^\tau H_0\mathbf{x}_0$ for $k=0,1,\ldots,N-1$.

Necessity. If any control sequence $\tilde{\mathbf{u}}=(\tilde{\mathbf{u}}_1,\tilde{\mathbf{u}}_2,\ldots,\tilde{\mathbf{u}}_{N-1})$ which minimizes the cost function $J(\mathbf{x}_0,\mathbf{u})$, then we have

$$J\left(\mathbf{x}_0,\tilde{\mathbf{u}}\right)=\sum_{k=0}^{N-1}E\left[\left(\tilde{\mathbf{u}}_k+L_k^+M_k\mathbf{x}_k\right)^\tau L_k\left(\tilde{\mathbf{u}}_k+L_k^+M_k\mathbf{x}_k\right)\right]-c\gamma+\mathbf{x}_0^\tau H_0\mathbf{x}_0, \tag{59}$$

for $k=0,1,\ldots,N-1$. The above equality implies that

$$\sum_{k=0}^{N-1}E\left[\left(\tilde{\mathbf{u}}_k+L_k^+M_k\mathbf{x}_k\right)^\tau L_k\left(\tilde{\mathbf{u}}_k+L_k^+M_k\mathbf{x}_k\right)\right]=0,\quad k=0,1,\ldots,N-1. \tag{60}$$

Since $L_k \geq 0$, we get the following equivalent condition:

$$L_k\left(\tilde{\mathbf{u}}_k + L_k^+ M_k \mathbf{x}_k\right) = 0, \quad k = 0, 1, \ldots, N-1. \tag{61}$$

We see that $\tilde{\mathbf{u}}_k$ solves the following equation:

$$L_k \tilde{\mathbf{u}}_k + L_k L_k^+ M_k \mathbf{x}_k = 0, \quad k = 0, 1, \ldots, N-1. \tag{62}$$

By using Lemma 3 with $L = L_k$, $M = I$, $N = -L_k L_k^+ M_k \mathbf{x}_k$, it is easy to verify that

$$LL^+ NMM^+ = N. \tag{63}$$

Then, we obtain the solution of (62) with

$$\tilde{\mathbf{u}}_k = -L_k^+ M_k \mathbf{x}_k + Z_k - L_k^+ L_k Z_k, \quad Z_k \in \mathbf{R}^m, \ k = 0, 1, \ldots, N-1. \tag{64}$$

As in (35), the optimal control can be represented by

$$\mathbf{u}_k = -\left(L_k^+ M_k + Y_k - L_k^+ L_k Y_k\right)\mathbf{x}_k + Z_k - L_k^+ M_k Z_k, \quad k = 0, 1, \ldots, N-1. \tag{65}$$

□

5. Numerical Example

In this section, application of Theorem 9 to solve constraint optimal control problem is illustrated. We present a two-dimensional indefinite LQ problem with terminal state constraint for discrete-time uncertain systems. A set of specific parameters of the coefficients are given as follows:

$$\begin{aligned}
\mathbf{x}_0 &= \begin{pmatrix} 0 \\ 1 \end{pmatrix}, \\
c &= 2.0408, \\
N &= 2, \\
A_0 &= \begin{pmatrix} 1 & 0 \\ 1 & 1 \end{pmatrix}, \\
A_1 &= \begin{pmatrix} 2 & 0 \\ 0 & 1 \end{pmatrix}, \\
B_0 &= \begin{pmatrix} 2 \\ 1 \end{pmatrix}, \\
B_1 &= \begin{pmatrix} 1 \\ 0 \end{pmatrix}, \\
\lambda_0 &= -\frac{\sqrt{3}}{2}, \\
\lambda_1 &= \frac{\sqrt{3}}{2}.
\end{aligned} \tag{66}$$

The state weights and the control weights are as follows:

$$\begin{aligned}
Q_0 &= \begin{pmatrix} -1 & 0 \\ 0 & 1 \end{pmatrix}, \\
Q_1 &= \begin{pmatrix} -1 & 0 \\ 0 & -1 \end{pmatrix}, \\
Q_2 &= \begin{pmatrix} 0 & 0 \\ 0 & 0 \end{pmatrix}, \\
R_0 &= -1, \\
R_1 &= -2.
\end{aligned} \tag{67}$$

Note that, in this example, the state weight Q_0 is negative semidefinite, Q_1 is negative definite, and Q_2 is positive semidefinite and the control weights R_0 and R_1 are negative definite.

In order to find the optimal controls and optimal cost value of this example, we have to solve the following equations:

$$\begin{aligned}
H_k &= Q_k + \left(1 + \frac{1}{3}\lambda_k^2\right) A_k^\tau H_{k+1} A_k - M_k^\tau L_k^+ M_k, \\
& L_k L_k^+ M_k - M_k = 0, \\
L_k &= R_k + \left(1 + \frac{1}{3}\lambda_k^2\right) B_k^\tau H_{k+1} B_k \geq 0, \\
M_k &= \left(1 + \frac{1}{3}\lambda_k^2\right) B_k^\tau H_{k+1} A_k, \quad k = 0, 1, \\
H_2 &= Q_2 + \gamma I, \\
X_{k+1} &= \left(1 + \frac{1}{3}\lambda_k^2\right)\left(A_k X_k A_k^\tau + A_k X_k K_k^\tau B_k^\tau \right. \\
&\quad \left. + B_k K_k X_k A_k^\tau + B_k K_k X_k K_k^\tau B_k^\tau\right), \\
&\qquad k = 0, 1, \ X_0 = \mathbf{x}_0 \mathbf{x}_0^\tau.
\end{aligned} \tag{68}$$

Firstly, we have

$$X_0 = \mathbf{x}_0 \mathbf{x}_0^\tau = \begin{pmatrix} 0 & 0 \\ 0 & 1 \end{pmatrix}. \tag{69}$$

Then, we get $\gamma = 2$ by solving (68), and we obtain

$$H_2 = Q_2 + \gamma I = \gamma \begin{pmatrix} 1 & 0 \\ 0 & 1 \end{pmatrix} = \begin{pmatrix} 2 & 0 \\ 0 & 2 \end{pmatrix}. \tag{70}$$

Secondly, by applying Theorem 9, we obtain the optimal feedback control and optimal cost value as follows.

For $k = 1$, we obtain

$$L_1 = R_1 + \left(1 + \frac{1}{3}\lambda_1^2\right) B_1^T H_2 B_1 = 0.5 > 0,$$
$$M_1 = \left(1 + \frac{1}{3}\lambda_1^2\right) B_1^T H_2 A_1 = (5, 0),$$
$$H_1 = Q_1 + \left(1 + \frac{1}{3}\lambda_1^2\right) A_1^T H_2 A_1 - M_1^T L_1^+ M_1 = \begin{pmatrix} -41 & 0 \\ 0 & 1.5 \end{pmatrix}. \quad (71)$$

The optimal feedback control is $\mathbf{u}_1 = K_1 \mathbf{x}_1$, where

$$K_1 = -L_1^+ M_1 = (-10, 0). \quad (72)$$

For $k = 0$, we obtain

$$L_0 = R_0 + \left(1 + \frac{1}{3}\lambda_0^2\right) B_0^T H_1 B_0 = 0.875 > 0,$$
$$M_0 = \left(1 + \frac{1}{3}\lambda_0^2\right) B_0^T H_1 A_0 = (1.875, 1.875),$$
$$H_0 = Q_0 + \left(1 + \frac{1}{3}\lambda_0^2\right) A_0^T H_1 A_0 - M_0^T L_0^+ M_0 = \begin{pmatrix} -54.3929 & -2.1429 \\ -2.1429 & -1.1429 \end{pmatrix}. \quad (73)$$

The optimal feedback control is $\mathbf{u}_0 = K_0 \mathbf{x}_0$, where

$$K_0 = -L_0^+ M_0 = (-2.1429, -2.1429). \quad (74)$$

Finally, the optimal cost value is

$$V(\mathbf{x}_0) = \mathbf{x}_0^T H_0 \mathbf{x}_0 - c\gamma = -5.2245. \quad (75)$$

6. Conclusion

We have considered the indefinite LQ optimal control with terminal state constraint involving state and control dependent uncertain noises. We first transform the uncertain LQ optimal control problem into a deterministic LQ optimal control problem. By means of the matrix maximum principle, we have presented a necessary condition for the existence of optimal linear state feedback control. Besides, we have proved the well-posedness of the indefinite LQ constraint problem by applying the technique of completing squares. For further work, we will consider discrete-time indefinite LQ optimal control model with inequality constraint.

Conflict of Interests

The authors declare that there is no conflict of interests regarding the publication of this paper.

Acknowledgments

This work is supported by the National Natural Science Foundation of China (no. 61203050) and the Natural Science Research Project of the Education Bureau of Anhui Province (China) (no. KJ2015A076).

References

[1] R. E. Kalman, "Contributions to the theory of optimal control," *Boletin de la Sociedad Matematica Mexicana*, vol. 5, no. 2, pp. 102–119, 1960.

[2] W. M. Wonham, "On a matrix Riccati equation of stochastic control," *SIAM Journal on Control and Optimization*, vol. 6, pp. 681–697, 1968.

[3] S. Xie, Z. Li, and S. Wang, "Continuous-time portfolio selection with liability: mean-variance model and stochastic LQ approach," *Insurance: Mathematics & Economics*, vol. 42, no. 3, pp. 943–953, 2008.

[4] S. P. Chen, X. J. Li, and X. Y. Zhou, "Stochastic linear quadratic regulators with indefinite control weight costs," *SIAM Journal on Control and Optimization*, vol. 36, no. 5, pp. 1685–1702, 1998.

[5] X. Li and X. Y. Zhou, "Indefinite stochastic LQ controls with Markovian jumps in a finite time horizon," *Communications in Information and Systems*, vol. 2, no. 3, pp. 265–282, 2002.

[6] A. Beghi and D. D'Alessandro, "Discrete-time optimal control with control-dependent noise and generalized Riccati difference equations," *Automatica*, vol. 34, no. 8, pp. 1031–1034, 1998.

[7] H. Yang, "Study on stochastic linear quadratic optimal control with quadratic and mixed terminal state constraints," *Journal of Applied Mathematics*, vol. 2013, Article ID 674327, 8 pages, 2013.

[8] W. Zhang and G. Li, "Discrete-time indefinite stochastic linear quadratic optimal control with second moment constraints," *Mathematical Problems in Engineering*, vol. 2014, Article ID 278142, 9 pages, 2014.

[9] B. Liu, *Uncertainty Theory*, Springer, Berlin, Germany, 2nd edition, 2007.

[10] B. Liu, *Uncertainty Theory: A Branch of Mathematics for Modeling Human Uncertainty*, Springer, Berlin, Germany, 2010.

[11] X. Chen, "American option pricing formula for uncertain financial market," *International Journal of Operations Research*, vol. 8, no. 2, pp. 27–32, 2011.

[12] Y. Gao, "Uncertain models for single facility location problems on networks," *Applied Mathematical Modelling*, vol. 36, no. 6, pp. 2592–2599, 2012.

[13] Z. Qin and S. Kar, "Single-period inventory problem under uncertain environment," *Applied Mathematics and Computation*, vol. 219, no. 18, pp. 9630–9638, 2013.

[14] B. Zhang and J. Peng, "Uncertain programming model for uncertain optimal assignment problem," *Applied Mathematical Modelling*, vol. 37, no. 9, pp. 6458–6468, 2013.

[15] H. Ke, T. Su, and Y. Ni, "Uncertain random multilevel programming with application to production control problem," *Soft Computing*, vol. 19, no. 6, pp. 1739–1746, 2015.

[16] Y. Zhu, "Uncertain optimal control with application to a portfolio selection model," *Cybernetics and Systems*, vol. 41, no. 7, pp. 535–547, 2010.

[17] L. Sheng and Y. Zhu, "Optimistic value model of uncertain optimal control," *International Journal of Uncertainty, Fuzziness and Knowledge-Based Systems*, vol. 21, supplement 1, pp. 75–87, 2013.

[18] H. Yan and Y. Zhu, "Bang-bang control model for uncertain switched systems," *Applied Mathematical Modelling*, vol. 39, no. 10-11, pp. 2994–3002, 2015.

[19] B. Liu, "Some research problems in uncertainty theory," *Journal of Uncertain Systems*, vol. 3, no. 1, pp. 3–10, 2009.

[20] R. Penrose, "A generalized inverse of matrices," *Mathematical Proceedings of the Cambridge Philosophical Society*, vol. 51, no. 3, pp. 406–413, 1955.

[21] Y. Zhu, "Functions of uncertain variables and uncertain programming," *Journal of Uncertain Systems*, vol. 6, no. 4, pp. 278–288, 2012.

[22] M. Athans, "The matrix minimum principle," *Information and Computation*, vol. 11, pp. 592–606, 1967.

22

H_∞ Optimal Inversion Feedforward and Robust Feedback Based 2DOF Control Approach for High Speed-Precision Positioning Systems

Chao Peng,[1,2] Chongwei Han,[2] Jianxiao Zou,[1] and Guanghui Zhang[2]

[1]*School of Mechanical Engineering, University of Electronic Science and Technology of China, Chengdu, Sichuan 611731, China*
[2]*Northwest Institute of Mechanical & Electrical Engineering, Xianyang, Shanxi 712099, China*

Correspondence should be addressed to Chao Peng; pcddiy@163.com

Academic Editor: Petko Petkov

This paper proposed a novel H_∞ optimal inversion feedforward and robust feedback based two-freedom-of-freedom (2DOF) control approach to address the positioning error caused by system uncertainties in high speed-precision positioning system. To minimize the H_∞ norm of the positioning error in the presence of model uncertainty, a linear matrix inequality (LMI) synthesis approach for optimal inversion feedforward controller design is presented. The specification of position resolution, control width, robustness, and output signal magnitude imposed on the entire 2DOF control system are taken as optimization objectives of feedback controller design. The robust feedback controller design approach integrates with feedforward controller systematically and is obtained via LMI optimization. The proposed approach was illustrated through a simulation example of nanopositioning control in atomic force microscope (AFM); the experiment results demonstrated that the proposed 2DOF control approach not only achieves the performance specification but also could improve the positioning control performance compared with H_∞ mixed sensitivity feedback control and inversion-based 2DOF control.

1. Introduction

The performance of high speed-precision positioning system such as piezoelectric and precision motor actuated system is required to meet the specification in scanning probe microscopy (SPM) [1, 2], microelectromechanical system [3, 4], optical-hard disk drive [5, 6], and so on. The positioning control performance could be characterized by positioning resolution, control bandwidth, and system robustness [1–6]. Various feedback control approaches have been studied and demonstrated that it could render requirements on positioning resolution and robustness satisfied at low frequencies [7–9]. However, the positioning resolution and control bandwidth of feedback control are limited for the Bode Sensitivity Integral theorem [10, 11]. It has been demonstrated that the feedforward control could increase the feedback control bandwidth [12, 13]. To improve the system positioning control performance, considerable 2DOF control approaches which combine the feedforward controller and feedback controller have been studied [14–18].

The inversion control technology could achieve exact trajectory tracking and precision positioning for linear and nonlinear system without system uncertainty [19, 20]. The stable-inversion [21], preview-based optimal inversion [22], and inversion-based iterative control [23] approach have improved the practicality of inversion technology. However, the positioning control performance of inversion feedforward control is limited by modeling uncertainties and disturbances [24]. Thus, recently, the inversion feedforward based 2DOF control approaches which utilize the robust feedback control to compensated the limitation of inversion feedforward controller have been presented and demonstrated that they could achieve good positioning performance [25–27]. However, the feedforward controller and feedback controller in the above 2DOF control approaches are designed separately; the feedback controller is designed by conventional

H_∞ mixed sensitivity feedback control approach without taking the effect of feedforward controller into account. There exist challenges in inversion feedforward based 2DOF control: (1) inversion feedforward controller design optimization with considering the system uncertainty and (2) how to integrate the feedforward controller and feedback controller design systematically to obtain the desired requirement positioning resolution, control bandwidth, and robustness of the entire 2DOF control system.

To minimize the adverse effect caused by system uncertainties and make the requirements of performance of the entire 2DOF control system satisfied, a novel H_∞ optimal inversion feedforward and robust feedback based 2DOF control approach is proposed in this paper. The main contribution of this paper is as follows. (1) A new LMI representation for H_∞ optimal inversion feedforward controller design which takes minimizing the H_∞ norm of the positioning error caused by feedforward control in presence of the system uncertainty as optimization objective. (2) A 2DOF control positioning performance optimization problem based on H_∞ mixed sensitivity is formulated and a LMI optimization based robust feedback controller design approach integrating with feedforward controller systematically. The proposed 2DOF control design approach is evaluated through the experiment.

The rest of this paper is organized as follows. The design objective of H_∞ optimal inversion feedforward and robust feedback based 2DOF control is formulated in Section 2. In Section 3, the LMI based H_∞ optimal inversion feedforward controller design approach is presented. The design of robust feedback controller is given in Section 4. The proposed 2DOF control approach is illustrated through AFM simulation experiment in Section 5. Finally, the conclusion is discussed in Section 6.

2. Problem Formulation

2.1. System Description. Consider the SISO LTI system with parameter uncertainty:

$$G(s) = p(s,0) - \frac{n(s,\theta)}{d(s,\theta)}, \qquad (1)$$

where $n(s,\theta)$ and $d(s,\theta)$ are polynomials of Laplace variable s and $\theta = \{\theta_1, \theta_2, \ldots, \theta_q\} \in \mathbb{R}^q$ is the uncertain parameter vector that parameterizes the transfer function $G(s)$.

Defining the nominal parameter vector of θ as $\theta_m = \{\theta_{m1}, \theta_{m2}, \ldots, \theta_{mq}\}$, the nominal model of the system could be denoted as $G_m(s) = p(s, \theta_m)$.

We assume that θ is known to lie in a box domain Ω which is defined as

$$\Omega = \left\{\theta \mid \theta_i = \theta_{m_i} + \Delta\theta_i,\ \Delta\theta_i \in [a_i, b_i],\ i = 1, 2, \ldots, q\right\}, \qquad (2)$$

where $\Delta\theta_i$ are the variation of θ_i.

Assumption 1. System $G(s)$ is invertible, and the relative degree of $G(s)$ is at least two (i.e., has at least two more poles than zeros). System $G(s)$ and its inverse $G^{-1}(s)$ are hyperbolic; that is, $G(s)$ does not have poles or zeros on the imaginary axis of the complex plane.

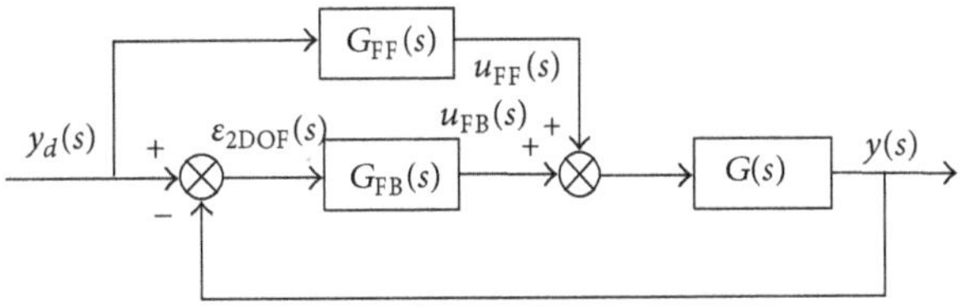

FIGURE 1: Block diagram of 2DOF control system.

Remark 2. The requirements that the system $G(s)$ is invertible and hyperbolic and that the relative degree of $G(s)$ is more than one are needed for computation [21] and robustness [22] of the exact inverse.

2.2. H_∞ Optimal Inversion Feedforward and Robust Feedback Based 2DOF Control. Consider the 2DOF control system shown in Figure 1. In this figure, $G(s)$ is the transfer function of a LTI plant as described in (1), $G_{FF}(s)$ and $G_{FB}(s)$ are feedforward controller and feedback controller, respectively. The signal $y_d(s)$ represents the desired outputting and $u(s)$ represents the input to the plant $G(s)$. The signal $y(s)$ represents the actual output of the entire 2DOF control system.

The transfer function from the desired trajectory $y_d(s)$ to actual outputs $y(s)$ and $G_{2DOF}(s)$ is given by

$$G_{2DOF}(s) = [G_{FF}(s) + G_{FB}(s)]\, G(s)\, S(s), \qquad (3)$$

where $S(s)$ is the feedback sensitivity function; that is,

$$S(s) = (1 + G(s) G_{FB}(s))^{-1}. \qquad (4)$$

The outputting error of entire 2DOF control system $\varepsilon_{2DOF}(s)$ can be decoupled as multiplication of the feedforward path outputting error $\varepsilon_{FF}(s)$ and the feedback sensitivity function $S(s)$:

$$\begin{aligned}\varepsilon_{2DOF}(s) &= 1 - G_{2DOF}(s) = (1 - G(s) G_{FF}(s))\, S(s) \\ &\triangleq \varepsilon_{FF}(s)\, S(s).\end{aligned} \qquad (5)$$

Remark 3. By Bode Sensitivity Integral theorem, the small feedforward path output error could improve the bandwidth of feedback control and output error. If $G_{FF}(s) = G^{-1}(s)$, the output error could be zero; however, it is impossible to find the exact inverse of the system for the modeling uncertainty.

The control block and output error of the 2DOF control system are now presented. The design goal of H_∞ optimal feedforward and robust feedback based 2DOF control design is to make the following two optimization objects satisfied.

(1) Find an H_∞ optimal robust inversion-based feedforward controller $G^*_{FF}(s)$ to minimize the H_∞ norm of feedforward path output error in the presence of system uncertainties; that is,

$$G^*_{FF}(s) = \arg\min_{G_{FF}} \sup_{\theta \in \Omega} \left\|\varepsilon_{FF}(s)\right\|_\infty . \qquad (6)$$

(2) After determining the feedforward controller $G^*_{\mathrm{FF}}(s)$, find a robust feedback controller $G^*_{\mathrm{FB}}(s)$ to minimize the H_∞ norm of transfer function $\Phi(G_{\mathrm{FB}})$; that is,

$$G^*_{\mathrm{FB}}(s) = \arg\min_{G_{\mathrm{FB}}} \sup_{\theta\in\Omega} \|\Phi(G_{\mathrm{FB}})\|_\infty, \tag{7}$$

where

$$\|\Phi(G_{\mathrm{FB}})\|_\infty = \left\| \begin{matrix} W_p(s) S_{2\mathrm{DOF}}(s) \\ W_u(s) S_{2\mathrm{DOF}}(s) G_{\mathrm{FB}}(s) \\ W_t(s) T_{2\mathrm{DOF}}(s) \end{matrix} \right\|_\infty. \tag{8}$$

$W_p(s)$ denotes the weighting function for output bandwidth and output error limitation, $W_t(s)$ denotes the weighting function for robustness performance of system, $W_u(s)$ denotes the weighting function for magnitude of output signal of feedback controller, and $S_{2\mathrm{DOF}}(s)$ and $T_{2\mathrm{DOF}}(s)$ are the transfer function from desired trajectory $y_d(s)$ to output error and $\varepsilon_{2\mathrm{DOF}}(s)$ and $y_d(s)$ to actual output $y(s)$.

3. H_∞ Optimal Inversion Feedforward Controller Design

This section is devoted to H_∞ optimal inversion feedforward controller design. At first feedforward controller based on H_∞ optimal inverse of $G(s)$ is designed by LMI optimization, and then the feedforward controller is implemented by preview-based inversion [21].

3.1. H_∞ Optimal Inversion of System. A modulation function $a(s)$ is introduced to inversion feedforward controller design as follows:

$$G_{\mathrm{FF}}(s) = a(s) G_m^{-1}(s) \triangleq a(s)\widehat{G}(s). \tag{9}$$

Consider Assumption 1; $\widehat{G}(s)$ is improper; that is, the poles number is less than zeros number. Assume $a(s)$ is proper; thus, inversion feedforward controller also could be decoupled as proper element $\widehat{G}_{\mathrm{impr}}(s)$:

$$\begin{aligned} G_{\mathrm{FF}}(s) &= a(s)\widehat{G}(s) = a(s)\frac{N_m(s)}{\mathrm{den}_m(s)} \\ &= a(s)\frac{A_m(s)}{\mathrm{den}_m(s)} B_m(s) \triangleq a(s)\widehat{G}_{\mathrm{pr}}(s) B_m(s) \\ &\triangleq a(s)\widehat{G}_{\mathrm{pr}}(s)\widehat{G}_{\mathrm{impr}}(s) \triangleq \widehat{G}_a(s)\widehat{G}_{\mathrm{impr}}(s), \end{aligned} \tag{10}$$

where $\mathrm{den}_m(s)$ and $N_m(s)$ are the denominator and numerator of $\widehat{G}(s)$, respectively, The order of $A_m(s)$ is equivalent to the order of $\mathrm{den}_m(s)$. Thus, the relative degree of $\widehat{G}_{\mathrm{pr}}(s)$ is zero and the order of $\widehat{G}_{\mathrm{impr}}(s)$ is equal to the relative degree of nominal model $G_m(s)$.

The transfer function of feedforward path output error $\varepsilon_{\mathrm{FF}}(s)$ could be rewritten as

$$\begin{aligned} \varepsilon_{\mathrm{FF}}(s) &= 1 - G_{\mathrm{FF}}(s) G(s) = 1 - \widehat{G}_a(s)\widehat{G}_{\mathrm{impr}}(s) G(s) \\ &\triangleq 1 - \widehat{G}_a(s)\widetilde{G}(s). \end{aligned} \tag{11}$$

Remark 4. Because the order of $\widehat{G}_{\mathrm{impr}}(s)$ is equal to the relative degree of nominal model $G_m(s)$, the relative degree of $\widetilde{G}(s)$ is equal to zero. The modulation function $a(s)$ could be obtained as $a(s) = \widehat{G}_a(s)/\widehat{G}_{\mathrm{pr}}(s)$.

Now, a proper transfer function $\widetilde{G}(s)$ whose relative degree is zero can be obtained by multiplying improper element $\widehat{G}_{\mathrm{impr}}(s)$ with $G(s)$. Consider the parameter uncertainty domain of $G(s)$, the parameters of $\widetilde{G}(s)$ can also be known to lie in a box uncertainty domain, and the state space matrices of $\widetilde{G}(s)$ which depend on θ [28, 29]. The sate space representation of $\widetilde{G}(s)$ can be given as

$$\begin{aligned} \dot{\widetilde{x}} &= \widetilde{A}(\theta)\widetilde{x} + \widetilde{B}(\theta)\widetilde{u}, \\ \widetilde{y} &= \widetilde{C}(\theta)\widetilde{x} + \widetilde{D}(\theta)\widetilde{u}. \end{aligned} \tag{12}$$

The state space representation of $\widehat{G}_a(s)$ is given as

$$\begin{aligned} \dot{\widehat{x}}_a &= \widehat{A}_a\widehat{x}_a + \widehat{B}_a\widehat{u}_a, \\ \widehat{y}_a &= \widehat{C}_a\widehat{x}_a + \widehat{D}_a\widehat{u}_a. \end{aligned} \tag{13}$$

Now, the state space representation of the feedforward path output error $\varepsilon_{\mathrm{FF}}(s)$ can be written as

$$\begin{aligned} \dot{x}_{\varepsilon_{\mathrm{FF}}} &= A_{\varepsilon_{\mathrm{FF}}} x_{\varepsilon_{\mathrm{FF}}} + B_{\varepsilon_{\mathrm{FF}}} u_{\varepsilon_{\mathrm{FF}}}, \\ y_{\varepsilon_{\mathrm{FF}}} &= C_{\varepsilon_{\mathrm{FF}}} x_{\varepsilon_{\mathrm{FF}}} + D_{\varepsilon_{\mathrm{FF}}} u_{\varepsilon_{\mathrm{FF}}}, \end{aligned} \tag{14}$$

where the state space matrices are given by

$$\begin{aligned} A_{\varepsilon_{\mathrm{FF}}}(\theta) &= \begin{bmatrix} \widetilde{A}(\theta) & \widetilde{B}(\theta)\widehat{C}_a \\ 0 & \widehat{A}_a \end{bmatrix}, \\ B_{\varepsilon_{\mathrm{FF}}}(\theta) &= \begin{bmatrix} \widetilde{B}(\theta)\widehat{D}_a \\ \widehat{B}_a \end{bmatrix}, \\ C_{\varepsilon_{\mathrm{FF}}}(\theta) &= \begin{bmatrix} -\widetilde{C}(\theta) & -\widetilde{D}(\theta)\widehat{C}_a \end{bmatrix}, \\ D_{\varepsilon_{\mathrm{FF}}}(\theta) &= 1 - \widetilde{D}(\theta)\widehat{D}_a. \end{aligned} \tag{15}$$

The objective of the inversion feedforward controller design in (6) could be transformed to find an optimal function; that is,

$$\begin{aligned} \widehat{G}^*_a(s) &= \min_{\widehat{G}_a} \sup_{\theta\in\Omega} \left(\|\varepsilon_{\mathrm{FF}}(s)\|_\infty\right) \\ &= \min_{\widehat{G}_a} \sup_{\theta\in\Omega} \left(\left\|1 - \widehat{G}_a(s)\widetilde{G}(s)\right\|_\infty\right). \end{aligned} \tag{16}$$

The bounded real lemma (BRL) [30] will be utilized for solving the optimal $\widehat{G}_a(s)$. The LMI representation lemma is given as follows.

Lemma 5. *A proper element in inversion feedforward controller $G_{FF}(s)$ as in (10) is the solution to the minimization problem in (16) whose state space matrices $\widehat{A}_a, \widehat{B}_a, \widehat{C}_a$, and $\widehat{D}_a$ and a symmetric matrix $V > 0$ could minimize the scale*

γ in presence of uncertain parameters, subjected to following parameterized LMI, that is,

$$\min_{\widehat{A}_a,\widehat{B}_a,\widehat{C}_a,\widehat{D}_a,V}\left(\max_{\theta}(\gamma)\right) \tag{17}$$

subject to

$$\begin{bmatrix} VA_{\varepsilon_{FF}}(\theta)+A_{\varepsilon_{FF}}^T(\theta)V & VB_{\varepsilon_{FF}}(\theta) & C_{\varepsilon_{FF}}^T(\theta) \\ B_{\varepsilon_{FF}}^T(\theta)V & -\gamma^2 I & D_{\varepsilon_{FF}}^T(\theta) \\ C_{\varepsilon_{FF}}(\theta) & D_{\varepsilon_{FF}}(\theta) & -I \end{bmatrix} < 0, \quad \theta\in\Omega. \tag{18}$$

Proof. Consider the uncertainty domain Ω and bounded real lemma; if there exist a positive-definite symmetric matrix V and matrices $\widehat{A}_a$, $\widehat{B}_a$, $\widehat{C}_a$, and $\widehat{D}_a$ which could render the parameterized LMIs in (18) satisfied, H_∞ norm of feedforward path positioning error is less than γ in the presence of uncertain parameters θ; that is, $\|\varepsilon_{FF}(s)\|_\infty < \gamma$, $\theta\in\Omega$.

Thus, if matrices $\widehat{A}_a$, $\widehat{B}_a$, $\widehat{C}_a$, and $\widehat{D}_a$ and a given $\widehat{G}_a(s)$ and V could satisfy the optimization objective in (17), a solution for minimizing problem in (16) $\widehat{G}_a^*(s)$ exists. This completes the proof. □

The solution to the optimization problem in (6) could be given by the following theorem.

Theorem 6. *Assume the H_∞ optimal inversion feedforward controller $G_{FF}^*(s)$ with form as (10) is $G_{FF}^*(s)=\widehat{G}_a^*(s)\widehat{G}_{impr}(s)$.*

(a) The controller $G_{FF}^*(s)$ that could solve the minimization problem in (6) exits, if there exists a solution $(X,Y,Z,U,W_{11},\widehat{V}_{11})$ where $W_{11}=W_{11}^T>0$, $\widehat{V}_{11}=\widehat{V}_{11}^T>0$ for $\theta\in\Omega$ for the following optimization:

$$\min_{X,Y,Z,U,W_{11},\widehat{V}_{11}}\left(\max_{\theta}(\gamma)\right) \tag{19}$$

subject to

$$\begin{bmatrix} \widetilde{A}(\theta)\widehat{V}_{11}^T+\widehat{V}_{11}\widetilde{A}^T(\theta) & \widetilde{A}(\theta)W_{11}+\widetilde{B}(\theta)Z+X+\widehat{V}_{11}\widetilde{A}^T(\theta) & \widetilde{B}(\theta)U+Y & -\widehat{V}_{11}\widetilde{C}^T(\theta) \\ * & \widetilde{A}(\theta)W_{11}+\widetilde{B}(\theta)Z+\left(\widetilde{A}(\theta)W_{11}\right)^T+\left(\widetilde{B}(\theta)Z\right)^T & \widetilde{B}(\theta)U & -W_{11}\widetilde{C}^T(\theta)-Z^T\widetilde{D}^T(\theta) \\ * & * & -\gamma^2 I & I-\left(\widetilde{D}(\theta)U\right)^T \\ * & * & * & -I \end{bmatrix} < 0. \tag{20}$$

(b) The proper element in $G_{FF}^*(s)$, $\widehat{G}_a^*(s)$, is given as follows:

$$\begin{aligned} \widehat{A}_a^* &= V_{12}^{-1}V_{11}X\left(W_{12}^T\right)^{-1}, \\ \widehat{B}_a^* &= V_{12}^{-1}V_{11}Y, \\ \widehat{C}_a^* &= Z\left(W_{12}^T\right)^{-1}, \\ \widehat{D}_a^* &= U. \end{aligned} \tag{21}$$

Considering the partition of $A_{\varepsilon_{FF}}$ and $B_{\varepsilon_{FF}}$, we introduce a partition of V and its inverse $W \triangleq V^{-1}$:

$$\begin{aligned} V &= \begin{bmatrix} V_{11} & V_{12} \\ V_{12}^T & V_{22} \end{bmatrix}, \\ W &= \begin{bmatrix} W_{11} & W_{12} \\ W_{12}^T & W_{22} \end{bmatrix}. \end{aligned} \tag{22}$$

Since $WV=I$,

$$\begin{aligned} W_{11}V_{11}+W_{12}V_{12}^T &= I, \\ W_{11}V_{12}+W_{12}V_{22} &= 0. \end{aligned} \tag{23}$$

Define a full ranked and square matrix, which will be used in following matrix inequalities transformation:

$$F \triangleq \begin{bmatrix} V_{11}^{-1} & 0 \\ W_{11} & W_{12} \end{bmatrix} \triangleq \begin{bmatrix} \widehat{V}_{11} & 0 \\ W_{11} & W_{12} \end{bmatrix}. \tag{24}$$

To obtain a feasible solution which could render the LMIs in (18) satisfied and consider the partition of V in (22), we introduce full ranked matrices X, Y, Z, and U and assume $\widehat{A}_a$, $\widehat{B}_a$, $\widehat{C}_a$, and $\widehat{D}_a$ have following form:

$$\begin{aligned} \widehat{A}_a &= V_{12}^{-1}V_{11}X\left(W_{12}^T\right)^{-1}, \\ \widehat{B}_a &= V_{12}^{-1}V_{11}Y, \\ \widehat{C}_a &= Z\left(W_{12}^T\right)^{-1}, \\ \widehat{D}_a &= U. \end{aligned} \tag{25}$$

Multiplying the inequalities in (18) by diag$\{F, I, I\}$ *and* diag$\{F^T, I, I\}$ *on the left and on the right, respectively, we obtain the following LMI:*

$$\begin{bmatrix} FVA_{\varepsilon_{FF}}(\theta)F^T + FA_{\varepsilon_{FF}}^T(\theta)VF^T & FVB_{\varepsilon_{FF}}(\theta) & FC_{\varepsilon_{FF}}^T(\theta) \\ B_{\varepsilon_{FF}}^T(\theta)VF^T & -\gamma I & D_{\varepsilon_{FF}}^T(\theta) \\ C_{\varepsilon_{FF}}(\theta)F^T & D_{\varepsilon_{FF}}(\theta) & -\gamma I \end{bmatrix} < 0. \tag{26}$$

Now, the LMI item in (26) could be written as

$$\begin{aligned} FVA_{\varepsilon_{FF}}(\theta)F^T &= \begin{bmatrix} \widetilde{A}(\theta)\widehat{V}_{11}^T & \widetilde{A}(\theta)W_{11} + \widetilde{B}(\theta)Z + X \\ \widetilde{A}(\theta)\widehat{V}_{11}^T & \widetilde{A}(\theta)W_{11} + \widetilde{B}(\theta)Z \end{bmatrix}, \\ FVB_{\varepsilon_{FF}} &= \begin{bmatrix} \widetilde{B}(\theta)\widehat{D}_a + V_{11}^{-1}V_{12}\widehat{B}_a \\ \widetilde{B}(\theta)\widehat{D}_a \end{bmatrix} = \begin{bmatrix} \widetilde{B}(\theta)U + Y \\ \widetilde{B}(\theta)U \end{bmatrix}, \\ FC_{\varepsilon_{FF}}^T &= \begin{bmatrix} -\widehat{V}_{11}\widetilde{C}^T(\theta) \\ -W_{11}\widetilde{C}^T(\theta) - Z^T\widetilde{D}^T(\theta) \end{bmatrix}, \\ D_{\varepsilon_{FF}} &= I - \widetilde{D}\widehat{D}_a = I - \widetilde{D}U. \end{aligned} \tag{27}$$

Replacing the LMI items in (26) with (27), we obtain the LMI as in (20). This completes the proof.

(b) It is obvious that $\widehat{G}_a^*(s)$ as (25) is the solution to (16) if the matrix variables X, Y, Z, U, W_{11}, and $\widehat{V}_{11}$ are the solution for (19); W_{12} and V_{12}, could be deduced from (23). This completes the proof.

Now, the H_∞ optimal inversion feedforward controller which satisfies the optimization problem in (6) can be obtained by Theorem 6.

3.2. Time Domain Implementation of Feedforward Controller. Note that the obtained H_∞ optimal inversion feedforward controller as $\widehat{G}_a^*(s)\widehat{G}_{\text{impr}}(s)$ is improper; it will be realized by preview-based inversion [22]. The design steps are given as follows.

(1) The improper inversion feedforward controller can be transformed to proper form as $\widehat{G}_a^*(s)$ by a transformed desired outputting $\widehat{y}_d(s)$ which carries the preview knowledge of $y_d(s)$. The transformed desired outputting $\widehat{y}_d(s)$ is obtained by multiplying the improper element $\widehat{G}_{\text{impr}}(s)$ and desired outputting $y_d(s)$. The output of feedforward controller $u_{\text{FF}}^*(s)$ is given as follows:

$$\begin{aligned} u_{\text{FF}}^*(s) &= G_{\text{FF}}^*(s)\,y_d(s) = \widehat{G}_a^*(s)\,\widehat{G}_{\text{impr}}(s)\,y_d(s) \\ &= \widehat{G}_a^*(s)\,\widehat{y}_d(s). \end{aligned} \tag{28}$$

(2) By the proper $\widehat{G}_a^*(s)$ and $\widehat{y}_d(s)$, the minimal state space realization of the H_∞ optimal inversion feedforward controller $G_{\text{FF}}^*(s)$ can be given as

$$\begin{aligned} \dot{x}_{\text{FF}}(t) &= A_{\text{FF}}x_{\text{FF}}(t) + B_{\text{FF}}\widehat{y}_d(t), \\ y_{\text{FF}}(t) &= C_{\text{FF}}x_{\text{FF}}(t) + D_{\text{FF}}\widehat{y}_d(t). \end{aligned} \tag{29}$$

(3) Finally, the time domain representation of output of the optimal feedforward controller $u_{\text{FF}}^*(t)$ can be obtained as

$$u_{\text{FF}}^*(t) = C_{\text{FF}}\int_{-\infty}^{t} e^{A_{\text{FF}}(t-\tau)B_{\text{FF}}}\,\widehat{y}_d(\tau)\,d\tau + D_{\text{FF}}\widehat{y}_d(t). \tag{30}$$

By the above step, the H_∞ optimal inversion is implemented in time domain.

4. Robust Feedback Controller Design

This section will discuss the solution to the robust feedback controller design problem in (7). The H_∞ mixed sensitivity synthesis scheme of 2DOF control system is shown in Figure 2.

By Figure 2, the sensitivity transfer functions of the entire 2DOF control system can be represented as follows:

$$\begin{aligned} S_{\text{2DOF}}(s) &= (1 - G_{\text{FF}}(s)\,G(s))\,S(s), \\ T_{\text{2DOF}}(s) &= (G_{\text{FF}}(s)\,G(s) + G_{\text{FB}}(s)\,G(s))\,S(s). \end{aligned} \tag{31}$$

Integrating with the feedforward which has been determined by design approach proposed in Section 3 and the weighting functions $W_p(s)$, $W_u(s)$, and $W_t(s)$, which impose the requirements for positioning resolution performance, robustness of the entire 2DOF control system, and output signal magnitude of feedback controller, find a feedback controller $G_{\text{FB}}^*(s)$ which could make the optimization problem in (7) satisfied. The design problem of feedback controller could be transformed to seek a controller $G_{\text{FB}}^*(s)$ which could minimize the H_∞ norm of transfer function from desired trajectory y_d to weighting output $T_{y_dZ}(s)$ as in Figure 2; that is,

$$G_{\text{FB}}^*(s) = \min_{G_{\text{FB}}}\sup_{\theta\in\Omega}\left\|T_{y_dZ}(s)\right\|_\infty. \tag{32}$$

The transfer function matrix of P in Figure 2 is given as follows:

$$P(s) = \begin{bmatrix} W_p(s)\left[1 - G(s)\,G_{\text{FF}}(s)\right] & -W_p(s)\,G(s) \\ 0 & W_u(s) \\ W_t(s)\,G(s)\,G_{\text{FF}}(s) & W_t(s)\,G(s) \\ 1 - G(s)\,G_{\text{FF}}(s) & -G(s) \end{bmatrix}. \tag{33}$$

The state space representation of P could be given as

$$\begin{aligned} \dot{x}_p &= A_p(\theta)\,x_p + B_p(\theta)\,y_d + B_{p_2}(\theta)\,u_{\text{FB}}, \\ Z &= C_{p_1}(\theta)\,x_p + D_{p_{11}}(\theta)\,y_d + D_{p_{12}}(\theta)\,u_{\text{FB}}, \\ \varepsilon_{\text{2DOF}} &= C_{p_2}(\theta)\,x_p + D_{p_{21}}(\theta)\,y_d. \end{aligned} \tag{34}$$

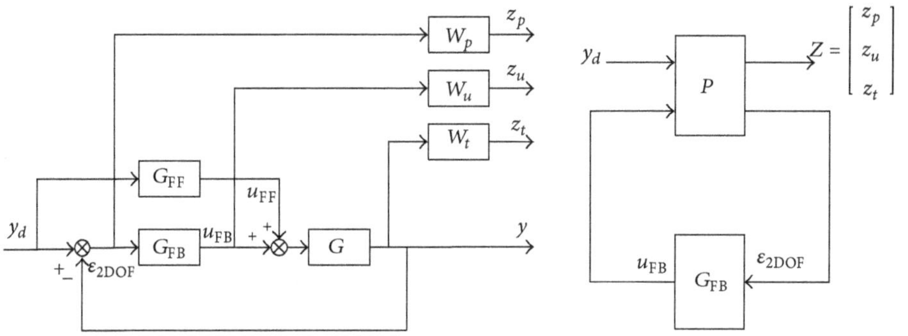

FIGURE 2: H_∞ mixed sensitivity synthesis scheme.

The state space representation of feedback controller G_{FB} is given as follows:

$$\begin{aligned} \dot{x}_{FB} &= A_{FB}x_{FB} + B_{FB}\varepsilon_{2DOF}, \\ u_{FB} &= C_{FB}x_{FB} + D_{FB}\varepsilon_{2DOF}. \end{aligned} \tag{35}$$

Theorem 7. *Consider the system as (34), if there exist symmetric matrices X and Y and matrices with appropriate dimension $\widehat{A}, \widehat{B}, \widehat{C}$, and $\widehat{D}$ that could satisfy the following optimization problem of positive scale γ, that is,*

$$\min_{X,Y,\widehat{A},\widehat{B},\widehat{C},\widehat{D}} (\max(\gamma)) \tag{36}$$

subject to

$$\begin{bmatrix} A_p(\theta)X + XA_p^T(\theta) + B_{p_2}(\theta)\widehat{C} + \left(B_{p_2}(\theta)\widehat{C}\right)^T & \widehat{A}^T + \left(A_p(\theta) + B_{p_2}(\theta)\widehat{D}C_{p_2}(\theta)\right) & B_{p_1}(\theta) + B_{p_2}(\theta)\widehat{D}D_{p_{21}}(\theta) & \left(C_{p_1}(\theta)X + D_{p_{12}}(\theta)\widehat{C}\right)^T \\ * & A_p^T(\theta)Y + YA_p(\theta) + \widehat{B}C_{p_2}(\theta) + \left(\widehat{B}C_{p_2}(\theta)\right)^T & YB_{p_1}(\theta) + \widehat{B}D_{p_{21}}(\theta) & \left(C_{p_1}(\theta) + D_{p_{12}}(\theta)\widehat{D}C_{p_2}(\theta)\right)^T \\ * & * & -\gamma I & \left(D_{p_{11}}(\theta) + D_{p_{12}}(\theta)\widehat{D}D_{p_{21}}(\theta)\right)^T \\ * & * & * & -\gamma I \end{bmatrix} < 0, \quad \begin{bmatrix} X & I \\ I & Y \end{bmatrix} > 0 \tag{37}$$

for $\theta = \theta^* \in \Omega$; the feedback controller which is the solution to the optimization problem

$$\min_{G_{FB}} \sup_{\theta=\theta^*} \left\|T_{y_dZ}(s)\right\|_\infty \tag{38}$$

could be given as

$$\begin{aligned} D_k &= \widehat{D}, \\ C_k &= \left(\widehat{C} - D_kC_{p_2}(\theta^*)X\right)\left(M^T\right)^{-1}, \\ B_k &= N^{-1}\left(\widehat{B} - YB_{p_2}(\theta^*)D_k\right), \\ A_k &= N^{-1}\left[\widehat{A} - Y\left(A(\theta^*) + B_{p_2}(\theta^*)D_kC_{p_2}(\theta^*)\right)X\right] \\ &\quad \cdot \left(M^T\right)^{-1} - B_kC_{p_2}(\theta^*)X\left(M^T\right)^{-1} \\ &\quad - N^{-1}YB_{p_2}(\theta^*)C_k, \end{aligned} \tag{39}$$

where N and M are deduced from $MN^T = I - XY$.

Proof. By BRL and the theorem in [30, 31], there exists a feedback controller such that $\|T_{y_dZ}(s)\|_\infty < \gamma$ for a certain parameter θ^*, if the LMIs in (37) hold, and the feedback controller is given in (39) for a system P with certain parameter θ^*. It is obvious that if there exists a solution of the minimization problem for γ, the feedback controller which is given in (39) is the solution for the minimization problem in (38). This completes the proof. □

Feedback controller G_{FB}^* which satisfies the optimization problem in (7) could be found by the following procedure.

Step 1. Discretize the uncertain parameters θ_i in box domain Ω, set $\theta_{ij} = \theta_{m_i} + (b_i - a_i)(j-1)/m$, $(1 \le j \le m+1;\ m > 1)$, and initialize the parameter vector set $\Theta = \{\Theta_k \mid \Theta_k = (\theta_{1k_1}, \theta_{2k_2}, \ldots, \theta_{qk_q})\}$; $(k_1, k_2, \ldots, k_q \in \{1, 2, \ldots, m\})$; $k \in \{1, 2, k \in \{1, 2, \ldots, m^q\}\}$, $k = 1$.

Step 2. By Theorem 7, obtain the feedback controller G_{FB_k}, which could satisfy the optimization problem $\min_{\Theta_k} \|T_{y_dZ}(s)\|_\infty$ and define $\tau = \min_{\Theta_k} \|T_{y_dZ}(s)\|_\infty$.

Step 3. Define $\lambda_{kl} = \min_{G_{FBk},\Theta_k} \|T_{y_dZ}(s)\|_\infty$. By BRL, we can obtain λ_{kl}. If $\tau < \lambda_{kl}$, for $l \ne k$, $l = 1, 2, \ldots, m^q$, then $G_{FB}^* = G_{FB_k}$ and stop.

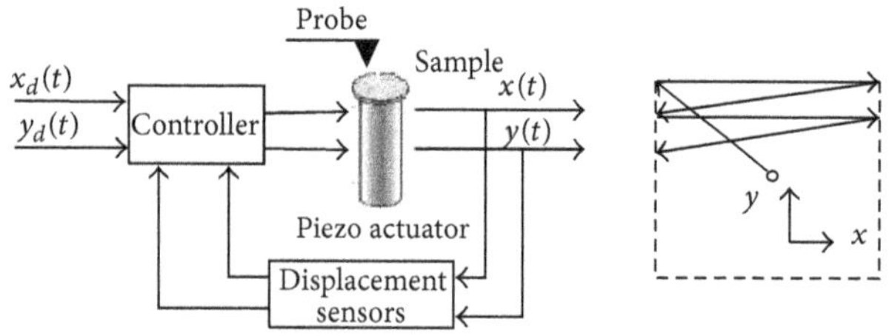

FIGURE 3: AFM operation diagram and scanning path.

Step 4. Else $k = k + 1$, go to Step 2.

5. Implementation and Experiment

In this section, we will conduct a simulation example of nanopositioning in atomic force microscope (AFM) operation to illustrate the positioning control performance through the proposed 2DOF control design approach.

5.1. Experimental System Description. The AFM system utilizes piezo actuator to enable the x-y axis scanning of the AFM probe relative to the sample surface during AFM imaging as shown in Figure 3. The output error will cause AFM image distortion and sample or probe damaging during scanning. Thus, the positioning control performance of the piezo actuator is important in AFM which measures the surface properties.

To demonstrate the proposed 2DOF control design approach, we will take the x-axis scanning motion control in the simulation experiment in MATLAB Simulink. The model of piezo actuator always is identified by several frequency response measurements. According to frequency response data, an approximate transfer function is obtained, which has parameter uncertainties compared with actual system. Considering the uncertainties, a second-order model of piezo actuator is given as follows [32]:

$$G(s) = \frac{4.019 \times 10^6}{s^2 + \theta_1 s + \theta_2}. \tag{40}$$

The variation bound is given as follows:

$$\theta_{m_i} + \Delta\theta_{i\min} \le \theta_i \le \theta_{m_i} + \Delta\theta_{i\max}, \quad (i = 1, 2), \tag{41}$$

where $\theta_{m_1} = 14.78$, $\Delta\theta_1 \in [0.1, 2]$; $\theta_{m_2} = 2.713 \times 10^5$, $\Delta\theta_2 \in [0.05 \times 10^5, 0.1 \times 10^5]$.

The frequency responses of piezo actuator are shown in Figure 4.

5.2. Design of the Feedforward Controller. The inversion feedforward controller $G_{\mathrm{FF}}(s)$ is selected as follows:

$$G_{\mathrm{FF}}(s) = a^*(s)\widehat{G}(s) = \widehat{G}_a^*(s)\widehat{G}_{\mathrm{impr}}(s). \tag{42}$$

The improper element $\widehat{G}_{\mathrm{impr}}(s)$ could be written as

$$\widehat{G}_{\mathrm{impr}}(s) = \frac{s^2 + 14.78s + 2.713 \times 10^5}{4.019 \times 10^6}. \tag{43}$$

The transfer function $\widetilde{G}(s)$ in (11) and its state matrices in (12) could be given as follows:

$$\begin{aligned}
\widetilde{G}(s) &= \widehat{G}_{\mathrm{impr}}(s)G(s) = \frac{s^2 + 14.78s + 2.713 \times 10^5}{s^2 + \theta_1 s + \theta_2}, \\
\widetilde{A} &= \begin{bmatrix} 0 & 1 \\ -\left(2.713 \times 10^5 + \Delta\theta_2\right) & -\left(14.78 + \Delta\theta_1\right) \end{bmatrix}, \\
\widetilde{B} &= \begin{bmatrix} 0 \\ 1 \end{bmatrix}, \\
\widetilde{C} &= \begin{bmatrix} -\Delta\theta_2 & -\Delta\theta_1 \end{bmatrix}, \\
\widetilde{D} &= 1.
\end{aligned} \tag{44}$$

By using Theorem 6 and LMI toolbox in MATLAB, the H_∞ optimal inversion feedforward controller is obtained as follows which could render

$$\begin{aligned}
G_{\mathrm{FF}}^*(s) = \widehat{G}_a^*(s)\widehat{G}_{\mathrm{impr}}(s) &= \left(\frac{0.9979s + 0.0188}{s + 0.0287}\right) \\
&\cdot \left(\frac{s^2 + 14.78s + 2.713 \times 10^5}{4.019 \times 10^6}\right).
\end{aligned} \tag{45}$$

The output of H_∞ optimal inversion feedforward controller u_{FF}^* could be obtained by the obtained $G_{\mathrm{FF}}^*(s)$ and preview inversion design steps as in (28)–(30).

5.3. Design of the Feedback Controller. At first, we specify the weighting function. It is noted that the tracking error could be achieved to around 30% by H_∞ optimal inversion feedforward controller. We chose the weighting function for control bandwidth and positioning error limitation as $W_p(s) = 80/(1 + 0.01s)$ which specifies the positioning error to be less than 2.5% below frequency at 125 Hz. The frequency response of $W_p(s)$ is shown in Figure 5.

Considering system robustness requirement and the parameter uncertainty domain, the weighting function for system robustness performance is chosen as $W_t(s) = (s + 1)/(10s + 8)$. The frequency response of $W_t(s)$ is shown in Figure 6.

Considering the limitation of the controller output signal, the weighting function for magnitude of output signal of feedback controller is chosen to be a constant, $W_u(s) = 1/50$.

Through design procedure of robust feedback controller and selecting $\tau = 0.2$, we obtained the feedback controller by using LMI toolbox in MATLAB. It is given as follows:

$$G_{\mathrm{FB}} = \frac{p_{\mathrm{FB}}(s)}{n_{\mathrm{FB}}(s)}, \tag{46}$$

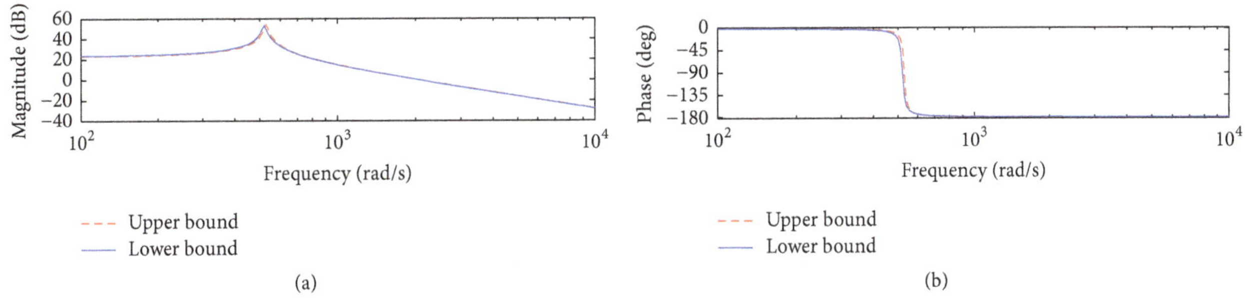

FIGURE 4: The frequency response of piezo actuator.

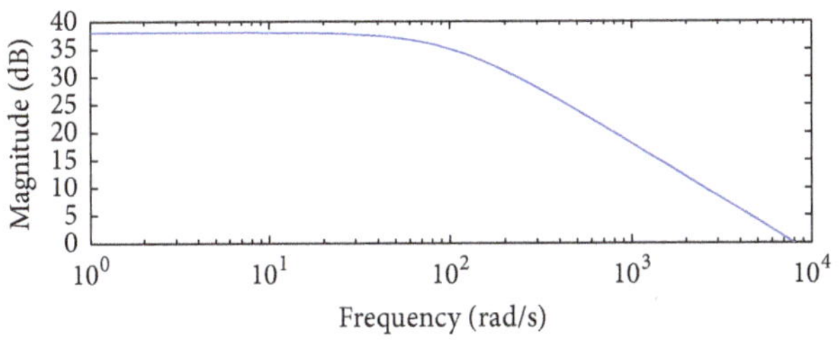

FIGURE 5: The frequency response of $W_p(s)$.

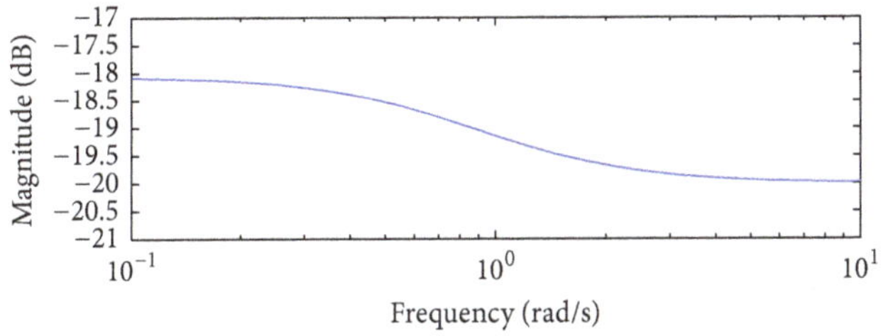

FIGURE 6: The frequency response of $W_t(s)$.

where

$$p_{\mathrm{FB}}(s) = 11.76s^6 + 4.10\times 10^7 s^5 + 3.47\times 10^{11} s^4 + 9.81\times 10^{14} s^3 + 1.08\times 10^{15} s^2 + 2.36\times 10^{14} s + 130.4,$$
$$n_{\mathrm{FB}}(s) = s^6 + 2.40\times 10^5 s^5 + 1.04\times 10^{10} s^4 + 4.67\times 10^{12} s^3 + 7.61\times 10^{11} s^2 + 1.81\times 10^{10} s + 5.59. \quad (47)$$

By H_∞ mixed sensitivity design approach without considering the feedforward controller and the above-selected weighting function, we obtained the controller by using MATLAB H_∞ mixed sensitivity synthesis command "hinf". It is given as follows:

$$G_H = \frac{2.63\times 10^6 s^3 + 4.09\times 10^7 s^2 + 7.12\times 10^{11} s + 5.70\times 10^{11}}{s^4 + 7.35\times 10^4 s^3 + 1.32\times 10^9 s^2 + 1.32\times 10^{11} s + 1.05\times 10^{11}}. \quad (48)$$

TABLE 1: Comparison of positioning control error.

Frequency	$e_{\max}$ (um)			E_{mean} (um)		
	(1)	(2)	(3)	(1)	(2)	(3)
5 HZ	0.0019	0.0027	0.0138	0.0002	0.0003	0.0073
50 HZ	0.0034	0.0061	0.0258	0.0003	0.0034	0.0148
125 HZ	0.0057	0.0103	0.0447	0.0008	0.0054	0.0303

5.4. Experiment and Result Discussion. The AFM x-axis positioning control simulation experiment which is tracking triangular trajectory at three different frequencies (5 Hz, 50 Hz, and 125 Hz) was conducted. For comparison, the positioning performances of three control approaches were tested by simulation experiment. The approaches are as follows: (1) the proposed H_∞ optimal inversion feedforward and robust feedback 2DOF control approach, that is, $(G^*_{\mathrm{FF}} + G^*_{\mathrm{FB}})$, (2) exact inversion feedforward-robust feedback based 2DOF control approach, that is, $(G_{\mathrm{FF}m} + G_H)$, and (3) H_∞ mixed sensitivity feedback control only, that is, G_H. The positioning control performance was evaluated by positioning maximum error and mean error, which are described as follows:

$$e_{\max} = \max_{k\in[1,n]} \left(y_d(k) - y(k)\right),$$
$$E_{\text{mean}} = \sqrt{\frac{\sum_{k=1}^{n}\left(y_d(k) - y(k)\right)^2}{n}}. \quad (49)$$

Figure 7 shows the positioning control simulation results and positioning errors by using different control approaches. The maximum positioning errors and mean errors for triangular trajectory scanning at different frequencies are shown in Table 1. The range of triangular trajectory was 1 μm. As shown, at scan rate 5 Hz, these three control approaches could achieve precision positioning; maximum positioning errors are 0.0019 μm, 0.0027 μm, and 0.0138 μm, respectively, and mean positioning errors are 0.0002 μm, 0.0003 μm, and 0.0073 μm, respectively. At scan rate 50 Hz, the proposed 2DOF control approach and the exact inversion feedforward and robust feedback based 2DOF control approach could obtain a better positioning control performance than H_∞ mixed sensitivity feedback control only. The maximum positioning errors obtained by approaches (1) and (2) are 0.0034 μm and 0.0061 μm which are as small as 13.2% and

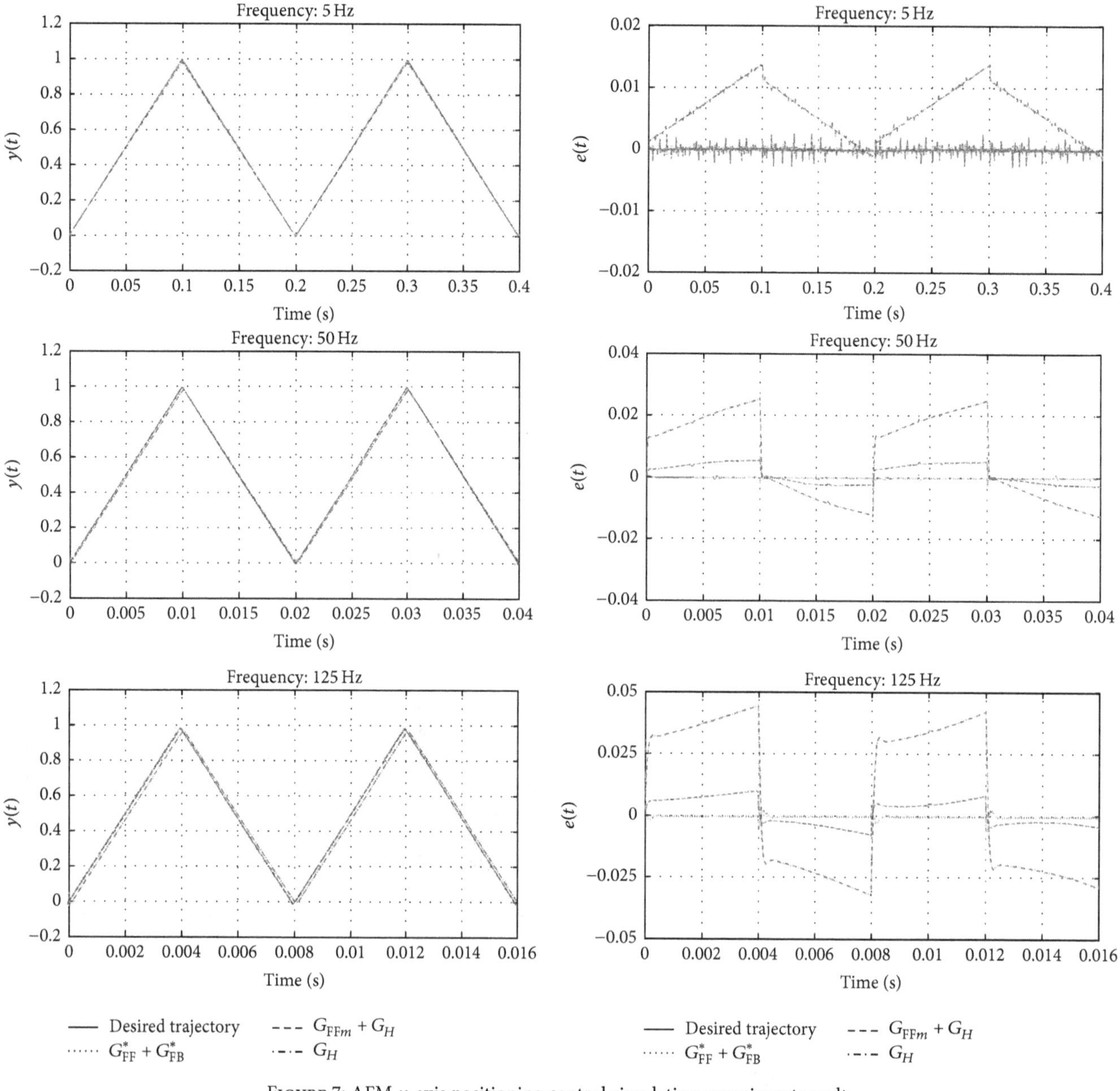

FIGURE 7: AFM x-axis positioning control simulation experiment result.

23.6% of that obtained by H_∞ mixed sensitivity feedback control. The positioning mean errors obtained by approaches (1) and (2) are 0.0003 μm and 0.0034 μm which are as small as 2% and 22% of that obtained by approach (3). At scan rate 125 Hz, only the proposed 2DOF control could obtain precision positioning. The maximum positioning error obtained by approach (1) is 0.0057 μm, which is as small as 55% of that obtain by approach (2) and is as small as 12.7% of that obtain by approach (3). The positioning mean error obtained by approach (1) is 0.0008 μm, which is as small as 14.8% of that obtained by approach (2) and is as small as 2.6% of that obtained by approach (3). Through the experiment result analysis, it is obvious that the proposed 2DOF control could achieve better positioning precision than the other two control approaches both in high speed scanning.

6. Conclusion

In this paper, a novel H_∞ optimal inversion feedforward and robust feedback based 2DOF control approach for high speed-precision positioning systems is proposed. In this approach, an H_∞ optimal inversion feedforward controller is designed to minimize the H_∞ norm of the tracking error in the presence of model uncertainty via linear matrix inequality (LMI) synthesis. Integrating with the feedforward controller systematically, a robust feedback controller is designed to render the requirements of positioning resolution, control width, robustness, and output signal magnitude imposed on the entire 2DOF control system satisfied via LMI optimization. The proposed control design approach is implemented in an AFM system x-axis positioning simulation experiment, and the experiment results demonstrated the effectiveness of the proposed 2DOF control approach.

Conflict of Interests

The authors declare that there is no conflict of interests regarding the publication of this paper.

Acknowledgment

This work is supported by China Postdoctoral Science Foundation under Grant 2014M560710.

References

[1] J.-C. Shen, W.-Y. Jywe, H.-K. Chiang, and Y.-L. Shu, "Precision tracking control of a piezoelectric-actuated system," *Precision Engineering*, vol. 32, no. 2, pp. 71–78, 2008.

[2] G. Schitter, F. Allgöwer, and A. Stemmer, "A new control strategy for high-speed atomic force microscopy," *Nanotechnology*, vol. 15, no. 1, pp. 108–114, 2004.

[3] M.-Y. Chen and J.-S. Lu, "High-precision motion control for a linear permanent magnet iron core synchronous motor drive in position platform," *IEEE Transactions on Industrial Informatics*, vol. 10, no. 1, pp. 99–108, 2014.

[4] B. E. Helfrich, C. Lee, D. A. Bristow et al., "Combined H_∞-feedback control and iterative learning control design with application to nanopositioning systems," *IEEE Transactions on Control Systems Technology*, vol. 18, no. 2, pp. 336–351, 2010.

[5] K. Yang, Y. Choi, and W. K. Chung, "On the tracking performance improvement of optical disk drive servo systems using error-based disturbance observer," *IEEE Transactions on Industrial Electronics*, vol. 52, no. 1, pp. 270–279, 2005.

[6] C. Du, L. Xie, J. N. Teoh, and G. Guo, "An improved mixed H_2/H_∞ control design for hard disk drives," *IEEE Transactions on Control Systems Technology*, vol. 13, no. 5, pp. 832–839, 2005.

[7] Z. Z. Liu, F. L. Luo, and M. A. Rahman, "Robust and precision motion control system of linear-motor direct drive for high-speed X-Y table positioning mechanism," *IEEE Transactions on Industrial Electronics*, vol. 52, no. 5, pp. 1357–1363, 2005.

[8] B. Yao and L. Xu, "Adaptive robust motion control of linear motors for precision manufacturing," *Mechatronics*, vol. 12, no. 4, pp. 595–616, 2002.

[9] A. C. Shegaonkar and S. M. Salapaka, "Making high resolution positioning independent of scan rates: a feedback approach," *Applied Physics Letters*, vol. 91, no. 20, Article ID 203513, 2007.

[10] J. S. Freudenberg, C. V. Hollot, R. H. Middleton, and V. Toochinda, "Fundamental design limitations of the general control configuration," *IEEE Transactions on Automatic Control*, vol. 48, no. 8, pp. 1355–1370, 2003.

[11] A. Karimi, M. Kunze, and R. Longchamp, "Robust controller design by linear programming with application to a double-axis positioning system," *Control Engineering Practice*, vol. 15, no. 2, pp. 197–208, 2007.

[12] T. Uchihashi, N. Kodera, H. Itoh, H. Yamashita, and T. Ando, "Feed-forward compensation for high-speed atomic force microscopy imaging of biomolecules," *Japanese Journal of Applied Physics*, vol. 45, no. 3, pp. 1904–1908, 2006.

[13] D. Croft and S. Devasia, "Vibration compensation for high speed scanning tunneling microscopy," *Review of Scientific Instruments*, vol. 70, no. 12, pp. 4600–4605, 1999.

[14] C. Lee, G. Mohan, and S. Salapaka, "2DOF control design for nanopositioning," in *Control Technologies for Emerging Micro and Nanoscale Systems*, vol. 413 of *Lecture Notes in Control and Information Sciences*, pp. 67–82, Springer, Berlin, Germany, 2011.

[15] G. Schitter, A. Stemmer, and F. Allgower, "Robust two-degree-of-freedom control of an atomic force microscope," *Asian Journal of Control*, vol. 6, no. 2, pp. 156–163, 2004.

[16] C. Lee and S. M. Salapaka, "Robust broadband nanopositioning: fundamental trade-offs, analysis, and design in a two-degree-of-freedom control framework," *Nanotechnology*, vol. 20, no. 3, Article ID 035501, 2009.

[17] K. Takanori, M. Yoshihiro, and I. Makoto, "LMI-based 2-degree-of-freedom controller design for robust vibration suppression positioning," *IEEJ Transactions on Industry Application*, vol. 131, no. 1, pp. 93–101, 2011.

[18] M. Araki and H. Taguchi, "Two-degree-of-freedom PID controllers," *International Journal of Control, Automation, and Systems*, vol. 1, no. 4, pp. 401–411, 2003.

[19] S. Devasia, D. Chen, and B. Paden, "Nonlinear inversion-based output tracking," *IEEE Transactions on Automatic Control*, vol. 41, no. 7, pp. 930–942, 1996.

[20] L. R. Hunt, G. Meyer, and R. Su, "Noncausal inverses for linear systems," *IEEE Transactions on Automatic Control*, vol. 41, no. 4, pp. 608–611, 1996.

[21] S. Devasia and B. Paden, "Stable inversion for nonlinear nonminimum phase time-varying system," *IEEE Transactions on Automatic Control*, vol. 43, no. 2, pp. 283–288, 1998.

[22] Q. Zou and S. Devasia, "Preview-based optimal inversion for output tracking: application to scanning tunneling microscopy," *IEEE Transactions on Control Systems Technology*, vol. 12, no. 3, pp. 375–386, 2004.

[23] Y. Yan, Y. Wu, Q. Zou, and C. Su, "An integrated approach to piezoactuator positioning in high-speed atomic force microscope imaging," *Review of Scientific Instruments*, vol. 79, no. 7, Article ID 073704, 9 pages, 2008.

[24] S. Devasia, "Should model-based inverse inputs be used as feedforward under plant uncertainty?" *IEEE Transactions on Automatic Control*, vol. 47, no. 11, pp. 1865–1871, 2002.

[25] Q. Zou, K. K. Leang, and E. Sadoun, "Control issues in high-speed AFM for biological applications: collagen imaging example," *Asian Journal of Control*, vol. 6, no. 2, pp. 164–178, 2004.

[26] C. G. L. Bianco and A. Piazzi, "A servo control system design using dynamic inversion," *Control Engineering Practice*, vol. 10, no. 8, pp. 847–855, 2002.

[27] K. K. Leang and S. Devasia, "Feedback-linearized inverse feedforward for creep, hysteresis, and vibration compensation in AFM piezoactuators," *IEEE Transactions on Control Systems Technology*, vol. 15, no. 5, pp. 927–935, 2007.

[28] M. Tahk and J. L. Speyer, "Modeling of parameter variations and asymptotic LQG synthesis," *IEEE Transactions on Automatic Control*, vol. 32, no. 9, pp. 793–801, 1987.

[29] G. Chesi, "Estimating the domain of attraction for uncertain polynomial systems," *Automatica*, vol. 40, no. 11, pp. 1981–1986, 2004.

[30] Y. He, M. Wu, and J.-H. She, "Improved bounded-real-lemma representation and H_∞ control of systems with polytopic uncertainties," *IEEE Transactions on Circuits and Systems II: Express Briefs*, vol. 52, no. 7, pp. 380–383, 2005.

[31] M.-N. Lee, J.-H. Moon, K. B. Jin, and M. J. Chung, "Robust H_∞ control with multiple constraints for the track-following system of an optical disk drive," *IEEE Transactions on Industrial Electronics*, vol. 45, no. 4, pp. 638–645, 1998.

[32] Q. Xu and Y. Li, "Model predictive discrete-time sliding mode control of a nanopositioning piezostage without modeling hysteresis," *IEEE Transactions on Control Systems Technology*, vol. 20, no. 4, pp. 983–994, 2012.

23

Quadcopter Aggressive Maneuvers along Singular Configurations: An Energy-Quaternion Based Approach

Ayman A. El-Badawy[1,2] and Mohamed A. Bakr[2]

[1]*Mechanical Engineering Department, Al-Azhar University, Cairo, Egypt*
[2]*Mechatronics Engineering Department, German University in Cairo, El Tagammoa El Khames, New Cairo, Cairo 11835, Egypt*

Correspondence should be addressed to Ayman A. El-Badawy; ayman.elbadawy@guc.edu.eg

Academic Editor: Xiao He

Automatic aggressive maneuvers with quadcopters are regarded as a highly challenging control problem. The aim is to tackle the singularities that exist in a vertical looping maneuver. Modeling singularities are resolved by writing the equations-of-motion of the quadcopter in quaternion form. Physical singularities due to underactuation are resolved by using an energy-based control. Energy-based control is utilized to overcome the uncontrollability of the quadcopter at physical singular configurations, for instance, when commanding the quadcopter to gain altitude while pitched at 90°. Three looping strategies (circular, clothoidal, and newly developed constant thrust) are implemented on a nonlinear model of the quadcopter. The three looping strategies are discussed along with their advantages and limitations.

1. Introduction

A looping maneuver (Figure 1) is executed when the centroid of the quadcopter moves along a circular path while pitching up to complete a 360° rotation.

Singular configurations are encountered while tracking the trajectories of a looping maneuver. Singular configurations are abnormal situations that should be avoided since they indicate either a malfunction of the mechanism, or a bad model [1]. Singular configurations are divided into two types: physical singular configurations and modelling singular configurations. Modelling singular configurations might reflect bad modelling decisions, for example, in a looping maneuver using Euler angles to model the quadcopter dynamics instead of quaternions [2]. Physical singular configurations reflect limitations in the design. For example, when the quadcopter's pitch angle approaches 90°, the controller loses the ability to command an acceleration in the vertical direction as the quadcopter is uncontrollable in this configuration. This is a limitation in the quadcopter's design since it has only 4 actuators in 6DOF (underactuated). At the physical singular configurations the controllability and the Jacobian matrices are rank deficient. Hence the quadcopter is uncontrollable.

Recently the design and implementation of control algorithms for aggressive maneuvers for unmanned aerial vehicles have been of interest to many research institutes and universities [3–13]. In the field of unmanned helicopters, human-piloted maneuvers are executed such that reference trajectories for the maneuver are extracted. These trajectories are then taught to the helicopters controller through reinforcement learning and so the maneuver can be replicated [3]. This approach implemented an entire air show of different maneuvers; however, this work addresses the challenge of designing trajectories in the full without the usage of human pilots.

Designing reference maneuver trajectories is not a simple task and so research has gone into the direction of reducing the complexity of the problem by dividing the flip trajectory into five steps (Acceleration, Start Rotation, Coast, Stop Rotate, and Recovery) [4–6]. Then these five steps were associated with five parameters. These five parameters were optimized to obtain the desired flips. This approach works on the optimization of the final state parameters and not the whole trajectory and thus it managed to perform flips and not a proper looping maneuver.

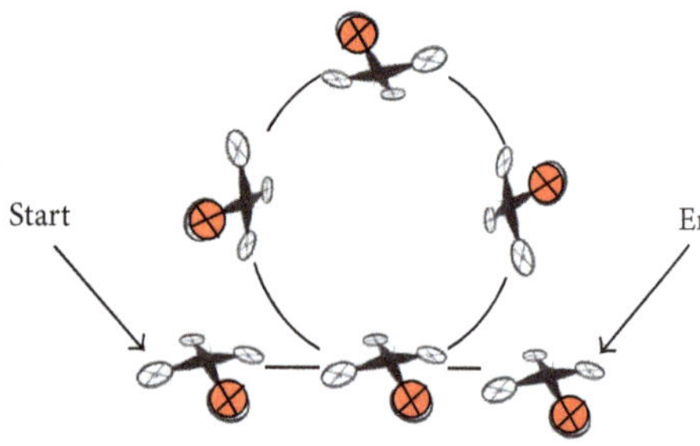

FIGURE 1: Looping maneuver.

In [7] the authors utilized geometric methods to define controllers that can achieve complex aerobatic maneuvers for a quadcopter. The quadcopter dynamics are modeled and expressed globally on the configuration manifold. It is coordinate-free, and therefore it overcomes modeling singularities. The authors managed to implement complex maneuvers such as recovering from being initially upside down. The paper considers the inverted orientation to be undesirable with a 180° attitude error. On the other hand, part of the maneuver in this work is to be inverted while executing an aggressive 360° loop.

The authors in [8] tackle the problem of generating real time trajectories for a quadcopter. These trajectories ensure safe passage through corridors. The authors discarded the small roll and pitch angle approximations as tight aggressive turns were required to maneuver the quadcopter in the corridors. On the other hand in this work the maneuver is more aggressive as it requires the pitch angle to change 360°. Moreover the looping maneuver done in this paper requires dealing with singularities, unlike tight turns where the pitch and roll angle are limited less than 90°.

The authors in [9] extend the flight envelope of the quadcopter by improving the accuracy of the aerodynamic model. The aerodynamic model is improved through the analysis of the blade flapping and thrust. The authors achieved high speeds and performed aggressive maneuvers. A stall turn maneuver is executed where a sudden pitch moment is applied, the quadcopter enters a steep climb, trading kinetic energy for height. At the peak of the maneuver where the velocity approaches zero, a yaw moment is applied to reverse the direction. However, in this paper the maneuver is more complicated where the quadcopter is required to trade kinetic energy with height while pitching up at the same time to execute the aggressive maneuver.

In [10] a complex aerobatic maneuver is approached by decomposing it into a sequence of discrete maneuvers. The author achieved safe switching between the maneuver segments while performing an autonomous backflip. The backflip maneuver is broken into three main stages: impulse, drift, and recovery. The maneuver initializes by rotating the quadcopter. Upon reaching the end of the first maneuver segment the motors are turned off for the drift mode, where the quadcopter rotates and falls under gravity. Finally, the recovery mode brings the quadcopter back to hover condition. The author in the paper depends on the inertia to pitch up and fall in the drift mode. On the other hand, in this work during the drift mode when the quadcopter has a pitch angle greater than 90° and less than 270° the motors are not turned off and energy is utilized to overcome the singular configurations and avoid falling due to gravity.

Research in [11–13] was successful in executing aggressive maneuvers as flying through narrow, vertical gaps such that the pitch angle of the quadcopter reaches 90°. Also the authors managed to perch on inverted surfaces with maximum inclination angle of 120°. This is done by designing trajectories and controllers defined by a sequence of segments. Each controller of each segment is then refined through iteration to account for errors in the dynamic model and noise in the sensors and actuators.

The goal of this work is to execute an aggressive 360° looping maneuver as shown in Figure 1. Performing such maneuver singularities that exist along the maneuver path has to be resolved. Modeling singularities were tackled by writing the equations-of-motion of the quadcopter in quaternion form. Energy-based control is then utilized to overcome the uncontrollability of the quadcopter at physical singular configurations. Three looping maneuvers (circular, clothoidal, and newly developed constant thrust) were implemented on a nonlinear model of the quadcopter. The three looping maneuvers were discussed along with their advantages and limitations.

2. Quadcopter Modelling and Control

2.1. Modelling. The inertial frame $I : \{O^I, X^I, Y^I, Z^I\}$ is defined by the ground, with gravity pointing in the positive Z^I direction. The body frame $B : \{O^B, X^B, Y^B, Z^B\}$ is defined by the orientation of the quadcopter, with the rotor axes pointing in the negative Z^B direction and the arms pointing in the x and y directions as shown in Figure 2.

The quad-rotor helicopter dynamic equations in quaternion form are found to be [14]

$$\begin{bmatrix}\ddot{x}\\ \ddot{y}\\ \ddot{z}\end{bmatrix} = \begin{bmatrix}0\\0\\g\end{bmatrix} + \mathbf{q}\otimes\begin{bmatrix}0\\0\\ \frac{1}{m}\end{bmatrix} U_1 \otimes \mathbf{q}^*,$$

$$\begin{bmatrix}\dot{p}\\ \dot{q}\\ \dot{r}\end{bmatrix} = \begin{bmatrix}\dfrac{\left(qr\left(I_z - I_y\right) + U_2 + I_p q\left(\Omega_{\text{front}} + \Omega_{\text{rear}} + \Omega_{\text{right}} + \Omega_{\text{left}}\right)\right)}{I_x}\\ \dfrac{\left(pr\left(I_x - I_z\right) + U_3 + -I_p p\left(\Omega_{\text{front}} + \Omega_{\text{rear}} + \Omega_{\text{right}} + \Omega_{\text{left}}\right)\right)}{I_y}\\ \dfrac{\left(pq\left(I_y - I_x\right) + U_4\right)}{I_z}\end{bmatrix}, \quad (1)$$

$$\dot{\mathbf{q}} = \frac{1}{2}\begin{bmatrix}0\\p\\q\\r\end{bmatrix}\otimes\mathbf{q}.$$

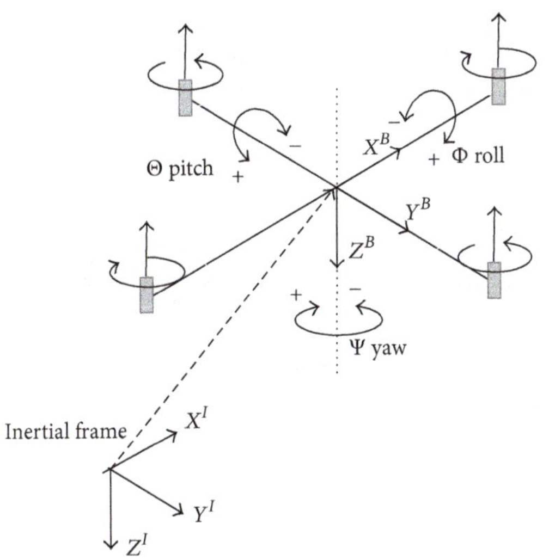

FIGURE 2: Coordinate systems.

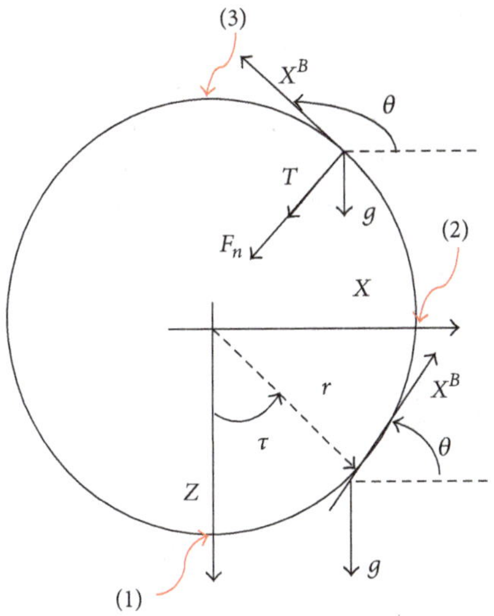

FIGURE 3: Perfect circular looping. In order to have sufficient energy for the quadcopter to traverse the interval between points (2) and (3) as shown in Figure 3, the quadcopter must build up enough energy at point (1). During the interval between points (2) and (3) the thrust vector only acts as a direction changer and does not contribute in increasing the altitude.

The x, y, and z represent the position of the centroid of the quadcopter relative to the inertial frame. The quadcopter is affected by the gravity shown by the gravitational acceleration g. The mass of the quadcopter is represented by m. I_x, I_y, and I_z are the moments of inertia about the X^B, Y^B, and Z^B axes, respectively. I_p is the propeller's moment of inertia about the Z^B axis. The quaternion $\mathbf{q}$ is a hyper complex number of rank 4, where $\mathbf{q} = [q_0 \; q_1 \; q_2 \; q_3]^T$. The quaternion units from q_1 to q_3 are called the vector part of the quaternion, while q_0 is the scalar part. The multiplication of two quaternions is done by the Kronecker product, denoted as $\otimes$ [14]. The quadcopter's p, q, and r represent the body-fixed angular velocity about the X^B, Y^B, and Z^B axes, respectively. The rotor propellers' speed are represented by Ω_{front}, Ω_{rear}, Ω_{right}, and Ω_{left}. Finally, U_1, U_2, U_3, and U_4 are the four control inputs of the quadcopter.

2.2. Controller Design. In order to track the desired trajectories a simple PD control law is designed for motions. For example, the roll angle control law is found to be

$$q_{1\text{err}} = [0 \;\; 1 \;\; 0 \;\; 0] \left(\mathbf{q}_c \otimes \mathbf{q}_a\right), \tag{2}$$

$$U_2 = K_p q_{1\text{err}} + K_d \dot{q}_{1\text{err}}, \tag{3}$$

where (2) is the quaternion difference between the command quaternion $\mathbf{q}_c$ and the actual quaternion $\mathbf{q}_a$. This difference gives the error quaternion [14]. The error quaternion is then multiplied by a vector to obtain the error quaternion "$q_{1\text{err}}$" that is related to the roll angle. Similarly, the pitch and yaw controllers are found to be

$$\begin{aligned} U_3 &= K_p q_{2\text{err}} + K_d \dot{q}_{2\text{err}}, \\ U_4 &= K_p q_{3\text{err}} - q_{3a} + K_d \dot{q}_{3\text{err}}. \end{aligned} \tag{4}$$

The control signal U_1 will be derived later while designing the trajectories.

2.3. Looping Trajectory Generation: An Energy-Based Approach. In order to approach the physical singularity problem looping trajectories are designed. These trajectories consist of a looping path and the thrust necessary to perform such loop. By calculating the controllability matrix at pitch angles 90° and 270° the system is uncontrollable at these points (physical singularity) and so the thrust calculated must be sufficient such that the quadcopter can cross these physical singularities using stored energy. In this section three types of looping paths are considered, a circular path, a clothoidal path, and a noncircular constant thrust path. The assumptions used for deriving the looping paths are as follows.

(1) The thrust vector is always directed normal to the path.

(2) The velocity is being commanded based on the energy conservation principle assuming a conservative system.

(3) The zero potential energy reference is the lowest point in the looping.

2.3.1. Perfect Circular Looping. Shown in Figure 3 is a side view of a perfect circular looping path.

By resolving the gravity and normalised thrust vector in the normal direction to the path, the normal force normalized by mass is found to be

$$F_n = T - g\cos(\tau). \tag{5}$$

By substituting (5) in Newton's second law in the normal direction, the normalised thrust is found to be

$$T = \frac{(V)^2}{r} + g\cos(\tau). \tag{6}$$

The term V^2/r is the centripetal acceleration and the $g\cos(\tau)$ is the gravity component in the normal direction. Since the quadcopter's X^B coordinate axis is always tangential to the path, therefore the path parameter τ can be replaced by the pitch angle θ. The magnitude of the velocity in (6) can be written using the conservation of energy principle:

$$V = \sqrt{\frac{2(L-U)}{m}}, \tag{7}$$

where L is the total energy and U is the potential energy of the quadcopter. Substituting (7) in (6) results in an equation relating the normalised thrust with the energy in the system:

$$T = \frac{2(L-U)}{mr} + g\cos(\theta). \tag{8}$$

From assumption (3) the potential energy is calculated using trigonometry to be

$$U = mgr(1-\cos(\theta)). \tag{9}$$

Substitute (9) in (8) and simplify

$$T = \frac{2L}{mr} + 3g\cos(\theta) - 2g. \tag{10}$$

The minimum total energy required is found to be

$$L > \frac{mr(-3g\cos(\theta)+2g)}{2}. \tag{11}$$

Note that the equality has been replaced by greater than since the rotational dynamics and thrust dynamics are coupled; therefore, at the peak of the looping, the thrust cannot be equal to zero.

The quadcopter must have enough energy to do the whole looping especially at the peak of the loop at $\tau = \theta = \pi$. Thus, the minimum total energy required to do a perfect circular looping must satisfy

$$L > \frac{5mrg}{2}. \tag{12}$$

The minimum total energy $5mrg/2$ from (12) is substituted in (10). The maximum value of the normalised thrust at the lowest point of the looping where the pitch angle is equal to zero. At pitch angle equal to zero the thrust normalised by mass is equal to $6g$. This thrust magnitude (normalised by mass) might be more demanding while designing the quadcopter. Therefore another approach is considered in order to perform the manuever. The clothoid looping is mainly used in roller coasters as it is characterised by lower load factor than a perfect circular looping [15].

2.3.2. Clothoid Looping. The clothoid looping is constructed from two clothoid curves as shown in Figure 4. λ is the path parameter, and E_t and E_n are the tangent and normal unit vectors. The first spiral curve extends from the start ($\lambda = 0$) to the peak ($\lambda = \sqrt{2}$) and the second spiral curve is a mirror

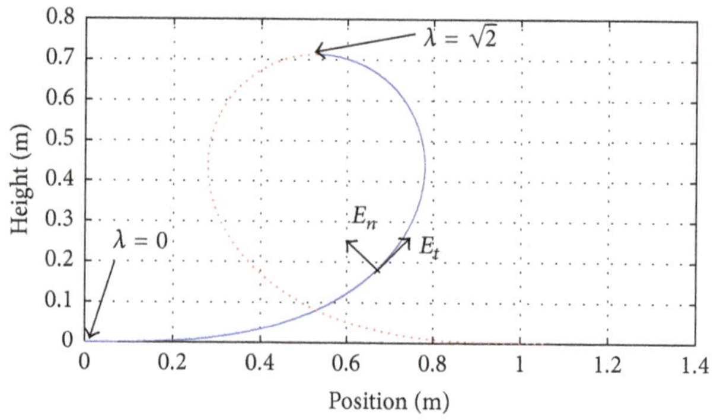

FIGURE 4: Clothoid looping.

of the first curve about the vertical axis passing through the peak.

The thrust analysis of the clothoid looping is done on one half of the loop (one clothoid curve). A clothoid curve is a member of the Euler spiral curves. It is a curved path whose curvature is linearly related to the arc length [15]. The parametric representation of the clothoid curve is found to be

$$\begin{aligned} x(\lambda) &= B\cdot C(\lambda), \\ z(\lambda) &= -B\cdot S(\lambda), \end{aligned} \tag{13}$$

where B is a constant which will determine the maximum height of the clothoid looping as it will be shown later. The functions $C(\lambda)$ and $S(\lambda)$ are the Fresnel integrals. They are expressed as definite integrals

$$C(\lambda) = \int_0^\lambda \cos\left(\frac{\pi}{2}t^2\right)dt, \tag{14}$$

$$S(\lambda) = \int_0^\lambda \sin\left(\frac{\pi}{2}t^2\right)dt, \tag{15}$$

where t is a dummy variable.

By differentiating (13) the speed along the clothoid curve is equal to "B" and so the path length is found to be

$$s = B\lambda. \tag{16}$$

The position vector $\overline{R}^I(s)$ of the first curve of the looping can be written as

$$\overline{R}^I(s) = \left(B\cdot C\left(\frac{s}{B}\right) \quad -B\cdot S\left(\frac{s}{B}\right)\right)^T. \tag{17}$$

The tangent unit vector is calculated by differentiating with respect to s [15] and thus it is determined to be

$$\overline{E}_t = \left(\cos\left(\frac{\pi s^2}{2B^2}\right) \quad -\sin\left(\frac{\pi s^2}{2B^2}\right)\right)^T. \tag{18}$$

The normal unit vector is calculated from the tangent unit vector and is found to be [15]

$$\overline{E}_n = \left(-\sin\left(\frac{\pi s^2}{2B^2}\right) \quad -\cos\left(\frac{\pi s^2}{2B^2}\right)\right)^T, \tag{19}$$

while the radius of curvature is

$$r = \frac{B^2}{\pi s}. \tag{20}$$

From the energy conservation principle, the magnitude of the velocity of one half of the clothoid looping can be written as

$$V^2 = V_0^2 - 2gz, \tag{21}$$

$$V = \sqrt{V_0^2 - 2gB \cdot S\left(\frac{s}{B}\right)}, \tag{22}$$

where V_0 is the initial velocity in the X^I direction. Same as the perfect circular path, if the normalised thrust and gravity vectors are resolved in the normal direction and then substituted in Newton's second law, the nomalised thrust is expressed as

$$T = \frac{(V)^2}{r} + g\cos(\tau). \tag{23}$$

The path parameter τ is interpreted as the path angle which is also equal to the quadcopter's pitch angle since the normalised thrust vector is always perpendicular to the path. Now all the equations are available to calculate the normalised thrust requirements to do a clothoid looping. By substituting (22), (20) and the path parameter from (19) in (23) the normalised thrust is found to be

$$T = \frac{\pi s V_0^2}{B^2} - \frac{2\pi s}{B} gS\left(\frac{s}{B}\right) + g\cos\left(\frac{\pi s^2}{2B^2}\right), \tag{24}$$

where the command input of the quadcopter can be written as:

$$U_1 = mT. \tag{25}$$

Since the quadcopter is limited to produce only positive thrust, therefore the thrust normalised by mass vector must be always greater than zero. The most critical point in the loop is at the peak at $\lambda = s/B = \sqrt{2}$, where the quadcopter must have enough energy to cross the peak of the looping. Therefore by substituting the $s/B = \sqrt{2}$, $S(\sqrt{2}) \approx 0.71$ (calculated numerically) and the thrust being greater than zero, the condition for the initial velocity is found to be

$$V_0 > \sqrt{1.65gB}. \tag{26}$$

Physically B is used to determine the maximum height of the clothoid looping. Since $S(\sqrt{2}) \approx 0.71$ and $z(t) = -B \cdot S(t)$ therefore the maximum height of the looping is approximately equal to $0.71B$. So the choice of B determines both the maximum height and the minimum initial velocity needed to perform the looping. The normalised thrust of (24) is plotted with the path length as shown in Figure 5.

Figure 5 shows two curves for the normalised thrust requirements to do one half of a clothoid looping. Both curves represent the same looping with the same maximum height ($B = 5$). The difference between them is the initial velocity.

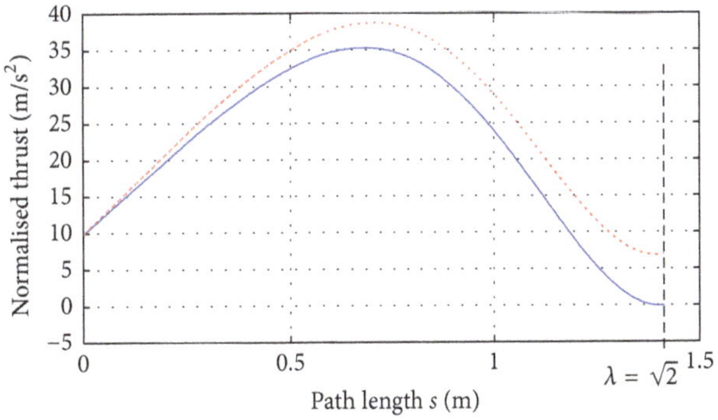

FIGURE 5: Normalised thrust requirements of one half of a clothoid loop.

Velocity 1 is the minimum initial velocity required to not to fall and velocity 2 is greater than velocity 1. Since B is the same, if the initial velocity is chosen higher than the minimum velocity, the centripetal acceleration along the path will increase. Therefore more normalised thrust is required in order to follow the looping path as shown in Figure 5. For the minimum velocity curve there is a linear increase of the normalised thrust from $1g$ to a maximum of $3.6g$. Based on design consideration this maximum thrust might be demanding but less than the circular path.

A second approach is then considered as shown next where the normalised thrust is specified and then the path is calculated.

2.3.3. Constant Thrust Looping: A New Trajectory Generation Approach. In this section the thrust is specified and then the path of the looping is determined. The thrust is assumed to be constant along the looping path.

The normalised thrust equation for an arbitrary path is shown in (23). Then by substituting the magnitude of the velocity with the conservation of energy equation, the normalised thrust is found to be

$$T = \frac{V_0^2 - 2gh}{r} + g\cos(\theta), \tag{27}$$

where h is the height which is equivalent to $-Z^I$ axis. Equation (27) contains three unknowns along the path, the height h, the radius of curvature r, and the pitch angle θ. The path is divided into arcs where a new pitch angle is calculated from the initial values of h and r along with the V_0 and T which are known. The new pitch angle is used to determine the final values of the arc which are used as the initial values for the next arc as it will be shown.

Figure 6 shows an arc of the looping path, where s is the path length and r is the radius of curvature. θ is the path parameter which is also equal to the pitch angle. Finally V is the velocity. The path length s is found to be

$$s = r\theta. \tag{28}$$

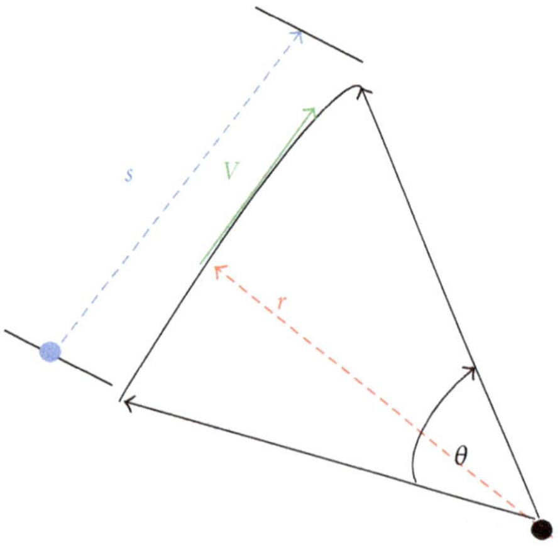

FIGURE 6: Constant thrust arc.

Then by taking the limit as the size of the arc approaches zero, the velocity is found to be

$$V = \frac{s}{t_{\text{step}}}, \tag{29}$$

where t_{step} is a small time step. Referring to the constant thrust algorithm shown in Algorithm 1 by substituting (28) in (29) the pitch angle calculation is shown in line (17). Thus, every time step the pitch angle is updated. Then using the energy conservation principle the new velocity is calculated as shown in line (19). Then position of the quadcopter is updated using the equations in line (22) and line (23). The final radius of curvature is calculated from the new velocity as shown in line (24). In Algorithm 1 the lines from (2) to (15) show the initialisations used to start the constant thrust looping algorithm.

The numerical technique presented here is Euler integration. The error in the solution is approximately proportional to the time step [16]. Therefore by decreasing the time step the error is minimized.

Figure 7 shows three looping paths at constant normalised thrust $3.5g$ but different initial velocities. Note that all the paths are plotted for the same period of time ($t = 12$ s).

As shown in Figure 7 at lower initial velocities, the number of loops per unit time increases since the normalised thrust is able to steer the quadcopter at a fast pitch rate due to the presence of less amount of energy in the system. Thus the rate of change of the pitch angle and the rate of change of the radius of curvature will decrease as the initial velocity increases. Increasing the initial velocity will increase the looping height and radius.

Alternatively shown next the normalised thrust is changing from one path to the other while the initial velocity is kept constant. Figure 8 shows three looping paths all at the same initial velocity ($V = 10\ \text{ms}^{-1}$) but different normalised thrust.

As shown in Figure 8 in the $2.5g$ and $3.5g$ looping there is enough normalised thrust to produce more centripetal acceleration capable of changing the direction of the velocity at a faster pitch rate. Thus, having a faster rate of change of the pitch angle will increase the rate of change of the local radius of curvature of the looping. As a result the quadcopter

(1) **Initialisation**;
(2) $g = 9.81$;
(3) $T = 14$; ▷ Thrust
(4) $t_{\text{step}} = 0.005$; ▷ time step
(5) $X = \text{zeros}(1, 2000)$; ▷ X-Array
(6) $Z = \text{zeros}(1, 2000)$; ▷ Z-Array
(7) $\theta = 0$;
(8) $\text{Energy} = \dfrac{V_{\text{initial}}^2}{2}$;
(9) $V_x = V_{\text{initial}}$;
(10) $V_z = 0$;
(11) $V = \sqrt{V_x^2 + V_z^2}$;
(12) $g_{\text{component}} = g\cos(\theta)$;
(13) $r = \dfrac{V^2}{T - g_{\text{component}}}$; ▷ radius of curvature
(14) $n = 1$;
(15) $t_{\text{end}} = 12$;
(16) **while** $t < t_{\text{end}}$ **do**
(17) $\theta_{n+1} = \dfrac{V * t_{\text{step}}}{r} + \theta_n$;
(18) $g_{\text{component}} = g\cos(\theta)$;
(19) $V = \sqrt{2 * (\text{Energy} - g * Z_n)}$;
(20) $V_z = V\sin(\theta)$;
(21) $V_x = V\cos(\theta)$;
(22) $Z_{n+1} = Z_n + t_{\text{step}} * V_z$;
(23) $X_{n+1} = X_n + t_{\text{step}} * V_x$;
(24) $r = \dfrac{V^2}{T - g_{\text{component}}}$;
(25) $t_{n+1} = t_n + t_{\text{step}}$;
(26) $n = n + 1$;
(27) **end**

ALGORITHM 1: Constant thrust.

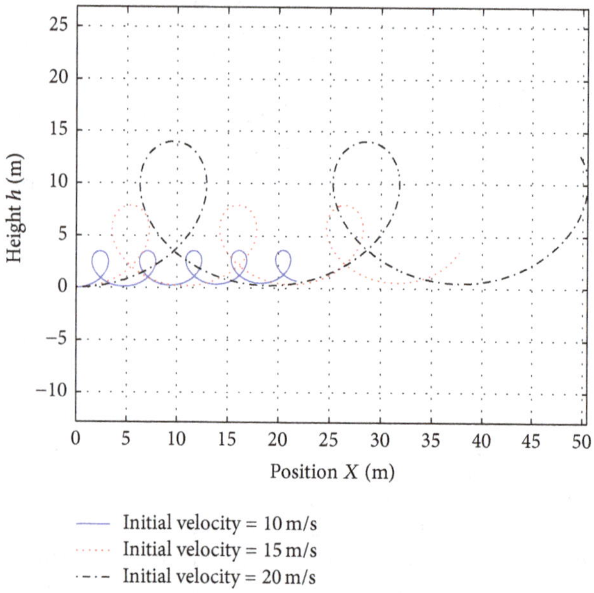

FIGURE 7: Varying the initial velocity at $3.5g$ normalised thrust for the same period of time.

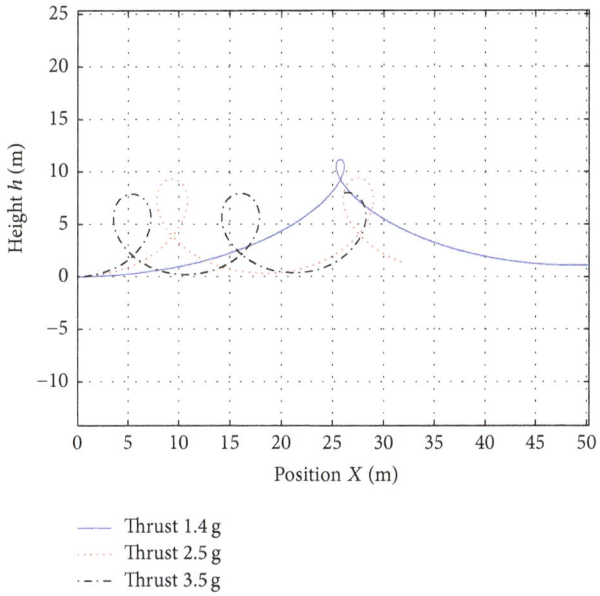

FIGURE 8: Varying the nomalised thrust at equal initial velocities for the period of time ($t = 12$ s).

will have enough kinetic energy at the peak of the looping in order to move sideways and form a looping shape. The $1.4g$ path is not really considered as a looping it is rather a flip. All three paths start with the same initial kinetic energy, the difference in shape is specified according to how much of the initial kinetic energy is transferred to potential energy and how much remains as kinetic energy at the peak of the looping.

3. Simulation

In this section the tracking performance of the 6DOF quadcopter model is evaluated with the looping commands. The block diagram shown in Figure 9 illustrates the simulation process.

As shown in Figure 9, the initial velocity is input along with the B for clothoidal commands or T for the constant thrust commands. The trajectories generated are then transformed into quaternions and the control input U_1. The error quaternion q_{err} is calculated from the quaternion difference between the command q_c and actual quaternion q_a. The error quaternion is then input to a PD controller and then the control signals are fed to the dynamics.

Since the perfect circular path is very demanding, therefore the perfect circle will not be implemented. The model parameters that are used in all simulations are shown in Table 1.

3.1. Clothoid Looping Commands. The following parameters were plugged in the clothoidal looping trajectory generation section and so the command thrust required and the command pitch angle are determined. The B is set to be equal to 5. The initial velocity is set to be the minimum initial velocity (refer to clothoid section) equal to $12.5\,\text{ms}^{-1}$.

TABLE 1: Looping parameters.

I_x	$7.5 * 10^{-3}\,\text{m}^4$
I_y	$7.5 * 10^{-3}\,\text{m}^4$
I_z	$1.3 * 10^{-2}\,\text{m}^4$
I_p	$6 * 10^{-5}\,\text{m}^4$
m	0.65 Kg

The time to perform the maneuver is 3 s. The proportional gain K_p and the derivative gain K_D are tuned to be equal 5 and 10, respectively.

The results are shown in Figures 10, 11, 12, and 13. Figures 10 and 11 show the command and state quaternion components q_0, q_2 as well as the command and state pitch angles.

As shown in Figure 10 the PD controller allows the state quaternion to track the command smoothly without any overshoot. Moreover due to the absence of any disturbances the tracking error is minimum. Note that the usage of quaternions eliminates the angular singularities. The quaternion components q_1 and q_3 are equal to zero since they are related to the roll and yaw angles.

Figure 11 is just a transformation of Figure 10 to illustrate the change in the pitch angle of the quadcopter along the loop. As shown in Figure 11 the PD control in the pitch angle controller allows the pitch to track the command smoothly without any overshoot.

Figure 12 shows the command normalised thrust over mass calculated from the clothoidal looping section.

As shown in Figure 12 the thrust curve is symmetrical. The first half of the curve represents the first spiral curve of the clothoid loop and the second represents the second spiral curve. Note that the normalised thrust reaches a maximum value of $4.5g$. An important point to be noticed is that the normalised thrust cannot reach zero at the peak of the looping since the pitching motion and the thrust are coupled. Therefore, if the thrust reaches zero at the peak, the quadrotor will lose its ability to pitch up. Finally Figure 13 shows a visualisation of the quadcopter doing a clothoidal looping.

As shown in Figure 13 the black dashed loop is the reference path and the blue solid loop is the actual loop. The quadcopter is first commanded to hover to 10 m. The blue loop is not identical to the black loop. This is due to the presence of the quadcopter dynamics. The pitch angle dynamics and the position dynamics are not identical, since the pitch dynamics are much faster than the position dynamics, the quadcopter did not track the desired trajectory. The addition of the damping term in the pitch angle controller slows down the pitch angle dynamics. Figure 14 shows the effect of the removal of the D-controller from the pitch angle controller.

As shown in Figure 14 the removal of the damping gain caused the pitch angle to change faster than the position dynamics, thus, leading the quadcopter to exit the maneuver at a higher altitude.

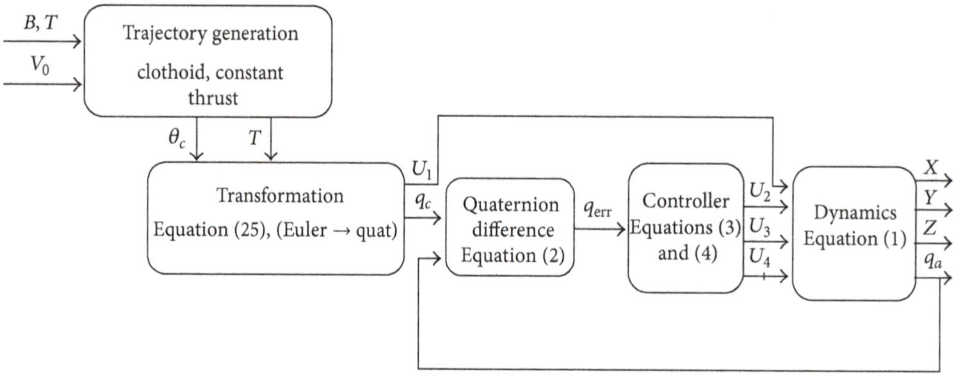

FIGURE 9: System block diagram.

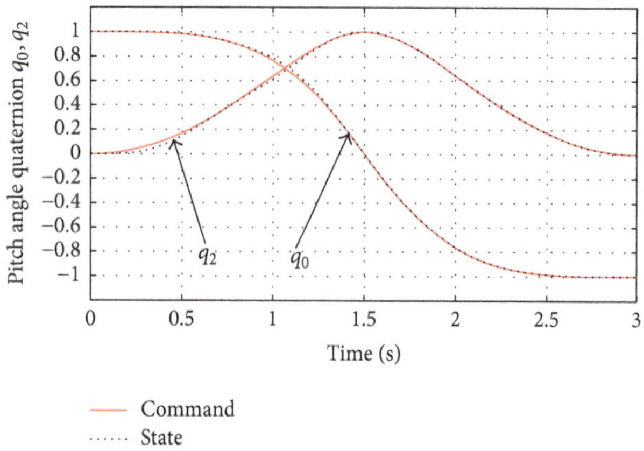

FIGURE 10: Command and state quaternion for a clothoid loop.

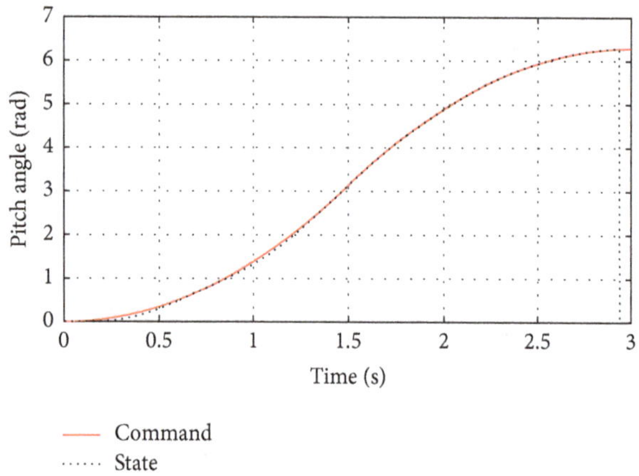

FIGURE 11: Command and state pitch angle for a clothoid loop.

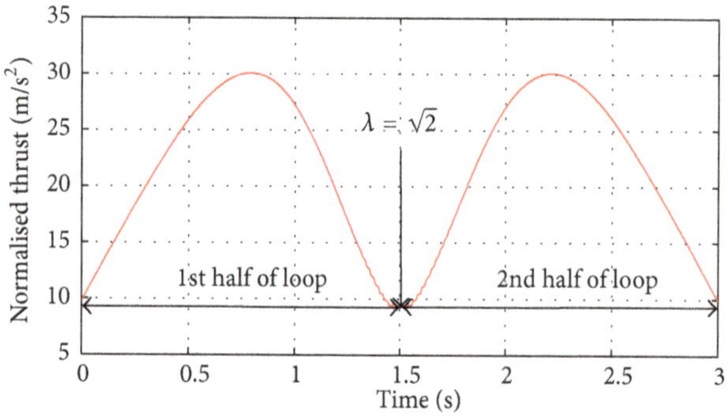

FIGURE 12: Command normalised thrust for a clothoid loop.

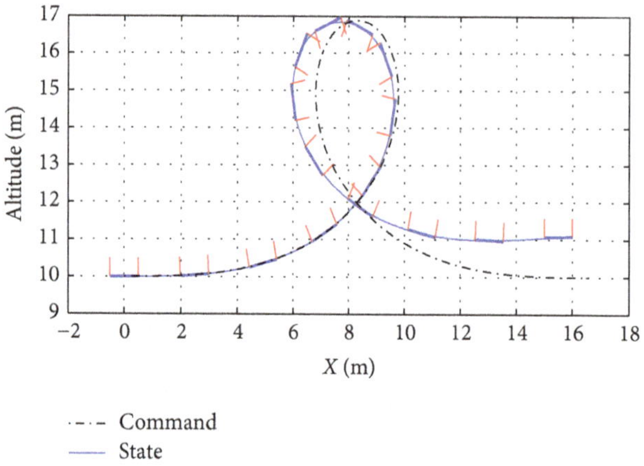

FIGURE 13: Quadcopter visualisation with PD controller.

3.2. Constant Thrust Looping. The following parameters were plugged in the constant looping thrust equations and so the command pitch angle is determined. The constant normalised thrust chosen for this simulation is $T = 20\,\text{m/s}^2$. The initial velocity is set to be 10 m/s. The time to perform the maneuver is 2.5 s. The proportional gain K_p and the derivative gain K_D are tuned to be equal to 5 and 10, respectively.

The results are shown in Figures 15, 16, and 17. Figures 15 and 16 show the command and state quaternion components q_0, q_2 as well as the command and state pitch angles.

As shown in Figure 15 the state quaternion tracks the command smoothly. The quaternion components q_1 and q_3 are equal to zero since they are related to the roll and yaw angles.

Due to the absence of external disturbance and the presence of a smooth PD controller, the state pitch angle

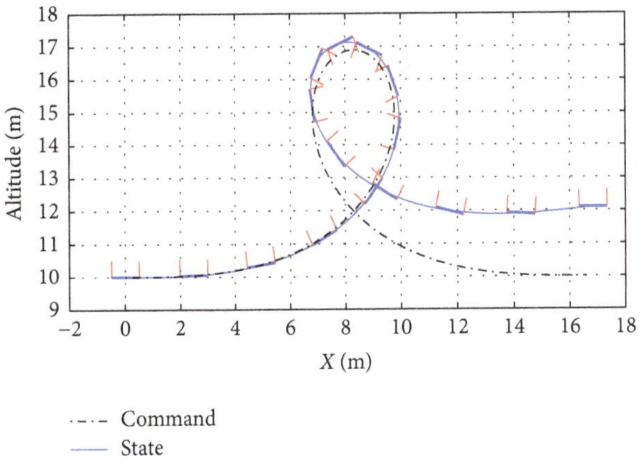

FIGURE 14: Quadcopter visualisation with P controller.

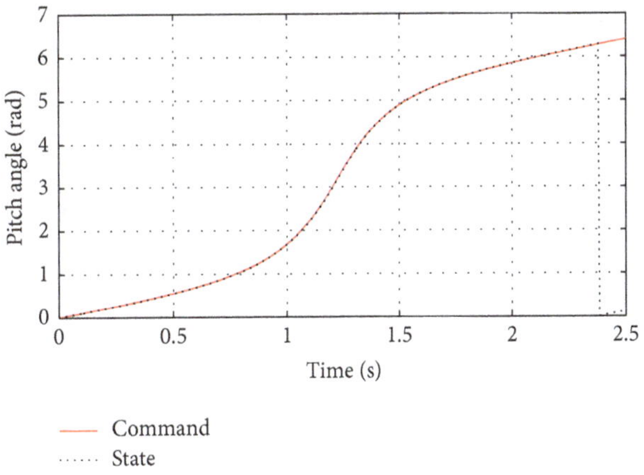

FIGURE 16: Command and state pitch angle for a constant thrust loop.

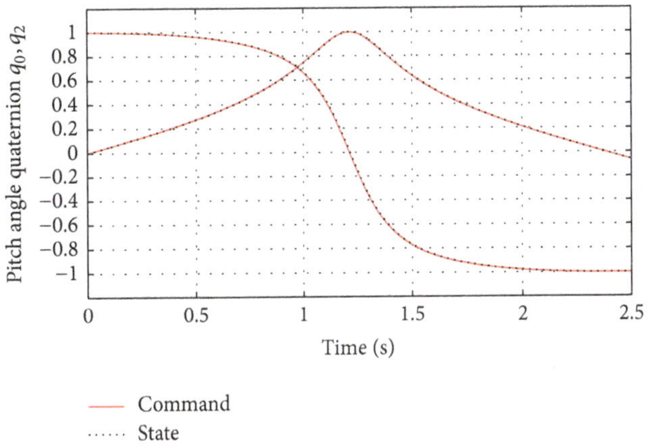

FIGURE 15: Command and state quaternion for a constant thrust loop.

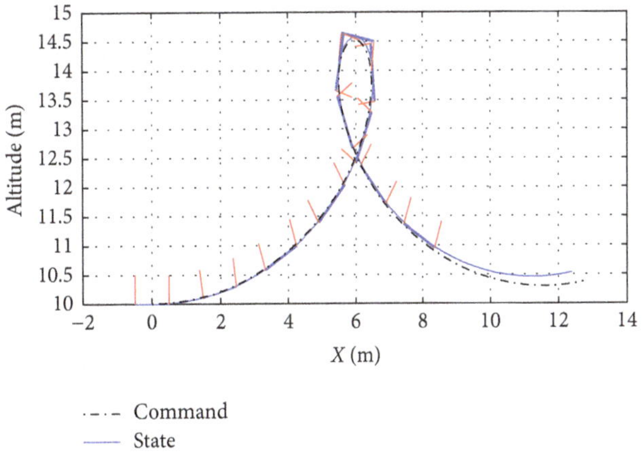

FIGURE 17: Visualisation of constant thrust loop.

tracks the command smoothly with minimum error as shown in Figure 16.

Figure 17 shows the visualisation of the quadcopter while performing the constant thrust looping.

As shown in Figure 17 the black dashed loop is the reference path and the blue solid loop is the actual loop. The quadcopter is first commanded to hover to 10 m. The blue loop nearly identical to the black loop. The position and pitch dynamics do not affect much the shape of the looping since in this case normalised thrust (acceleration) is always at steady state.

4. Conclusion

The trajectories designed using energy allowed the quadcopter to cross the physical singular configurations smoothly. The three looping paths and trajectories designed were found to be very promising in extending the flight envelope of the quadcopter in order to perform aggressive looping maneuver. The perfect circular looping is found to be the most demanding as it required an entry thrust equal to $6g$. A clothoid looping was found to be less demanding and as it requires normalised thrust equal to $3.5g$. Finally a new innovative approach is proposed where the thrust is specified and then the path is determined. This method used a constant thrust approach which might be the most applicable since the looping maneuver depends on the design specifications of the quadcopter. A quadcopter can perform a constant thrust looping at normalised thrust starting from $1.4g$.

Conflict of Interests

The authors declare that there is no conflict of interests regarding the publication of this paper.

References

[1] E. J. Haug, *Computer Aided Kinematics and Dynamics of Mechanical Systems*, vol. 1, Allyn & Bacon, Boston, Mass, USA, 1989.

[2] J. Diebel, "Representing attitude: Euler angles, unit quaternions, and rotation vectors," *Matrix*, vol. 58, p. 1516, 2006.

[3] P. Abbeel, A. Coates, and A. Y. Ng, "Autonomous helicopter aerobatics through apprenticeship learning," *The International Journal of Robotics Research*, vol. 29, no. 13, pp. 1608–1639, 2010.

[4] S. Lupashin, A. Schöllig, M. Sherback, and R. D'Andrea, "A simple learning strategy for high-speed quadrocopter multi-flips," in *Proceedings of the IEEE International Conference on Robotics and Automation (ICRA '10)*, pp. 1642–1648, IEEE, Anchorage, Alaska, USA, May 2010.

[5] S. Lupashin and R. D'Andrea, "Adaptive openloop aerobatic maneuvers for quadrocopters," in *Proceedings of the 18th IFAC World Congress*, pp. 2600–2606, Milan, Italy, August-September 2011.

[6] S. Lupashin and R. D'Andrea, "Adaptive fast open-loop maneuvers for quadrocopters," *Autonomous Robots*, vol. 33, no. 1-2, pp. 89–102, 2012.

[7] T. Lee, M. Leok, and N. H. McClamroch, "Control of complex maneuvers for a quadrotor UAV using geometric methods on SE (3)," http://arxiv.org/abs/1003.2005.

[8] D. Mellinger and V. Kumar, "Minimum snap trajectory generation and control for quadrotors," in *Proceedings of the IEEE International Conference on Robotics and Automation (ICRA '11)*, pp. 2520–2525, Shanghai, China, May 2011.

[9] H. Huang, G. M. Hoffmann, S. L. Waslander, and C. J. Tomlin, "Aerodynamics and control of autonomous quadrotor helicopters in aggressive maneuvering," in *Proceedings of the IEEE International Conference on Robotics and Automation (ICRA '09)*, pp. 3277–3282, IEEE, Kobe, Japan, May 2009.

[10] J. H. Gillula, H. Huang, M. P. Vitus, and C. J. Tomlin, "Design of guaranteed safe maneuvers using reachable sets: autonomous quadrotor aerobatics in theory and practice," in *Proceedings of the IEEE International Conference on Robotics and Automation (ICRA '10)*, pp. 1649–1654, IEEE, Anchorage, Alaska, USA, May 2010.

[11] D. Mellinger, N. Michael, and V. Kumar, "Trajectory generation and control for precise aggressive maneuvers with quadrotors," *The International Journal of Robotics Research*, vol. 31, no. 5, pp. 664–674, 2012.

[12] D. Mellinger, M. Shomin, and V. Kumar, "Control of quadrotors for robust perching and landing," in *Proceedings of the International Powered Lift Conference*, pp. 205–225, 2010.

[13] M. Piedmonte and E. Feron, "Aggressive maneuvering of autonomous aerial vehicles: a human-centered approach," in *Proceedings of the International Symposium of Robotics Research*, vol. 9, pp. 413–420, London, UK, 2000.

[14] E. Fresk and G. Nikolakopoulos, "Full quaternion based attitude control for a quadrotor," in *Proceedings of the 12th European Control Conference (ECC '13)*, pp. 3864–3869, Zurich, Switzerland, July 2013.

[15] R. Müller, "Roller coasters without differential equations—a Newtonian approach to constrained motion," *European Journal of Physics*, vol. 31, no. 4, pp. 835–848, 2010.

[16] J. C. Butcher, *Numerical Methods for Ordinary Differential Equations*, vol. 1, John Wiley & Sons, New York, NY, USA, 2003.

24

Longitudinal Motion Control of AUV Based on Fuzzy Sliding Mode Method

Duo Qi, Jinfu Feng, and Jian Yang

Aeronautics and Astronautics Engineering College, Air Force Engineering University, Xi'an 710038, China

Correspondence should be addressed to Duo Qi; qi33song@sina.com

Academic Editor: Lifeng Ma

According to the characteristics of AUV movement, a fuzzy sliding mode controller was designed, in which fuzzy rules were adopted to estimate the switching gain to eliminate disturbance terms and reduce chattering. The six-degree-of-freedom model of AUV was simplified and longitudinal motion equations were established on the basis of previous research. The influences of first-order wave force and torque were taken into consideration. The REMUS was selected to simulate the control effects of conventional sliding mode controller and fuzzy sliding mode controller. Simulation results show that the fuzzy sliding mode controller can meet the requirements and has higher precision and stronger antijamming performances compared with conventional sliding mode controller.

1. Introduction

Nowadays the ocean space is an important competition field of military and economic powers in the world, and many countries are vigorously developing deep sea exploration technology. As an intercrossed subject of ocean engineering and robot technology, autonomous underwater vehicles (AUV) are playing increasingly significant roles in underwater activities, such as offshore oil exploitation, underwater target search, marine science research, and military application [1–5].

The stable and efficient control of AUV is very difficult for its inherent highly nonlinear, uncertain hydrodynamic parameters and external disturbances. Wang et al. [6] adopted S-surface control method to simulate the heading control and depth control of a mini AUV, and, furthermore, they simulated long distance traveling following a planned path. The results showed that the AUV has good spatial maneuverability and verified the feasibility and reliability of control method. Ma and Cui [7] proposed a robust path-following control method for nonlinear and underactuated AUV based on a fuzzy hybrid control strategy. Jia et al. [8] presented a nonlinear iterative sliding mode controller based on the virtual guide method. It can decrease the static error and overshoot and achieves high tracking precision. Sahu and Subudhi [9] developed an adaptive control law for the AUV to track the desired trajectory and verified the stability of the controller using Lyapunov's direct method. The simulation results demonstrate that the controller is feasible for the tracking of uncertain parameters model. Lapierre and Soetanto [10] designed a new backstepping controller, which can get rid of the limits of initial conditions and make the tracking error converge to zero.

The sliding mode control has been successfully applied to dynamic positioning and motion control of underwater vehicle [11, 12], due to its simple algorithm, robustness against modeling imprecision, and external disturbances. However, the discontinuous switching characteristics of sliding mode control will cause chattering, which not only affects the control accuracy but also degrades the system performance and even severely damages the control units. Many researchers have put forward solutions to eliminate the chattering phenomenon from different angles, such as adaptive method [13], neural network method [14], feedback linearization method [15], and fuzzy method [16]. According to the experience, a proper switching gain can reduce chattering [17]. The fuzzy control has many advantages; for example, it needs no accurate mathematical model and has good robustness.

TABLE 1: Motion mode and attitude parameters of AUV.

Degree of freedom	Motion modes	Force/torque (in the body-fixed coordinate)	Linear velocity/angular velocity (in the body-fixed coordinate)	Location/Euler angles (in the earth-fixed coordinate)
1	Back/forward (movement along the x-axis)	F_x	u	x
2	Sway (movement along the y-axis)	F_y	v	y
3	Lift/dive (movement along the z-axis)	F_z	w	z
4	Roll (rotation along the x-axis)	M_x	p	ϕ
5	Pitch (rotation along the y-axis)	M_y	q	θ
6	Yaw (rotation along the z-axis)	M_z	r	ψ

The fuzzy sliding mode control combines the advantages of sliding mode control and fuzzy control and can make discrete control signals continuous to reduce chattering effectively.

The main contribution of this paper is to design a fuzzy sliding mode controller for the longitudinal motion control of AUV with the consideration of first-order wave force and torque. Based on the conventional sliding mode control, fuzzy rules are adopted to estimate the switching gain to eliminate disturbance terms and reduce chattering. The simulation results show that the fuzzy sliding mode controller can meet the requirements. Compared with conventional sliding mode controller, it has higher precision and stronger antijamming performances, which has good practical values.

2. Longitudinal Motion Model of AUV and Wave Disturbance

Earth-fixed coordinate and AUV body-fixed coordinate are shown in Figure 1. Six-DOF kinematic modes and attitude parameters are defined in the coordinate system as shown in Table 1.

According to the parameters in Table 1, we define vectors as follows: $\eta_1 = (x, y, z)^T$, $\eta_2 = (\phi, \theta, \psi)^T$, $\eta = (\eta_1, \eta_2)^T$, $\tau_1 = (F_x, F_y, F_z)^T$, $\tau_2 = (M_x, M_y, M_z)^T$, $\tau = (\tau_1, \tau_2)^T$, $V = (u, v, w)^T$, $\omega = (p, q, r)^T$, $\nu = (V, \omega)^T$, the center of gravity position is $r_G = (x_g, y_g, z_g)^T$, and the center of buoyancy position is $r_B = (x_B, y_B, z_B)^T$.

The longitudinal motion equations of AUV with respect to the body-fixed moving frame are described by a set of nonlinear differential equations as follows:

$$\begin{aligned} m\left(\dot{u} + wq - x_g q^2 + z_g \dot{q}\right) &= F_x, \\ m\left(\dot{w} - uq - z_g q^2 - x_g \dot{q}\right) &= F_z, \\ I_{yy}\dot{q} + m\left[z_g\left(\dot{u} + wq\right) - x_g\left(\dot{w} - uq\right)\right] &= M_y. \end{aligned} \tag{1}$$

Ignore all the high-order terms under the condition of low speed; the mathematical model above can be simplified as

$$\begin{aligned} m\left(\dot{u} + z_g \dot{q}\right) &= F_x, \\ m\left(\dot{w} - Uq - x_g \dot{q}\right) &= F_z, \\ I_{yy}\dot{q} + m\left[z_g \dot{u} - x_g\left(\dot{w} - uq\right)\right] &= M_y, \end{aligned} \tag{2}$$

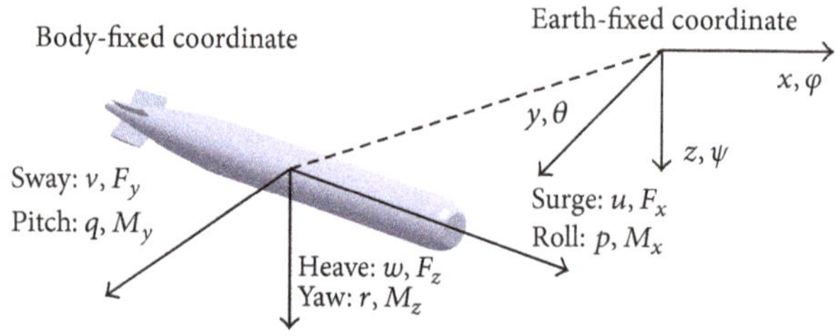

FIGURE 1: Reference coordinates and 6-DOF coordinates of AUV.

where

$$\begin{aligned} F_x &= X_{\dot{u}}\dot{u} + X_u u + X_q q + X_\theta \theta, \\ F_z &= Z_{\dot{w}}\dot{w} + Z_{\dot{q}}\dot{q} + Z_w w + Z_q q + Z_{\delta_s}\delta_s, \\ M_y &= M_{\dot{w}}\dot{w} + M_{\dot{q}}\dot{q} + M_w w + M_q q + M_{\delta_s}\delta_s, \end{aligned} \tag{3}$$

and $X_{\dot{u}}, Z_{\dot{w}}, M_{\dot{w}} \ldots$ are hydrodynamic parameters.

According to the practical situation, we suppose that the AUV moves with a constant velocity $u = u_0$, and the longitudinal motion equations of AUV can be described further [18]:

$$\begin{bmatrix} \dot{z} \\ \dot{\theta} \end{bmatrix} = \begin{bmatrix} \cos\theta & 0 \\ 0 & 1 \end{bmatrix} \begin{bmatrix} w \\ q \end{bmatrix} - \begin{bmatrix} u_0 \sin\theta \\ 0 \end{bmatrix} = A_1 \begin{bmatrix} w \\ q \end{bmatrix} - A_{\theta 1}, \tag{4}$$

$$\begin{aligned} \begin{bmatrix} \dot{w} \\ \dot{q} \end{bmatrix} &= M^{-1} u_0 \begin{bmatrix} Z_w & Z_q + m \\ M_w & M_q \end{bmatrix} \begin{bmatrix} w \\ q \end{bmatrix} \\ &+ M^{-1} {u_0}^2 \begin{bmatrix} Z_\delta \\ M_\delta \end{bmatrix} \delta_s + M^{-1} \left[\left(W - B_0\right)\cos\theta \right. \\ &\left. - \left(x_g W - x_B B_0\right)\cos\theta - \left(z_g W - z_B B_0\right)\sin\theta\right] \\ &+ M^{-1} \begin{bmatrix} Z_d \\ M_d \end{bmatrix} = A_2 \begin{bmatrix} w \\ q \end{bmatrix} + A_\delta \delta_s + A_m x_\theta + A_d, \end{aligned} \tag{5}$$

where $M^{-1} = \begin{bmatrix} m - Z_{\dot{w}} & -Z_{\dot{q}} \\ -M_{\dot{w}} & I_y - M_{\dot{q}} \end{bmatrix}$.

The first-order wave force mainly affects the AUV during its moving near water surface. The first-order wave force, which is of high frequency and periodical, has amplitude

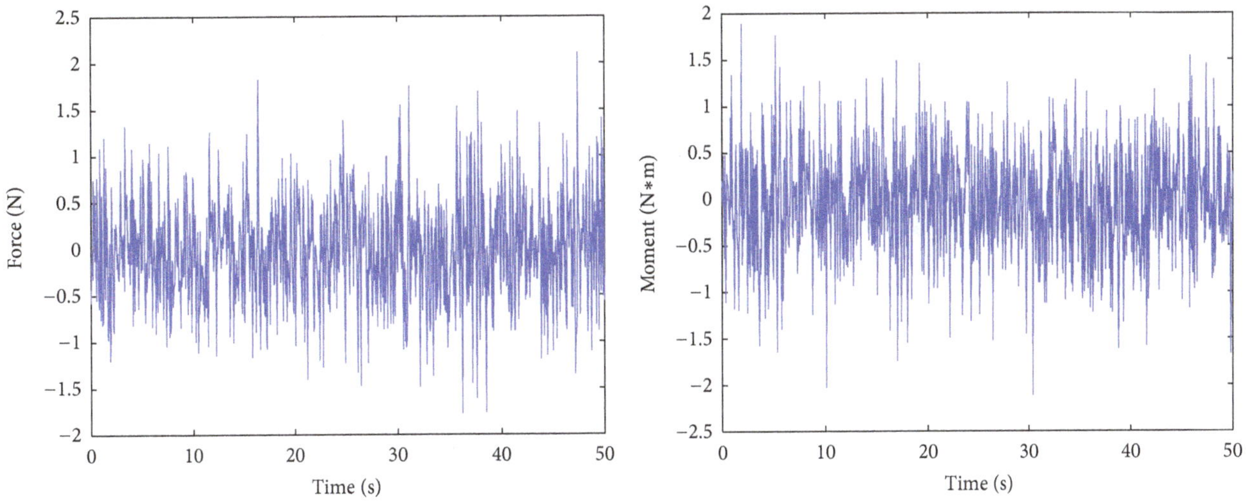

FIGURE 2: Wave force and wave torque.

proportional to wave height. In this paper, the Hirom approximation formula is adopted to calculate the first-order wave force and torque, and the concrete form is

$$Z = \left(780 - 145 \sum F \sin \omega_a t\right) \sum F \sin \omega, \qquad M = 1070F \cos \omega_a t, \tag{6}$$

where $F = a\omega^2 e^{-\omega^2(H+h(t)/g)}$; $\omega_a = -\omega(1 - \omega U \cos \beta/g)$; $a_i = \sqrt{2S(\omega)\delta\omega}$; $h(t)$ is the wave height; g is gravitational acceleration; ω is wave frequency; β is wave to course angle; $h(t)$ is the distance between the AUV and sea surface. The first-order wave force and torque under the conditions that $H = 10$ m, $u = 2$ m/s, and $\beta = 0^\circ$ are shown in Figure 2, respectively.

From (5) and (6), the model can be expressed as

$$\dot{x} = Ax + Bu + D, \tag{7}$$

where state vector is $x = [z\ \theta\ w\ q]^T$, input is $u = \delta_s$, $A = \begin{bmatrix} O & A_1 \\ O & A_2 \end{bmatrix}$, $B = \begin{bmatrix} O \\ A_\delta \end{bmatrix}$, δ_s is rudder angle, and D is the sum of disturbance terms and uncertain terms with wave force and torque.

3. Design of Fuzzy Sliding Mode Controller

The control of AUV, a typical underactuated system, is difficult in complex and variable underwater environment. Sliding mode control has the characteristic of discontinuity which forces the system to make a small range and high frequency sliding motion along a certain state. When the system is in the sliding mode, the control plant is invariant to uncertain parameters and disturbance. However, the invariance comes at the cost of high chattering, and it severely impacts the practical application of sliding mode control. It is one of the most effective ways to determine the switching gain in order to reduce the chattering by fuzzy method.

Let us suppose that the desired target state is $x_d = [z_d\ \theta_d\ w_d\ q_d]^T$, and the control error is defined as follows:

$$e = x - x_d. \tag{8}$$

The control target is to find a design of u to minimize the control error. We choose the switching function as

$$s = Ce, \tag{9}$$

where $C = [c_1, c_2, c_3, 1]$, which satisfies the Hurwitz stability condition.

With uncertain disturbance, we define the sliding mode control law as

$$u = u_{eq} + u_s, \tag{10}$$

where u_{eq} is the equivalent control and u_s is the switching control.

Let us suppose that, after a period of time, the system reaches sliding mode surface, and, in an ideal situation, the control system will meet

$$s = Ce = 0, \qquad \dot{s} = C\dot{e} = 0. \tag{11}$$

The equivalent control is

$$u_{eq} = -(CB)^{-1}(CAx) + (CB)^{-1}C\dot{x}_d. \tag{12}$$

We set the switching controller as

$$u_s = k\,\mathrm{sgn}(s), \tag{13}$$

TABLE 2: Fuzzy table of output.

$s\dot{s}$	NB	NM	NS	ZO	PS	PM	PB
Δk	NB	NM	NS	ZO	PS	PM	PB

where k is switching gain; sgn(s) is sign function. Consider

$$\operatorname{sgn}(s)=\begin{cases}+1, & s>0,\\ 0, & s=0,\\ -1, & s<0,\end{cases}$$

$$\begin{aligned}s\dot{s}&=s[C\dot{e}]=s[C(Ax+Bu+D-\dot{x}_d)]=s(CAx\\&+CBu+CD-C\dot{x}_d)=s\big(CAx\\&+CB\big[-(CB)^{-1}(CAx)+(CB)^{-1}C\dot{x}_d+k\operatorname{sgn}(s)\big]\\&+CD-C\dot{x}_d\big)=s(CAx-CAx+C\dot{x}_d\\&+CBk\operatorname{sgn}(s)+CD-C\dot{x}_d)=s(CBk\operatorname{sgn}(s)\\&-CD).\end{aligned} \tag{14}$$

Let $k=-\varepsilon(CB)^{-1}$, where $\varepsilon>\max(|CD|)$; thus $s\dot{s}<0$.

The reachability is verified. When the system enters into the sliding mode, it is effective to deal with the uncertain disturbance.

Appropriate switching gain can reduce the chattering efficiently. $s\dot{s}$ is the input, and Δk is the output. Define both the fuzzy sets of $s\dot{s}$ and Δk as [NB, NM, NS, ZO, PS, PM, PB], the universe is $[-3,-2,-1,0,1,2,3]$, and membership function adopts the triangle function. Fuzzy control rules are as follows:

(1) If $s\dot{s}>0$, k should be increased.

(2) If $s\dot{s}<0$, k should be decreased.

It is known that when the system goes out of the sliding mode surface, the switching gain should be increased to make the system reach the surface as quickly as possible; otherwise, if the system is on the sliding mode surface, the switching gain should be increased to reduce chattering. Fuzzy table of output is shown in Table 2.

Centroid method is adopted to achieve defuzzification, in which the enclosed area centroid by membership function and x-axis is calculated. We define μ as the fuzzy set of variable x; then the defuzzification result is

$$\varepsilon=\frac{\int x\mu\,dx}{\int \mu\,dx}. \tag{15}$$

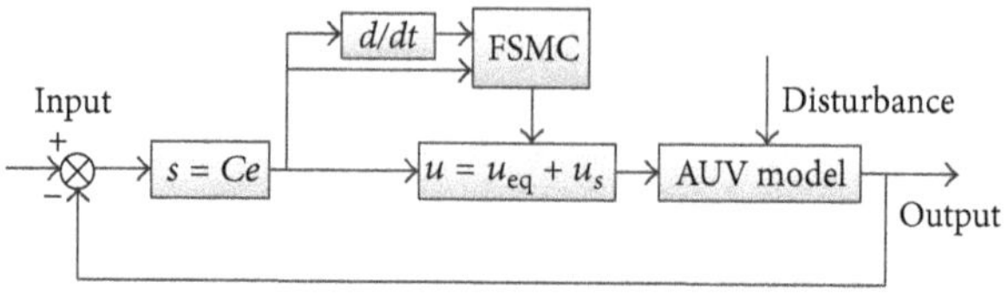

FIGURE 3: Diagram of fuzzy sliding mode control system.

We define the Lyapunov function as

$$\begin{aligned}V&=\frac{1}{2}s^2,\\ s\dot{s}&=s[C\dot{e}]=s[C(Ax+Bu+D-\dot{x}_d)]=s(CAx\\&+CBu+CD-C\dot{x}_d)=s\big(CAx\\&+CB\big[-(CB)^{-1}(CAx)+(CB)^{-1}C\dot{x}_d+k\operatorname{sgn}(s)\big]\\&+CD-C\dot{x}_d\big)=s(CAx-CAx+C\dot{x}_d\\&+CBk\operatorname{sgn}(s)+CD-C\dot{x}_d)=s(CBk\operatorname{sgn}(s)\\&-CD).\end{aligned} \tag{16}$$

Let $k=-\varepsilon(CB)^{-1}$, where $\varepsilon>\max(|CD|)$; thus $s\dot{s}<0$. Asymptotically stable condition is met.

4. Analysis of Simulation Result

The process of AUV longitudinal motion control is simulated to validate the effectiveness of this sliding mode control. The simulation environment is Matlab (R2011a)/Simulink, and the external disturbance is normal sea condition. Choose REMUS as the control plant. Hydrodynamic and physical parameters of REMUS when the AUV moves underwater are as follows:

$$\begin{aligned}&M_{\dot{q}}=-5.30\,\text{kg·m}^2/\text{rad};\ Z_{\dot{q}}=-2.24\,\text{kg·m/rad};\\&M_{\dot{w}}=-2.35\,\text{kg·m};\ Z_{\dot{w}}=-47.9\,\text{kg};\\&M_q=-23.2\,\text{kg·m/rad};\ Z_q=-26.6\,\text{kg/rad};\\&M_w=15.9\,\text{kg};\ Z_w=-45.6\,\text{kg/m};\\&M_\delta=-6.51\,\text{kg/rad};\ Z_\delta=-6.51\,\text{kg/(m·rad)};\\&x_g=0;\ y_g=0;\ z_g=0;\\&x_B=0;\ y_B=0;\ z_B=-0.02\,\text{m};\\&W=363\,\text{N};\ B_0=371\,\text{N}.\end{aligned}$$

The diagram of fuzzy sliding mode control system is shown in Figure 3.

Initial depth of AUV is $Z_0=0$ m, and target depth is $Z=10$ m; initial pitch angle is $\theta_0=0.18$ rad, and target pitch angle is $\theta_0=0.01$ rad; initial pitch angle rate is $q_0=-0.01$ rad/s, and initial velocity along the z direction $w_0=0.15$ m/s. The switching function matrix is $C=\begin{bmatrix}25 & 10 & 8 & 1\end{bmatrix}$.

First, the sine wave response is used to test the performance of the controller, and the simulation result is shown in

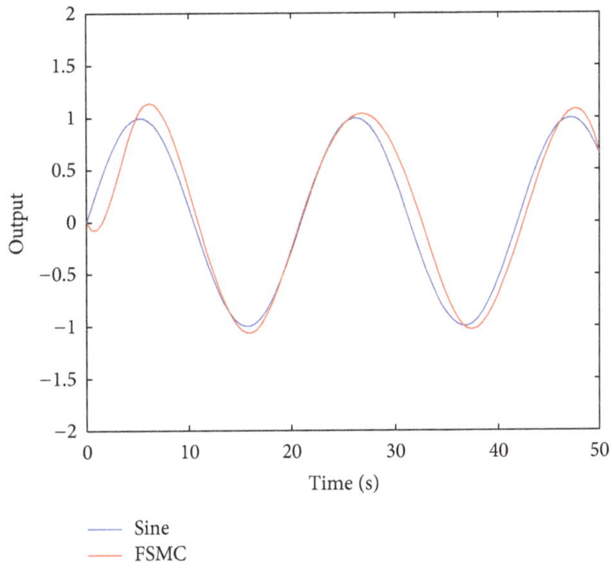

FIGURE 4: Sine wave response.

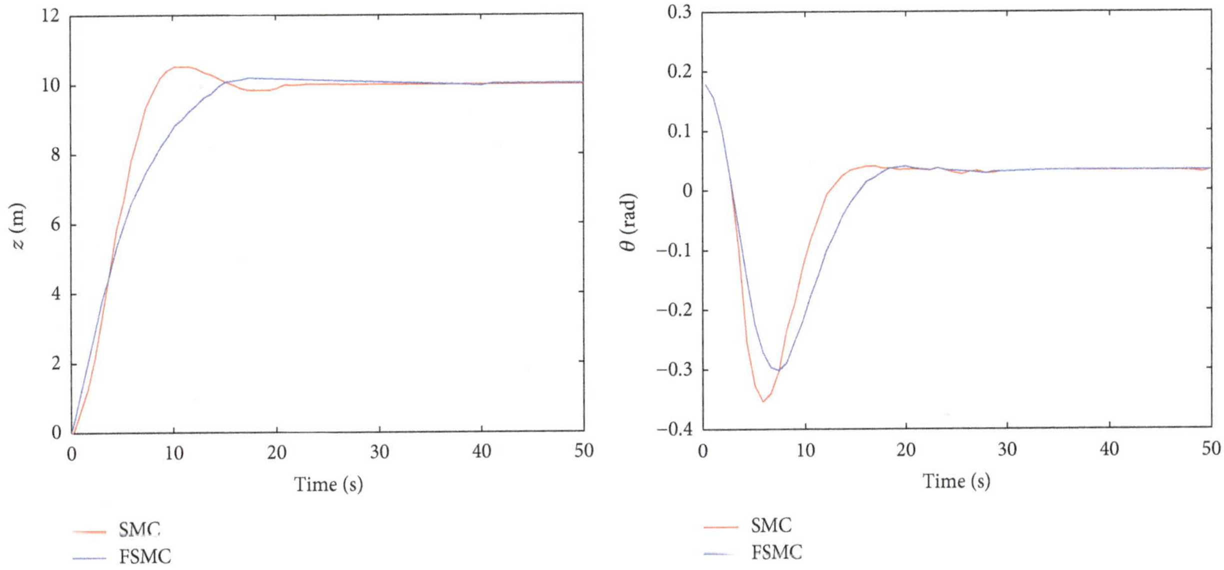

FIGURE 5: Changing curves of depth and pitch angle.

Figure 4. It can be seen that the fuzzy sliding mode controller has a good performance to track a varying state trajectory.

The conventional sliding mode control (SMC) and fuzzy sliding mode control (FSMC) are adopted, respectively, to control the longitudinal motion of AUV. The depth changing curves and pitch angle changing curve are shown in Figure 5, and the velocity changing along z direction curve and pitch angle changing rate are shown in Figure 6.

As shown in the figures, under the normal sea conditions, both control methods are robust and can meet the requirements, which means that they can reach the control target and keep the system stable. Compared with conventional sliding mode control, fuzzy sliding mode control has a smaller overshoot and shorter adjusting time. Overshoot of SMC is about 18% while that of FSMC is less than 2%; adjusting time of SMC is 25.53 s and that of FSMC is 19 s. In addition, FSMC reduces the chattering phenomenon and control error, and the overall performance of FSMC is superior to that of SMC.

Figure 7 shows the changing curve of rudder angel. SMC has a longer adjusting time and bigger overshoot. A slight chattering will appear when the system reaches stable state. FSMC can adjust the horizontal rudder angles in a short time to reach steadiness without chattering.

5. Conclusions

A fuzzy sliding mode controller is designed to improve the control precision and antijamming capability of AUV in this paper, which combines the sliding mode control and fuzzy

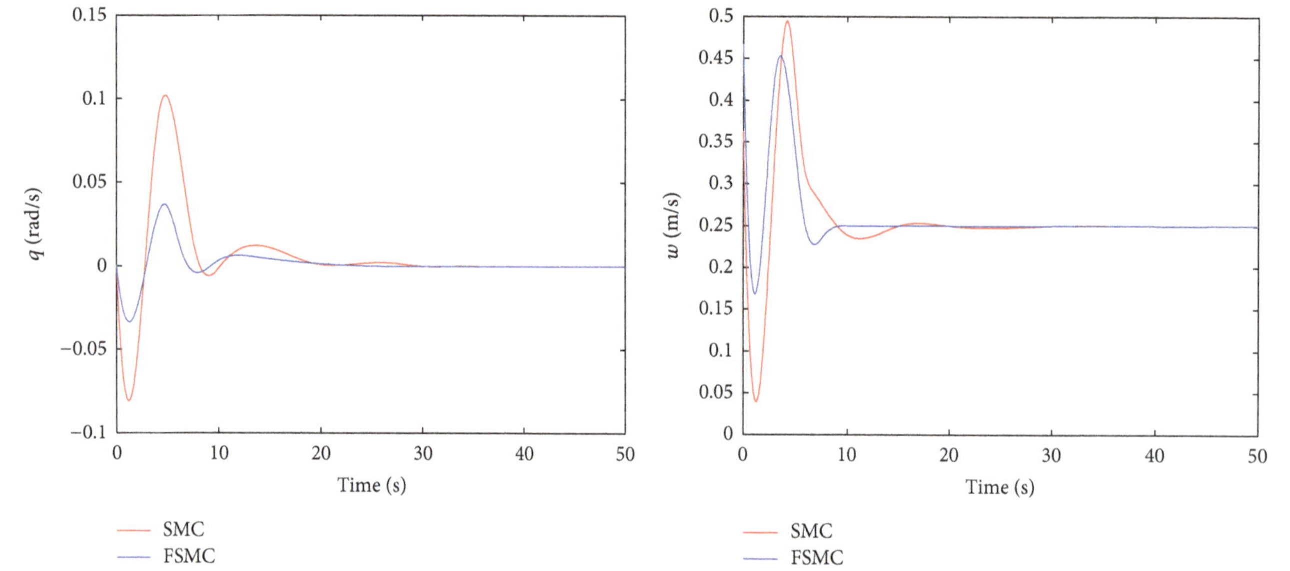

FIGURE 6: Changing curves of pitch rate and heave velocity.

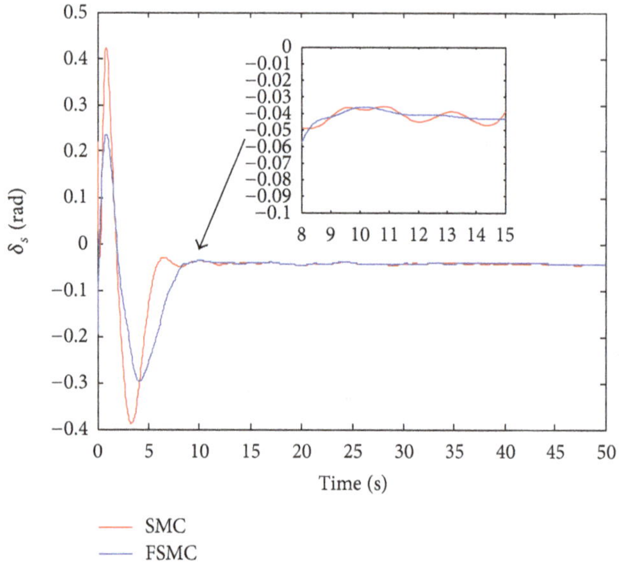

FIGURE 7: Changing curves of rudder angle.

control. Fuzzy rules are adopted to estimate the switching gain to eliminate disturbance terms and reduce chattering. The simulation results show that the fuzzy sliding mode controller can meet our requirements and has a higher precision and stronger antijamming performances compared with conventional sliding mode controller. In the further research, membership functions will be taken into consideration to improve the performance of fuzzy sliding mode controller.

Conflict of Interests

The authors declare that there is no conflict of interests regarding the publication of this paper.

References

[1] A. L. Forrest, B. Laval, M. J. Doble et al., "AUV measurements of under ice thermal structure," in *Proceedings of the MTS/IEEE OCEANS*, pp. 1–10, IEEE Press, Quebec City, Canada, September 2008.

[2] B.-H. Jun, J.-Y. Park, F.-Y. Lee et al., "Development of the AUV 'ISiMI' and a free running test in an ocean engineering basin," *Ocean Engineering*, vol. 36, no. 1, pp. 2–14, 2009.

[3] J. Petrich, C. A. Woolsey, and D. J. Stilwell, "Planar flow model identification for improved navigation of small AUVs," *Ocean Engineering*, vol. 36, no. 1, pp. 119–131, 2009.

[4] J. Jiang, B. Song, G. Pan, and M. Chang, "Study on design of shape and hydrodynamic layout for ultra-low-speed AUV," *Torpedo Technology*, vol. 19, no. 5, pp. 321–324, 2011 (Chinese).

[5] T. Li, D. Zhao, Z. Huang, and S. Su, "A method for self-estimating the depth of maneuvering AUV based on the grey particle filter," *Journal of National University of Defense Technology*, vol. 35, no. 5, pp. 185–190, 2013 (Chinese).

[6] B. Wang, L. Wan, Y.-R. Xu, and Z.-B. Qin, "Modeling and simulation of a mini AUV in spatial motion," *Journal of Marine Science and Application*, vol. 8, no. 1, pp. 7–12, 2009.

[7] L. Ma and W.-C. Cui, "Path following control of autonomous underwater vehicle based upon fuzzy hybrid control," *Control Theory & Applications*, vol. 23, no. 3, pp. 341–346, 2006.

[8] H. M. Jia, L. J. Zhang, X. Q. Cheng, X. Q. Bian, Z. P. Yan, and J. J. Zhou, "Three-dimensional path following control for an underactuated UUV based on nonlinear iterative sliding mode," *Acta Automatica Sinica*, vol. 38, no. 2, pp. 308–314, 2012.

[9] B. K. Sahu and B. Subudhi, "Adaptive tracking control of an autonomous underwater vehicle," *International Journal of Automation and Computing*, vol. 11, no. 3, pp. 299–307, 2014.

[10] L. Lapierre and D. Soetanto, "Nonlinear path-following control of an AUV," *Ocean Engineering*, vol. 34, no. 11-12, pp. 1734–1744, 2007.

[11] Y. F. Peng, "Robust intelligent sliding model control using recurrent cerebellar model articulation controller for uncertain

nonlinear chaotic systems," *Chaos, Solitons & Fractals*, vol. 39, no. 1, pp. 150–167, 2009.

[12] D. Kim, H.-S. Choi, J.-Y. Kim, J.-H. Park, and N.-H. Tran, "Design of an underwater vehicle-manipulator system with redundancy," *International Journal of Precision Engineering and Manufacturing*, vol. 16, no. 7, pp. 1561–1570, 2015.

[13] X.-Y. Luo, Z.-H. Zhu, and X.-P. Guan, "Chattering reduction adaptive sliding-mode control for nonlinear time-delay systems," *Control and Decision*, vol. 24, no. 9, pp. 1429–1435, 2009 (Chinese).

[14] L. Zhang, Y. Pang, Y. Su, and Y. Liang, "HPSO-based fuzzy neural network control for AUV," *Journal of Control Theory and Applications*, vol. 6, no. 3, pp. 322–326, 2008.

[15] L.-X. Pan, H.-Z. Jin, and L.-L. Wang, "Robust control based on feedback linearization for roll stabilizing of autonomous underwater vehicle under wave disturbances," *China Ocean Engineering*, vol. 25, no. 2, pp. 251–263, 2011.

[16] S.-H. Ryu and J.-H. Park, "Auto-tuning of sliding mode control parameters using fuzzy logic," in *Proceedings of the American Control Conference (ACC '01)*, vol. 1, pp. 618–623, IEEE, Arlington, Va, USA, June 2001.

[17] F.-J. Lin and W.-D. Chou, "An induction motor servo drive using sliding-mode controller with genetic algorithm," *Electric Power Systems Research*, vol. 64, no. 2, pp. 93–108, 2003.

[18] J. Cao, Y. Su, and J. Zhao, "Design of an adaptive controller for dive-plane control of a torpedo-shaped AUV," *Journal of Marine Science and Application*, vol. 10, no. 3, pp. 333–339, 2011.

25

Some New Generalized Retarded Gronwall-Like Inequalities and Their Applications in Nonlinear Systems

Haiyong Qin,[1] Xin Zuo,[2] and Jianwei Liu[2]

[1]*School of Mathematics, Qilu Normal University, Jinan, Shandong 250013, China*
[2]*Department of Automation, China University of Petroleum (Beijing), Changping, Beijing 102249, China*

Correspondence should be addressed to Haiyong Qin; qhymath@hotmail.com

Academic Editor: Petko Petkov

The Gronwall inequalities are of significance in mathematics and engineering. This paper generalizes the Gronwall-like inequalities from different perspectives. Using the proposed inequalities, the difficulties to discuss the controllability of integrodifferential systems of mixed type can be solved. Meanwhile, two examples as their applications are also given to show the effectiveness of our main results.

1. Introduction

Integral inequalities provide a powerful and important tool in the study of qualitative properties of solutions of nonlinear differential, integral, and integrodifferential equations, as well as in the modeling of science and engineering problems (see [1]). One of the most famous inequalities of this type is known as "Gronwall's inequality," "Bellman's inequality," or "Gronwall-Bellman's inequality" (see [2, 3]). Recently, the celebrated Gronwall inequality and its generalizations play increasingly important roles in the qualitative analysis of differential, integral, and integrodifferential equations. Based on the different purposes, many researchers put their efforts in exploring new inequalities and their applications in many fields, and many useful Gronwall-like integral inequalities have been established in various problems (see [4–17]).

Lipovan [18] proved a Gronwall-like inequality, and in order to show its applications, Lipovan applied his main results to the qualitative analysis of solutions to certain integral equations, functional differential equations, and retarded differential equations. Ye et al. [19] gave a generalized Gronwall inequality with singularity which can be applied to weakly singular Volterra integral equations and fractional integral and integrodifferential equations. Liu [20] proved a comparison result, which is widely known later and always used to provide explicit bounds on solutions and estimate on noncompactness. In addition, some existence theorems of solutions and iterative approximation of the unique solution for the nonlinear integrodifferential equations of mixed type are obtained. However, it is worth mentioning that it is difficult to deal with integrodifferential systems which include a Fredholm operator in nonlinearity unless powerful integral inequalities are established.

In this paper, we prove a generalization of the Gronwall inequality. As an application, we show that the inequality can be applied to the controllability analysis of abstract control system and existence analysis. Sufficient conditions ensuring the controllability of certain impulsive integrodifferential system of mixed type are obtained. The main difficulties from the Fredholm operator can be overcome.

The rest of the paper is organized as follows. In Section 2, we present some preliminaries and lemmas and prove some generalized Gronwall-like inequalities. In Section 3, we discuss the controllability of impulsive integrodifferential systems of mixed type in a Banach space as an application. In Section 4, we give another example to illustrate the application of our main results vividly. Finally, conclusions are given in Section 5.

2. Integral Inequalities

In this section, we present a generalization of the Gronwall-like inequality which can be called a comparison result in many literatures (see [20]). Unless otherwise stated, we denote $\mathbb{R}_+ = (0, +\infty)$, $\mathbb{R}^+ = [0, +\infty)$ in this paper.

Lemma 1. *Suppose that $u, f, g \in C([t_0, T), \mathbb{R}^+)$. Let $w \in C(\mathbb{R}^+, \mathbb{R}^+)$ be nondecreasing with $w(u) > 0$ for $u > 0$ and let $\alpha, \beta \in C^1([t_0, T), [t_0, T))$ be nondecreasing with $\alpha(t), \beta(t) \le t$ on $[t_0, T)$. If*

$$u(t) \le m_1(t) + m_2(t) \int_{\alpha(t_0)}^{\alpha(t)} f(s) w(u(s)) ds + m_3(t) \int_{\beta(t_0)}^{\beta(t)} g(s) w(u(s)) ds, \quad (1)$$

where $m_i(t)$ is a nonnegative, continuous function defined on $t_0 \le t < T$, and there exist nonnegative constants M_i such that $m_i(t) \le M_i$ $(i = 1, 2, 3)$, then, for $t_0 \le t < t_1$, one has

$$u(t) \le G^{-}\left(G(M_1) + M_2 \int_{\alpha(t_0)}^{\alpha(t)} f(s) ds + M_3 \int_{\beta(t_0)}^{\beta(t)} g(s) ds\right), \quad (2)$$

where $G(r) = \int_1^r (1/w(s)) ds$, $r > 0$, $t_1 \in (t_0, T)$ is chosen so that

$$G(M_1) + M_2 \int_{\alpha(t_0)}^{\alpha(t)} f(s) ds + M_3 \int_{\beta(t_0)}^{\beta(t)} g(s) ds \in \mathrm{Dom}(G^{-}), \quad (3)$$

for all t lying in the interval $[t_0, t_1)$.

Proof. Noting the conditions we imposed, we have

$$u(t) \le M_1 + M_2 \int_{\alpha(t_0)}^{\alpha(t)} f(s) w(u(s)) ds + M_3 \int_{\beta(t_0)}^{\beta(t)} g(s) w(u(s)) ds. \quad (4)$$

Let us denote

$$U(t) = M_1 + M_2 \int_{\alpha(t_0)}^{\alpha(t)} f(s) w(u(s)) ds + M_3 \int_{\beta(t_0)}^{\beta(t)} g(s) w(u(s)) ds. \quad (5)$$

Obviously, we have $U(t_0) = M_1$ and

$$U'(t) = M_2 f(\alpha(t)) w(u(\alpha(t))) \alpha'(t) + M_3 g(\beta(t)) w(u(\beta(t))) \beta'(t). \quad (6)$$

Since $\alpha(t) \le t$ and $\beta(t) \le t$ on $[t_0, T)$, then

$$U'(t) \le M_2 f(\alpha(t)) w(U(t)) \alpha'(t) + M_3 g(\beta(t)) w(U(t)) \beta'(t). \quad (7)$$

By the definitions of G, we obtain that

$$\frac{d}{dt} G(U(t)) \le M_2 f(\alpha(t)) \alpha'(t) + M_3 g(\beta(t)) \beta'(t) \quad (8)$$

integrate both sides, and we conclude that

$$G(U(t)) \le G(M_1) + M_2 \int_{\alpha(t_0)}^{\alpha(t)} f(s) ds + M_3 \int_{\beta(t_0)}^{\beta(t)} g(s) ds. \quad (9)$$

Because G^{-} is increasing on $\mathrm{Dom}(G^{-})$, we get

$$u(t) \le G^{-}\left(G(M_1) + M_2 \int_{\alpha(t_0)}^{\alpha(t)} f(s) ds + M_3 \int_{\beta(t_0)}^{\beta(t)} g(s) ds\right). \quad (10)$$

□

Remark 2. Next, we shall show that Lemma 1 generalizes some existing results:

(1) For $m_1(t) \equiv k$, $m_2(t) \equiv 1$, and $m_3(t) \equiv 0$, we obtain theorem in [18]. Further supposing that $\alpha(t) \equiv t$, we get the celebrated Bihari's inequality.

(2) Set $w(u) \equiv u$. Note that $G(u) = \int_1^{\infty} (1/s) ds = \infty$; then the previous result (2) holds.

(3) Compared with Theorem 1 in [19], this lemma has a different range of applications.

Corollary 3. *Suppose that $u, f, g \in C([t_0, T), \mathbb{R}^+)$. Let $\alpha, \beta \in C^1([t_0, T), [t_0, T))$ be nondecreasing with $\alpha(t), \beta(t) \le t$ on $[t_0, T)$. If*

$$u(t) \le M_1 + M_2 \int_{\alpha(t_0)}^{\alpha(t)} f(s) u(s) ds + M_3 \int_{\beta(t_0)}^{\beta(t)} g(s) u(s) ds, \quad (11)$$

where M_i $(i = 1, 2, 3)$ is nonnegative constants, then, for $t_0 \le t < T$, one has

$$u(t) \le M_1 \exp\left(M_2 \int_{\alpha(t_0)}^{\alpha(t)} f(s) ds + M_3 \int_{\beta(t_0)}^{\beta(t)} g(s) ds\right). \quad (12)$$

Remark 4. It is easy to get the following results.

(1) Assume that $M_2 = 1$ and $M_3 = 0$; we know that corollary in [18] is valid.

(2) With $M_3 = 0$ and $\alpha(t) = t$, we obtain the celebrated Gronwall-Bellman inequality.

Theorem 5. *Suppose that u, f, $g \in C([t_0, T), \mathbb{R}^+)$. Let $w \in C(\mathbb{R}^+, \mathbb{R}^+)$ be nondecreasing with $w(u) > 0$ for $u > 0$ and let $\alpha, \beta \in C^1([t_0, T), [t_0, T))$ be nondecreasing with $\alpha(t), \beta(t) \le t$ on $[t_0, T)$. If*

$$\begin{aligned} u(t) \le M_1 + M_2 \int_{\alpha(t_0)}^{\alpha(t)} f(s)\, w(u(s))\, ds \\ + M_3 \int_{\beta(t_0)}^{\beta(t)} g(s)\, w(u(s))\, ds \\ + M_4 \int_{t_0}^{T} [u(s)]^{\lambda} ds, \end{aligned} \tag{13}$$

where M_i $(i = 1, 2, 3, 4)$ are nonnegative constants, $0 \le \lambda < 1$, and $\int_1^{\infty}(1/w(s))ds = \infty$, then, for $t_0 \le t < T$, one has

$$u(t) \le m^{-}(0), \tag{14}$$

where $m(s) = (2s - MM_1)^{1-\lambda} - s^{1-\lambda} - (1-\lambda)MM_4T$, M is a positive constant, and $m^{-}(\cdot)$ represents the inverse of $m(\cdot)$.

Proof. By Lemma 1 and Remark 2 (2), for $t_0 \le t < T$, there exists a constant $M > 0$ such that

$$u(t) \le M\left(M_1 + M_4 \int_{t_0}^{T} [u(s)]^{\lambda} ds\right). \tag{15}$$

Define

$$\begin{aligned} p(t) \\ = M\left(M_1 + M_4 \int_{t_0}^{t} [u(s)]^{\lambda} ds + M_4 \int_{t_0}^{T} [u(s)]^{\lambda} ds\right); \end{aligned} \tag{16}$$

we have

$$\begin{aligned} p(t_0) = MM_1 + MM_4 \int_{t_0}^{T} [u(s)]^{\lambda} ds, \\ p'(t) \le MM_4 [p(t)]^{\lambda}. \end{aligned} \tag{17}$$

Integrating from t_0 to t, we get

$$[p(t)]^{1-\lambda} - [p(t_0)]^{1-\lambda} \le (1-\lambda) MM_4 (t - t_0); \tag{18}$$

then

$$p(t) \le \left\{[p(t_0)]^{1-\lambda} + (1-\lambda) MM_4 (t - t_0)\right\}^{1/(1-\lambda)}. \tag{19}$$

Since

$$\begin{aligned} 2p(t_0) - MM_1 = p(T) \\ \le \left\{[p(t_0)]^{1-\lambda} + (1-\lambda) MM_4 (T - t_0)\right\}^{1/(1-\lambda)}, \end{aligned} \tag{20}$$

we can deduce that

$$(2p(t_0) - MM_1)^{1-\lambda} - [p(t_0)]^{1-\lambda} \le (1-\lambda) MM_4 T. \tag{21}$$

Let

$$m(s) = (2s - MM_1)^{1-\lambda} - s^{1-\lambda} - (1-\lambda) MM_4 T. \tag{22}$$

Observe that $m \in C([MM_1/2, \infty], \mathbb{R}^+)$ and

$$\begin{aligned} m\left(\frac{MM_1}{2}\right) = -\left(\frac{MM_1}{2}\right)^{1-\lambda} - (1-\lambda) MM_4 T < 0, \\ \lim_{s \to 0} \frac{m(s)}{s^{1-\lambda}} = 2^{1-\lambda} - 1 > 0. \end{aligned} \tag{23}$$

It is easy to get that there exists a s_0 such that $m(s_0) = 0$; then $p(t_0) \le s_0$. Therefore

$$u(t) \le p(t_0) \le m^{-}(0), \quad t \in [t_0, T). \tag{24}$$

The proof is completed. □

Remark 6. (i) If $M_1 = M_4 = 0$, we have $m^{-}(t_0) = 0$; that is, $u(t) \equiv 0$.

(ii) Generally speaking, the spectral radius of Fredholm operators should not be less than one. However, there is no doubt that here the above inequality is satisfied as a particular case.

3. Controllability of Differential Systems of Mixed Type

In this section, we shall give an application to show that the proposed inequalities are useful in investigating the existence of mild solutions and controllability of differential systems of mixed type. Unfortunately, since the spectral radius of Fredholm operators should not be less than one, the inequality used in previous paper may be not suitable (see [21–26]). Therefore, more powerful integral inequalities should be established to solve the problem. In order to illustrate this problem, we consider the following impulsive integrodifferential system in a Banach space:

$$\begin{aligned} x'(t) = A(t)x(t) + f(t, x(t), (Sx)(t), (Tx)(t)) \\ + (Bu)(t), \quad t \in I = [0, b], \\ \Delta x(t_i) = I_i(x(t_i)) = x(t_i^{+}) - x(t_i^{-}), \\ i = 1, 2, \ldots, s, \\ x(0) = x_0, \end{aligned} \tag{25}$$

where operators S and T are defined as follows:

$$\begin{aligned} (Sx)(t) = \int_0^t k(t, s, x(s))\, ds, \\ (Tx)(t) = \int_0^b h(t, s, x(s))\, ds. \end{aligned} \tag{26}$$

$A(t)$ is a family of linear operators which generates an evolution operator

$$\begin{aligned} G : \Delta = \{(t, s) \in [0, b] \times [0, b] : 0 \le s \le t \le b\} \\ \longrightarrow L(\mathbb{X}), \end{aligned} \tag{27}$$

where $L(\mathbb{X})$ is the space of all bounded linear operators in $\mathbb{X}$ and $\mathbb{X}$ is a Banach space. Assume that $k \in C[\Delta \times \mathbb{X}, \mathbb{X}]$ and $h \in C[I \times I \times \mathbb{X}, \mathbb{X}]$. f is continuous. $0 = t_0 < t_1 < t_2 < \cdots < t_s < t_{s+1} = b$. $I_i \in C[\mathbb{X}, \mathbb{X}]$, $(i = 1, 2, \ldots, s)$ are impulsive functions, and $x(t_i^+)$ and $x(t_i^-)$ represent the right and the left limits of $x(t)$ at $t = t_i$, respectively. $B \in L[U, \mathbb{X}]$ is a bounded linear operator and the control function $u(\cdot)$ is given in $L^2[I, U]$ and U is a Banach space. Set $T_r = \{x \in \mathbb{X} \mid \|x\| \le r\}$ and $B_r = \{x \in \mathrm{PC}[I, \mathbb{X}] \mid \|x\| \le r\}$. $\mathrm{PC}[I, \mathbb{X}] = \{x \in C[I, \mathbb{X}] \mid x \text{ is continuous on } (t_i, t_{i+1}),\ i = 0, 1, \ldots, s,\ x(t_i^-) = x(t_i), \text{ and the right limit } x(t_i^+) \text{ exists}, i = 1, 2, \ldots, s\}$. Obviously, $\mathrm{PC}[I, \mathbb{X}]$ is a Banach space with the norm $\|x\|_{\mathrm{PC}} = \sup_{t\in I}\{\|x(t)\|\}$.

Suppose that the following hypotheses are satisfied.

(H_1) $A(t) : D(A) \to \mathbb{X}$ is a family of linear operators, generating an equicontinuous evolution system $\{G(t,s) : (t,s) \in \Delta\}$; that is, $(t,s) \to \{G(t,s)x : x \in B_r\}$ is equicontinuous for $t > 0$ and for all bounded subsets B_r.

(H_2) For any $r > 0$, f is uniformly continuous on $I \times T_r \times T_r \times T_r$ and I_i $(i = 1, 2, \ldots, s)$ are bounded on T_r. There exist functions $b_p \in C[I, \mathbb{R}^+]$ $(p = 1, 2, 3, 4)$ and k^*, $h^* > 0$ such that

$$\begin{aligned} \|f(t,x,y,z)\| &\le b_1(t) + b_2(t)\|x\|^{\lambda} + b_3(t)\|y\| \\ &\quad + b_4(t)\|z\|, \\ \|k(t,s,x)\| &\le k^* \|x\|^{\lambda}, \\ \|h(t,s,x)\| &\le h^* \|x\|^{\lambda}, \end{aligned} \tag{28}$$

where $0 \le \lambda < 1$. Define $b_p = \max\{b_p(t) \mid t \in I\}$.

(H_3) The linear operator $W : L^2[I, U] \to \mathbb{X}$ is defined by

$$Wu = \int_0^b G(b,s) Bu(s)\, ds. \tag{29}$$

(i) W has an invertible operator W^{-1} which takes values in $L^2[I,U]/\mathrm{Ker}W$ and there exist positive constants L_B and L_W such that $\|B\| \le L_B$ and $\|W^{-1}\| \le L_W$;

(ii) there exists $K_W \in C[I, \mathbb{R}^+]$ such that, for any bounded set $H \subset \mathbb{X}$,

$$\alpha\left(\left(W^{-1}H\right)(t)\right) \le K_W(t)\,\alpha(H). \tag{30}$$

Define $K_W = \max\{K_W(t) \mid t \in I\}$; $\alpha(\cdot)$ represents the Kuratowski noncompactness measure.

(H_4) There exist $l_q \in C[I, \mathbb{R}^+]$ $(q = 1, 2)$ such that, for any equicontinuous set, $D \subset B_r$, such that

$$\begin{aligned} &\alpha\left(f(t, D(s), (TD)(s), (SD)(s))\right) \\ &\quad \le l_1(t)\,[\alpha(D(s))]^{\lambda} + l_2(t)\,\alpha((TD)(s)). \end{aligned} \tag{31}$$

Define $l_q = \max\{l_q(t) \mid t \in I\}$.

Theorem 7. *Assume that conditions (H_1)–(H_4) hold. Then the system (25) is controllable.*

Proof. Using (H_3) (i), for every $x \in \mathrm{PC}[I, X]$, without loss of generality, define the control

$$\begin{aligned} u_{0x}(t) &= W^{-1}\Bigg[x_1 - G(b,0)x_0 \\ &\quad - \int_0^b G(b,s) f(s, x(s), (Tx)(s), (Sx)(s))\, ds\Bigg](t), \\ u_{jx}(t) &= W^{-1}\Bigg[x_1 - G(b,0)x_0 \\ &\quad - \int_0^b G(b,s) f(s, x(s), (Tx)(s), (Sx)(s))\, ds \\ &\quad - \sum_{i=1}^{j} G(t,t_i) I_i(x(t_i))\Bigg](t), \end{aligned} \tag{32}$$

where $j = 1, \ldots, s$. Define operator Q as follows:

$$\begin{aligned} (Qx)(t) &= G(t,0)x_0 + \int_0^t G(t,s)\left(f + Bu_{sx}\right)(s)\, ds \\ &\quad + \sum_{0<t_i<t} G(t,t_i) I_i(x(t_i)); \end{aligned} \tag{33}$$

clearly, using the control $u_{sx}(t)$, the fixed point of operator Q is a solution of the system (25), and $x_1 = (Qx)(b)$; that is, system (25) is controllable. From the conditions we imposed, it is easy to get that operator Q is continuous.

Set $\Omega_0 = \{x \in \mathrm{PC}[I, \mathbb{X}] \mid x = \lambda Qx,\ 0 \le \lambda \le 1\}$. Assume that there exists $\lambda_0 \in [0,1]$ such that $\overline{x}(t) = \lambda_0(Q\overline{x})(t)$. Next, we shall use the method of piecewise discussion.

(i) When $t \in [0, t_1]$,

$$\overline{x}(t) = \lambda_0 G(t,0)x_0 + \lambda_0 \int_0^t G(t,s)\left(f + Bu_{0\overline{x}}\right)(s)\, ds. \tag{34}$$

From Ji et al. [23], we know that there exists $L_G > 0$ such that $\|G(t,s)\| \le L_G$ for any $(t,s) \in I \times I$. Thus

$$\begin{aligned} \|\overline{x}(t)\| &\le L_G\|x_0\| + L_G\int_0^t \left\|\left(f + Bu_{0\overline{x}}\right)(s)\right\| ds \\ &\le L_G\|x_0\| + L_G\int_0^t \|f\|\, ds \\ &\quad + L_G L_B \int_0^t \left\|u_{0\overline{x}}(s)\right\| ds, \end{aligned} \tag{35}$$

where

$$\int_0^t \|f\| ds$$
$$\leq \int_0^b b_1(s)\, ds + b_2 \int_0^t \|\overline{x}(s)\|^\lambda ds$$
$$+ b_3 k^* t_1 \int_0^t \|\overline{x}(s)\|^\lambda ds + b_4 h^* t_1 \int_0^b \|\overline{x}(s)\|^\lambda ds, \quad (36)$$
$$\|u_{0\overline{x}}\|_{L^2}$$
$$= \left\| W^{-1} \left[x_1 - G(b,0)\, x_0 - \int_0^b G(b,s)\, f\, ds \right] \right\|_{L^2}$$
$$\leq L_W \left[\|x_1\| + L_G \|x_0\| + L_G \int_0^b \|f\|\, ds \right].$$

Let $u(t) = \|\overline{x}(t)\|$; then $u(t) \in C[[0,t_1],\mathbb{R}^+]$; we have

$$u(t) \leq L_G \|x_0\| + L_G L_B b \left[L_W \|x_1\| + L_W L_G \|x_0\| \right]$$
$$+ L_G \left[\int_0^b b_1(s)\, ds + b_2 \int_0^t [u(s)]^\lambda ds \right.$$
$$\left. + b_3 k^* t_1 \int_0^t [u(s)]^\lambda ds + b_4 h^* t_1 \int_0^b [u(s)]^\lambda ds \right] \quad (37)$$
$$+ L_G^2 L_B L_W b \left[\int_0^b b_1(s)\, ds + b_2 \int_0^b [u(s)]^\lambda ds \right.$$
$$\left. + b_3 k^* t_1 \int_0^b [u(s)]^\lambda ds + b_4 h^* t_1 \int_0^b [u(s)]^\lambda ds \right].$$

Since $\int_1^\infty (1/s^\lambda) ds = \infty$, then by Theorem 5, there exists a constant C_0 such that $u(t) \leq C_0$, $t \in [0,t_1]$; that is, there exists a constant C_0 independent of u such that $u(t) \leq C_0$, $t \in [0,t_1]$. The above inequality implies that $\|\overline{x}(t)\| \leq C_0$. From (H_2), there also exists a constant $M_0 > 0$ independent of $\overline{x}$ such that

$$\left\| f\left(t, \overline{x}(t), \overline{x}'(t), (T\overline{x})(t), (S\overline{x})(t)\right) \right\| \leq M_0,$$
$$\left\| I_1\left(\overline{x}(t_1)\right) \right\| \leq M_0, \quad (38)$$
$$\forall t \in [0,t_1], \quad \|\overline{x}(t)\| \leq C_0.$$

Thus $\|\overline{x}(t_1^+)\| = \|\overline{x}(t_1) + I_1(\overline{x}(t_1))\| \leq C_0 + M_0$.

(ii) When $t \in (t_1,t_2]$,

$$\overline{v}(t) = \begin{cases} \overline{x}(t), & t \in (t_1,t_2], \\ \overline{x}(t_1^+), & t = t_1. \end{cases} \quad (39)$$

Then $\overline{v}(t) \in C[[t_1,t_2],\mathbb{X}]$ and

$$\overline{v}(t) = \lambda_0 G(t,0)\, x_0 + \lambda_0 \int_0^{t_1} G(t,s) \left(f + B u_{1\overline{v}}\right)(s)\, ds$$
$$+ \lambda_0 \int_{t_1}^t G(t,s) \left(f + B u_{1\overline{v}}\right)(s)\, ds \quad (40)$$
$$+ \lambda_0 G(t,t_1)\, I_1\left(x(t_1)\right).$$

From results (38) and similar to the proof of (i), we can know that there exists $C_1 > 0$ that does not depend on $\overline{x}$ such that $\|\overline{v}(t)\| \leq C_1$, $t \in [t_1,t_2]$. So $\|\overline{x}(t)\| \leq C_1$, $t \in (t_1,t_2]$.

By the same method as above, we can prove that there exists a constant $C_s > 0$ that does not depend on $\overline{x}$ such that

$$\|\overline{x}(t)\| \leq C_s, \quad t \in (t_s, b]. \quad (41)$$

Let $C = \max\{C_i \mid 0 \leq i \leq s\}$; then $\|\overline{x}(t)\| \leq C$, $t \in I$. Thus Ω_0 is a bounded set in $PC[I,\mathbb{X}]$. Take $R > C$; let $\Omega = \{x \in PC[I,\mathbb{X}] \mid \|x\|_{PC} < R\}$; obviously Ω is a bounded open set in $PC[I,\mathbb{X}]$ and $\theta \in \Omega$. From the choice of R, we know that if $x \in \partial\Omega$ and $\lambda \in [0,1]$, we have $x \neq \lambda Qx$.

Let $H \subset \overline{\Omega}$ be a countable set and $H \subset \overline{co}(\{\theta\} \bigcup Q(H))$. By (H_1) and (H_2), it is easy to see that $Q(H)$ is equicontinuous on each $[t_i, t_{i+1}]$, $(i = 0,1,2,\ldots,s)$.

Next, we shall prove that $(QH)(t)$ is relatively compact for each $[t_i,t_{i+1}]$. In the same way, we discuss step by step as follows.

(i) When $t \in [t_0,t_1]$

$$\alpha(H(t)) \leq \alpha((QH)(t)) \leq \alpha\left(G(t,0)\, x_0 \mid x \in H\right)$$
$$+ \alpha\left(\int_0^t G(t,s) \left(f + B u_{0x}\right)(s)\, ds \mid x \in H\right)$$
$$\leq 2 L_G l_1 \int_0^t [\alpha(H(s))]^\lambda ds$$
$$+ 2 L_G l_2 k^* b \int_0^t \alpha(H(s))\, ds \quad (42)$$
$$+ L_G L_B \int_0^t \alpha\left(u_{0x} \mid x \in H\right) ds,$$
$$\alpha\left(u_{0x} \mid x \in H\right) \leq K_W \left(2 L_G l_1 \int_0^b [\alpha(H(s))]^\lambda ds \right.$$
$$\left. + 2 L_G l_2 k^* b \int_0^t \alpha(H(s))\, ds \right).$$

Let $m(t) = \alpha(H(t))$, $t \in [0,t_1]$; then $m(t) \in C[[0,t_1],\mathbb{R}^+]$. Thus

$$m(t) \leq 2 L_G l_1 \int_0^t [\alpha(H(s))]^\lambda ds$$
$$+ 2 L_G l_2 k^* b \int_0^t \alpha(H(s))\, ds$$

$$+2K_W L_G l_1 b \int_0^t [\alpha(H(s))]^\lambda ds + 2K_W L_G l_2 b k^* t \int_0^t \alpha(H(s))\, ds. \quad (43)$$

By Remark 6 (i), we have $m(t) \equiv 0, t \in [0, t_1]$. Thus $\alpha(H(t)) = 0$; that is, H is a relatively compact set in $PC[[0, t_1], \mathbb{X}]$. Since $I_1 \in C[\mathbb{X}, \mathbb{X}]$, $\alpha(I_1(H(t_1))) = 0$, then $H(t_1)$ is a relatively compact set in $\mathbb{X}$.

(ii) For $t \in (t_1, t_2]$, we know

$$\begin{aligned}
&\alpha(H(t)) \\
&\leq \alpha(G(t,0)x_0 \mid x \in H) \\
&\quad + \alpha\left(\int_0^t G(t,s)(f + Bu_{1x})(s)\, ds \mid x \in H\right) \\
&\quad + \alpha(I_1(x(t_1)) \mid x \in H) \\
&\leq 2L_G l_1 \int_0^t [\alpha(H(s))]^\lambda ds \\
&\quad + 2L_G l_2 k^* b \int_0^t \alpha(H(s))\, ds \\
&\quad + L_G L_B \int_0^t \alpha(u_{1x} \mid x \in H)\, ds.
\end{aligned} \quad (44)$$

Similar to the proof of (i), we can deduce that $m(t) \equiv 0$, $t \in [t_1, t_2]$. Therefore $\alpha(H(t)) = 0, t \in [t_1, t_2]$. In particular, $\alpha(H(t_2)) = 0$, so H is a relatively compact set in $C[[t_1, t_2], \mathbb{X}]$. Similarly, we can show that H is a relatively compact set in $C[[t_i, t_{i+1}], \mathbb{X}]$ $(i = 2, 3, \ldots, s)$. Thus H is a relatively compact set in $PC[I, \mathbb{X}]$.

In conclusion, we deduce that Q has at least one fixed point in Ω by the Mönch fixed point theorem; that is, system (25) has at least one mild solution in $PC[I, \mathbb{X}]$. Thus system (25) is controllable on I. □

4. The Uniqueness and Global Existence of Solutions

Consider the following integral equation:

$$u(t) = k(t) + \int_0^{\alpha(t)} f(s)\, w(u(s))\, ds + \int_0^{\beta(t)} g(s)\, w(u(s))\, ds, \quad (45)$$

where $k, f, g \in C(\mathbb{R}^+, \mathbb{R}^+)$ with $w(0) = 0$ and $\alpha, \beta \in C^1(\mathbb{R}^+, \mathbb{R}^+)$ are nondecreasing with $\alpha(t), \beta(t) \leq t$ on $\mathbb{R}\mathbb{R}^+$. Assume that (45) has a solution $u \in C([0, T], \mathbb{R}^+)$ on some maximal interval of existence $[0, T]$. Moreover, if $T < \infty$,

$$\limsup_{t \to T} u(t) = \infty. \quad (46)$$

Theorem 8. *Assume that*

$$\|w(x) - w(y)\| \leq Z(\|x - y\|) \quad (47)$$

with $Z \in C(\mathbb{R}^+, \mathbb{R}^+)$ nondecreasing, $Z(x) > 0$ for $x > 0$. If

$$\int_0^1 \frac{1}{Z(s)} ds = \int_1^\infty \frac{1}{Z(s)} ds = \infty, \quad (48)$$

then (45) has a unique solution defined on $\mathbb{R}^+$. Moreover, if k is bounded on $\mathbb{R}^+$ and if either α, β is bounded on $\mathbb{R}^+$ or $\int_0^\infty f(s)ds, \int_0^\infty g(s)ds < \infty$, then its solution is bounded on $\mathbb{R}^+$.

Proof. Suppose that, on some interval $[0, t_0]$, (45) has two solutions $u_1, u_2 \in C([0, t_0], \mathbb{R}^+)$; we obtain

$$\begin{aligned}
&u_1(t) - u_2(t) \\
&\quad = \int_0^{\alpha(t)} f(s)[w(u_1(s)) - w(u_2(s))]\, ds \\
&\quad + \int_0^{\beta(t)} g(s)[w(u_1(s)) - w(u_2(s))]\, ds.
\end{aligned} \quad (49)$$

Denote $u(t) = \|u_1(t) - u_2(t)\|$; we have

$$u(t) \leq \int_0^{\alpha(t)} f(s) Z(u(s))\, ds + \int_0^{\beta(t)} g(s) Z(u(s))\, ds. \quad (50)$$

Set

$$G(r) = \int_1^r \frac{1}{Z(s)} ds, \quad r > 0. \quad (51)$$

Then $G(0) = -\infty$ and $G(\infty) = \infty$. There exists $\epsilon > 0$, where

$$u(t) \leq \epsilon + \int_0^{\alpha(t)} f(s) Z(u(s))\, ds + \int_0^{\beta(t)} g(s) Z(u(s))\, ds, \quad 0 \leq t \leq t_0. \quad (52)$$

From Theorem 5, we know that

$$u(t) \leq G^-\left(G(\epsilon) + \int_0^{\alpha(t)} f(s) Z(u(s))\, ds + \int_0^{\beta(t)} g(s) Z(u(s))\, ds\right), \quad 0 \leq t \leq t_0. \quad (53)$$

From Remark 6, $u(t) \to 0, \epsilon \to 0$; then $u_1(t) = u_2(t)$, and the uniqueness of the solution can be obtained.

Next, we will show that the solution is global; that is, $T = \infty$, where T is the maximal time of existence. If $T < \infty$, we set $k_0 = \max_{0 \leq t \leq T}\{k(t)\}$, and we obtain that

$$u(t) \leq k_0 + \int_0^{\alpha(t)} f(s) Z(u(s))\, ds + \int_0^{\beta(t)} g(s) Z(u(s))\, ds, \quad 0 \leq t < T, \quad (54)$$

as $w(u(s)) = w(u(s)) - w(0) \leq Z(u(s))$ for $0 \leq t < T$. By Lemma 1 and $u \in C[[0,T], \mathbb{R}^+]$, we deduce that

$$u(t) \leq G^{-}\left(G(k_0) + \int_0^{\alpha(t)} f(s)\,ds + \int_0^{\beta(t)} g(s)\,ds\right), \quad 0 \leq t < T. \tag{55}$$

Since k is bounded on $\mathbb{R}^+$ and either α, β is bounded on $\mathbb{R}^+$ or $\int_0^{\infty} f(s)ds$, $\int_0^{\infty} g(s)ds < \infty$ satisfies; then $u(t)$ is bounded on $\mathbb{R}^+$. Thus the previous inequality (55) contradicts (46). Thus the global existence is proved. □

5. Conclusions

This paper generalizes a more general Gronwall-like inequality with a Fredholm operator. Using the proposed inequality, we solve a difficult problem in the research of the controllability of integrodifferential systems of mixed type in Banach space. Meanwhile, we also prove the uniqueness and global existence of solutions for a class of integral equations. Therefore, the results we obtained are very important and powerful tools. However, it should be more useful than we can imagine in qualitative properties of many other nonlinear problems, such as existence, estimation of solutions, dependence of solutions on parameters in nonlinear analysis, and control.

Conflict of Interests

The authors declare that there is no conflict of interests regarding the publication of this paper.

References

[1] B. G. Pachpatte, *Inequalities for Differential and Integral Equations*, Academic Press, New York, NY, USA, 1998.

[2] T. H. Gronwall, "Note on the derivatives with respect to a parameter of the solutions of a system of differential equations," *The Annals of Mathematics. Second Series*, vol. 20, no. 4, pp. 292–296, 1919.

[3] R. Bellman, "The stability of solutions of linear differential equations," *Duke Mathematical Journal*, vol. 10, pp. 643–647, 1943.

[4] Y. Jalilian and R. Jalilian, "Existence of solution for delay fractional differential equations," *Mediterranean Journal of Mathematics*, vol. 10, no. 4, pp. 1731–1747, 2013.

[5] O. Lipovan, "Integral inequalities for retarded Volterra equations," *Journal of Mathematical Analysis and Applications*, vol. 322, no. 1, pp. 349–358, 2006.

[6] W.-S. Wang, "A generalized retarded Gronwall-like inequality in two variables and applications to BVP," *Applied Mathematics and Computation*, vol. 191, no. 1, pp. 144–154, 2007.

[7] R. P. Agarwal, S. Deng, and W. Zhang, "Generalization of a retarded Gronwall-like inequality and its applications," *Applied Mathematics and Computation*, vol. 165, no. 3, pp. 599–612, 2005.

[8] S. Guo, I. Moroz, L. Si, and L. Han, "Several integral inequalities and their applications in nonlinear differential systems," *Applied Mathematics and Computation*, vol. 219, no. 9, pp. 4266–4277, 2013.

[9] Y.-H. Lan and Y. Zhou, "High-order $\mathscr{D}^{\alpha}$-type iterative learning control for fractional-order nonlinear time-delay systems," *Journal of Optimization Theory and Applications*, vol. 156, no. 1, pp. 153–166, 2013.

[10] Q. Feng and F. Meng, "Some new Gronwall-type inequalities arising in the research of fractional differential equations," *Journal of Inequalities and Applications*, vol. 2013, article 429, 2013.

[11] K. Zheng, W. Feng, and C. Guo, "Some new nonlinear weakly singular inequalities and applications to Volterra-type difference equation," *Abstract and Applied Analysis*, vol. 2013, Article ID 912874, 6 pages, 2013.

[12] B. Zheng, "Some new Gronwall-Bellman-type inequalities based on the modified Riemann-Liouville fractional derivative," *Journal of Applied Mathematics*, vol. 2013, Article ID 341706, 8 pages, 2013.

[13] H. Wang and B. Zheng, "Some new dynamic inequalities and their applications in the qualitative analysis of dynamic equations," *WSEAS Transactions on Mathematics*, vol. 12, no. 10, pp. 967–978, 2013.

[14] Y.-H. Kim, "Gronwall, Bellman and Pachpatte type integral inequalities with applications," *Nonlinear Analysis: Theory, Methods & Applications*, vol. 71, no. 12, pp. e2641–e2656, 2009.

[15] H. Ye and J. Gao, "Henry-Gronwall type retarded integral inequalities and their applications to fractional differential equations with delay," *Applied Mathematics and Computation*, vol. 218, no. 8, pp. 4152–4160, 2011.

[16] J. Wang, X. Xiang, W. Wei, and Q. Chen, "The generalized Gronwall inequality and its application to periodic solutions of integrodifferential impulsive periodic system on Banach space," *Journal of Inequalities and Applications*, vol. 2008, Article ID 430521, 22 pages, 2008.

[17] Y. Peng, X. Xiang, and W. Wei, "Nonlinear impulsive integro-differential equations of mixed type with time-varying generating operators and optimal controls," *Dynamic Systems and Applications*, vol. 16, no. 3, pp. 481–496, 2007.

[18] O. Lipovan, "A retarded Gronwall-like inequality and its applications," *Journal of Mathematical Analysis and Applications*, vol. 252, no. 1, pp. 389–401, 2000.

[19] H. Ye, J. Gao, and Y. Ding, "A generalized Gronwall inequality and its application to a fractional differential equation," *Journal of Mathematical Analysis and Applications*, vol. 328, no. 2, pp. 1075–1081, 2007.

[20] L. Liu, "Iterative method for solutions and coupled quasi-solutions of nonlinear integro-differential equations of mixed type in Banach spaces," *Nonlinear Analysis: Theory, Methods & Applications*, vol. 42, pp. 583–598, 2000.

[21] H. Qin and X. Zuo, "Controllability of nonlocal boundary conditions for impulsive differential systems of mixed type in banach spaces," in *Proceedings of the 10th IEEE International Conference on Control and Automation (ICCA '13)*, pp. 1015–1019, IEEE, Hangzhou, China, June 2013.

[22] Y. Li and H. Zhang, "The solutions of initial value problems for nonlinear second order integro-differential equations of mixed type in Banach spaces," *Journal of Shandong University (Natural Science)*, vol. 45, pp. 93–98, 2010 (Chinese).

[23] S. Ji, G. Li, and M. Wang, "Controllability of impulsive differential systems with nonlocal conditions," *Applied Mathematics and Computation*, vol. 217, no. 16, pp. 6981–6989, 2011.

[24] L. Liu, F. Guo, C. Wu, and Y. Wu, "Existence theorems of global solutions for non-linear Volterra type integral equations

in Banach spaces," *Journal of Mathematical Analysis and Applications*, vol. 309, no. 2, pp. 638–649, 2005.

[25] F. Guo, L. Liu, Y. Wu, and P.-F. Siew, "Global solutions of initial value problems for nonlinear second-order impulsive integro-differential equations of mixed type in Banach spaces," *Nonlinear Analysis. Theory, Methods & Applications*, vol. 61, no. 8, pp. 1363–1382, 2005.

[26] L. Liu, C. Wu, and F. Guo, "Existence theorems of global solutions of initial value problems for nonlinear integrodifferential equations of mixed type in banach spaces and applications," *Computers & Mathematics with Applications*, vol. 47, no. 1, pp. 13–22, 2004.

Permissions

All chapters in this book were first published in JCSE, by Hindawi Publishing Corporation; hereby published with permission under the Creative Commons Attribution License or equivalent. Every chapter published in this book has been scrutinized by our experts. Their significance has been extensively debated. The topics covered herein carry significant findings which will fuel the growth of the discipline. They may even be implemented as practical applications or may be referred to as a beginning point for another development.

The contributors of this book come from diverse backgrounds, making this book a truly international effort. This book will bring forth new frontiers with its revolutionizing research information and detailed analysis of the nascent developments around the world.

We would like to thank all the contributing authors for lending their expertise to make the book truly unique. They have played a crucial role in the development of this book. Without their invaluable contributions this book wouldn't have been possible. They have made vital efforts to compile up to date information on the varied aspects of this subject to make this book a valuable addition to the collection of many professionals and students.

This book was conceptualized with the vision of imparting up-to-date information and advanced data in this field. To ensure the same, a matchless editorial board was set up. Every individual on the board went through rigorous rounds of assessment to prove their worth. After which they invested a large part of their time researching and compiling the most relevant data for our readers.

The editorial board has been involved in producing this book since its inception. They have spent rigorous hours researching and exploring the diverse topics which have resulted in the successful publishing of this book. They have passed on their knowledge of decades through this book. To expedite this challenging task, the publisher supported the team at every step. A small team of assistant editors was also appointed to further simplify the editing procedure and attain best results for the readers.

Apart from the editorial board, the designing team has also invested a significant amount of their time in understanding the subject and creating the most relevant covers. They scrutinized every image to scout for the most suitable representation of the subject and create an appropriate cover for the book.

The publishing team has been an ardent support to the editorial, designing and production team. Their endless efforts to recruit the best for this project, has resulted in the accomplishment of this book. They are a veteran in the field of academics and their pool of knowledge is as vast as their experience in printing. Their expertise and guidance has proved useful at every step. Their uncompromising quality standards have made this book an exceptional effort. Their encouragement from time to time has been an inspiration for everyone.

The publisher and the editorial board hope that this book will prove to be a valuable piece of knowledge for researchers, students, practitioners and scholars across the globe.

List of Contributors

Antonello Baccoli and Alessandro Pisano
Department of Electrical and Electronic Engineering, University of Cagliari, 09123 Cagliari, Italy

Yi Zhang
Department of Electrical Engineering, North China University of Science and Technology, Tangshan 063000, China
The State Key Laboratory of Alternate Electrical Power System with Renewable Energy Sources, North China Electric Power University, Beijing 102206, China

Xiangjie Liu and Yujia Yan
The State Key Laboratory of Alternate Electrical Power System with Renewable Energy Sources, North China Electric Power University, Beijing 102206, China

Yang Yang, Yao Gang and Wang Rong
College of Civil Engineering, Chongqing University, Chongqing 400044, China

Wang Hengyu
Huadian New Energy Development Co., Gansu 730000, China

Shunya Nagai
The Graduate School of Advanced Technology and Science, Tokushima University, 2-1Minamijosanjima, Tokushima 770-8506, Japan

Hidetoshi Oya
The Institute of Technology and Science, Tokushima University, 2-1 Minamijosanjima, Tokushima 770-8506, Japan

Huimin Xu and Xiangjie Liu
The State Key Laboratory of Alternate Electrical Power System with Renewable Energy Sources, North China Electric Power University, Beijing 102206, China

Xuedong Zhang
School of Energy and Power Engineering, North China Electric Power University, Baoding 071003, China

C. H. Chai and Johari H. S. Osman
Faculty of Electrical Engineering, University Technology Malaysia, 81300 Skudai, Malaysia

Yin Zhao, Ying-kai Xia, Ying Chen and Guo-Hua Xu
School of Naval Architecture and Ocean Engineering, Huazhong University of Science and Technology, Wuhan 430074, China

Jesús U. Liceaga-Castro and Irma I. Siller-Alcalá
Departamento de Electrónica, Universidad Autónoma Metropolitana-Azcapotzalco, 02200 Ciudad de México, DF, Mexico

Eduardo Liceaga-Castro and Luis A. Amézquita-Brooks
CIIIA-FIME, Universidad Autónoma de Nuevo León, 66451 Monterrey, NL, Mexico

Tie Wang, Ruiliang Zhang and Jinxian Shen
Department of Vehicle Engineering, Taiyuan University of Technology, Taiyuan 030024, China

Guoxing Li and Fengshou Gu
Department of Vehicle Engineering, Taiyuan University of Technology, Taiyuan 030024, China
Centre for Efficiency and Performance Engineering, University of Huddersfield, Huddersfield HD1 3DH, UK

Jeang-Lin Chang and Tsui-Chou Wu
Department of Electrical Engineering, Oriental Institute of Technology, Banciao District, New Taipei City 220, Taiwan

Jin-Hua Ding and De-Quan Wang
School of Mechanical Engineering & Automation, Dalian Polytechnic University, Dalian 116034, China

Teng Gao
School of Mechanical Engineering & Automation, Dalian Polytechnic University, Dalian 116034, China
School of Control Science and Engineering, Dalian University of Technology, Dalian 116024, China

Jin-Yan Song
School of Information Engineering, Dalian Ocean University, Dalian 116023, China

Juanxiu Liu, Yifei Wu, Jian Guo and Qingwei Chen
School of Automation, Nanjing University of Science and Technology, Nanjing 210094, China

Adel Ouannas
Department of Mathematics and Computer Science, University of Tebessa, 12002 Tebessa, Algeria

Raghib Abu-Saris
Department of Health Informatics, College of Public Health and Health Informatics, King Saud Bin Abdulaziz University for Health Science, Riyadh 11481, Saudi Arabia

Lieping Zhang and Yu Zhang
Guangxi Key Laboratory of New Energy and Building Energy Saving, Guilin University of Technology, Guilin 541004, China
College of Mechanical and Control Engineering, Guilin University of Technology, Guilin 541004, China

Yingxiong Luo
College of Information Science and Engineering, Guilin University of Technology, Guilin 541004, China

Qixin Zhu
School of Mechanical Engineering, Suzhou University of Science and Technology, Suzhou 215009, China
School of Electrical and Electronic Engineering, East China Jiaotong University, Nanchang 330013, China

Kaihong Lu
School of Electrical and Electronic Engineering, East China Jiaotong University, Nanchang 330013, China

Yonghong Zhu
School of Mechanical and Electronic Engineering, Jingdezhen Ceramic Institute, Jingdezhen 333001, China

Hai-gang Xu, Yue-feng Liao and Xiao Han
Henan Mechanical and Electrical Engineering College, Xinxiang 453002, China

Xiaomeng Li, Zhanshan Zhao and Meixia Zhu
School of Computer Science & Software Engineering, Tianjin Polytechnic University, Tianjin 300387, China

Jing Zhang
School of Textiles, Tianjin Polytechnic University, Tianjin 300387, China
Tianjin Vocational Institute, Tianjin 300410, China

Awantha Jayasiri, Raymond G. Gosine and George K. I. Mann
Faculty of Engineering and Applied Science, Memorial University of Newfoundland, St. John's, NL, Canada A1B3X5

Peter McGuire
2C-CORE, Captain Robert A. Bartlett Building, Morrissey Road, St. John's, NL, Canada A1B3X5

Guohai Liu, Jun Yuan, Wenxiang Zhao and Yaojie Mi
School of Electrical and Information Engineering, Jiangsu University, Zhenjiang 212013, China

Chao Sun
College of Information Science and Engineering, Northeastern University, Shenyang, Liaoning 110819, China
College of Sciences, University of Science and Technology Liaoning, Anshan, Liaoning 114051, China

XiQin He
College of Sciences, University of Science and Technology Liaoning, Anshan, Liaoning 114051, China

FuLi Wang
College of Information Science and Engineering, Northeastern University, Shenyang, Liaoning 110819, China
State Key Laboratory of Synthetical Automation of Process Industries, Northeastern University, Shenyang, Liaoning 110819, China

Yuefen Chen
School of Science, Nanjing University of Science and Technology, Nanjing 210094, China
College of Mathematics and Information Science, Xinyang Normal University, Xinyang 464000, China

Minghai Yang
College of Mathematics and Information Science, Xinyang Normal University, Xinyang 464000, China

Jianxiao Zou
School of Mechanical Engineering, University of Electronic Science and Technology of China, Chengdu, Sichuan 611731, China

Chao Peng
School of Mechanical Engineering, University of Electronic Science and Technology of China, Chengdu, Sichuan 611731, China
Northwest Institute of Mechanical & Electrical Engineering, Xianyang, Shanxi 712099, China

Chongwei Han and Guanghui Zhang
Northwest Institute of Mechanical & Electrical Engineering, Xianyang, Shanxi 712099, China

Ayman A. El-Badawy
Mechanical Engineering Department, Al-Azhar University, Cairo, Egypt
Mechatronics Engineering Department, German University in Cairo, El Tagammoa El Khames, New Cairo, Cairo 11835, Egypt

Mohamed A. Bakr
Mechatronics Engineering Department, German University in Cairo, El Tagammoa El Khames, New Cairo, Cairo 11835, Egypt

Duo Qi, Jinfu Feng and Jian Yang
Aeronautics and Astronautics Engineering College, Air Force Engineering University, Xi'an 710038, China

Haiyong Qin
School of Mathematics, Qilu Normal University, Jinan, Shandong 250013, China

Xin Zuo and Jianwei Liu
Department of Automation, China University of Petroleum (Beijing), Changping, Beijing 102249, China

www.ingramcontent.com/pod-product-compliance
Lightning Source LLC
Chambersburg PA
CBHW080503200326
41458CB00012B/4069